Escaping the Bonds of Earth
The Fifties and the Sixties

Ben Evans

Escaping the Bonds of Earth

The Fifties and the Sixties

 Springer

Published in association with
Praxis Publishing
Chichester, UK

Ben Evans
Space Writer
Atherstone
Warwickshire
UK

SPRINGER–PRAXIS BOOKS IN SPACE EXPLORATION
SUBJECT *ADVISORY EDITOR*: John Mason, M.B.E., B.Sc., M.Sc., Ph.D.

ISBN 978-0-387-79093-0 Springer Berlin Heidelberg New York

Springer is a part of Springer Science + Business Media (*springer.com*)

Library of Congress Control Number: 2009925769

Cover design: Jim Wilkie
Copy editing: David Harland
Typesetting: BookEns Ltd, Royston, Herts., UK

Printed in Germany on acid-free paper

To Michelle

Contents

Illustrations

Author's preface

Overshadowed by the dark events of Vietnam, civil rights, the Kennedy and King murders, the Bay of Pigs and a close shave with nuclear holocaust, the Sixties will hopefully also be remembered by history as the decade in which humanity first ventured into the heavens. Men and a woman left Earth's atmosphere, spacewalked hundreds of kilometres above their home planet, rendezvoused and docked their ships together and travelled to the Moon for the first time. These triumphs, however, were tempered by tragedy: three astronauts asphyxiated in a launch pad fire, then a cosmonaut killed during his ill-fated descent to Earth. Still, by the end of the decade, both the United States and the Soviet Union had firmly established their presence in space. The excitement and euphoria which these years inspired were felt not just in America and Russia, but throughout the world. By the time Frank Borman, Jim Lovell and Bill Anders circled the Moon on Christmas Eve 1968, it is said that no fewer than a billion people back home were watching or listening.

This book explores the history of humanity's early exploration of space, beginning with the pioneering flight by Yuri Gagarin and ending with Apollo 8's circumnavigation of the Moon. It will, I hope, form the basis of a series to commemorate the first half-century of human exploration in space. By the time of that momentous anniversary in 2011, perhaps, the ongoing drive towards private spaceflight and 'space tourism' will begin to make human journeys into the heavens so commonplace that it will be impossible to catalogue them all! It is my most fervent wish that a further volume – covering the decade from 2011 – will be impossible to write, because men and women will be in space so often and human spaceflight will have changed from the realm of the few and the privileged to the realm of the many.

My intention in writing this volume was to convey some of my own enthusiasm for what was one of the most remarkable decades in human history. Born in 1976, sadly, I missed it all, and still await the chance to see my first manned lunar landing. Still, I have attempted to introduce the reader to some of the problems faced in the early days: from the basic questions of whether men could breathe, eat and avoid going mad in space, to more complex issues of the kinds of fuels and atmospheres to be used in rockets and spacecraft and the techniques needed to accomplish orbital rendezvous, docking and reaching the Moon. Many of the techniques pioneered by the trailblazing heroes of the Sixties continue to be used today by Shuttle and

International Space Station crews and close parallels can be drawn between Apollo lunar mission design and plans for the United States' proposed return to the Moon in 2020. However, in my mind, at least, the real achievement of that handful of early astronauts and cosmonauts is that they drew our attention away from petty problems on Earth and refocused it once more on the excitement of exploration, the thrill of discovery and the conquest of new frontiers. That legacy, that passion for adventure and that yearning to stretch our horizons, will surely drive the next generation of space explorers and inspire our next 50 years in space.

Ben Evans
Atherstone, February 2009

Acknowledgements

This book would not have been possible without the support of a number of individuals to whom I am hugely indebted. I must firstly thank my beautiful wife, Michelle, whom I married in August 2008, for her constant love, support and encouragement throughout the two long years it has taken to plan, research and write this manuscript. As always, she has been uncomplaining during the weekends and holidays when I sat up late typing on the laptop or poring through piles of books, old newspaper clippings, magazines, interview transcripts, press kits and websites. It is to her, with my love, that I wish to dedicate this book. My thanks also go to Clive Horwood of Praxis for his enthusiastic support and to David Harland for reviewing the manuscript. Others to whom I owe a debt of thanks include Sandie Dearn, Ken, Alex and Jonathan Jackson, Malcolm and Helen Chawner, Jemma Jessica James and Shaun Joicey, Vicky Farmer and Leanne and Jonathan O'Neill. To those friends who have supported and encouraged my fascination with all things 'space' over the years, many thanks: to Andy Salmon, Dave Evetts, Mike Bryce and Rob and Jill Wood. Finally, it has become customary for me to offer some grudging acknowledgement to the various pets which have entered into my extended family: the cat count presently numbers four (Simba, Sooty, Benjy and Bailey) and our own excitable pair of golden retrievers: biscuit-hunting Rosie and woodwork-chewing Milly.

1

From the East

CONTRASTS

The Sixties were a decade of contrasts. Their three thousand, six hundred and fifty-three days were marked by some of the most tumultuous, violent and devastating, yet far-reaching, inspiring and influential events in human history. They saw enormous political, social and cultural change and have been seen as a nostalgic era of peace and liberalism, overshadowed by a dark cloud of hatred, oppression and wanton excess. They began, ominously, under the longest shadow of the Cold War. Only days after the first man rocketed into space, a newly-elected United States president and a feisty Soviet premier locked horns over the fortunes of a young Cuban revolutionary, bringing the possibility of nuclear war onto an international stage.

As the decade wore on, that very same president was publicly cut down by an assassin's bullet – as, indeed, was his younger brother a few years later – and the Soviet leader was toppled from office in 1964, hours after bragging to the world of his nation's latest space triumph. Elsewhere, decades of servile colonialism drew to an end as a handful of African countries finally achieved independence from European mastery; some evolving into stable democracies fit for the modern world, others degenerating into corrupt and despotic dictatorships. Younger generations, inspired by the unequal conservative norms of the time, as well as an increasingly unpopular war brewing in Vietnam, cultivated a social revolution which swept across much of the western world.

Simmering discontent in America's black community boiled over with the 1968 murder of Martin Luther King in Memphis and, a year later, with the point-blank assassination of Black Panther Party co-founder Fred Hampton in Chicago. Meanwhile, Vietnam consumed ever-increasing numbers of lives on both sides – including, at My Lai, the infamous massacre of hundreds of unarmed civilians – and enforced conscription led to massive opposition, culminating in the 500,000-strong Moratorium protests in late 1969. Voting rights were questioned: why, asked American youth, should they die for their country if they were barred from casting at the ballot box? Remarkably, amid all this chaos and carnage, men visited the Moon and, as one observer told Apollo 8 astronaut Frank Borman, saved what was otherwise the darkest point of the decade.

On the fringes of Europe, equally divisive measures were being undertaken to forcibly separate eastern communists from the democratic west. Beginning in August 1961, less than a week after the Soviets launched their second cosmonaut, East German troops sealed borders and set about building a physical barrier between the eastern and western halves of Berlin. An initial barbed-wire fence was, by 1965, replaced by one of the most hated icons of the communist regime: the 157 km Berlin Wall. In spite of its clear violation of the Potsdam Agreement (which granted Britain, France and the United States a say over Berlin's post-war future) little effort was made to challenge the wall by force. Even President John Kennedy's administration acquiesced that its existence was "a fact of international life". Closed by chain fences, walls, minefields and manned by sharpshooters, the despised barrier would divide families, friends and communities for almost three decades.

As Soviets and Americans spacewalked outside their Earth-circling ships and raced to put a man on the Moon, efforts to promote democracy in eastern communist states, including Poland and Yugoslavia, but notably Czechoslovakia, came to nothing. The optimistic Prague reforms of Alexander Dubček in the spring of 1968 raised such alarm that 250,000 Warsaw Pact troops and thousands of Soviet tanks rumbled into the country to stifle any attempt to create a new nation of pluralism, tolerance and improved human rights. The invasion provoked widespread opposition both within Czechoslovakia – visibly expressed through the self-immolation of student Jan Palach in Wenceslas Square in January 1969 – and from beyond, even from within the Soviet Union itself. Three hundred thousand emigrations from Czechoslovakia to the west represented an exodus so high in number that it has not been seen since. Dubček himself was forced from office to ensure that, in future, his country would subordinate its interests to those of the Eastern Bloc.

Similar opposition and destruction, not merely of people and places, but of an entire way of life, commenced in 1966, as the abject failure of Chairman Mao's Five-Year Plan to bring lasting economic prosperity to China culminated in the rampages of the Red Guards and the abolition of the so-called 'Four Olds', believed to stand in the way of socialist progress. Over the next few years, old customs, old cultures, old habits and old ideas were systematically eradicated, as the old world was smashed in favour of the new. It should have granted the Chinese people their most extensive period of free speech yet seen; in reality, it was a 'freedom' severely impaired by Maoist ideology, military brutality and the biggest single attempt by a nation to eliminate its own identity ever seen in the modern age.

A revolution of a somewhat different kind came one night in February 1964, with the triumphant arrival in New York of four mop-topped Liverpudlians called the 'Beatles'; their appearance on the Ed Sullivan Show transformed them overnight into one of the few British acts at the time to achieve enormous success in the United States. The so-called 'British Invasion' was followed by an infusion of new musical talent from across the Atlantic: the Kinks, the Yardbirds, the Moody Blues, the

Rolling Stones and the Who. Yet, although the late John Peel once remarked that his distinctive Merseyside accent alone enabled him to break into American radio, the invasion was by no means restricted to music. British movies, characters and television series, from James Bond to Mary Poppins and 'A Hard Day's Night' to 'The Avengers', were met with great enthusiasm Stateside.

The British Invasion, though, formed only part of a wider 'counter-culture', which ran like a broad vein through the mid to late Sixties, encompassing demands for improved rights and freedoms for women, homosexuals and racial minorities. Rampant use of psychedelic drugs seemed to journey hand-in-hand with, and influence, the music, artwork, movies and attitudes of the time. Only months after three American astronauts died in a launch pad fire and a Soviet cosmonaut plunged to his death when his parachute failed, one of the defining moments of this counter-culture came with San Francisco's 1967 Summer of Love and the associated rise of the hippie movement. Two years later came Woodstock, although the infamous Tate-LaBianca killings of August 1969 provoked growing mistrust of the counter-culture and its lax morals. Indeed, the excesses of the period prompted Jefferson Airplane co-founder Paul Kanter to quip: "If you can remember anything about the Sixties, you weren't really there!"

It is fortunate, therefore, that another of the decade's most persistent themes will remain forever entrenched in human memory. After countless millennia spent staring up at the heavens and wondering what lay beyond the thin veil of our atmosphere, men – and, in 1963, a woman – finally broke free of their home planet. Some would spend many days circling Earth, others would open hatches and venture outside in pressurised space suits to work, still more would dock their spacecraft together and a few hardy souls would visit the Moon. By the end of the first decade of humanity's adventure in space, men would have left their footprints in lunar dust.

We would be naïve and foolish to suppose that Russia and America – communist and capitalist rivals – undertook these escapades for purely scientific and peaceful purposes, although undoubtedly both of these reasons played a part. The development of rockets capable of hurling humans into space emerged from a long-nurtured desire on both sides to send weapons across thousands of kilometres and drop them onto each other's cities. In fact, at an August 1961 press conference in Moscow to announce the flight of the second cosmonaut, Gherman Titov, a New York Tribune journalist was not interested in the scientific accomplishments of the mission, but rather in its military implications. Was Titov's Vostok 2 spacecraft, the journalist asked, capable of delivering bombs to pre-selected spots on Earth? The cosmonaut, with a hint of embarrassment, replied that it was not, but the question certainly demonstrated the reality that space was the new 'high ground' and would be exploited by both superpowers for their own ends.

VEIL OF SECRECY

On a bleak, featureless expanse of steppe, some 200 km east of the Aral Sea, lies a tiny junction on the Moscow-to-Tashkent railway, known as Tyuratam. In the local

Kazakh tongue, its name is roughly translatable as the gravesite of Tyura, beloved son of the great Mongol conqueror Genghis Khan, whose medieval empire spanned much of Asia. According to some sources, the place began as an ancient cattle-rearing settlement on the north bank of the Syr Darya River, although at least one Soviet-era journalist expressed preference for giving it a more modern origin, hinting at its foundation as recently as 1901 as an outpost to refill steam engines passing between Orenburg and Tashkent.

Its significance over the past half a century, though, cannot be disputed. It was from this sparsely populated region, five decades ago, that the first steps of a journey far more audacious, much longer and considerably more difficult than any the Great Khan could have envisaged were taken. It was this place that Gary Powers, following the line of railway tracks in his U-2 reconnaissance aircraft, tried to find when he was shot down in May 1960. It was this remote corner of old Soviet Central Asia – a region swarming with scorpions, snakes and poisonous spiders, whose climate is characterised by vicious dust storms, soaring summertime highs of 50°C and plummeting wintertime lows of –25°C – that a young man, clad in a bloated, pumpkin-orange suit and glistening white helmet, sat atop a converted ballistic missile, defied all the odds and took humanity's first voyage beyond the cradle of Earth.

One morning in April 1961, as five months of bitter snow and fierce, hurricane-strength blizzards yielded to the first murmurings of spring on the steppe, his kind achieved what had previously existed only in dreams. He rose from Earth, as Socrates once said, right to the top of the atmosphere and beyond and obtained a glimpse of the world from which he came. Contrary to some long-held expectations, the vista that Yuri Alexeyevich Gagarin beheld from 175–300 km high was not flat, nor could he discern any atlas-like lines dividing the countries, nor still, it is said, did he perceive any physical notion of God. Instead, he saw a beautiful, fragile oasis; a world iridescent with life and colour, encircled by what he described as "a very distinct and pretty blue halo" of an atmosphere, which almost merged into the blackness of space beyond. Flying at 28,000 km/h, in his journey of just 108 minutes, he somehow managed to plant fleeting images into his brain of rivers, islands, continents, forests and mountains. Never before had they been seen from so high by human eyes.

It is hardly surprising that the site from which he left Earth – known today simply as 'Gagarin's Start' and still used to blast humans into space – was kept under wraps by the Soviet government. Indeed, in the early Seventies, when American astronaut Tom Stafford asked to visit the site in readiness for the joint Apollo-Soyuz mission, he met stubborn resistance. The Soviets' desire to mislead and confuse prying westerners about this ultra-secret place was pursued to such an extent that even its name remained imprecise. Today, it is still variously known as Tyuratam, after the tiny railhead, or, more often, as Baikonur, which covers a wider and different geographical area. In fact, the town of Baikonur is more than 200 km from the launch base. For this reason, at a 1975 press conference, ABC News anchorman Jules Bergman expressed displeasure at the Baikonur name, pointing out that, despite its diminutive size, Tyuratam is actually closer.

Clad in space suit and helmet, the first man in space, Yuri Alexeyevich Gagarin, bids farewell to the genius who made his flight possible, Sergei Pavlovich Korolev.

Whatever one's preference, in February 1955 the site was chosen for a research and testing facility for the R-7 intercontinental ballistic missile; a missile developed by Sergei Korolev, the famed 'chief designer' of early Soviet spacecraft and rockets, originally to deliver huge warheads across distances of several thousand kilometres and, later, to send the first men into orbit. Assembly of the R-7 base – consisting of airports, rocket hangars, control blockhouses and the first of several colossal launch pads – was completed in a little over two years and, on 4 October 1957, one of these behemoths carried the world's first artificial satellite, Sputnik 1, aloft. Within a month, a living creature, the dog Laika, was boosted into space aboard Sputnik 2. However, not all missions were successful: two of Korolev's Mars-bound probes exploded shortly after liftoff and in October 1960 an R-16 missile misfired on the pad, destroying the launch complex in a conflagration which claimed the lives of almost 130 technicians, military officers, engineers and Marshal Mitrofan Nedelin. An American reconnaissance satellite, which overflew the site a day later, saw only a blackened smudge across the barren steppe. The awful truth of exactly what happened would not reach western ears for decades.

Ironically, only days before the R-16 disaster, Premier Nikita Khrushchev had boasted to the United Nations that Russia was producing intercontinental ballistic missiles "like sausages from a machine". On the fateful night of 23 October, just half an hour before its scheduled liftoff, the R-16 exploded, destroying the launch pad and breaking in half. Everyone in the vicinity of the inferno was either incinerated in the 3,000°C temperatures or succumbed to the missile's toxic propellants. Marshal Nedelin's remains were recognisable only from a Gold Star pinned to his uniform, whilst another man was identified from the height of his burned corpse. Although the R-16 was not directly connected to the R-7, which would be used for piloted missions, its loss caused an inevitable delay to the first manned space launch. In fact, many design organisations were involved with both the R-16 and R-7 and Nedelin himself chaired the State Commission for the man-in-space effort.

Consequently, in spite of its relative youth, the place already had historic and tragic attributes by the time Nikolai Kamanin, head of the newly-established cosmonaut team, arrived there in the spring of 1961 to oversee final preparations to send a man into space. The middle-aged Kamanin was one of the Soviet Air Force's most distinguished generals, having led air brigades and divisions during the Second World War. In 1934 he had received the coveted Hero of the Soviet Union accolade for his role in the daring rescue of the icebound steamship Chelyuskin on the frozen Chukchi Sea. Throughout the Sixties, as the cosmonauts' commander, Kamanin frequently disagreed with Korolev over differing policies, attitudes and requirements for the spacecraft and rockets, the men who would ride them and the often whimsical desires of the Soviet leadership. His memoirs are preserved in a series of quite remarkable diary entries, first published in 1995, which reveal a tough, bitter man who would blame his country's loss of the Moon race on Soviet engineers' unwillingness to give cosmonauts active control of their spacecraft.

Kamanin's diaries paint a portrait of a man who fought fiercely for 'his' cosmonauts and show the close relationship between them during their time together on the isolated Kazakh steppe. However, he has also been described by space analyst Jim Oberg as an "authoritarian space tsar, a martinet" and by Soviet journalist Yaroslav Golovanov as "a malevolent person ... a complete Stalinist bastard". Others, including cosmonaut Alexei Leonov, have proven more complimentary, seeing him as "very approachable" with a keen love of sports, especially tennis. Still, his brand of leadership, in most cases, successfully prepared the first generation of space explorers for their ventures into the heavens.

By the beginning of April 1961, wrote Kamanin, a number of obstacles remained to be overcome before a man could be launched. The basic design of his spacecraft had already been established and tested under the smokescreen name of 'Korabl-Sputnik' ('Spaceship-Satellite') which, between May 1960 and March 1961, had ferried dogs, rats, mice, flies, plant seeds, fungi and even a full-sized human mannequin famously nicknamed 'Ivan Ivanovich' into orbit. These missions evaluated everything from the spacecraft's habitability to the performance of its ejection seat. Some proved to be dismal failures: the retrorockets of one mission fired in the wrong direction, sending the capsule into a higher orbit, while a July 1960 attempt exploded seconds after liftoff, killing its two canine passengers, Chaika and Lisichka. Others, notably the flight of the dogs Belka and Strelka, were hugely successful.

The latter were launched at 11:44 am on 19 August, accompanied by mice, insects, plants, fungi, cultures, seeds of corn, wheat, peas, onions, microbes, strips of human skin and other specimens. Two internal cameras provided televised views of them throughout the day-long mission. At first, the images showed the dogs to be deathly still – Belka, in particular, squirmed uncomfortably and vomited during the fourth orbit – prompting medical chief Vladimir Yazdovsky to gloomily recommend no more than one circuit for the first manned flight. The dogs' return to Earth, however, was perfect and the capsule landed just 10 km from its intended spot in the Orsk region of the southern Urals. Belka and Strelka earned their places in history as the first living creatures recovered safely from orbit.

Upon examination, both were found to be in excellent condition, with no fundamental changes to their health. This data, together with the exemplary performance of the capsule's systems, provided encouragement that a Soviet man could be launched before the year's end. In fact, documentation from the Council of Chief Designers to the Central Committee of the Communist Party, produced around this time and finally declassified in 1991, revealed the formal timetable for sending a human pilot aloft. It recommended one or two more test flights in October and November 1960, before attempting a manned shot in December. Signed by ministerial heads, rather than the standard deputy ministers, the document clearly reflected how important the man-in-space effort was to the Soviet leadership.

Known as 'Vostok' ('East' or 'Upward Rising'), the machine that Yuri Gagarin, Gherman Titov and others would fly comprised a spherical cabin to house the cosmonaut and a double-cone-shaped instrument section. However, unlike the United States' man-in-space effort, the true form of Vostok remained hidden from the world and would not be revealed in its entirety until a full-scale model appeared at the Moscow Economic Exhibition in April 1965. Until then, the Soviet Union's propaganda apparatus continued to misinform western observers as to precisely what kind of spacecraft had placed the first man into orbit. Careful to maintain the ambiguity, Gagarin himself waxed lyrical, cryptically describing it as "more beautiful than a locomotive, a steamer, a plane, a palace and a bridge; more beautiful than all of these creations put together". His praise, though under-standable, was not especially helpful. In the four years before Vostok was finally unveiled, the world could rely only on brief clips from Soviet documentaries and scenes from the Moscow Parades, which variously showed a contraption with an attached rocket stage, payload shroud and even, in the case of Vostok 2, a pair of short, stubby wings.

Today, at the Tsiolkovsky Museum in Kaluga, just south-west of Moscow, Vostok is presented for what it was: a 4,730 kg monster of a spacecraft, some 4.4 m in length and 2.4 m wide. Its capsule – nicknamed 'the ball' or 'sharik' ('little sphere') by the cosmonauts – comprised a little over half of its total weight, rendering it so heavy that not only a hefty parachute, but also an impact-cushioning rocket, would be needed to bring a man safely to the ground. Since this additional weight would have pushed it above the R-7's payload capacity, Soviet designers incorporated an escape system to stabilise Vostok's own descent by parachute, then allowing the cosmonaut to eject at a relatively low altitude of 7 km and land under his own canopy. During his descent, he would separate from his seat and touch down at about 5 m/sec. The Vostok, on the other hand, would impact roughly twice as fast, easily sufficient to injure the cosmonaut had he remained inside.

Questions of whether or not cosmonauts remained aboard their capsules throughout descent and landing were by no means insignificant. In order for such flights to be officially recognised for their achievements – specifically by taking the World Aviation Altitude Record – the Fédération Aéronautique Internationale (FAI) required pilots to remain inside their machines from launch until landing. This would sidestep any accusation that they had been obliged to abandon their ships due to problems and ensure that their missions would be considered a successful 'first'. The reality that none of Vostok's fliers accompanied their capsules to the ground, but still made successful flights, was kept carefully hidden by the Soviets for many years.

In June 1960, Korolev took Gagarin, Titov and 18 other cosmonauts to his OKB-1 design bureau in Kaliningrad, north-east of Moscow, to see the first Vostoks in production. Their silvery spheres contained no aerodynamics, control surfaces or propulsion systems and, with their double-coned instrument sections, could only stand upright with the aid of metal frames. All of the cosmonauts were fighter pilots who could barely comprehend what they were seeing. All were enthusiastic to someday guide these machines through space, but none could understand how, without wings, they were supposed to do it.

Within the capsule, they found a tan-coloured rubber cladding, covering a myriad of wiring and piping, with little obvious instrumentation, save for a single panel containing switches, status indicators, a chronometer and a small globe representing Earth. Other systems would provide Vostok's fliers with temperature, pressure, carbon dioxide, oxygen and radiation readings, as well as ticking off each circuit of their home planet. The bulky ejection seat occupied much of the cabin. At its foot was the 'Vzor' ('Eyesight') periscope to indicate that the spacecraft was correctly oriented for atmospheric re-entry. It consisted of a central view, encircled by eight ports; when Vostok was perfectly centred with respect to the horizon, all eight would be lit up. Within reach of the cosmonaut was a small food locker, providing up to ten days' worth of supplies, and herein lay one of the problems that, in April 1961, stood in the way of future flights.

Attached to the base of the capsule was the double-coned instrument section, some 2.25 m long and 2.4 m wide and weighing 2,270 kg. Spread around the 'waist' between the two sections was a set of 16 spherical oxygen and nitrogen tanks for Vostok's life-support system. At the bottom was the TDU-1 retrorocket, which employed a self-igniting mixture of nitrous oxide and an amine-based fuel. Capable of delivering a total thrust of 1,614 kg with a specific impulse of 266 seconds, the device would operate for 45 seconds with a 275 kg propellant load, slowing Vostok by around 155 m/sec to permit atmospheric entry at the end of a mission.

As long as the retrorocket operated without incident, the best opportunities for bringing the capsule back to Earth, and back to Soviet territory, were after one orbital pass – as was planned for Gagarin's flight – or a full day later, somewhere in the midst of the 17th revolution. Admittedly, it could be fired at any other time, if necessary, but at the risk of bringing Vostok down within foreign borders or possibly

The Vostok spacecraft during construction in the assembly shop.

into the sea. However, in the worst-case scenario that the retrorocket should fail to fire at all, the cosmonaut may have had to remain in orbit for up to ten days until his spacecraft naturally decayed from orbit. Although food and water were made available for such a long flight, the capsule's lithium hydroxide canisters, meant to scrub exhaled carbon dioxide from the cabin, proved insufficient and in ground tests were expended within four days. Since Gagarin's mission would last only a couple of hours, Kamanin felt that this would not preclude the ability to launch him in April 1961, but believed a new approach would be needed for longer flights.

Other obstacles had arisen during the Korabl-Sputnik missions, among them the 800 kg ejection seat. Although the ejection of a dog named Chernushka from the Korabl-Sputnik 4 spacecraft on 9 March 1961 and that of another dog, Zvezdochka, together with the life-sized mannequin Ivan Ivanovich, from Korabl-Sputnik 5 two weeks later, were successful, sea-based trials proved harder to quantify. Despite these misgivings, after three successful tests on the ground and from an Il-28 aircraft, coupled with the Korabl-Sputnik results, the seat was declared ready.

Months earlier, in September 1960, Korolev had submitted his proposal for a human flight to the Central Committee of the Communist Party and received approval. His original plan was to stage the mission before the end of that year, but the failure of Korabl-Sputnik 3 on 1 December – in which the dogs Pchelka and Mushka were incinerated, said the Soviets, when their capsule re-entered the atmosphere at too steep an angle – cast doubt on this schedule. (In reality, Korabl-Sputnik 3 had suffered a failure of its TDU-1 retrorocket and was remotely destroyed lest it land in foreign territory.) Another attempt just three weeks later uncovered a rare anomaly with the R-7 booster itself, whose third stage ran out of thrust halfway to space, although both of its canine passengers were ejected safely ... landing thousands of kilometres off-course in a remote and inhospitable area of Siberia. Earlier in the year, another R-7 had exploded seconds after liftoff, ironically with many of the cosmonauts on hand to watch. "We saw how it could fly," Gherman Titov said darkly. "More important, we saw how it blows up."

By 7 April 1961, the final decision to go ahead with the mission was made and events moved rapidly. The Soviets were keenly aware that the United States' first attempt to put a man into space was imminent, perhaps as soon as 28 April, although Kamanin firmly believed that Vostok would beat them. By this time, Gagarin and Titov had been selected as the prime and backup candidates for the flight. Their last few days were spent undergoing refresher classes on spacecraft and rocket systems, including the troublesome ejection seats. At one meeting, Kamanin reminded them of the option to fire the seat manually in the event of an emergency. Titov expressed total confidence in the seat and felt that worrying about it was a waste of time. Gagarin, on the other hand, offered a more considered response, perhaps so as not to embarrass Titov or the seat's engineers, by pointing out that although his confidence was high, the manual option increased his chance of survival. He certainly knew how to play the game of cosmonaut politics.

It is difficult, though, to determine if one single event allowed the decision of Gagarin over Titov to be made. In their 1998 biography of Gagarin, Jamie Doran and Piers Bizony stressed that Kamanin himself had a hard time deciding between the two men. Only days before the mission, he noted in his diary that Titov completed his training more accurately than Gagarin, but that Titov's "stronger character" put him in a better position to fly Vostok 2, which was planned to spend a full day in space. Others have hinted that Korolev simply liked Gagarin from their first meeting, that his calmness under duress and even his respectful removal of his boots before entering a Vostok capsule in the OKB-1 workshop may have played a part. Of equal, perhaps overriding, significance from the Soviet government's point of view was the political need to favour a humble farmboy (Gagarin) over Titov, a more bourgeois teacher's son. Even the simplicity of character between Gagarin and Soviet premier Nikita Khrushchev, both of whom came from peasant roots, has been cited, together with Titov's perceived lack of charm, reserved personality and 'strangeness' when spouting from memory reams of poetry or quotes from tsarist literature.

Certainly, an early evaluation of Gagarin's personality, conducted in August 1960 by Soviet Air Force physicians and psychologists, was highly favourable of his 'other' talents. It read: "Modest: embarrasses when his humour gets a little too racy: high degree of intellectual development evident in Yuri; fantastic memory; distinguishes himself from his colleagues by his sharp and far-ranging sense of attention to his surroundings: a well-developed imagination: quick reactions: persevering, prepares himself painstakingly for his activities and training exercises, handles celestial mechanics and mathematical formulae with ease as well as excels in higher mathematics: does not feel constrained when he has to defend his point of view if he considers himself right: appears that he understands life better than a lot of his friends … " He seemed the perfect choice.

On 9 April, that choice was publicly made and filmed in colour by the official cameraman, Vladimir Suvorov. In reality, the decision had been made in secret the previous day, after which Kamanin had told both cosmonauts the outcome. Titov, he wrote, was visibly disappointed to have lost the flight, but both men mimed their way through 'spontaneous', though pre-rehearsed, speeches. Ironically, Suvorov's camera ran out of film halfway through the acceptance speech, forcing Gagarin to repeat everything, word for word.

At 5:00 am Moscow Time on 11 April, the enormous R-7 booster, carrying the Vostok 1 spacecraft – simply labelled 'Vostok', so as to give no hint to the world that it might be the first in a series of missions – left the main assembly building in a horizontal position atop a railcar. Korolev, who knew it intimately as his 'Semyorka' ('Little Seven'), accompanied it to the launch pad. Fuelled by liquid oxygen and kerosene, it consisted of a two-stage core, measuring 34 m long and 3 m in diameter and weighed 280,000 kg. Strapped around the core were four tapering boosters. Upon arrival at the pad, it was raised to a vertical position, ready for the arrival of its human passenger early the next morning. Liftoff was set for 9:07 am on 12 April, with plans calling for the jettison of the strap-on boosters two minutes into the flight, followed by insertion into low-Earth orbit at 9:18 am. Half an hour later, Vostok would orient itself in preparation for an automatic retrofire at 10:25 am, parachute deployment from the capsule at 10:43 am, ejection of Gagarin at 10:44 am and touchdown of both shortly thereafter. The entire flight would be shorter than one of today's Hollywood blockbusters.

That night, the prime and backup cosmonauts stayed in a cottage close to the pad, their every toss and turn monitored by strain gauges fitted to their mattresses to allow physicians to determine whether they experienced restful sleep. The results indicated that they did, but Gagarin would later admit to Korolev that he hardly rested at all and spent much of his time trying to remain perfectly still in bed, so that he would be declared well prepared to fly the following morning. Months later, he would joke with Kamanin that the only reason Titov did not ride Vostok 1 was because he had rolled over in his sleep. Korolev also slept very little. His major concern was that the R-7's third stage might fail during the ascent, perhaps dropping the spacecraft into the ocean near Cape Horn, an area notorious for its violent storms. Shortly before the launch, he demanded that a telemetry antenna be set up at Tyuratam to confirm the satisfactory operation of the third stage; if it worked as planned, the telemetry data would print out a string of 'fives' on tape, but if not, there would be a string of 'twos'.

At 5:30 am on what is now universally known as 'Cosmonautics Day', 12 April 1961, Korolev and his head of medical preparations, Vladimir Yazdovsky, woke Gagarin and Titov from their slumbers. After washing, shaving and a breakfast of meat puree and toast with blackcurrant jam, physicians glued sensor pads onto their torsos and sent them to the spacecraft assembly building to don their pumpkin-orange suits. These had been designed by Gai Severin, the Soviet Union's most accomplished manufacturer of attire and ejection seats for MiG fighter pilots. Severin utilised several key elements of earlier suit designs – including a tight fit around the legs to prevent blood from pooling into the lower torso and starving the supply to the brain – to protect Gagarin from the rapid acceleration of the R-7. The main layers of the suit, designed and built in only nine months, consisted of a blue-tinted rubber material, overlaid by the high-visibility orange coverall.

On launch morning, Titov donned his suit first in order to reduce Gagarin's time spent overheating inside his own uncomfortable garment. As he continued his own suiting-up, Gagarin realised for the first time that he was – or soon would be – the most famous man on Earth. In his heavily-censored account of the mission, 'The Road to the Stars', published later in 1961, he recalled technicians offering him slips of paper and work passes on which to scrawl his signature. Titov beheld this and wished Gagarin luck, although he was disappointed and, even as he rode the bus to the launch pad, considered his role that day as hopeless. "He was commanding the flight and I was his backup," Titov said later, "but we both knew, 'just in case' wasn't going to happen. What could happen at this late stage? Was he going to catch the flu between the bus and the launch gantry? Break his leg? It was all nonsense! We shouldn't have gone out to the launch pad together. Only one of us should have gone." Vladimir Yazdovsky, who was also aboard the bus and gave him the order to remove his suit as soon as Gagarin was strapped inside Vostok, recalled Titov's tension. Admittedly, Titov would fly Vostok 2 in a few months' time, but history would not recall the Second Cosmonaut with the same clarity as the First.

Much tradition surrounds Gagarin's trip to the pad, including, famously, his need to relieve himself through his suit's urine tube against the tyres of the bus. Unable to share the going-away custom of kissing three times on alternate cheeks, he and Titov merely clanked their helmets together in a sign of solidarity. Korolev gave Gagarin a tiny pentagon of metal – a duplicate of a plaque flown on the Luna 1 probe two years before – and wryly suggested that, someday, perhaps the First Cosmonaut could pick up the original from the Moon's dusty surface.

After reaching the top of the gantry, Gagarin manoeuvred himself into his ejection seat. Senior engineer Oleg Ivanovsky and chief test pilot Mark Gallai tightened his harnesses and plugged his suit hoses into Vostok's oxygen supply. After giving him a good-luck tap on his helmet, Ivanovsky motioned for Gagarin to lift his faceplate to inform him of three significant numbers.

Since the beginning of their adventure in space, the Soviets had operated their machines through on-board systems controlled exclusively from the ground and, for a time, considered that manned missions should be undertaken in the same way. What would happen, wondered the medical experts, if a cosmonaut went mad in orbit, overcome by a profound sense of separation from his home planet, or even attempted to defect to the west, deliberately bringing his capsule down onto foreign soil? Guidance, it was decided, must be automatic.

However, if the cosmonaut's sanity and devotion to the Motherland could be demonstrated, and if he needed to assume command, a six-digit keypad was provided to unlock Vostok's navigation system, disengage it from the automatic controls and allow him to fly the ship. The three-digit combination for the keypad would be radioed to Gagarin if ground staff considered him sane enough to take over. Yet the logic was questionable: what would happen if Vostok lost attitude control or its radio went dead and communications were impossible? Instead, a face-saving measure was adopted, whereby the three-digit code would be kept in a sealed envelope aboard the spacecraft, ready for the cosmonaut to open if needed. Gagarin's ability to open the envelope, type in the numbers and activate the keypad would supposedly 'prove' that he had not lost his mind and was fully aware of his actions.

"It was a dangerous comedy," Ivanovsky recalled later, "part of the silly secrecy we had in those days." The envelope, though, had to be placed somewhere within easy reach, should he need to get to it in an emergency and a mentally unstable Gagarin could easily have opened it if he really wanted to do so. Mark Gallai, whose role included supervising the training of the Vostok cosmonauts, agreed, pointing out that all were qualified military pilots with experience of flying high-altitude, nighttime missions. The chance of them going mad was considerably less likely than suffering a radio failure. Consequently, betraying an official state secret and theoretically putting himself at risk of a lengthy prison spell, Ivanovsky told Gagarin the three digits: 1-2-5. To his surprise, the cosmonaut smiled and replied that Kamanin had already given him the combination!

Assisted by Tyuratam's chief of rocket troops, Vladimir Shapovalov, and two launch pad staff members, the next task for Ivanovsky and Gallai was to seal Gagarin inside his capsule for liftoff. At this point, a problem reared its head. A series of electrical contacts encircling Vostok's hatch should have registered a signal – known as 'KP-3' – to Korolev and his control team in the nearby blockhouse, informing them that it was secure. Furthermore, the signal was supposed to confirm that explosive charges around the hatch could jettison it at a millisecond's notice in the event of an emergency and enable Gagarin to eject. On the gantry, the contacts seemed fine and the enormous hatch – which "weighed about a hundred kilos and was a metre wide," according to Ivanovsky – was manhandled into place and the laborious process of screwing its 30 bolts began. No sooner had they finished, the launch pad's telephone rang. No KP-3 signal had been received, barked Korolev, and he demanded that they unscrew, remove, then reseal the hatch. After more fiddling, it was done. This time, thankfully, the KP-3 signal came back clearly.

It was now less than 40 minutes away from the projected 9:07 am launch. Ivanovsky, Gallai and the remaining personnel left the pad for the nearest control bunker. Vladimir Suvorov, determined to seize the most important photo opportunity of his career, stayed out in the open and would record some of the 20th century's most remarkable imagery as the first man headed into space. Elsewhere, Gherman Titov was midway through stripping off his own space suit and his neckpiece was halfway over his head when the attending technicians disappeared to watch the launch. Meanwhile, alone in his tiny capsule, the young Soviet Air Force senior lieutenant, who had celebrated his 27th birthday and the birth of his second daughter a few weeks earlier, could scarcely have believed that his humble, peasant-stock roots in the small Russian village of Klushino could possibly have brought him this far.

'STAR SAILORS'

A little more than a year earlier, Gagarin had been just one of hundreds of Soviet military pilots who received unusual instructions to undergo classified briefings and physical and psychological tests as part of an entirely new, and mysterious, aviation project. The search for the world's first spacefarers began in earnest in May 1959, when representatives of the armed forces, the scientific community and the design bureaux met at the Soviet Academy of Sciences under the supervision of Vice-President Mstislav Keldysh to discuss methods of selecting the most suitable candidates for Earth-orbital missions. Aviators, rocketeers and even car racers were considered in the early days, but, at length, bowing to Soviet Air Force pressure, Keldysh agreed to narrow the selection criteria to qualified pilots from this branch of the military.

Despite its obvious vested interest in wanting to have 'its' fliers taking the first manned spacecraft beyond the atmosphere, the logic was inescapable: Soviet Air Force pilots had proven themselves under exposure to hypoxia, high pressures and varying G loads and had undergone rigorous ejection-seat and parachute training. In addition to their flying experience, candidates would only be admissible if they could meet the height and weight requirements of the Vostok spacecraft: they needed to be no taller than 1.75 m and weigh no more than 72 kg. Moreover, anticipating that they would be embarking on lengthy careers as 'cosmonauts' (the word literally means 'star sailor'), the age limit was set between 25–30 years old.

Throughout 1959, groups of physicians were sent to a number of air bases in the western Soviet Union and by August the selection teams had the records of more than 3,000 pilots ready for inspection. Most of these were eliminated at a fairly early stage, on the basis of not meeting the height, weight, age or medical criteria; some, indeed, were dropped for bronchitis, angina, gastritis and colitis, renal and heptic colic and pathological cardiac shifts. The remainder were then systematically interviewed from early September, still unaware of exactly what the so-called 'special flights' project entailed. Three thousand soon became a little over two hundred, who were despatched in groups of about 20 for further tests at the Central Scientific Research Aviation Hospital in Moscow. In addition to more interviews, the candidates were spun in stationary seats to assess their vestibular apparatus, placed in low-pressure barometric chambers and wrung through a centrifuge to evaluate their performance under high-gravity loads. Original plans, it seemed, called for seven or eight pilots, but Sergei Korolev insisted on tripling this number, for no other reason than because he wanted a larger team than the United States' seven-strong Mercury group.

In January 1960, Konstantin Vershinin, commander-in-chief of the Soviet Air Force, formally signed plans to establish a centre for cosmonaut training in Moscow. Although it was under the control of physicians, the Air Force General Staff eventually assumed command of cosmonaut affairs, under Nikolai Kamanin. It was Kamanin who formally approved a final shortlist of 20 Air Force cosmonaut candidates in late February: Ivan Anikeyev, Pavel Belyayev, Valentin Bondarenko, Valeri Bykovsky, Valentin Filatyev, Yuri Gagarin, Viktor Gorbatko, Anatoli

Kartashov, Yevgeni Khrunov, Vladimir Komarov, Alexei Leonov, Grigori Nelyubov, Andrian Nikolayev, Pavel Popovich, Mars Rafikov, Georgi Shonin, Gherman Titov, Valentin Varlamov, Boris Volynov and Dmitri Zaikin. Their ages ran from just 23 in Bondarenko's case to as old as 34 in that of Belyayev; this criteria was waived in a couple of instances out of respect for their exemplary performance during testing. Some of them would become the most famous names in spaceflight history, whilst others would disappear into anonymity ... and, in a few cases, disgrace.

On 7 March, the 20 cosmonauts were given their welcoming speech by Vershinin at the Central Scientific Research Aviation Hospital. A week later, after settling their affairs at their respective air bases, they began training with Vladimir Yazdovsky's first class in aerospace medicine. The following four months were consumed by a mixture of in-depth lectures and an intense physical fitness regime, the latter of which included two hours daily of intensive calisthenics at the Central Army Stadium in Moscow. Parachute training was also conducted in the Saratov region, near Engels, using a converted An-2 aircraft and within six weeks each of the candidates had made between 40–50 jumps over water and land, from high and low altitudes and during daytime and nighttime.

It was partway through this training regime that the Air Force began exploring a number of more suitable sites for the cosmonauts to continue their preparations. Two possible places were identified: one in Balashikha and the other close to the Tsiolkovsky Railway Station in Shelkovo. The latter was eventually chosen in recognition of its isolated location, large area and proximity to Korolev's OKB-1 design bureau, the Academy of Sciences and the Monino airfield. The new cosmonauts and their training staff relocated to the new site at the end of June 1960 and the suburb itself, near Shchyolkovo, some 30 km north-east of Moscow, came to be known as Zelenyy ('Green'). Nowadays, it has become world-famous as 'Zvezdny Gorodok' (variously 'Star City' or 'Starry Township').

With the appearance of the new cosmonaut town came the development of the first spacecraft simulators in which they could work. Known as 'TDK-1' and built by the M.M. Gromov Flight Research Institute, it received its first trainees – Gagarin, Kartashov, Nikolayev, Popovich, Titov and Varlamov, nicknamed 'The Vanguard Six' – in the summer of 1960. The make-up of this group changed almost immediately, however, when a reddening was discovered on Kartashov's spine, diagnosed as haemorrhages and he was dropped from the Six. He was eventually dismissed from the cosmonaut team in April 1962. Shortly after Kartashov's removal, Varlamov was involved in a swimming accident in which he displaced a cervical vertibra and disqualified himself from consideration. Their places were taken by Bykovsky and Nelyubov. The 'new' Six formed a cadre who would vie for the chance to become the Soviet Union's first man in space. Indeed, with the exception of Nelyubov, they would fly five of the six Vostok missions. First among them, of course, was Yuri Gagarin, whom many cosmonauts felt had been a strong contender right from the first time he met Korolev. His journey to the stars had begun, rather strangely, as a child saboteur.

FARMBOY

After checking his safety, the boy crept over to a pile of Panzer batteries and began dropping fistfuls of soil into their accumulator caps. On other occasions, he might deliberately muddle up their chemicals, pouring them into the wrong compartments, so that SS commanders would return furiously to Klushino at nightfall to complain about their tanks' dead batteries. Sometimes, he would even shove potatoes deep inside the exhaust pipes of German military cars. It was the autumn of 1942. The Nazi invasion had begun a year before, but, after scoring several major successes, they had now been drawn so deep into Soviet territory that the harsh Russian winter prompted a lengthy retreat. The Smolensk district, some 160 km west of Moscow and containing Klushino, lay directly in the path of the Nazi fallback. It was here that the young boy, Yuri Alexeyevich Gagarin, born on 9 March 1934, saw war for the first time. Yet although his childlike attempts at sabotage certainly helped his village's struggle against the German occupiers, he didn't do it for patriotism. He did it to avenge himself on the Devil.

'The Devil' was a red-haired Bavarian known only as Albert, whose job included collecting flat batteries from German vehicles and replenishing them with acid and purified water. The Klushino children had already used broken glass to burst their tyres and, one day, in retaliation, Albert tried to murder Gagarin's younger brother, six-year-old Boris, by stringing him from an apple tree with a woollen scarf. The attempt failed, but Albert still managed to evict the entire family from their home. Later, the two elder Gagarin children were abducted and taken to Poland, their father was beaten and their mother slashed with a scythe. It was just the start of what Gagarin's fellow cosmonaut Alexei Leonov would later describe as "a very dark period for our country".

Following the expulsion of the Germans from Klushino in March 1944, the family eventually moved to nearby Gzhatsk, building a new home, and as Gagarin entered his teens he and Boris learned to read from Russian military manuals. However, it was witnessing a wartime dogfight between two Soviet Yak fighters and a pair of German Messerschmitts that kindled Gagarin's interest in aviation, garnered a fascination with mathematics and physics and led to model aircraft clubs and maddening demands for his father to build him miniature gliders. By 1950, he had applied – but was not accepted – to study at the College of Physical Culture in Leningrad, hoping to become a gymnast or sportsman. Instead, he took his second option as an apprentice foundryman at the Lyubertsy Steel Plant in Moscow. His work record led directly to training at the newly-built technical school in Saratov on the Volga River and, while there, he saw a notice for an aeroclub, which he promptly joined.

After graduating from Saratov in 1955, and by now with considerable flying experience in an antiquated Yak-18 trainer through the aeroclub, Gagarin was recommended by his instructor for the pilots' school in Orenburg. This required him to sign up as a Soviet Air Force cadet, which neither of his parents found particularly appealing, and one of his Orenburg instructors, Yadkar Akbulatov, considered him by no means a piloting prodigy and felt that Gagarin may have failed the school altogether if he could not land without bouncing on his tyres. Ultimately, Akbulatov recalled later, Gagarin solved his problem by placing a cushion under his seat to gain a clearer line of sight with the runway. Never again, it is said, did he fly without the benefit of a cushion.

Whilst at Orenburg, Gagarin met his future wife, Valentina, whom he married in October 1957, only three weeks after the launch of Sputnik 1. By this time, he had begun flight training in high-performance MiG-15 jets, successfully qualified with outstanding grades from the pilot's school and earned a lieutenant's commission with a posting to the Nikel base at the northern tip of Murmansk. During his time at this remote place, 300 km north of the Arctic Circle, he flew MiG-15 reconnaissance missions, enduring terrible weather conditions, with ice on his control surfaces and snow blindness proving an almost daily hazard. Then, in October 1959, with his wife and two-year-old daughter desperate to leave the sub-zero setting, something happened that would change his life forever.

At first, the recruiting team that arrived at Nikel sought interviews with the pilots, in which mysterious questions were posed about flying in "more modern planes", then "flying something completely new" and later "long-distance rocket flights". As the pilots were winnowed into smaller groups, they were sent to the Bordenko Military Hospital in Moscow for extensive medical and psychological screening. Finally, after more than 2,000 candidates had been poked and prodded, on 8 March 1960 a cadre of 20 pilots was selected from throughout the Soviet Union to fly the first missions into space. Most were twentysomethings – the intention being that they would be hired whilst young to prepare them for careers as cosmonauts – and each was less than 170 cm tall and below 70 kg in weight to meet Vostok's limitations.

For the remainder of 1960, the trainees undertook academic and physical work in Moscow's major scientific establishments, including the Zhukovsky Academy of Aeronautical Sciences on Leningradsky Prospekt and the Institute of Medical and Biological Problems, near Petrovsky Park. Gagarin and the others were sealed, individually, inside isolation chambers for days at a time, with the barest of provisions, as pressure levels were alternately raised and lowered and the trainees given mind-numbing tasks to test their ability to overcome the tedium of lengthy flights. Towards the end of one 24-hour session, Gagarin even tried to alleviate his boredom by conjuring up songs about the electrodes taped to his chest. At other times, the cosmonauts were whirled in a centrifuge – sometimes peaking at 13 G – which caused their breathing to become laboured, their facial muscles to twist and contort and, said Gagarin, "the blood in my veins felt as heavy as mercury". Still more torture was effected through oxygen-starvation experiments, which made the more 'normal' pilot-training activities, such as parachute jumps, a blessed relief.

On 6 January 1961, six of the cosmonauts – Gagarin, Titov, Valeri Bykovsky, Andrian Nikolayev, Pavel Popovich and Grigori Nelyubov – were selected for final examinations prior to the announcement of who would fly first. Eleven days later, the six men were each sealed inside a Vostok simulator for almost an hour apiece to describe the equipment and operations to be conducted during each phase of a mission. Their understanding of how to orient the spacecraft for a manual retrofire and properly egress the capsule was scrutinised. Gagarin, Titov, Nikolayev and

Popovich were rated "outstanding" by the examiners, with Nelyubov and Bykovsky a close "good". Written examinations followed, after which all six were proclaimed fit to fly, with Gagarin, Titov and Nelyubov categorised as the best candidates.

Three days before launch, Gagarin was chosen and Titov would later agree that from an image-conscious Soviet perspective, he was indeed the perfect choice. "Look at his biography," Titov recalled years later. "Yuri Gagarin is an ordinary man from Smolensk. During the war, his brother and sister were driven away. After graduating, he went to a factory school and became a foundry worker, then followed the Saratov industrial specialised school, an air club and a flight school. He was a lad who made his dream come true all by himself, without a father and mother to help him." In fact, on 12 April 1961, as Gagarin waited atop Korolev's latest Little Seven, ready for his chance to make history, his parents, brothers and sister had absolutely no knowledge of what he was about to do. By nightfall, Yuri Alexeyevich Gagarin would be the most famous man on the planet.

"LET'S GO!"

His voice exhibited clarity, calmness and confidence when Korolev called over the radio an hour before launch for a status update. To alleviate the boredom, Pavel Popovich, the cosmonaut on duty in the control bunker, arranged for some music to be played. This stopped abruptly at 8:51 am as Korolev gruffly announced that, with a little over 15 minutes to go, it was time for Gagarin to seal his gloves and close his visor. Unlike their counterparts in the United States, no '5-4-3-2-1' countdown was followed; the R-7 was simply fired at the appointed time. As a result, the last few seconds before humanity's first voyage into space were almost anti-climatic. With steady rhythm, Korolev barked out in turn: "Launch key to 'go' position ... Air purging ... Idle run," and, finally, at 9:07 am, "Ignition!" The gantry's supporting arms sprang clear from the sides of the rocket as its 20 kerosene-fed engines, with an explosive yield of 880,000 kg, roared to full power.

Gagarin would later describe the initial sensation of liftoff as "an ever-growing din", albeit no louder than the sounds he had experienced flying high-performance jets at Nikel. His helmet muffled much of the R-7's rumble, although the vibrations remained apparent. At some point, a few seconds after leaving Earth, he yelled the immortal words "Poyekhali!" ("Let's go!") over the radio circuit. Within two minutes, as the G loads began to build, he found it increasingly difficult to speak and would later compare the sensation to the stress of a harsh turn in a MiG. The pressure lifted momentarily as the R-7's four strap-on boosters burned out and separated; after a brief pause, the central core of the rocket accelerated and the G loads began to rise again. By three minutes into the flight, at 9:10 am, the Little Seven's nose fairing was jettisoned from around Vostok's ball, giving Gagarin his first glimpse of the dark blue sky and the clear curvature of Earth as he reached the edge of space.

The R-7's core, now exhausted of propellant, finally fell away some five minutes into the flight, leaving Gagarin reliant on an upper-stage engine, which inserted him

Vostok 1 ascends from Tyuratam atop Korolev's Little Seven. Note the opened petals of the 'tulip' launch platform.

into orbit promptly at 9:18 am, exactly as Korolev had planned. Unknown to the cosmonaut, the core had actually burned for longer than anticipated, leaving Vostok in an orbit with an apogee – high point – of 327 km, rather than the intended 230 km. It added a little more height to Russia's already-won World Aviation Altitude Record. As Muscovites arrived at work and the western world still slumbered – figuratively and literally – a new era had begun. Gagarin, whose heart rate soared from 66 to 158 beats per minute during ascent, had won the first lap of the space race.

Although the noise and intense vibration of the R-7 was now gone, it was succeeded by the steady murmur of fans, ventilators, pumps and the hiss of static in his ears. His first experience of the state of weightlessness, properly termed 'microgravity', was hampered by the fact that he was tightly strapped into his ejection seat. However, he had purposely carried a small Russian doll as a gravity indicator and watched as it floated comically in midair. So too did a notepad and pencil. In fact, this would not be the doll's only experience of space travel: in 1991, cosmonaut Musa Manarov would carry it on his mission to commemorate the 30th anniversary of Gagarin's flight.

As the capsule slowly rotated, to avoid wasting propellant on unnecessary manoeuvres and also to ensure that various sections of Vostok did not become too hot or cold, the First Cosmonaut had his first opportunity to view the world through the Vzor. In his official statement, made in Moscow a few days later, he would describe "a smooth transition from pale blue, blue, dark blue, violet and absolutely black ... a magnificent picture". He started jotting down observations in his notepad, but after the pencil floated away he turned instead to the on-board tape recorder to log his thoughts, describing weightlessness as "not at all unpleasant" and confirming that the food and drink were good. Two Russian schoolgirls, who would sample that same food in less than two hours' time, would be inclined to disagree.

Ten minutes after launch, Vostok was heading eastwards in the direction of Siberia and Gagarin gradually drifted out of radio range with Tyuratam, to be picked up in turn by other listening posts at Novosibirsk, Kolpashevo, Khabarovsk and the easternmost station on Soviet soil, Yelizovo, near Petropavlovsk on the Kamchatka Peninsula. To attract Gagarin's attention, each station transmitted musical themes to him: Muscovite tunes, 'Waves of the Amur', Baglanova songs and others. "During that early period," Alexei Leonov wrote, "the Mission Control Centre at Yevpatoriya in the Crimea ... was still under construction, so communication and control from the ground ... were performed from the radio stations above which the spacecraft flew." Leonov had been sent to the remote Yelizovo site a few days earlier as its cosmonaut representative and, to underline the secrecy, had no idea if Gagarin or Titov was aboard Vostok. Then, as the spacecraft passed over Kamchatka at around 9:21 am, he saw the first crude television images from the cabin. "I could not make out his facial features," he wrote, "but I could tell from the way he moved that it was Yuri."

Leonov had been instructed not to initiate communications with Vostok unless given permission to do so and he duly remained silent. However, within moments, Gagarin radioed a request for details of his "flight path", keeping the orbital nature of the mission – for now – a secret from western ears. "The radio operator by my side did not realise his finger was depressing the button that opened the radio link ... when he turned to speak to me," Leonov wrote. The open mike broadcast the words of both Leonov and the radio operator straight up to Vostok. As soon as he heard Leonov tell the operator that "everything is going fine", Gagarin responded with "Give my regards to Blondin", fair-haired Leonov's nickname.

The journey continued, cutting diagonally across the Pacific Ocean from volcanic Kamchatka's land of fire and ice into darkness and toward the sleeping Americas. "The transition into Earth's shadow," Gagarin would explain to a Moscow press conference a few days later, "took place very rapidly. Darkness comes instantly and nothing can be seen. The exit from the Earth's shadow is also rapid and sharp." The unexpected suddenness of the shift from orbital daytime to nighttime led to some mutterings that the flight was a fraud; a testament, clearly, to how limited humanity's understanding of space travel was at the time. Midway through the darkness, at 9:26 am, Vostok rose above the horizon of the Electronic Intelligence (ELINT) station on the Alaskan island of Shemya, giving the United States its first awareness that a Soviet man was indeed in orbit. Admittedly, American long-range radars had detected the R-7's launch, but it was Shemya which confirmed without doubt that live dialogue was ongoing with an Earth-circling cosmonaut.

By 9:32 am, a stable orbit had been achieved – reaching a maximum apogee of 327 km and dipping to a perigee of 169 km, inclined 64.95 degrees to the equator – and, shortly thereafter, Gagarin began his passage across Hawaii, then out over the South Pacific. Following several requests, he was assured by the Khabarovsk station via long-range radio at 9:53 am that his orbit was satisfactory.

Minutes later, precisely on the hour, as he crossed the Strait of Magellan, Radio Moscow announced the electrifying news: "The world's first spaceship, Vostok, with a man on board, was launched into orbit from the Soviet Union on 12 April 1961. The pilot-navigator of the satellite-spaceship Vostok is a citizen of the USSR, Flight Major Yuri Gagarin. The launching of the multi-stage space rocket was successful and, after attaining the first escape velocity and the separation of the last stage of the carrier rocket, the spaceship went into free flight on a round-the-Earth orbit. According to the preliminary data ... "

The effect throughout the world, and particularly in the United States, was dramatic. Although it was known that the mission was underway virtually since its launch, and with certainty since Shemya confirmed Gagarin's dialogue, and even though the imminence of the flight was not unexpected, the event sent shockwaves through the Kennedy administration. Before retiring to bed the previous night, the president had displayed a sense of foreboding and his science advisor, Jerome Wiesner, had gone so far as to prepare a statement for the press. Elsewhere, in Florida, NASA's press officer John 'Shorty' Powers was awakened in the middle of the night by an alert journalist who had just heard of Gagarin's launch. Powers, unfortunately, had not. With clear irritation in his voice and uttering words he would regret, he yelled down the phone: "What is this? We're still asleep down here!" The journalist could not resist exploiting the figurative irony of Powers' words. The United States and its own effort to put a man in space had indeed been caught off-guard. Next morning came the front-page headline: 'Soviets Put Man In Space. Spokesman Says US Asleep.'

It was not only the world that expressed shock and disbelief at Gagarin's achievement; even those closest to him – his family – had little or no inkling of what had happened. Only his wife Valentina knew that he would be flying into space, although he had told her that the mission was planned for 14 April, so as not to worry her. He informed his parents that he was going on a business trip and, according to his sister, would be travelling "very far". Upon learning of the flight, his mother's reaction was to buy a train ticket to visit her daughter-in-law in Moscow and help care for their children. His father, meanwhile, was working on the collective farm near Klushino and, after hearing that someone called Major Yuri Alexeyevich Gagarin was in space, he responded that it could not be his son, who was "only a senior lieutenant". The enormity of what his son had done became apparent when he headed to the local soviet for more information and was quickly pushed onto the phone with a Communist Party official in Gzhatsk. Within hours, Alexei Gagarin and his sons were answering an endless stream of calls from journalists within and beyond the Soviet Union.

High above Earth, at 10:25 am, just off the coast of western Africa, near Angola, Vostok's retrorocket fired automatically to begin its re-entry into the atmosphere. Gagarin would never need to touch his controls, nor open the envelope, nor worry about the three magic numbers. Eight thousand kilometres still remained to be covered before landing back in the Soviet Union. However, all did not go according to plan. Ten seconds after retrofire, as intended, the four metal straps holding the instrument section and the capsule together severed, but an electrical cable failed to separate. The cable contained a thick bundle of wiring which provided power and data to the capsule.

For ten minutes, the two remained connected, with the unneeded instrument section trailing behind and causing the capsule to experience wild gyrations as high as 30 degrees per second. The incident, which was not revealed by the Soviets, but which Gagarin hinted at in his state-sanctioned report, was potentially disastrous. The capsule was weighted in such a way that it would naturally rotate to point the thickest part of its heat shield into the direction of travel. If the instrument section did not separate cleanly, the capsule could not assume its correct attitude and properly bear the brunt of re-entry heating. Gagarin could burn alive.

Fortunately, at 10:35 am, as the spacecraft's meteoric descent path neared Egypt, the cable finally burned through and separated the instrument section. After slinging the capsule away with a spin so severe that Gagarin almost lost consciousness, it finally attained aerodynamic equilibrium with its heat shield positioned accurately. It was later revealed that the complications arose when a retrorocket valve failed to close properly, allowing some fuel to escape from the combustion chamber. As a result of this loss, the engine cut off a second too early, slowing Vostok at a less-than-expected rate and preventing a normal shutdown command from being issued. In the absence of this command, the engine's propellant lines remained open and pressurised gas and oxidiser continued to escape through the nozzle and served to induce the wild gyrations.

Although the engine was ultimately cut off by a timer, its lack of delivered thrust caused the spacecraft's control system to scrub the primary sequence to separate the capsule from the instrument section. The notes of the cosmonauts' physician, Yevgeni Karpov, auctioned by Sotheby's in March 1996, revealed his concern and, indeed, analysts have speculated that – had the world known of these problems – the Sixties space race might have slowed dramatically.

Gagarin himself made one brief reference to the incident. "The craft began to revolve and I told ground control about it," he wrote. "The turning I had worried about soon stopped … " Little else would be known or even suspected for almost three decades. However, in his testimony before the top-secret State Commission on 13 April, whose panel included both Korolev and Kamanin, he described feeling the intensity of the capsule's oscillations and an audible sound of 'cracking', possibly from the structure of Vostok itself or from the expansion of the thermal cladding material. Through the Vzor, he saw the bright crimson glow of ionised gases rushing past the capsule and would later estimate that the G forces exceeded 10 and "data on the control gauges started to look blurry … starting to turn grey in my eyes".

Eventually, as he heard denser air whistling past the capsule and saw blue – not black – sky outside, he braced himself for ejection. By this point, Vostok had crossed back into Soviet territory, on the Black Sea coast, near Krasnodar. Gagarin's ejection procedure was supposed to be fully automatic, triggered when the on-board sensors registered an outside atmospheric pressure consistent with an altitude of 7 km. He may, however, have felt that the oscillations were too severe to risk it and it would appear that he initiated the sequence manually. With a tremendous roar and rush of air, the hatch above his head blew away and the ejection seat's rocket propelled him from the capsule at terrifying speed. Protected from frigid high-altitude temperatures of –30°C by his suit, Gagarin descended to Earth; his main parachute opened successfully, followed by its backup, and the First Cosmonaut hit the ground under two canopies.

The location at which his feet touched terra firma, in a field some 26 km south-west of the town of Engels in the Saratov region, near Smelovka, are today marked by a 12 m obelisk and plaque, inscribed with the legend 'Y.A. Gagarin Landed Here'. The formal marker was placed there on 14 April 1961. However, the historic nature of the event had already led someone to erect a small commemorative signpost on the spot, instructing potential trespassers not to remove it and announcing the time of his landing as 10:55 am Moscow Time. Less than two hours had elapsed since Gagarin's launch from Tyuratam.

Tractor driver Yakov Lysenko heard, but could not see, the ejection sequence as a loud 'crack' in the sky. Seconds later, he saw the Vostok capsule descending under its own automatically-deployed parachute and immediately returned to Smelovka to raise the alarm. A hastily-assembled search party was greeted by what Lysenko would later describe as a "very lively and happy" Gagarin, who identified himself to them as "the first space man in the world". Farmer Anna Takhtarova also recalled the strange sight of the orange-clad cosmonaut telling them not to be afraid and asking for a telephone to call Moscow.

Shortly thereafter, the Soviet military, under General Andrei Stuchenko, arrived in force to take over the recovery effort. Gagarin had been promoted to the rank of major during his flight and was greeted as such by one of the local officers, Major Gasiev. "It was a complete invasion force," Yakov Lysenko said later of the military's arrival. "They didn't allow us to get too close. They're very strange people." It was understandable, at least from Stuchenko's perspective. He had already been told in no uncertain terms by an anonymous Kremlin official the previous day that on his head, literally, lay the responsibility to safely recover Gagarin. Stuchenko obviously wanted to leave nothing to chance.

The Vostok capsule hit the ground a couple of kilometres from the cosmonaut himself, since his high-altitude ejection had caused them to drift apart. For years, the spacecraft's landing site was officially one and the same with Gagarin's own, to avoid FAI suspicions that both had not touched down together. However, the Vostok site is known with certainty, thanks to a group of children who happened to be playing in a meadow near the banks of a tributary to the Volga River. They saw the capsule land. Schoolgirls Tamara Kuchalayeva and Tatiana Makaricheva described the dent it left in the soft earth and related how the boys clambered inside and began handing out and trying the tubes of space food. "Some of us were lucky and got chocolate," Makaricheva recalled of the unusual mid-morning snacks. "The others got mashed potatoes. I remember tasting some [of this] and spitting it out." Kuchalayeva agreed that she would not eat it again. Gagarin's 108-minute adventure, it seemed, made him a hero not only for being the first to survive space, but also for being the first to survive the tastelessness of space food.

REACTION

Twenty minutes after the R-7 blasted off, Nikolai Kamanin boarded an An-12 aircraft, bound for the industrial city of Kuibishev, today's Samara. With him were Gherman Titov, Mark Gallai and a substantial delegation from Tyuratam. Whilst airborne, they learned of Gagarin's landing in Saratov and toasted his success with cognac. The cosmonaut, meanwhile, had already spoken with Nikita Khrushchev by telephone from Engels, before heading on to Kuibishev. On the outskirts of the city, in a special dacha on the banks of the Volga, Gagarin was given a medical examination and a day's rest before his journey to Moscow. The mission was over. Shortly, his new life as an international celebrity would begin.

But not yet. On 13 April, whilst still secluded in the Kuibishev dacha, he underwent his official, two-and-a-half-hour interview by the Vostok State Commission; the only opportunity for 'the truth' about the mission to be revealed, behind closed doors, to Korolev, Kamanin and other high-ranking officials. Although he undoubtedly described the problem with the instrument section, it remains unclear as to why this was not properly resolved in time for the next flight, Vostok 2, other than the possibility that changes were implemented, but failed to work. Meanwhile, efforts to secure the World Aviation Altitude Record had already led sports official Ivan Borisenko to hurriedly get the First Cosmonaut's signature on FAI documents within hours of landing. In his fictitious account of the proceedings, published in 1978, Borisenko would recall "dashing up to the descent module, next to which stood a smiling Gagarin". The reality that capsule and cosmonaut landed a couple of kilometres apart was kept closely guarded.

After his day on the Volga, during which time he also played billiards with Titov and described his experiences, Gagarin flew to Moscow on the morning of 14 April aboard an Ilyushin-18. He had already rehearsed the half-hour speech that he would deliver to Khrushchev – in which Nikolai Kamanin played the role of the Soviet leader – but could hardly have anticipated the sheer outpouring of adoration for him. On the outskirts of the capital, a squadron of seven MiG fighters intercepted his aircraft and escorted him down Lenin Prospekt, Red Square and along Gorky Street to Vnukovo Airport, where the Il-18 touched down just 100 m from Khrushchev's flower-bedecked reception stand. The premier congratulated Gagarin, announcing that "you have made yourself immortal because you are the first man to penetrate space". Following the party line and successfully currying immense favour with Khrushchev, the First Cosmonaut responded by challenging the "other countries" to try to catch up with superior Soviet technology.

However, partly due to sour grapes, but mainly because of the intense mistrust that the Russians had themselves created through their ridiculous secrecy, some observers in those 'other countries' were already doubting that the mission had happened at all or – at the very least – that it had not occurred precisely as reported. The Soviet campaign of misinformation became evident when their sports officials filled in the FAI paperwork to register Gagarin's flight on 30 May 1961. The name and co-ordinates of the launch site, they wrote, were 'Baikonur' at '47°22′00″N, 65°29′00″E', whereas in reality the site was close to Tyuratam at 45°55′12.72″N, 63°20′32.32″E, a considerable distance to the south-west. Indeed, speculation abounded as late as July 1961 over conflicting reports, obscure photographs and a lack of reliable eyewitnesses. Other suspicions lingered over whether Gagarin landed in his capsule or by parachute. On 17 April, just five days after the mission, a correspondent for the London Times wrote that "no details have been given about the method of landing" and revealed that, when questioned at a press conference, Gagarin had "skated over the question".

Nonetheless, within hours of the flight, NASA Administrator Jim Webb appeared on American television to congratulate the Soviets and express his disappointment, but also to offer reassurance that Project Mercury – the United States' own man-in-space effort – would not be stampeded into a premature speeding-up of its schedule. His remarks did little to dampen the fury of the House Space Committee, which verbally roasted both Webb and his deputy, Hugh Dryden, on 13 April. It made no difference; the Soviets had won the first lap of the space race and John Kennedy, still only months into his presidency, had to respond with something spectacular. Faced with persistent questions from Congress as to why the United States should remain in second place to Russia in space, together with a perceived 'gap' in missile-building technology, Kennedy knew that Project Mercury's first manned flight would not even match, let alone surpass, Vostok's achievement. Indeed, it was unlikely that a single-orbit piloted mission could be attempted before the end of 1961, so temperamental was the new Atlas launch

Rumour has abounded for years that this is Gagarin's sharik after touchdown. It was said to have been so badly damaged that it required extensive repairs before it could be placed on display.

vehicle needed to achieve such a feat. By that time, the Soviets could well have pushed their lead even further.

A goal on a longer-term basis, with an above-average chance of success, was crucial for America's young president. On 14 April, Kennedy called an informal brainstorming session with several aides to discuss suitable space goals. Landing a man on the Moon emerged as the best option to draw the Soviets into a race which the United States could conceivably win. Not only would it convey a message to the world of American technological prowess, but it would clearly beat Russia. A few days later, however, circumstances on a Caribbean island just south-east of Florida made Kennedy's need for something – anything – to bolster his administration even more urgent.

THE BAY OF PIGS

Deep within the Gulf of Cazones, on the southern coast of Cuba, is a place known as Bahía de Cochinos. In English, 'cochinos' is sometimes translated as 'pigs', although this may be erroneous and could refer instead to a species of triggerfish. In mid-April 1961, events at this small, nondescript place – the 'Bay of Pigs' – would lead to a major diplomatic incident between the United States, Russia and the newly-established pro-communist regime of Fidel Castro on the island. It would leave the Kennedy administration, still reeling from Yuri Gagarin's flight, severely embarrassed and, in the eyes of socialists, would significantly raise the profile of both the Soviet Union and Communism.

The roots of the debacle had actually been laid during the presidency of Kennedy's predecessor, Dwight Eisenhower, in March 1960. A year after Castro had come to power with his own brand of revolutionary rule, the CIA had begun secret efforts to train and equip a force of up to 1,500 Cuban exiles, with the intention of invading the island and overthrowing the dictator. Initial plans sought to land a brigade close to the old colonial city of Trinidad, some 400 km south-east of Havana, where the population was known to generally oppose Castro's regime.

Already, the dictator was beginning to align himself with the Soviet Union, agreeing in February 1960 to buy Russian oil and expropriating the American-owned refineries in Cuba when they refused to process it. The Eisenhower administration promptly cut diplomatic ties with the fledgling nation, which only served to strengthen Castro's links with the Soviets. When Eisenhower reduced Cuba's sugar import quota in June 1960, Castro responded by nationalising $850 million-worth of American property and businesses. Although some of his policies proved popular among the Cuban poor, they alienated many former supporters of the revolution and precipitated over a million migrations to the United States.

In February 1961, less than a month after his inauguration, an opportunity presented itself for Kennedy to topple Castro: the Cuban armed forces possessed Soviet-made tanks and artillery, together with a formidable air force, including A-26 Intruder medium-range bombers, Harrier Sea Fury fighter-bombers and T-33 Shooting Star jets – surely a tangible threat to the United States' security. As these plans were being thrashed out, the landing site for the anti-Castro brigade was changed to an area in Matanzas Province, 200 km south-east of Havana, at the Bay of Pigs. The exiles' chance of success here was limited still further by warnings from senior KGB agents, by loose talk in Miami and by the interrogation of over 100,000 Cuban suspects, which gradually exposed the plans for the invasion.

Of critical importance to these plans was Operation Puma, which sought to undertake 48 hours of air strikes, eliminating Castro's air force and ensuring that the exiles – known as 'Brigade 2506' – could land safely at the Bay of Pigs. This failed when additional waves of air support were cancelled; Kennedy wanted the invasion to appear as if engineered wholly by the Cuban exiles and not by his own government. For this reason, he had insisted the landing site be moved from Trinidad to the Bay of Pigs – the former was a popular resort and would undoubtedly grab unwanted headlines if the invasion should fail. It was a fatal mistake. Trinidad was actually an ideal spot: in addition to the broadly anti-Castro sentiment of its people, it offered excellent port facilities, armaments and was close to the Escambray Mountains, an anti-communist rebel stronghold. In order to maintain the ability of his administration to claim 'plausible deniability' and avoid admitting that it was actually an American-financed operation, Kennedy doomed the invasion to failure.

On 17 April 1961, two days after the first bombing run and still under the impression that they could rely upon several more waves of decisive air cover, over 1,500 Cuban exiles landed at the Bay of Pigs in four chartered transport ships. They were joined by a pair of CIA-owned infantry craft, together with supplies, ordnance and equipment. The hope that they would find support in the local populace, however, proved fruitless. Cuban militia had already contained the Escambray rebels, Castro had executed several key suspects thought to be involved in the plot and troops were waiting at the Bay of Pigs. The hard-fighting exiles, by now aware that they would not receive effective air support and were likely to lose, were forced back to the beach. By the time the fighting ended on 21 April, 68 exiles were dead, together with four American pilots, and the remainder captured. Some would be executed and over 1,100 imprisoned. After lengthy negotiations, the latter were released 20 months later in exchange for $53 million in food and medicine from the United States.

The fiasco proved extremely embarrassing for the Kennedy administration and was quickly followed by the forced resignations of the CIA director, his deputy and the deputy director of operations. Although he admitted responsibility for the bungled invasion, as the fighting in Cuba drew to an end, on 20 April, Kennedy refined his plans to draw the Soviets into a space race and perhaps gain more credibility for his government. "Is there any space programme," he asked Vice-President Lyndon Johnson in one of the 20th century's most influential memos, "that promises dramatic results in which we could win? Do we have a chance of beating the Soviets by putting a laboratory in space or a trip around the Moon or by a rocket to land on the Moon or by a rocket to go to the Moon and back with a man?" His motives, of course, were chiefly political, but he was clearly pinning his colours to the space flag.

One of the main personalities approached by Johnson as he weighed up the options was the famed rocket scientist Wernher von Braun, who, in a 29 April memo, felt that the "sporting chance" of sending a three-man crew around the Moon before the Soviets was somewhat higher than putting an orbital laboratory aloft. Others, including Secretary of Defense Robert McNamara, would even push for a landing on Mars, although his motivations for such a proposal have been questioned. Von Braun, who had designed Nazi Germany's infamous V-2 missile before coming to the United States in 1945 as a key player in its rocketry and space programmes, felt that a lunar landing was the best option, since "a performance jump by a factor of ten over their present rockets is necessary to accomplish this feat. While today we do not have such a rocket, it is unlikely that the Soviets have it". The rocket to which von Braun referred, known as Saturn, remained in the early planning stages, but a commitment to its development had been one of the conditions he had applied before agreeing to join NASA in October 1958. "With an all-out crash effort," he told Johnson, "I think we could accomplish this objective in 1967–68."

Von Braun's judgement won the day for Johnson. Three weeks later, still smarting from Bay of Pigs humiliation, Kennedy delivered the speech which – perhaps more than any other – would truly define his presidency.

A DAY IN ORBIT

Gherman Titov's dismay at having lost the chance to fly first in space was tempered somewhat by the realisation that his own orbital journey in August 1961 would be more than ten times longer. In fact, one of the reasons cited by Nikolai Kamanin for deferring the poetry-loving teacher's son from the first Vostok to the second had been his greater physical endurance to handle a longer period in the peculiar state of weightlessness. Ironically, Titov's response to this environment would earn him the unenviable record of becoming the first person to suffer space sickness.

The plans for a lengthy mission had been sketched out earlier in 1961, with Sergei Korolev wanting a cosmonaut to spend 24 hours aloft. Kamanin, together with many cosmonauts and their physicians, believed such an endeavour to be too ambitious – "too adventurist", he wrote in his diary – and advocated a shorter, three-orbit mission, lasting around five hours and landing in the eastern Soviet Union. Korolev rejected it. His opposition was based on sound judgement: a recovery during this period, and specifically between the second and seventh orbits, would not be possible, since retrofire would need to occur whilst in Earth's shadow. If this happened, Vostok 2's solar orientation sensor could not function reliably.

With typical single-minded determination, Korolev ordered his deputy, Konstantin Bushuyev, to lay plans for a 24-hour flight. His unwavering effort, devotion and – to a great extent – obstinacy was the result of a hard, driven, thankless life of service to the Soviet Union by a man of pure technical genius. Born in 1907 in the central Ukraine, Korolev's interest in aviation and rocketry emerged at a young age. Under Stalin's regime, with its ingrained fear of the power of the individual, there was little opportunity for the 'intelligentsia' to prosper and, as a highly respected and brilliant engineer, Korolev quickly found himself arrested and sentenced to ten years of hard labour in the Siberian gulag. The Nazi invasion of 1941, however, prompted his release to support the war effort. Subsequently, Korolev set to work developing an arsenal of rockets and missiles which he hoped could someday transport instruments into the high atmosphere and, eventually, into space. His masterpiece, the R-7, though principally intended for the Soviet military as a ballistic missile, would indeed eventually put satellites and men into orbit. By giving it this dual-purpose use, he kept his military critics quiet by satisfying their needs and his own.

Nikita Khrushchev's regime proved generally supportive of Korolev and his projects, but for different reasons: the Presidium – later known as the Politburo – was far more interested in the glamour, political and military impact of spacegoing rocketry than purely upon scientific advancement. Indeed, on the evening of 11 April 1961, when Korolev informed Khrushchev of the final preparations to launch Vostok, the Soviet leader's exasperated response amounted to a demand for him to "get on with it". Even in the wake of Gagarin's triumph, Korolev – still officially a state secret and known only as the 'chief designer' to the outside world – received no congratulation, was barred from wearing his medals and even had to thumb a lift into Moscow when his antiquated Chaika limousine broke down. Later efforts by the Nobel Prize Committee to create an award for the anonymous chief designer fell on deaf ears in the Soviet leadership.

It is astonishing, therefore, that the resilience of the man – whose sufferings in Stalin's gulag had left him physically weakened – was so high in the light of so little tangible reward. Under pressure from Korolev to fly the 24-hour mission, Kamanin reluctantly acquiesced, but imposed a condition that a manually-implemented retrofire could be conducted between the second and seventh orbits if the cosmonaut felt unwell. Opposition to the long flight, though, remained strong. As late as June 1961, Soviet Air Force officers, physicians and cosmonauts felt more comfortable with a three-orbit mission and the dispute remained unsettled until Korolev took his plan all the way to Leonid Smirnov, head of the State Committee for Defence Technology, who opted in favour of spending 24 hours in space. Later that month, the Vostok 2 State Commission convened, named Titov as the prime cosmonaut and Andrian Nikolayev as his backup.

A summertime launch was highly desirable for Korolev. Already, since Gagarin's pioneering flight, no fewer than two American astronauts had ventured into space, albeit on 15-minute suborbital 'hops' from Florida into the Atlantic Ocean. Khrushchev wanted a summertime launch, too, but at a specific point and possibly for specific reasons. In mid-July, he summoned Korolev to his Crimean vacation home and hinted strongly that Vostok 2 should fly no later than 10 August; some observers have since speculated that this was deliberately engineered to provide propaganda cover for the initial steps to build the Berlin Wall just a few days later. "While it was not the first case in which Khrushchev had suggested a particular time for a specific launch," wrote Asif Siddiqi in 'Challenge to Apollo', "it was clearly the first occasion in which the launch of a mission was timed to play a major role in the implementation of Soviet foreign policy."

The western world knew little of the Vostok 2 plans, of course, and persistent rumours abounded that other Soviet efforts to put men – and a woman – into space had backfired and ended disastrously. One notable example suggested that a woman had been launched on 16 May, a few weeks after Gagarin, but that her re-entry had been delayed, perhaps due to damage incurred by her Vostok capsule's heat shield. A decision to come home on 23 May, due to dwindling oxygen supplies, was apparently taken ... and on the 26th, the state-run Tass news agency announced the return of a large unmanned satellite, which burned up during re-entry. More notorious was a story penned by the pro-communist London newspaper The Daily Worker, which revealed, two days before Gagarin flew, that a renowned Soviet test pilot had been killed during his return from space. Such stories did not seem to go away for many years and, as late as 1979, the British Interplanetary Society suggested that the son of

Russian aircraft designer Sergei Ilyushin had flown before Gagarin, but had landed "badly shaken" and had "been in a coma ever since".

These rumours have since been shown for what they are, but they certainly demonstrate the lack of knowledge of exactly what was going on behind the Iron Curtain at this time. Indeed, news of Titov's impending launch did not reach even the keenest western ears until 5 August. Late that evening, the Agence-France-Presse issued a cable from Moscow, reporting that further 'rumours' from the Soviet capital hinted at a manned launch within 24 hours.

Early the following morning, Titov headed for Gagarin's Start in an old eggshell-blue bus and rode the elevator to the capsule that would be his home for more than a day. Following an unfortunate incident during training, he had been reminded, only half-jokingly, by his fellow cosmonauts not to get his parachute lines entangled after ejecting from Vostok 2 or "they would be forced to expel him from the corps". As he was being strapped in, Titov was handed a notepad and pencil to log his experiences during the flight. Seconds after 9:00 am Moscow Time, American radar installations detected the launch of the R-7, although President Kennedy had been informed the previous night that a second Soviet manned shot was imminent.

Two hours into the flight, at 10:45 am, Radio Moscow's famous wartime announcer Yuri Levitan boomed out the details for a listening world. Vostok 2 was in an orbit of 178–257 km, inclined 64.93 degrees to the equator. For the first time, and undoubtedly for propaganda purposes rather than in the interests of 'true' openness, Tass revealed the radio frequencies on which the cosmonaut was transmitting his reports. These appeared in the state-run Pravda newspaper on the morning of 7 August, together with details that the 143.625 MHz voice transmitter was frequency-modulated with a frequency deviation of plus or minus 30 kHz; obviously a clear invitation to western radio enthusiasts to listen in. Following the doubts over the authenticity of Gagarin's mission, this would eliminate any suggestion that Titov's flight might be a fake. In fact, listening posts in western Europe, including the Meudon Observatory, near Paris, heard the cosmonaut's voice within two hours of liftoff, as did Reuters' monitoring station outside London. The BBC also picked up an announcement from Titov, in which he provided details about Vostok's cabin temperature – a pleasant 22°C, he reported – together with his personal callsign, 'Oriel' ('Eagle').

In his post-flight press conference, held a few days later, he would describe candidly the acceleration, noise and vibrations during the launch as having been endurable. Weightlessness, though, posed a different challenge. "The first impression," he told a packed Moscow State University auditorium on 13 August, "was that I was flying with my feet up. After a few seconds, however, everything returned to normal. The Sun shone through the illuminators and there was so much light inside the cabin that I could turn off the artificial illumination. When the Sun did not shine directly into the illuminators, it was possible for me to observe, simultaneously, the Earth – which was illuminated by the Sun – and the stars above, which were sharp and bright little points on a very black sky."

Although he was undoubtedly impressed by his view of the heavens, Titov's experience of the microgravity environment would be somewhat different. Shortly after reaching orbit, he began to feel disorientated and uncomfortable and, even before beginning his sleep period at around 6:30 pm he exhibited symptoms of vertigo: dizziness, nausea, headaches. Titov, the teacher's son from the village of Verkhnie Zhilino, in the Altai region, would be the first of many to suffer from the condition known as 'space sickness'.

TEACHER'S SON

To this day, Gherman Stepanovich Titov remains the youngest person ever to have flown into space, a record he has held for almost five decades. On 6 August 1961, he was just a month shy of his 26th birthday. Born on 11 September 1935, he was named Gherman – an unusual name for a Russian – by his father, in honour of a favourite Pushkin character from 'The Queen of Spades'. Titov's own love of literature, though, went far beyond the inspiration for his name: in his cosmonaut days, he was well-known for quoting long reams of poetry or fragments from stories or novels. Jamie Doran and Piers Bizony have hinted that, in the "egalitarian workers' and peasants' paradise" that was the old Soviet Union, this may have harmed his chances of becoming the first man in space. Unlike Pushkin, whose liberal views and influence on generations of Russian rebels led the Bolsheviks to consider him an opponent to bourgeois literature, Titov's pride, love of poetry and reading and a "suspicion of class" bestowed on him by his learned father made him somewhat less appealing to Nikita Khrushchev's regime than Yuri Gagarin.

His breakthrough to reach the hallowed ranks of the first cosmonaut team in March 1960 came about through his excellence as a MiG fighter pilot. Titov had entered the Ninth Military Air School at Kustanai in Kazakhstan in 1953, transferring to the Stalingrad Higher Air Force School two years later, where he commenced military flight training. Following qualification, in September 1957 he was attached to two different Air Guard regiments in the Leningrad Military District and subsequently became a Soviet Air Force pilot in the Second Leningrad Aviation Region. His selection as a cosmonaut, he would recall more than three decades later, was almost a fluke, with the answers he gave to the physicians and psychologists bordering on arrogance. He seemed non-committal in his interviews even when the subject of "flying sputniks" in orbit was broached. However, he said, "I was curious about how it would be to fly a sputnik and I was told that I had been called to Moscow. I went to Moscow and I was enrolled into the cosmonauts' team".

Titov's selection was lucky in another way, too. At the age of 14, he had crashed his bicycle and shattered his wrist. Instead of revealing the injury to his parents, he nursed it secretly, unwilling to show any sign of weakness, particularly as he had already signed up for elementary training at aviation school. During his time as a cadet, fearful that his injury would be discovered, Titov bluffed them by performing early-morning exercises on a set of parallel bars, until his damaged wrist appeared as good as the other. When he underwent intensive X-rays for the cosmonaut selection in 1960, the medical staff found nothing amiss. Only years after his Vostok 2 flight, when they learned of the injury, did they tell him that his recruitment would never have been sanctioned if they had known.

SPACE SICKNESS

Titov's long-hidden wrist injury may not have been detected, but almost immediately after reaching orbit another condition would become readily apparent. He would secure the unenviable record of becoming the first person to suffer from Space Adaptation Syndrome (SAS) – 'space sickness' – which is today known to affect around half of all space travellers. Research over the past five decades has generally concluded that it is a nauseous malaise, somewhat akin to motion sickness, which typically lasts no more than two or three days of a mission. In Titov's case, it manifested itself in sensations of disorientation and discomfort, coupled with feelings of dizziness and recurrent headaches. "As soon as the [R-7's] third stage split, I felt turned upside down," he recounted years later. "I couldn't understand why I felt this way. Then I saw the Earth begin to turn slower in my eyes. Three or four minutes later, this feeling of being in the upside-down position went away." He managed 'lunch' at 9:30 am and supper at 2:00 pm, consuming puree, bread, pâté, green peas and meat and washing it down with blackcurrant juice, but when he tried to eat again during his sixth orbit, he vomited.

Sleep brought mixed blessings. Around ten and a half hours into the mission, and roughly an hour after bidding goodnight to flight controllers, his pulse rate dropped significantly from around 88 beats per minute to as low as 53. At about the same time, he awoke and was surprised to find his arms floating in midair, due to the absence of gravity. He tucked them under a security belt. "Once you have your arms and legs arranged properly," he recalled in his state-sanctioned autobiography, "space sleep is fine. I slept like a baby." Actually, Titov overslept by some 35 minutes, waking at 2:37 am, still feeling unwell, but recovering towards the end of his 12th orbit to eat breakfast. He would describe the food, including sausages and cold coffee with milk, which he had also been obliged to eat on the ground as a familiarisation exercise, as "joyless". On Earth, however, flight controllers were sufficiently concerned to scrub future Vostok missions until an explanation for Titov's strange reaction to weightlessness could be found.

Even today, five decades later, explanations and countermeasures for the condition remain imprecise. It appears to be aggravated by the subject's ability to move around freely in the microgravity environment, with over 60 per cent of Space Shuttle astronauts reporting the complaint, and appears to be more prevalent in 'larger' spacecraft. The cramped nature of the Vostok, Voskhod, Mercury and Gemini capsules made the sickness virtually unknown in the early Sixties and Titov's unexpected response, together with other factors, may have prompted the decision not to fly him in space again. Indeed, following his mandatory appearance at Lenin's Tomb in Moscow after the mission, he was quietly whisked away to hospital for tests to determine if he was sick.

Modern thinking postulates that the influence of weightlessness on the vestibular apparatus – the workings of the inner ear, which control balance – could present a possible root cause. This disorientation arises, it is theorised, when sensations from the eyes and other areas of the body conflict with those from the vestibular apparatus and with information stored in the brain as a result of a lifetime spent in 'normal' terrestrial gravity. Over a few days, a 'repatterning' of the central memory network occurs, such that unfamiliar sensations from eyes and ears begin to be correctly interpreted and adjustment to the new environment can commence. Today, motion sickness medicines have been shown to help counter it, but are rarely used, with most space fliers expressing preference to adapt naturally over a few days in orbit, rather than risk starting their missions in a drowsy state.

Of course, at the time of Titov's flight, this was unknown. A one-day mission, with little opportunity to fully adapt to weightlessness, complicated his reaction to the environment still further. In the wake of Vostok 2, Nikolai Kamanin and others would notice the cosmonaut's increasingly hyperactive and undisciplined behaviour – his love of women, excessive drinking and fast cars got Titov into hot water with his superiors on many occasions – and led to mutterings that it could have been triggered by his exposure to weightlessness. Kamanin penned his concerns in a July 1964 diary entry, although he was later assured that other Vostok fliers, some of whom spent as many as five days in orbit, exhibited no such personality changes.

To be fair, it would appear that Titov and a few other cosmonauts – Yuri Gagarin included – had gotten themselves into trouble purely by exploiting the fame which had suddenly befallen them and which was completely at odds with their previous restricted lives under Soviet communism. Titov was reprimanded repeatedly within the first year after Vostok 2: riding a motorcycle during a parade in Romania, consorting with prostitutes, an 'incident' with his female chauffeur, speeding, drunkenness, leaving a satchel of highly-classified papers in his unattended car and a hit-and-run traffic accident. His arrogance, too, was a constant concern, with demands for his own jet, involvement in decision-making processes and wanting to take his wife with him on foreign tours. It would be embarrassing to publicly disgrace him, but as the Sixties wore on it became increasingly unlikely that Titov would fly again. The original plan to hire pilots in their twenties to develop them as 'career' cosmonauts seemed to have partially backfired. In his diary, Kamanin revealed his frustrations and admitted that the public role of the cosmonaut was far larger than had been envisaged; more academic training was needed and Titov, Gagarin and others were enrolled into university before commencing further mission preparations.

Naturally, official Tass communiqués yielded no indication of any sickness whilst in orbit. Titov's pulse rate during his second orbital pass was given as 88 beats per minute and he was quoted during his fourth and fifth circuits as feeling "fine", "completely comfortable" and enduring weightlessness "in an excellent manner". When the cosmonaut's relaxed, smiling face was broadcast to a Soviet television audience during his fifth orbit it revealed little of the discomfort he was experiencing. By the end of his tenth orbit, having covered a distance of over 410,000 km, the communiqués boasted that he had travelled "more than the distance to the Moon". Aside from space sickness, Titov's mission was almost a complete success, with the exception of a malfunctioning heater, which allowed the cabin temperature to drop to a chilly 6.1°C. During his very first orbit, he took the manual controls of his capsule and checked out its systems, relayed greetings to the United States and received a congratulatory call from Khrushchev, who promoted him from captain to major and upgraded his status in the Communist Party from that of a candidate to a full member.

Gazing Earthward through the Vzor orientation device, Titov recalled how small his home planet appeared, even though he had regularly flown MiG fighters at high altitude, and was amazed at how quickly a circuit of the globe was completed. Travelling at some 28,000 km/h, his day-long mission orbited 17 times, during which he recorded ten minutes of film with a professional-quality Konvas movie camera and took photographs with hand-held Zritel cameras. "Flying over North America, I sent my greetings … and, 80 minutes later, I sent my greetings to the people of the Soviet Union, so it takes an hour and a half to circle the Earth," he said later. "It's very impressive. That was the brightest impression. I had the feeling that our Earth is a sand particle in the Universe, comparable to a particle of sand on the shore of the ocean. Here we live, and used to threaten each other with nuclear bombs, and I thought that no matter to what society you belong and what your relation is, you have to understand that we are all spacemen and the Earth is our spacecraft and we have to work here like spacemen do." The sentiment among many in the United States, however, was still one of fear of the capabilities of this unknown communist empire. "It makes me sick to the stomach", one American military officer growled to a Time magazine writer.

Although Titov's feelings were undoubtedly sincere, the distrust of Khrushchev's regime was not helped by its excessive secrecy. Even the appearance of Vostok itself remained unknown: not for four more years would the shape of the spacecraft be unveiled to the world and Soviet misinformation gained yet more notoriety in October 1961 when the propaganda film *To The Stars Again* was aired about Titov's mission. In it, the orbital motion of Vostok 2 was illustrated by a model fitted with stubby wings! According to a Soviet source quoted in the 18 December issue of the magazine 'Missiles and Rockets', the wings were "connected … with manoeuvres Major Gherman Titov reportedly carried out during the last orbits of his 17-orbit flight, when he was in the denser regions of the atmosphere". The wings led to a plethora of rumours in the west: was Vostok being developed as a military spacecraft, observers wondered, or perhaps as a platform to manoeuvre in space and conduct orbital rendezvous exercises? Later images, published in July 1962, even showed Mercury-style heat shielding and retrorockets attached to the spacecraft.

Titov's retrofire occurred automatically at 9:41 am on 7 August, high above south-western Africa, and he was able to observe "with great interest," he told the packed auditorium at Moscow State University a few days later, "the bright illumination of the air, which enveloped the spaceship during its re-entry into the denser layers of the atmosphere". Vostok 2's return was not perfect, though, since its capsule and instrument section also remained attached together and were only separated by aerodynamic heating. It would also become apparent that Titov landed dangerously close to a railway line, which led to the inclusion of representatives of the Soviet rail authorities on future State Commissions before launches. Half an hour after retrofire, the cosmonaut ejected and descended, like Gagarin, by parachute. He landed at 10:18 am, in a ploughed field near Krasniy Kut in the Saratov district, not far from his predecessor's own touchdown site. The mission had lasted 25 hours and 18 minutes and Titov had travelled more than 700,000 km. Despite his sickness in orbit, he was described as exhibiting "a fit of euphoria" after landing and on his return flight to Kuibishev for debriefing, he alarmed the medical staff by opening and downing a beer in complete violation of the rules.

Suspiciously, Titov would later recount that it was his choice to either ride his Vostok to the ground or parachute out when he had descended to a sufficiently low altitude. "As already reported," he told the Moscow State University audience, "the structure of the cosmic ship and its systems for landing provided the following two landing methods: landing by remaining inside the spaceship or by means of ejection of the pilot's seat from the spaceship and descent by parachute. I was permitted to select my own way of landing. Contrary to Gagarin's method of landing the ship, I decided to try out the second method." Clearly, Titov's need to reinforce to the world that Gagarin had landed with his ship and had not ejected highlighted doubts in the west that were still prevalent over precisely how the first two cosmonauts returned to Earth. As a result of admitting not landing in his ship, which Nikolai Kamanin later confirmed when he filled in the FAI paperwork in March 1962, Titov's one-day flight would not hold the official record for spaceflight duration. That would go instead to American astronaut John Glenn, who had completed a five-hour orbital voyage in February of that year.

For Nikita Khrushchev's regime, which just a few days later would prove instrumental in the construction of the Berlin Wall, the Vostok 2 flight represented yet another tangible example of communist superiority over the capitalist west. Titov and his young wife, Tamara, were paraded through Moscow to Red Square, where the Second Cosmonaut saluted atop Lenin's Tomb and helicopters rained tiny, multi-coloured pictures of him into the streets. As he embarked on the public appearances circuit, he began to divulge the first minor details that weightlessness, although it "does not interfere with man's capacity for work", did leave him with uneasy sensations in his inner ear. Psychological unease posed another obstacle, through homesickness, although this would be the only 'sickness' to which Titov would officially admit. "I knew that there was something in the nature of homesickness called nostalgia," he said, "but, up there, I found there is also homesickness for the Earth. I don't know what it should be called, but it does exist."

THE WALL RISES

One week after Titov's mission, on 13 August 1961, East German troops sealed borders inside Berlin and began the construction of what would become a permanent barrier around the three western sectors of the city. Since the end of the Second World War, the former territory of Nazi Germany had been divided into four occupation zones – controlled, respectively, by the United States, Britain, France and the Soviet Union – and, although it lay deep within the Russian sector, the old capital was itself divided into four areas. Within the next few years, increasing tensions led the American, British and French sectors to become consolidated into the Federal Republic of Germany, together with 'West' Berlin, while the Soviet region became known as the German Democratic Republic, with 'East' Berlin. Under the direction of Joseph Stalin and his foreign minister Vyacheslav Molotov, East German border defences were substantially improved during the Fifties, ostensibly to prevent the free movement of western agents in the GDR, and a barbed-wire fence was erected. However, the border remained open, albeit with restrictions, to traffic.

The closure of this border with a more permanent barrier in August 1961 was carried out by East German authorities, with no direct Soviet involvement, although some observers have seen Nikita Khrushchev's insistence on launching his nation's second cosmonaut only days earlier as propaganda cover for constructing the wall. It was built slightly inside East German territory to ensure that it did not intrude on any sector of West Berlin and such was the tenacity with which it was assembled that streets were torn up and rendered impassable, railway stations closed and tram lines cut. The result was that West Berlin, now completely surrounded, became an isolated enclave in the hostile territory of the Russian zone. Officially, the East German perspective of the wall's purpose was as a means of protecting the "new and more beautiful life" of the socialist republic from western "imperialists and militarists". In reality, it was a means of forcible separation and ensured that defections would be eliminated; if necessary, by deadly force.

Indeed, East German guards were encouraged to regard anyone attempting to escape as a traitor and shoot them, "even when the border is breached in the company of women and children". Despite these risks, as East Germans saw the barbed-wire monstrosity rise, some took their chance. On 15 August, a young guard named Conrad Schumann leapt over into West Berlin and was driven away at high speed by a waiting car. Other escapees jumped from apartment windows, used hot-air balloons, dug tunnels, flew ultralights and slid along aerial wires. One even drove a sports car at full speed through the barrier.

Inside West Berlin, meanwhile, mass demonstrations, led by Mayor Willy Brandt, who vehemently criticised the United States for failing to respond, achieved little. President Kennedy, speaking a few weeks earlier, had acknowledged that he could only hope to protect West Germans; to attempt to do the same for East Germans would lead to an embarrassing failure. In spite of the fact that the wall's very presence violated the post-war Potsdam Agreement, Kennedy's administration later told the Soviet government that the barrier was now "a fact of international life" and refused to challenge it by force.

Nonetheless, a show of force to provide at least some visible reassurance for West Berliners was needed. On the afternoon of 19 August, General Lucius Clay – Kennedy's special advisor and mastermind of the 1948 Berlin Airlift – and Vice-President Lyndon Johnson arrived in the city. Early the following morning, a column of nearly 500 armoured vehicles and 1,500 troops left the Helmstedt-Marienborn checkpoint, arriving in Berlin just before midday, under the wary watch of East German police. Clay and Johnson met the force, which paraded through the streets, and duly left the city on 21 August under the command of General Frederick Hartel and a 4,200-strong brigade. For the next three and a half years, American battalions would rotate into West Berlin by autobahn at three-monthly intervals to demonstrate Allied rights to the city.

The wall, however, remained … and grew. In June 1962, shortly before two more cosmonauts pulled off the Soviet Union's next space spectacular, work started on a second, parallel fence. This lay slightly further inside East German territory and effectively created a no-man's land between the two barriers. It would become notorious as the wall's 'death strip': paved with raked gravel to make footprints easy to spot, it afforded would-be defectors no cover, was filled with tripwires and provided a clear line of fire for the ever-watchful East German guards. Two months later, on 17 August, the new wall claimed its first victim. In full view of hundreds of witnesses, together with the western media, 19-year-old bricklayer Peter Fechter was shot as he tried to escape into West Berlin. It was only the beginning. By the time the wall finally fell in November 1989, 133 'official' – and probably far more – defectors would be murdered attempting to gain their freedom.

SPACE DUET

Six months after Gherman Titov's flight, the gap in space achievement between the Soviet Union and the United States had begun to close slightly. On 20 February 1962, astronaut John Glenn embarked on a five-hour mission which not only completed three circuits of the globe, but also – since he landed inside his ship – established a new, FAI-approved record for the longest 'confirmed-successful' flight to date. Surprisingly, not until August would Soviet cosmonauts again journey into the heavens, mainly because Sergei Korolev's design bureau had its hands full preparing the Zenit spy satellite, based upon Vostok technology, for launch. Nonetheless, his manned spaceflight ambitions had by no means stagnated. As early as 15 August 1961, he had proposed to Nikolai Kamanin the launch of three manned Vostoks at 24-hour intervals, with the first mission spending three days aloft, the others around two days apiece. Although Kamanin had agreed in principle, he felt that Korolev's November 1961 target for the joint endeavour was too ambitious and refused to endorse such a schedule. Further, in the wake of Titov's space sickness episode, he did not feel that the next flight should exceed two days.

Korolev was furious, particularly since missions limited to two days and an inability to launch more than one Vostok per day meant that only two spacecraft, rather than three, could be in orbit at the same time. However, a dual mission could be better handled by the Soviet tracking and rescue networks than a triple-Vostok flight. Plans for the dual mission thus proceeded and, in October 1961, cosmonaut Andrian Nikolayev, fully-suited, went through a three-day-long simulated 'flight' in a spacecraft mockup. Ongoing problems with Zenit, however, made a November launch impossible and Korolev began to push instead for Vostoks 3 and 4 to fly in December or January. This, too, was opposed by Kamanin, on grounds that weather conditions would be too severe to support adequate launch and recovery operations. Additionally, several technical hurdles contributed to make launches before the spring untenable: trials of a new parachute and space suit were not going well, the environmental system was still not perfected and problems with Vostok's temperature controls needed further attention.

Meanwhile, government official Dmitri Ustinov, chairman of the Military-Industrial Commission and primary manager of all Soviet missile and space projects during this period, demanded that the joint flight be launched by 12 March 1962 in direct response to John Glenn's mission. In his diary, Kamanin fumed at the knee-jerk reactions of the Soviet leadership, whose months of inactivity, followed by an insistence to launch in an impossibly short time span, was seriously hampering the nation's space effort. He had reason for yet more anger on 22 February, when it was decided to allow Vostoks 3 and 4 to each spend three full days in space, although Korolev admitted that he would only push for the extra day if both cosmonauts remained healthy. Those cosmonauts were confirmed as Andrian Grigoryevich Nikolayev and Pavel Romanovich Popovich, both Soviet Air Force officers; the former a major, the latter a lieutenant-colonel. Their backups would be Valeri Bykovsky and Vladimir Komarov.

Even pressure from higher-ups like Ustinov was insufficient to avoid the postponement of the joint flight beyond March. Technical problems and a Zenit launch failure delayed the manned shot until at least the second week of April. An additional delay until the end of May was enforced by the launch of the Cosmos 4 spacecraft and, finally, the explosion of a Zenit rocket seconds after liftoff on 22 June pushed the Nikolayev-Popovich flight into late July.

Kamanin remained opposed to the idea of three-day missions, as did many of the cosmonauts and their training teams. In his diary, he substantiated his fears by noting the results of a recent American-West German experiment which had confined five men to a fallout shelter for five days. The results had revealed gradual physical and mental deterioration, which Kamanin and Gherman Titov felt could also apply to cosmonauts if flight durations were extended prematurely. By June 1962, as the Vostok 3 and 4 launches slipped further into the summer, some Soviet engineers predicted that the Americans would be in a position to stage an 18-orbit mission – eclipsing that of Titov – before the year's end. Kamanin disagreed, speculating (correctly, as it turned out) that such a capability would be beyond them until mid-1963. However, none of this deviated from the fact that a new mission was needed to prove that the Soviets remained in the game.

By the time the State Commission finally convened in mid-July, Korolev had received Khrushchev's blessing for a full three-day mission, with Kamanin representing the sole voice of dissent. Low-Earth orbit, however, was not a quiet place at the time. Only days earlier, on 9 July, the United States' Starfish Prime test had detonated a thermonuclear warhead 400 km above Johnston Island in the Pacific Ocean, inadvertently disabling several satellites and leading to serious concerns about increased levels of radiation in low-Earth orbit. In fact, according to a 1963 article in the Journal of Geophysical Research, Starfish Prime created a man-made radiation belt of high-energy electrons which persisted for years. Soviet scientists were convinced that Vostok missions lasting less than five days should still be safe and would not expose their cosmonauts to abnormally high doses of radiation. Still, clearly savouring the propaganda value of the event, the Kremlin politely asked the United States to refrain from further nuclear tests whilst their men were in orbit.

As August opened, the experiments and plans for the joint mission were concluded. One of them called for observations of the final stages of the Vostoks' R-7s to be conducted shortly after separation and demonstrate the cosmonauts' ability to properly orient their spacecraft. This latter exercise would require both Nikolayev and Popovich to execute very slow manoeuvres of just 0.06 degrees per second, followed by 'barbecue rolls' of 0.5 degrees per second to maintain the correct thermal balance across all spacecraft surfaces. Late on the evening of 10 August, following a two-hour delay to check fasteners on Nikolayev's ejection seat, the first R-7, with Vostok 3 secreted in its nose, was rolled out to Gagarin's Start in readiness for launch the following morning. By this time, it had been confirmed that, assuming Nikolayev remained healthy, his flight would be extended to a record-breaking four days in orbit.

STARFISH PRIME

On 9 July 1962, an event which would have important ramifications for two Soviet cosmonauts and an American astronaut got underway on Johnston Island in the Pacific Ocean. It was part of a joint effort between the Defense Atomic Support Agency (DASA) and the Atomic Energy Commission (AEC) and was known under its umbrella designation of 'Operation Dominic'. Its objective, nicknamed 'Starfish Prime', was to detonate a thermonuclear warhead some 400 km above Earth's surface. The testing of such devices in both the Pacific and Nevada in 1962–63 sought to evaluate new weapons designs, their effects and their reliability. Its timing was crucial: in August 1961, Nikita Khrushchev had announced the end of a three-year moratorium and resumed Soviet weapons testing a few weeks later, when the most powerful hydrogen bomb ever built – nicknamed 'Ivan' by the Russians – was detonated over the Novaya Zemlya archipelago in the Arctic Ocean.

Both it and Operation Dominic would arouse much condemnation and, indeed, despite their own efforts, the Soviets would request diplomatic assurances that the Americans refrain from their nuclear weapons testing whilst cosmonauts Nikolayev and Popovich were in orbit. Starfish Prime was one of five tests conducted by the United States 'in outer space', although an initial launch attempt on 20 June 1962 had failed when its Thor carrier missile experienced an engine malfunction, crashed, showered the Johnston area with radioactive metal and caused "slight" nuclear contamination.

Less than three weeks later, atop another Thor, the test commenced. Its warhead produced an explosive yield equivalent to 1.4 megatons of TNT and, although too high and too far beyond the 'sensible' atmosphere to create a fireball, triggered a whole host of problems back on Earth. Across Hawaii, for example, some 1,500 km to the east of Johnston Island, the effects of its electromagnetic pulse caused 300 street lights to fail, television sets and radios to malfunction, burglar alarms to sound in unison and power lines to fuse. On the westernmost Hawaiian island of Kauai, telephone links with the other islands were severed when a microwave link burned out and much of the Pacific skyline was lit by an eerie, man-made aurora, lasting for more than seven minutes. Observers in the far-off Samoan Islands, more than 3,200 km from the blast, even recorded the aurora on film.

According to a witness of the test, whose remarks were published in August, "a brilliant white flash burned through the clouds" at the stroke of 9:00 am, "rapidly changing to an expanding green ball of irradiance extending into the clear sky above the overcast. From its surface extruded great white fingers, resembling cirrostratus clouds, which rose to 40 degrees above the horizon in sweeping arcs turning downward toward the poles and disappearing in seconds to be replaced by spectacular concentric cirrus-like rings moving out from the blast at tremendous initial velocity, finally stopping when the outermost ring was 50 degrees overhead ... All this occurred, I would judge, within 45 seconds. As the greenish light turned to purple and began to fade at the point of burst, a bright red glow began to develop on the horizon ... expanding inward and upward until the whole eastern sky was a dull, burning-red semicircle ... "

Despite the risk, 'rainbow bomb parties' were offered by Hawaiian hotels on their rooftops to view the effects of the Starfish Prime detonation. By the end of July, however, circumstances had changed markedly. Another test, 'Bluegill Prime', suffered an engine failure on the pad. It was detonated, after ignition, by range safety officers, completely destroying both it and the launch facility. More than three months of repairs – and extensive decontamination – would follow. The largest nuclear weapons testing project ever conducted by the United States would also be its last in the high atmosphere. In August 1963 and coming into effect just two months later, the Limited Test Ban Treaty was signed by the United States, the Soviet Union and Britain to restrict all future detonations to underground. For a time, at least, it helped to enforce a dramatic slowdown of the arms race between the superpowers.

AN IRON MAN AND "A PUZZLE"

When he was shortlisted as a candidate for the first man in space in January 1961, Andrian Grigoryevich Nikolayev was described by his examiners as "the quietest" of the six finalists. By the time he launched into orbit on 11 August 1962 to begin the longest manned space mission to date, he had earned another nickname: 'Iron Man', due to his astonishing stamina and ability to sit alone in an isolation chamber, without stimulus or awareness of the passage of time, for no less than four whole days. Born on 5 September 1929 on a collective farm in the village of Sorseli in the forested Chuvash region of the Volga River valley, he was one of four children and discovered a love of aviation when, aged eight, he visited a nearby airfield. One story from his early years tells how he clambered into the branches of a tree and announced that he intended to fly from it; fortunately, local villagers changed his mind and persuaded him to come down.

Following his father's death in 1944, his intention was to support his family, although this was opposed by his mother, who wanted him to gain a full education. Nikolayev entered medical school, then tried his hand at forestry, serving as a lumberjack and timber camp foreman for a time, before joining the Soviet Army. He initially trained as a radio operator and machine gunner, demonstrating "composure under stress" when he crashed a flamed-out jet in a field rather than bailing out. Undoubtedly, this was a contributory factor in his selection as a cosmonaut trainee, along with Yuri Gagarin, Gherman Titov and 17 others in March 1960. A bachelor at the time of Vostok 3, he is famously said to have kissed his girlfriend goodbye at the foot of the launch pad. That 'girlfriend' – 25-year-old Valentina Tereshkova – would not only become his wife a little over a year later, but would also become the first woman to venture into space.

In stark contrast to the quiet, reserved nature of Nikolayev, that of Vostok 4 cosmonaut Pavel Romanovich Popovich has been described as considerably more extroverted. He was also the only member of the 1960 cosmonaut group to have flown a 'high-performance' aircraft, having piloted the MiG-19. Interestingly, although he was shortlisted among the final six candidates for the first Vostok mission, his examiners labelled him "a puzzle" and mysteriously attributed his behaviour to "secret family problems". A lieutenant-colonel at the time of the flight, he became the most senior-ranking cosmonaut yet to reach orbit. Born on 5 October 1930 in Uzyn, within the Kiev Oblast in the north of the then-Ukrainian Soviet Socialist Republic, Popovich is today revered as the first ethnic Ukrainian spacefarer.

During his early teens, Popovich apparently so loathed the Nazi occupation that he refused to learn German at school, instead stuffing cotton into his ears and being expelled as a result. He was, it is said, even dressed in old frocks and passed off as a girl by his mother to avoid being sent away to Nazi labour camps. After the Second World War, Popovich worked as a herdsman, before achieving a diploma from a technical school in the Urals and entering the Soviet Air Force. Whilst assigned as a fighter pilot in Siberia, he met his future wife, Marina, a woodcutter's daughter who became a high-ranking officer and engineer. She was also an accomplished stunt pilot and outspoken UFO researcher, which her husband, too, later embraced. In fact, in 1984, after his retirement from the cosmonaut corps, Popovich headed the Soviet Academy of Sciences' UFO Commission. Like Titov, he was a voracious reader, an admirer of Hemingway and Stendhal and often quoted the works of the Soviet poets Sergei Yesenin and Vladimir Mayakovsky. In the isolation chamber, he proved very much the opposite of steely Nikolayev: he was more light-hearted and jocular, often relieving the tedium by dancing and singing operatic arias with such gusto that scientists and engineers gathered to listen.

FLYING FALCON, SOARING EAGLE

When Radio Moscow announced Nikolayev's successful 11:24 am launch on 11 August, observers could be forgiven for wondering what this latest Soviet mission might entail. The answer became clear when Popovich roared aloft from Gagarin's Start in an R-7 he had christened 'Swallow' at 11:02 the following morning and, within an hour of reaching orbit, he had established visual line of sight with Vostok 3. In his post-flight debriefing, Nikolayev would recount that, despite positioning his ship in the correct attitude, he had been unable to see Popovich's launch from space. Over the following days, despite the paucity of reliable information from Tass, western analysts set to work plotting the two ships' radio signals and estimated that they were flying some 120 km apart. This led to increased speculation about whether the Soviets had trumped the United States again by achieving what the Americans were not expected to achieve until their two-man Gemini spacecraft flew in 1964: rendezvous in orbit. If, indeed, they had achieved this remarkable feat of celestial mechanics, a Soviet man on the Moon by the middle of the decade was entirely possible. "Once they have achieved orbital rendezvous," the British Interplanetary Society's Kenneth Gatland said at the time, "they have taken a vital step toward lunar flight."

Strictly speaking, what had been achieved was not rendezvous. The wording of the official communiqués allowed the interpretation that the spacecraft had manoeuvred to reduce their initial separation, but the reduction was due to orbital dynamics, after which the range increased again. Nonetheless, inserting two manned spacecraft into similar orbits at the same time was a feat that the United States could not hope to match for several years and the propaganda value of this was fully exploited. As announced, the purpose of the missions by Nikolayev and Popovich, callsigned 'Falcon' and 'Golden Eagle', respectively, was to check "contact" between two spacecraft flying in similar orbits. Although the minimum range is believed to have been 6.5 km, one account claimed that the cosmonauts had been able to see each other through their capsules' portholes! Hence their "contact" seems to have been exclusively visual or by radio.

Indeed, the Sohio tracking station in Cleveland, Ohio, reported that after their initial close proximity, the two spacecraft drifted more than 2,800 km apart. "We're convinced that if they had the proper equipment, they could have touched," the station's supervisor was later quoted as saying by Time magazine. Yet the dual mission did induce some concern in the Pentagon. "If the Russians can send Colonel Popovich up to look at Major Nikolayev," said one officer, "they can go up and look at one of our birds. Why, they could knock out those delicate instruments in some of our satellites by hitting them with almost anything." Others were more cautious, pointing out that the interception of one Vostok by another was simplified by the fact that both had launched from the same pad.

Unlike Gherman Titov, who had experienced space sickness shortly after reaching orbit, neither Nikolayev nor Popovich appeared to be affected by the ailment. Their spirits seemed high as they congratulated each other over a shortwave channel which linked their two ships and even engaged in a three-way radio conversation with

Sergei Korolev and his wife with the cosmonauts selected in March 1960 and a number of their trainers. Included in this portrait are five of the six Vostok fliers. Gagarin is the only person wearing a tie; Titov stands directly behind Korolev's wife; Popovich sits at the far left on the front row; Nikolayev is second from the left on the second row; and Bykovsky is second from the right, also on the second row. These cosmonauts, and, of course, Korolev himself, were responsible for some of the most remarkable triumphs in mankind's early conquest of space.

fellow cosmonaut Yuri Gagarin at the control centre. Elsewhere on the ground, after the official announcement had been made, Muscovites gathered in their hundreds in the streets to listen as loudspeakers blared out the news of the latest Soviet spectacular in space. Possibly in an effort to show how much more 'roomy' Vostok was compared to the Americans' cramped Mercury capsules, it was revealed that Nikolayev released his shoulder straps and floated 'around' the cabin.

Further details trickled out with the suggestion that he had worried about bumping into things as he moved around; Popovich, too, it was claimed, accidentally banged his head whilst floating across the cabin. Obviously, after the dimensions of Vostok were revealed to the world in April 1965, it became clear that neither cosmonaut had much room in which to move and the stories were simply a clever game designed to keep western listeners guessing about the spacecraft's true size.

The Soviets clearly only revealed what they wanted the outside world to know and, indeed, when American networks asked to plug in on televised images of Nikolayev and Popovich via the Telstar communications satellite, they were politely refused. Instead, Soviet embassies released photographs of the cosmonauts at play with their families, at seaside resorts, riding pedal boats and even one of Nikolayev sniffing poppies. The two men, meanwhile, worked methodically through their detailed experimental programmes, photographing and – in Nikolayev's case – filming Earth in colour for the first time. They checked their ships' systems, monitored communications, verified guidance . . . and even found time to request the latest football scores. They chatted, too, about their food. Instead of the toothpaste-tube-like fare that Titov had endured, they were provided with packed meals: small, bite-sized chunks of veal cutlet and chicken, together with sandwiches and pastries. A disappointed Nikolayev, upon learning that Popovich had a small piece of dried fish in his food locker, asked for some; to which Popovich gamely invited him to "come a little closer and we'll share what we've got".

Medical personnel hoped that the packed meals, which were more 'normal' than those consumed by Titov, might help avoid a recurrence of space sickness. "It was just as pleasant as a good restaurant," Nikolayev would recall after landing and it would appear that neither cosmonaut experienced any of the dizziness, nausea or headaches suffered by Vostok 2's pilot. In their post-flight debriefings, both would explain that they moved their heads sharply from right to left with no ill effects. However, an unfortunate misunderstanding appears to have curtailed Popovich's mission. It was becoming clear that occurrences of space sickness might be linked to the reactions of individual cosmonauts, rather than as a result of long missions. Consequently, before launch, the cosmonauts had been given the callsign "observing thunderstorms" – "groza" – to report to ground controllers if they felt unwell and desired an immediate return to Earth. Unluckily for Popovich, it would seem that he really was observing enormous thunderheads over the Gulf of Mexico and made an innocent remark about them. Within hours, and just a few minutes after Nikolayev's own landing, Popovich's mission was over and he was back on Soviet soil.

Although he had tried to explain, whilst still in orbit, that he really was observing meteorological thunderstorms and was not ill, neither Nikolai Kamanin nor Yuri Gagarin wanted to take the risk, suspecting that he had experienced an attack of nausea, made the transmission and later relented, not wishing to admit to any weakness. To be fair, problems with Vostok 4's life-support system had already caused the cabin temperature to plummet to just 10°C and some officials were pushing to bring the cosmonaut home on his 49th orbit. Additionally, since he had only expected to remain aloft for three days, Popovich had not conserved his on-board provisions with the same tenacity as Nikolayev; still, he remained cheerful, active and eager to complete a lengthy mission.

It is ironic that these events should have transpired because, by 13 August, optimism was high that both cosmonauts were sufficiently healthy to complete four-day missions: the only voice of dissent came from Kamanin, still fearful of the effects of long-duration flights on the human body. One such effect began to manifest itself midway through Nikolayev's flight when he vented his frustration on personnel at a Soviet tracking station, who had provided him with incorrect timing information. "You were wrong by five minutes," he barked. "Please give me a new time recording now. Can't you hear what I say? Start the timing, for heaven's sake!" Tension and fatigue, it seemed, were something even the Iron Man could not avoid.

Nikolayev's feet touched Earth at 9:52 am Moscow Time on 15 August, in the hilly desert country close to the coal-mining city of Karaganda in north-central Kazakhstan, some 2,400 km south-east of Moscow. He was followed, barely seven minutes later, by Popovich, who landed a few kilometres away and at the same (48th) parallel; a similar landing principle would be adopted during the joint flight of Vostoks 5 and 6 in June 1963. Some observers have speculated that the reason was to deploy recovery forces in an east-to-west pattern. Both re-entries appeared to be smooth and not as eventful as those of Gagarin and Titov, with Nikolayev commenting only that his capsule "revolved randomly on reaching the denser atmosphere" and he experienced deceleration forces of 8–9 G. He also recounted that, although there were boulders in the landing zone, he was able to guide his parachute successfully and touch down in a clear area. After recovery, the two now-bearded cosmonauts were reunited, greeting each other, it is said, with embraces, kisses and spontaneous song. As they munched watermelon and chatted with locals in a crowded Kazakh rest house, their sole complaint was that the heat and discomfort of the desert was greater than it had been in space.

OCTOBER CRISIS

Eight weeks after the joint flight of Nikolayev and Popovich, the Cold War took its most dramatic turn. It was a turn which has since been regarded as the moment at which the Soviet Union and the United States came closest to full-scale nuclear conflict. It began on 14 October 1962, when reconnaissance photographs taken by a U-2 aircraft revealed the construction of ballistic missile bases in Cuba. President John Kennedy regarded it as a significant threat to both his own nation and others in the western hemisphere and, moreover, a pro-communist country in Latin America was unthinkable at this tense time. Events had not been helped by the president's own half-hearted support of a group of Cuban exiles to topple Fidel Castro's regime in April 1961. This had prompted the revolutionary dictator to declare his fledgling nation a socialist republic, ally himself openly with the Soviet Union and begin efforts to modernise Cuba's military infrastructure.

Kennedy's other attempts at aggression included Operation Mongoose, which sought to destabilise and overthrow Castro through a series of (ultimately unsuccessful) covert military activities, together with continuous reconnaissance overflights and harassment from the United States' Guantanamo naval base. In April 1962, Nikita Khrushchev agreed to supply surface-to-air and surface-to-surface missiles to Cuba for coastal defence and, a few weeks later, began installing Soviet-controlled nuclear weapons on the island. This emplacement was a clear response to Kennedy's own installation of 15 Jupiter missiles at Izmir in Turkey, all of which were aimed at cities in the western Soviet Union, including Moscow. By late July, more than 60 Russian ships had reached Cuba and fears grew that an American invasion was imminent. A joint Congressional resolution had already authorised the use of force if their interests were threatened and a huge military exercise, Operation Ortsac, was planned in the Caribbean for October.

Although it was widely known that the United States possessed a far larger nuclear arsenal than the Soviet Union, the close proximity of missiles stationed in Cuba would reduce the warning time of any launches to little or none. Meanwhile, between June and October, Soviet merchant ships delivered two dozen launch pads, 42 missiles, 45 nuclear-tipped warheads, 42 Il-28 bombers, a 40-strong MiG regiment, two Anti-Air Defence divisions and three mechanised infantry units to the island. Kennedy had initially expressed scepticism that Khrushchev would risk provoking an international confrontation: even in late August, he told Congress that the Cuban missiles were for defensive, not offensive, purposes. Further reassurances to this effect came from Soviet ambassador Anatoli Dobrynin and from Khrushchev himself.

The military build-up, however, continued. On 8 October, Cuban President Osvaldo Dorticós Torrado told the United Nations General Assembly that, if his country was attacked, it would protect itself. "We have sufficient means," he warned, "with which to defend ourselves; we have, indeed, our inevitable weapons, which we would have preferred not to acquire and which we do not wish to employ." A few days later, a U-2 flight identified the construction of a missile base at San Cristóbal in western Cuba; Kennedy first saw the photographs on 16 October and duly assembled the National Security Council's executive committee, which concluded on three possible courses of action: an air assault on the bases, a full-scale military invasion or a naval blockade to prevent the arrival of more weapons. Although the Joint Chiefs of Staff supported the second option, believing that the Soviets would not stop the United States from conquering Cuba, Kennedy was more cautious. He feared that "if they don't take action in Cuba, they certainly will in Berlin". An air attack, too, would be unwise, giving the Soviets an ideal excuse to take West Berlin.

The naval blockade – or at least a more selective 'quarantine', focusing solely on the weapons – was also endorsed by Secretary of Defense Robert McNamara. Under international law, it would represent an act of war, but Kennedy felt that the Soviets would not be provoked into an attack by a mere blockade. Despite a meeting with Soviet foreign affairs minister Andrei Gromyko, who assured him again that there were no offensive weapons in Cuba, Kennedy moved to obtain approval from the Organisation of American States (OAS) for military action. In support of the quarantine, which would occur in international waters, Argentina, Venezuela, the Dominican Republic, Colombia and Trinidad and Tobago offered destroyers, air force and other military units, escort ships and the use of naval bases.

In a telegram despatched late on 24 October, Khrushchev urged Kennedy to "understand that the Soviet Union cannot fail to reject the arbitrary demands of the United States", regarding the blockade as "an act of aggression" which Russian ships would be ordered to ignore. Kennedy's response was that his hand had been forced after repeated reassurances that no offensive weapons existed in Cuba and he implored his Soviet counterpart to "take necessary action to permit a restoration of the earlier situation". That same night, the Joint Chiefs instructed the Strategic Air Command to lift its defence readiness condition to 'DEFCON 2' – the second-highest state of military alertness – and the directive was deliberately transmitted uncoded to ensure that Soviet intelligence agents could understand it. The first efforts to enforce the blockade – intercepting the tanker Bucharest and boarding the Lebanese freighter Marcula – began the next day.

Tensions were raised still further when Kennedy authorised the loading of nuclear missiles onto American military aircraft and ordered low-level reconnaissance missions over Cuba to increase from twice daily to once every two hours. Secret negotiations began shortly thereafter to seek a diplomatic resolution to the crisis, with suggestions that the Soviets remove their weapons under the supervision of the United Nations and Fidel Castro publicly announce his refusal to accept them in future. In exchange, it was proposed that the United States would promise never to invade Cuba. The response from Kennedy was that his government was "unlikely to invade" if the missiles were removed. Early on 26 October, Khrushchev broadcast a message on Radio Moscow and wired Kennedy to the same effect: offering to remove the Cuban weapons in exchange for the United States dismantling their own Jupiter missile emplacements in Turkey.

Around midday, the crisis deepened when a U-2, flown by Rudolph Anderson, was shot down by one of the Cuban missiles; later, several Crusader aircraft on low-flying reconnaissance missions were also fired upon. Despite an earlier decision to order an attack under such circumstances, Kennedy reserved judgement. He was keen to accept Khrushchev's offer, although the Turkish government was unhappy about relinquishing the Jupiter defences and such a trade would effectively undermine relations with a NATO ally. Ultimately, and after much deliberation, Kennedy agreed secretly to remove all missiles set in Turkey and respect Cuban sovereignty, in exchange for Khruschev's removal of all weapons from the Caribbean island. However, the decision to remove the Turkish batteries was not made public at the time, with the effect that Khrushchev appeared to be the loser. His apparent 'retreat' from a situation which many observers felt that he had started – coupled with his perceived inability to handle international crises – has been cited as one of the factors in his overthrow two years later.

A WOMAN IN SPACE

By the time Andrian Nikolayev and Pavel Popovich returned from their joint flight in August 1962, it was becoming clear to the western world that the Soviet space effort focused solely on scoring one 'spectacular' after another. Unlike the American programme, which was proving to be more gradual, yet had a longer-term aim with its Apollo Moon landing project, the Soviet leadership seemed uninterested in the exploration of space. Theirs was a programme exploited by Nikita Khrushchev's regime purely for the political, military and propaganda advantages that it offered. One further advantage was first tabled by Sergei Korolev towards the end of 1961 and involved sending the first woman into space. On 30 December, the Central Committee of the Communist Party authorised the selection of five female cosmonauts and, by March 1962, after a lengthy evaluation and screening process, the candidates arrived at a male-dominated training facility, just outside Moscow, to begin preparations for the next Vostok stunt.

The women – Tatiana Kuznetsova, Valentina Ponomaryova, Irina Solovyeva, Valentina Tereshkova and Zhana Yerkina – were put through precisely the same training regime as the men. Their advantage was aided by the apparent disinterest of the United States in selecting female astronauts of its own. "At that time in America, women tried to make their way into the Mercury programme," Ponomaryova recalled in an interview years later. "They had not been invited, but some first-class women pilots began to act on their own. They reached the vice-president with their request to be allowed to participate in the space programme. Nothing came out of it, but since the Americans did not hide anything, some publications about this appeared in the press. Thus the decision [not] to create a women's group was made at the top." It was the perfect propaganda coup for a socialist state: 'proving' that women were, on the face of it at least, equal to men.

In spite of the obvious political nature of the decision, the women chosen all had rudimentary flying or parachuting expertise and Nikolai Kamanin envisaged a training programme of five or six months to prepare them for orbital Vostok missions. According to Ponomaryova, they "were selected through aviation clubs in the European part of the Soviet Union. They mostly selected sports parachute jumpers, since in the Vostok spacecraft the cosmonaut had to land on a parachute. Parachute jumping is a complex skill and therefore to train a novice in a short time is impossible. I had been trained as a pilot and had only eight jumps. I was a third-category jumper; in comparison to the sports master Irina Solovyeva's 800 jumps, my eight jumps were nothing". However, despite checking the documents of around 200 female aviation sports candidates, Kamanin was presented with only 58 'suitable' candidates, of whom five were finally selected.

"When we arrived," said Ponomaryova, "we were enrolled as privates of the Soviet Air Force. We found ourselves in a military unit, in which we became an alien part, with our different characters and different concepts. Our commanders had great difficulty dealing with us, since we did not understand the requirements of the service regulations and we did not understand that orders had to be carried out. Military discipline in general was, for us, an alien and difficult concept. Specialists from Korolev's bureau visited us and gave lectures on the Vostok spacecraft [and] specialists from other organisations also gave lectures."

Intensive instruction in rocket technology, navigation techniques and astronomy also formed part of the syllabus. Like their male counterparts, each woman endured repeated runs in the centrifuge, together with daily physical and vestibular training, parachute jumps and flights in two-seater MiG-15 jets to prepare them as much as possible for the weightless environment. "There were many special devices for stimulating and training the vestibular system," Ponomaryova said later, including "rotating chairs, electric current, chairs on unstable surfaces and so on. 'Real' weightlessness was simulated with flights, first on fighter planes and later on huge, specially-designed flying laboratories. Weightlessness there lasted 20–40 seconds, just enough to notice that a pencil sharpener was floating in front of you."

To evaluate their responses to weightlessness, each woman had to repeat phrases, write sentences, draw shapes and attempt to eat food from toothpaste-like tubes. However, one aspect in which they were not prepared was the actual operation of Vostok, since by this point there was a high level of confidence in the automatic controls. Years later, Ponomaryova admitted that, although weight issues may have been a contributory factor in Korolev's decision to minimise on manual controls, they were secondary to a sheer lack of understanding of how a human would behave in space. Although Gherman Titov had performed his tasks well, his unexpected sickness underlined this lack of knowledge. Spacecraft designer and Voskhod cosmonaut Konstantin Feoktistov later argued that the role of the human was to conduct research, not become a "servant to the machine". Ponomaryova, and doubless many other cosmonauts, felt the opposite: in emergencies, it would be imperative to allow the pilot a chance to control their ship.

By the winter of 1962, the training of the five women was complete and, following their final examinations, they were asked by Kamanin if they wished to become regular Soviet Air Force officers. All five accepted and were commissioned as junior lieutenants. On 19 November, when the final selections were made, Ponomaryova actually scored the best test results, but, it is said, did not offer the 'proper' replies to examiners' questions. When asked "What do you want from life?" she is reported to have responded with "I want to take everything it can offer". Valentina Tereshkova, on the other hand, spoke of her ambition "to support irrevocably the Communist Party", which may have contributed greatly to her eventual selection as the first woman in space. Ponomaryova later suspected that her own habit as a smoker, her somewhat aggressive feminism and her failure to vocally support communist ideals had been frowned upon by the selection board. Certainly, in his diary, Kamanin felt that Ponomaryova had "the most thorough theoretical preparation" and was "more talented than the others", yet he admitted her to be "arrogant, self-centred, exaggerates her abilities and does not stay away from drinking, smoking and taking walks".

Kamanin definitely favoured Tereshkova, although she had not received the highest marks, and considered her the best candidate and virtually a female version of Gagarin. His exact words were that "she is a Gagarin in a skirt". He recommended Irina Solovyeva ("the most objective of all") as her backup, with

Ponomaryova and the "persistently improving" Zhana Yerkina as options for later Vostoks. Kuznetsova had missed so much training that she did not take the final exams. In particular, she had performed poorly in both the pressure chamber and the centrifuge.

Plans for the flight changed considerably in the months leading up to Tereshkova's June 1963 launch. Early options, tabled in November 1962, included sending the female cosmonaut into orbit for three or four days or flying a dual mission with two female cosmonauts – much like that of Nikolayev and Popovich – or launching a man and a woman in separate Vostoks. The latter option was finally chosen by the Central Committee on 21 March 1963, in which cosmonaut Valeri Bykovsky would fly Vostok 5 on a record-breaking mission of between five and seven days. During his time aloft, he would be joined in orbit by Tereshkova, aboard Vostok 6, for about two days. Several weeks later, at the end of April, Korolev and – surprisingly – also Kamanin were pushing for Bykovsky to attempt a mission of eight to ten days, extending Vostok's life-support system and supplies to their limit. If successful, such a long-duration stunt would place the Soviet Union at least two years ahead of the Americans.

Still, there was some opposition to flying one woman, rather than two, particularly from the Soviet Air Force, which felt there was insufficient time to train Bykovsky for the other Vostok. Moreover, the shelf-life of spacecraft hardware expired in July 1963, necessitating a springtime or early summertime launch, and efforts to authorise ten more capsules for production had come to nothing. Plans were already underway to prepare a series of modified Vostoks, known as 'Voskhod', which would involve launching crews of two or three cosmonauts, crammed inside the tiny cabins, and conducting spacewalks. In March 1963, the Ministry of Defence categorically opposed building further Vostoks and it was decided that the Bykovsky-Tereshkova joint flight would be the last of the series. Kamanin, who now had four fully-qualified female cosmonauts, was livid that a dual-female mission had seemed the likely outcome and was now being eliminated at virtually the last moment.

Tereshkova, to be fair, was an enthusiastic flier and parachutist, but it was not just her careful replies to questions, or even her active membership of the Young Communist League, which endeared her to Nikita Khrushchev. Her occupation was a seamstress; an ordinary factory worker. She represented a plain Russian girl who would provide him with the chance to show that, under socialism, anybody could fly into space. Unlike the United States, which had scarcely considered female candidates during the selection of its Mercury astronaut team, Tereshkova would not be an elite intellectual, nor a professional fighter or test pilot, but instead would possess the kind of 'common touch' that Khrushchev liked. It was risky gamble.

THE PERFECT CANDIDATE?

Valentina Vladimirovna Tereshkova's parents provided the almost-perfect socialist background that Khrushchev wanted to present to the outside world. Her father, a tractor driver, had fought in the Russian Army as a sergeant and tank commander, dying in the Finnish Winter War when Tereshkova was two years old. Following her historic mission, incidentally, she was asked about possible ways in which the Soviet Union could demonstrate its gratitude to her: she requested a search to be conducted for the exact location of her father's death. This was duly done and a monument stands today in Lemetti, on the Russian side of the border with Finland, to commemorate Vladimir Tereshkov.

After her father's death, her mother single-handedly raised three children whilst working at the Krasny Perekop cotton mill. The young Tereshkova, who had been born on 6 March 1937 in the village of Bolshoye Maslennikovo, on the Volga River in the Yaroslavl Oblast of the western Soviet Union, did not commence her formal schooling until she was ten years old. She worked variously making coats, serving as an apprentice in a tyre factory and finally joined her mother and sister in 1955 as a loom operator at the cotton mill. She continued her academic studies in tandem, taking correspondence courses and eventually graduating from the Light Industry Technical School.

Her interest in aviation crystallised with membership of the Yaroslavl Air Sports Club, in which she quickly proved herself to be a skilful amateur parachutist, completing her first jump aged just 22. By the time she was picked as a cosmonaut candidate in March 1962, she had no fewer than 126 jumps under her belt and it was this achievement, coupled with the need for Vostok fliers to parachute out of their capsules during descent, which aided her selection. Her family and friends knew nothing of her plans: even her mother was under the impression that Tereshkova would be undertaking 'special studies' for a women's precision skydiving team. In fact, the first that Yelena Tereshkova knew about her daughter's achievement was on the day of her launch, via Radio Moscow.

Tereshkova's technical qualifications, admittedly, were lower than those of her male counterparts in the cosmonaut team, but her role as an active Party member – she had been the secretary of her local Young Communist League in 1961 – together with a war-hero father certainly brought her to the attention of the selection board and, finally, Khrushchev.

She was also well-liked by Kamanin as being "suitably feminine" and modest and, indeed, would demonstrate such attributes in an article entitled 'Women in Space', published by an American journal some years later. "I believe a woman should always remain a woman," she wrote, "and nothing feminine should be alien to her. At the same time, I strongly feel that no work done by a woman in the field of science or culture … however vigorous or demanding, can enter into conflict with her ancient 'wonderful mission' to love and be loved and with her craving for the bliss of motherhood." She was also doggedly determined in her efforts. Although she did not score the highest of the five candidates in her exams, her consistent effort prompted Yuri Gagarin to once comment that "she tackled the job stubbornly and devoted much of her own time to study, poring over books and notes in the evening". That was Tereshkova. Modest. Determined. Hard working. A good communist. To Khrushchev, she was ideal; the perfect candidate.

"ROMANOV AND JULIET"

Before February 1963, it seemed that the likely outcome of the woman-in-space project would involve two female cosmonauts, launched 24 hours apart in the March-April timeframe, with each spending two or three days aloft. Valentina Tereshkova was considered the most likely candidate for Vostok 5, with Valentina Ponomaryova following her aboard Vostok 6. Within weeks, the picture had changed considerably. A meeting of the Presidium of the Communist Party on 21 March killed off the plans and insisted instead that only one woman should fly and a male cosmonaut – a Soviet Air Force lieutenant-colonel named Valeri Fyodorovich Bykovsky, born on 2 August 1934 in Pavlovsky Posad, near Moscow – would be rushed into last-minute refresher training for the other Vostok. By inserting Bykovsky into the mix at such a late stage, the joint flight would have to be delayed; but, due to the limited shelf-life of the Vostok hardware, it would still have to fly before July 1963.

Like Gagarin, Titov, Nikolayev and Popovich before him, Bykovsky had been selected as a cosmonaut trainee in March 1960 and had a similar background. He had been one of the final six candidates for the first Vostok mission and, although his examination scores were good, it was noted that he made few important or substantial contributions to group discussions. His performance on Vostok 5, though, would be nothing short of exemplary and would mark the beginning of an illustrious, three-flight cosmonaut career, as well as securing a record which still stands to this day for the longest solo space mission in history. His backup, according to the final crew selections made on 10 May 1963, was Boris Volynov.

Only days earlier, Nikolai Kamanin had written in his diary that the joint endeavour would occur within the following six-week period, but noted that both Bykovsky and Volynov needed to make a few more parachute jumps and additional training runs in the Vostok simulator before being cleared for the mission. Two support cosmonauts, Alexei Leonov and Yevgeni Khrunov, were also listed as requiring more centrifuge training, pushing the launch until no earlier than the second week of June. Then, when the State Commission convened on 4 June, concerns were raised that wind speeds for 7 June were expected to be around 15–20 m/sec, exceeding the uppermost limit at which an R-7 could be launched. The failure of a command radio line, which required three or four days to repair, enforced an additional delay. A 10 June liftoff also became untenable due to the intensity of increased solar flare activity and the Crimean Observatory warned that risks would remain high for several days.

Even on 14 June, when Bykovsky finally clambered aboard Vostok 5, circumstances were far from perfect. Firstly, controllers reported that the ultra-shortwave transmitters on the spacecraft were refusing to operate properly; it was decided that, to avoid halting the launch preparations, the mission would rely upon its shortwave transmitters instead. Then a stuck pin in the ejection hatch caused a 30-minute delay and, finally, shortly before liftoff, the indicator light on the control panel for the R-7's third stage failed to light up as it should have done. This problem was traced to a failure in the stage's gyroscope instrumentation unit. Korolev was furious and began a heated argument, in front of other colleagues, with Viktor Kuznetsov, the man responsible for the gyroscope systems. Indeed, it was not a minor problem: the failure could have had a devastating impact on the mission and, if not fixed quickly, Vostok 5 would have been postponed another day, the rocket drained of its propellants and returned to the assembly building. Readying it for another attempt would have pushed the spacecraft itself past its shelf-life and derailed both flights. In light of these facts, it is perhaps not surprising that the hot-headed Korolev reacted as he did.

Fortunately, one of Kuznetsov's deputies announced that his engineers could replace the offending unit within a couple of hours and Bykovsky was kept aboard Vostok 5. At length, the cosmonaut – callsigned 'Hawk' – finally headed for space at 2:58:58 pm Moscow Time. Sadly, a less-than-nominal performance by the R-7 immediately reduced his mission from an expected ten-day orbital decay, which would have ensured eight days aloft, to an eight-day decay and a flight of around five or six days. The final orbit achieved was 162–209 km, inclined 65 degrees to the equator. Indeed, Bykovsky would recount in his post-mission debriefing that "the engine noise of the launch vehicle was weak", adding that, after separation from the rocket, he noticed a lot of frost particles which rendered it difficult for him to properly orient Vostok 5 to view the expended third stage. Nonetheless, for Khrushchev, the launch was ideal, corresponding as it did with a visit to Moscow by Harold Wilson, then-leader of Britain's Labour opposition. When asked by Wilson how many cosmonauts were in space this time, Khrushchev could not help but gleefully reply: "Only one ... so far!"

By the early hours of 16 June, it was clear to many western radio enthusiasts that something extraordinary was about to happen at Tyuratam, and Tereshkova was duly launched aboard Vostok 6 at 12:29:52 pm Moscow Time. Her liftoff, she reported later, was "excellent" and her experience of weightlessness presented no problems. In fact, despite her pulse rate soaring to 140 during the elevator ride up the gantry to the capsule, she seemed to have handled the ascent into orbit better than Nikolayev or Popovich, according to her biomedical readings. Original plans had called for her to launch five days into Bykovsky's mission, thus enabling them to both land on the same day, but the changes to the Vostok 5 duration called for a liftoff on his third orbital day instead. Both would then land on 19 June.

After insertion into space, Tereshkova's orbital plane caused the two Vostoks to draw towards each other for a few minutes, twice daily, with the closest approach of about 5 km. Some three hours after launch, Tereshkova's voice could be heard radioing her callsign 'Seagull' and she was certainly in direct voice contact with Bykovsky by 7:50 pm Moscow Time. Next day, ground controllers experienced some difficulty trying to contact her and even asked Bykovsky to help at one point. According to radio transmissions picked up by a Japanese station in Chiba, east of Tokyo, the Vostok 5 cosmonaut replied that he had attempted to contact Tereshkova without success, but did not think there was any reason to worry. Besides, the two Vostoks did not have a direct line-of-sight with one another on 17 June. All did appear to be well, in fact, and it appeared that Tereshkova was trying to contact the ground, but had been prevented from doing so, perhaps by being tuned to the wrong reception channel or, more likely, due to a problem with her receiver. At 5:58 pm Moscow Time on the 17th, the Enköping station in Sweden picked up a message from her, expressing that she felt "fine" and that all was well aboard Vostok 6. She even tried to communicate via rudimentary Morse code, which analysts described as "inexperienced". Fortunately, by 9:00 pm, she could be heard talking to ground controllers and asked to speak to "Number 20", the code for Sergei Korolev himself.

In his diary, Nikolai Kamanin wrote that communications in general were good and Bykovsky even reported that "she is singing songs for me", evidently hinting that the loss of contact did not prove a significant concern. It was speculated by the official Soviet media that she had simply fallen asleep. Indeed, she spoke to Nikita Khrushchev, her televised image was broadcast throughout the Soviet Union and she was able to undertake most of her planned scientific research programme. Although she experienced difficulty when changing camera films and could not reach her biological experiments to activate them, she was able to record images of terrestrial cloud cover and terrain. Using light filters, she observed Earth's horizon over the poles – describing it as "a light blue, beautiful band" – and the Moon on several occasions. Photographs returned from her mission and Bykovsky's flight suggested that it was possible to determine structures in the stratosphere and their data recorded two aerosol layers between 11.4 and 19.4 km above the surface.

Reports soon emerged, though, that Khrushchev's gamble of flying an 'ordinary' Russian girl, albeit one with over a hundred parachute jumps to her credit, was not entirely successful. Various accounts of the mission hinted that Tereshkova was unwell during the early stages of her flight and apparently vomited during her third day aloft; she also appeared tired and weak in televised images. She reported nagging pains in her right shin, pressure points from the helmet on her shoulder and left ear and irritation caused by biomedical sensors attached to a headband. In fact, both she and Bykovsky felt that future cosmonauts would be considerably more comfortable if they were permitted to remove their suits during missions. This recommendation proved ironic on the very next Soviet spaceflight, in October 1964, when three cosmonauts would fly without any space suit protection whatsoever.

Flannels, said Tereshkova, were too small and not moist enough to clean her face, there was no provision to clean her teeth or freshen up her mouth and she reported that she did not consume 40 per cent of her food. This could not be confirmed because she apparently gave away the rest of her food to onlookers at the landing site. Indeed, she complained that the bite-sized chunks of bread were too dry and, although she enjoyed the fruit juices and cutlet pieces, she began to crave Russian black bread, potatoes and onions by the end of the flight.

Bykovsky was also experiencing discomfort. An undisclosed problem with his spacecraft's waste management system, perhaps a spillage, had made conditions inside the Vostok 5 cabin very unpleasant and possibly contributed to the further shortening of the mission to just under five days. He also commented that the fan of his space suit's oxygen supply tended to cut off whenever he released himself from his seat and that this posed "a real problem". Nonetheless, he seemed to have enjoyed his time in orbit and performed many observations of terrestrial objects and places.

"I couldn't see Volgograd," he reported in his post-flight debriefing. "It was clouded over. I could make out islands easily and recognised Leningrad, the Nile and Cairo. At sea, I could see the wakes of ships and large barges; in Norway, the fjords and mountain summits. At night, through the Vzor, I could see lightning flashes and cities over South America. I saw aircraft contrails over France." Tereshkova, too, reported seeing fires in South America and the twinkling of city lights at night. Today, with almost five hundred individuals having journeyed into space, such sights are commonplace. In the summer of 1963, however, they were novel, hard to comprehend and truly remarkable.

RETURN TO EARTH

Described by some western observers as resembling a tougher-looking Ingrid Bergman, rumours would abound for decades that Tereshkova's mission had not gone well and, even, that she handled her return to Earth badly. Still, with radioed guidance from Gagarin, Titov and Nikolayev, she had managed to orient her spacecraft by manual control in a 20-minute-long experiment just before retrofire and successfully held the correct re-entry attitude. At 10:34:40 am Moscow Time on 19 June, the retrofire command was sent to Vostok 6 and executed 20 minutes later. However, the cosmonaut herself did not call out each event as required: she did not report a successful solar orientation, nor did she report the progress of the retrofire procedure, nor even the jettisoning of her spacecraft's instrument section. In fact, the only information coming in to the control centre was downlinked telemetry data.

Tereshkova ejected on time, but apparently broke a mission rule by opening her helmet faceplate and gazing upwards; she was hit by a small piece of falling metal, it is said, which cut her face. Barely missing a lake in the violently-gusting wind, she landed at 11:20 am, some 620 km north-east of Karaganda, on the 53rd parallel. The bland fare she had eaten over the past three days was replaced, thanks to kindly locals, by fermented milk, cheese, flat cakes and bread, although this ruined the physicians' chances of properly analysing her dietary intake. Three hours later, at 2:06 pm, after ejecting from Vostok 5, Bykovsky touched down 540 km to the north-west of the mining city and on the same parallel as Tereshkova. Unfortunately, in a worrying recurrence of the Vostok 1 and 2 problems, his instrument section failed to separate cleanly from the capsule, causing wild gyrations until the aerodynamic heating of re-entry finally burned the restraining straps away.

Little of this was evident in Bykovsky's post-mission report. "The solar orientation for retrofire worked correctly," he recalled, "and the braking engine fired for 39 seconds. Immediately after shutdown of the engine, the capsule separated from the service module. There were no big G forces during re-entry. There was a powerful explosion when the cabin hatch blew off" – it would appear that Bykovsky, too, 'chose' to eject – "and I was ejected from the capsule in my seat two seconds later ... " After landing, he was greeted by a man on horseback and a car which drove him to the charred Vostok 5, lying a couple of kilometres away. With a mission elapsed time of some 119 hours, just shy of five full days, he had set a new record for the longest solo manned spaceflight, which still remains unbroken. Tereshkova, too, had firmly ground all of America's Mercury astronauts beneath her heel in terms of space time: her 48 orbits of Earth and 70 hours aloft soundly surpassed all six of their missions combined.

Khrushchev loved it. A few days later, in Moscow, he declared that, unlike bourgeois society's emphasis on women as representing the weaker sex, the Soviet system permitted them to prosper and, literally, reach for the stars. Although the reality was decidedly more insincere, Tereshkova, for her part, agreed with him. "Since 1917," she said, "Soviet women have had the same prerogatives and rights as men. They share the same tasks. They are workers, navigators, chemists, aviators, engineers ... and now the nation has selected me for the honour of being a cosmonaut. On Earth, at sea and in the sky, Soviet women are the equal of men." Many observers in the west agreed. The wife of Philip Hart, the Democratic senator for Michigan, remarked that Russia was providing its women with a chance that American women simply did not have. Others, including anthropologist Margaret Mead, added that "the Russians treat men and women interchangeably. We treat men and women differently".

The purely propagandist nature of the mission would be shown up, however, by the fact that no more Soviet women would venture into space until 1982. Even a decade after Tereshkova's flight, fellow cosmonaut Alexei Leonov told an interviewer that her results indicated that "for women, flying in space is a hard job and they can do other things down here. After training, she will be 28 or 29 and if she is a good woman, she will have a family by then". Clearly, the real motivation of the Soviets in putting a woman into orbit was different from how it was presented to – and, surprisingly, accepted by – the rest of the world. Andrian Nikolayev, who married Tereshkova in a lavish ceremony presided over by Khrushchev at the Moscow Wedding Palace on 3 November 1963, shared Leonov's sentiment. "The mission programme makes big demands on her, especially if she is married," Nikolayev said, "so nowadays we keep our women here on Earth. We love our women very much; we spare them as much as possible. However, in the future, they will surely work on board space stations, but as specialists – as doctors, as geologists, as astronomers and, of course, as stewardesses!"

Whatever Tereshkova's own opinions on the matter of future female space travellers, she dived with vigour into her post-flight life on the international speaking circuit, touring India, Pakistan, Mexico, the United States, Cuba, Poland, Bulgaria and elsewhere, assuming dozens of ceremonial posts and moving into the office of the president of the Committee of Soviet Women on Pushkin Square in Moscow. Like the other cosmonauts, she was honoured as a Hero of the Soviet Union, together with the Order of Lenin and the Gold Star Medal. Five months after Vostok 6, amid much pomp and circumstance, she wed Nikolayev, previously the only bachelor cosmonaut. Apparently, their romance had developed during training – one story told that she kissed him farewell at the foot of the Vostok 3 gantry – and at the wedding, a beaming Khrushchev himself gave the bride away. The truth, it seemed, was very different. Although a daughter, Yelena, was born to the couple in June 1964, becoming the first child whose parents had both flown into space,

Nikolayev and Tereshkova were not even living together by the end of that same year. They divorced in 1982. To this day, speculation exists as to whether or not their marriage represented a genuine love match or a cynical ploy engineered by Khrushchev.

Tereshkova herself has always argued vehemently against allegations that she performed poorly during her flight, saying only that she suffered from fatigue and lack of sleep. Sergei Korolev had already been heard to mutter that he would not deal with "broads" again and, at a private interview with her on 11 July 1963, he expressed severe displeasure with her performance. His anger was shared by his deputy, Vasili Mishin, who claimed that she had been "on the edge of psychological instability". In fact, whilst drafting the official press release a few weeks earlier, the head of medical preparations, Vladimir Yazdovsky, had suggested including a paragraph to explain Tereshkova's "overwhelming emotions", coupled with tiredness and a sharply reduced ability to complete all of her assigned tasks. He was persuaded not to do so by Kamanin. Other rumours hint that she experienced significant menstruation whilst in orbit and at one stage became hysterical and began crying uncontrollably until she was scolded by Korolev over the radio.

Two decades later, in June 1983, shortly before the launch of America's first woman into space, Time magazine speculated that perhaps Tereshkova's poor preparation for the mission had contributed to the Soviets' decision not to fly female cosmonauts more frequently. Despite these apparent slurs on her ability, it was revealed in 2004 that an error in Vostok 6's control software had made the spacecraft ascend from orbit instead of descending; Tereshkova had noticed the fault during her first day aloft and had reported it to Korolev. She then calmly entered data to repair the mistake and landed safely. None of this was made public for more than four decades.

After her flight, Tereshkova would remain for many years a member of the cosmonaut corps, though in a purely honorific capacity, with little chance of another flight. She would study at the Zhukovsky Air Force Academy, graduating in October 1969 with distinction as a professional engineer, before learning of the dissolution of the female cosmonaut team later that year. To be fair, she and her female colleagues were never truly considered 'regular' members of the corps and the availability of flight assignments for them, realistically, ended shortly after Vostok 6. The entire woman-in-space effort was, for Korolev, simply a means of currying favour with Khrushchev: providing him with yet another propaganda coup to upstage the Americans, in exchange for signing off plans for the 'real' space programme to continue. That space programme would take two paths after Vostok: an improved version of the capsule (Voskhod) would fly from 1964 onwards with crews of up to three cosmonauts for long-duration and spacewalking exercises and an entirely new (Soyuz) spacecraft would be inaugurated shortly thereafter to pave the way for Earth-circling space stations and manned journeys to the Moon.

2

Monkeys to Men

HEROES

An expectant hush descended over the journalists in the conference room of the Dolley Madison House, opposite Lafayette Park in downtown Washington, DC. It was the afternoon of 9 April 1959. Backstage, clad in civilian suits, two of them wearing bow ties, stood the United States' first team of astronauts – the 'Mercury Seven'. At 2:00 pm Eastern Standard Time, the presiding officer, Walt Bonney, finally spoke. "Ladies and gentlemen," he began, "in about 60 seconds, we will give you the announcement you have all been waiting for: the names of the seven volunteers who will become the Mercury astronaut team."

Those volunteers – Scott Carpenter, Gordo Cooper, John Glenn, Virgil 'Gus' Grissom, Wally Schirra, Al Shepard and Donald 'Deke' Slayton – had been chosen after three months of careful screening of 110 experienced combat and test pilots. The list had gradually been reduced through a series of invasive and, in many cases, degrading medical and psychological evaluations. The pilots had been split into three groups, two of which were summoned secretly to the Pentagon in January 1959 for briefings on the effort to send a man into space. Afterwards, the 69 candidates from the first two groups had been asked if they wished to volunteer for further tests and, to the great surprise of many on the selection board, more than 90 per cent of them agreed. The third group, unneeded, was never called up. It had been assumed that the pilots would be so entrenched in their military careers that shifting to a civilian space project would hold little appeal. Already, some of the United States' most accomplished test pilots – Chuck Yeager among them – had poured scorn on the idea of rocketing men into space atop converted ballistic missiles. Even Al Shepard's father (himself a former military officer) expressed concern that his son had made the wrong decision. He would eat his words years later when Shepard not only became the first American in space, but also the fifth man to walk on the Moon.

Sixty-nine candidates rapidly dwindled to just thirty-two, who reported to Randy Lovelace's aerospace medicine clinic in Albuquerque, New Mexico, for additional evaluations. Over the course of a week, every spot on the bodies of Shepard and the

The Mercury Seven. From the left are Gordo Cooper, Wally Schirra (partially obscured), Al Shepard, Gus Grissom, John Glenn, Deke Slayton and Scott Carpenter. Behind them is a Mercury capsule, which all of them but Slayton flew into space.

others was sampled, measured, poked and prodded, with scarcely a muscle, bone or gland left untouched. Throats were scraped, stool and semen samples taken, electricity zapped into hands and intensely uncomfortable 'steel eels' inserted into rectums. Wally Schirra would later call Lovelace's tests "an embarrassment, a degrading experience … sick doctors working on well patients".

Survivors of the clinic devised a tradition of inviting the newer candidates to dinner at a local Mexican restaurant. At one of these gatherings, the veterans each had at their feet a jug of urine, which they had been obliged to collect for medical purposes during their stay. One evening, accidentally, Gus Grissom knocked over his jug, but, thanks to the quick-thinking crowd of test pilots, was provided with a ready solution: to order more beer. Several rounds, and a number of trips to the lavatory, later, Grissom's jug had its required amount of urine …

Still more tests followed at the Aeromedical Laboratory of Wright-Patterson Air Force Base in Dayton, Ohio, where the pilots withstood cold water pumped into their ears, sat for hours in overheated saunas, endured soundproofed and darkened isolation rooms, blew up balloons until they were out of breath, walked on treadmills until their heart rates soared to 180 beats per minute and were photographed from every conceivable angle and into every conceivable orifice. Many perceived the whole thing as excessive and a waste of time. "I'd flown combat missions and done operational test flying for 17 years by that point," wrote Deke Slayton. "The fact that I'd survived should have told them all they needed to know about stress. At least putting me in the blackout chamber, they let me catch a nap!"

Not only were the selectors looking for the most physically unbreakable men, they were also scrutinising their reactions to the tests and the testers. Would they crack, psychologically, under the unknown stresses imposed by the mysterious space environment? Personality questions prompted them to explore their individual motivations for wanting to become astronauts, their concerns about their health, their frustrations, their 'thoughts' and even whether their desires to fly jets and rockets arose from feelings of sexual inadequacy. The seven men eventually chosen, in addition to their combat and test flying credentials, were all highly intelligent – with IQs of between 131 and 141 – but, said psychologist George Ruff, all were "oriented toward action, rather than thought".

A MAN IN SPACE

Initially dubbed 'Project Astronaut' – a term later dropped because it placed too much emphasis on 'the man', rather than 'the mission' – the effort and its search for volunteers was carried out under the auspices of the newly-founded National Aeronautics and Space Administration (NASA), a government body established by President Dwight Eisenhower in the autumn of 1958. It represented the combined parts of the National Advisory Committee for Aeronautics (NACA), which, since 1915, had employed thousands of personnel at several research centres across the United States to design newer, better and faster aircraft. These included the Bell-built X-1 vehicle, in which Chuck Yeager first broke the sound barrier in 1947. However, in addition to taking NACA's old resources, the new NASA also assumed control of the United States Army's Jet Propulsion Laboratory (JPL) in Pasadena, California, and absorbed its ongoing aeronautical, rocketry and man-in-space projects.

Proposals for a civilian agency of this type had been made in the summer of 1957, during the International Geophysical Year, and led to a formal report, submitted to James Killian, chair of Eisenhower's Science Advisory Council, that December. At around the same time, NACA Director Hugh Dryden, responding to the Soviet launch of Sputnik 1, also felt that "an energetic programme of research and development for the conquest of space" was acutely needed. In March 1958, Killian added weight to the proposal and suggested that a new agency should be based on a "strengthened and redesignated NACA", utilising all of its 7,500 employees, $300 million-worth of research assets and $100 million annual budget, "with a minimum of delay". Later that same month, Eisenhower outlined his administration's future aims in space: to explore, to support national defences, to bolster the United States' prestige and to advance scientific achievement. Future projects would begin with preliminary experiments, followed by automated exploration, then limited manned missions, robotic planetary flights and, eventually, journeys to the Moon and Mars.

A civilian space agency had already won the support of Eisenhower, who distrusted the significant role that the military was playing in space affairs, and the bill for its creation was quickly pushed through Congress, thanks to the efforts of Senator Lyndon Johnson. Eisenhower signed the National Aeronautics and Space Act on 29 July 1958 and NASA officially came into being on the first day of October, headquartered at the Dolley Madison House until better facilities could be found. Its first administrator, Keith Glennan, was a former Case Western University president, and one of his earliest official tasks, on 17 December, was to announce America's man-in-space effort to the nation and publicly give it a name: 'Project Mercury'. Designed by NACA aerodynamicist Max Faget, the spacecraft would employ a truncated cone, sitting on a dish-shaped heat shield, to be launched on either a suborbital trajectory atop a Redstone missile or an orbital flight aboard an Atlas rocket. Project Mercury, however, was not the only man-in-space effort: for at least two years beforehand, the military had harboured its own plans. One of the most prominent of these, cultivated by the United States Air Force, was dubbed, somewhat unimaginatively, 'Man In Space Soonest' (MISS).

In July 1957, the Air Force's Scientific Advisory Committee arranged through the Los Angeles-based Rand Corporation to hold a two-day conference to discuss state-of-the-art space projects. Six months later, in the wake of Sputnik 1, a panel of scientists led by Edward Teller concluded that there was no technical reason why the Air Force could not launch a man into space within two years and an abbreviated plan was set in motion to explore the feasibility of placing a vehicle into orbit atop a converted Atlas. Contracts to build mockups of the spacecraft were awarded to North American Aviation and General Electric in March 1958 and, with a sense of great urgency, plans were implemented for an effort initially called 'Man In Space', then, from June, an accelerated 'Man In Space Soonest'.

Animal-carrying flights, read the proposal, would be attempted in 1959, followed by a manned mission in October 1960 and lunar landings as early as 1964. The five-year project would cost $1.5 billion. MISS, a ballistic capsule measuring 1.8 m in diameter and about 2.4 m long, would be fully automated and capable of supporting a single astronaut for up to 48 hours in orbit. Interestingly, the astronaut would lie supine in a contoured couch which could be rotated according to the direction of the G forces building up during ascent and re-entry.

Two camps existed over which missile – Atlas or Thor – should be used to loft the MISS spacecraft into orbit; the former was considered too unreliable and, moreover, would subject its astronaut to around 20 G, beyond the limits of human tolerance, in an abort situation. A two-stage Atlas, on the other hand, could provide a shallower re-entry flight path and reduce this to a survivable 12 G, but others expressed preference for a modified Thor intermediate-range ballistic missile, fitted with a Nomad fluorine-hydrazine upper stage. Eventually, on 2 May 1958, detailed designs for the MISS spacecraft, its operational procedures and the decision to employ the

Thor missile, were forwarded to Air Force Headquarters, with a first manned launch tentatively scheduled for October 1960.

However, it was felt in many quarters that development of the Thor-Nomad would take longer than planned, perhaps requiring 30 test launches and causing massive cost overruns. Consequently, the Air Force's undersecretary was convinced that using a modified Atlas as the launch vehicle could cut the project costs below the $100 million mark. Unfortunately, this would also mean cutting the orbital altitude achievable by MISS from 275 km to just 185 km, essentially putting it out of range of the tracking network for much of its flight. Still, on 15 June 1958, the Atlas was brought on board, the project's budget descended to $99.3 million and the first manned launch was targeted for April 1960.

Ten days later, the Air Force selected test pilots Robert Walker, Scott Crossfield, Robert Rushworth, William Bridgeman, Alvin White, Iven Kincheloe, Robert White, Jack McKay and – notably – Neil Armstrong as candidates to fly the MISS spacecraft. Arriving on the scene ten months before the Mercury Seven and almost two years before the first Soviet cosmonaut team was chosen they represented the first 'astronaut' selection in history. These astronauts would have been little more than passengers, inspiring the denigration of Chuck Yeager and others that the early space fliers were 'spam in a can', riding relatively simple ballistic capsules and parachuting to a water landing in the vicinity of the Bahamas.

Within weeks, however, Eisenhower's plan to create a civilian space agency had developed into legislation and Brigadier-General Homer Boushey of Air Force Headquarters announced that the Bureau of the Budget was blocking the further release of funds for MISS. A chance remained to make the project a reality if its costs could be kept below an impossible $50 million ceiling in 1959, although this would have pushed the first mission into the spring of 1962. Eisenhower's ingrained distrust of military involvement in the human spaceflight effort, coupled with the fact that the soon-to-be-formed NASA would not be spending more than $40 million on its own man-in-space project for 1959, signalled the final death-knell for MISS. By the third week of August 1958, Eisenhower assigned NASA specific responsibility for developing and carrying out manned space missions and $53.8 million, set aside for Air Force projects, including MISS, was transferred from the Department of Defense to the civilian agency. It has been speculated that, had it gone ahead with the required level of funding, it is quite possible that MISS would have beaten the Soviets into space, orbiting a man sometime in 1960.

At around the same time, the Army was planning its own, simpler, man-in-space effort, initially called 'Man Very High' (with Air Force participation, utilising the Manhigh gondola design) and, later, 'Project Adam'. This had been the brainchild of Wernher von Braun, designer of the V-2 missile and among a handful of German rocketry experts brought to the United States in the wake of the Second World War. Had it gone ahead, its proponents claimed, it would have reached space even 'sooner' than the Soonest. Utilising a converted Redstone missile, it would have placed a capsule onto a ballistic, suborbital trajectory, probably similar to that followed by the first two manned Mercury missions. The Army's astronaut would have been housed inside an ejectable cylinder, 1.2 m wide and 1.8 m long, which itself would have been encased inside the Redstone's nosecone. The rather tongue-in-cheek justification for the project was as a step towards improving techniques of troop transportation, although Hugh Dryden scornfully remarked that "tossing a man up in the air and letting him come back ... is about the same technical value as the circus stunt of shooting a young lady from a cannon". By July 1958, the Army was told that Project Adam's impracticability meant that it would not receive its requested $12 million of funding.

Meanwhile, the Navy, not to be outdone, proposed its own Manned Earth Reconnaissance (MER) initiative. This would have taken the form of a cylindrical spacecraft with a sphere at each end. After launch atop a two-stage booster, the spherical ends of the vehicle would expand laterally along two structural, telescoping beams to form a delta-winged, inflated glider with a rigid nose. The astronaut would then be able to make a controlled re-entry and water landing. Several studies were undertaken, including one jointly between Convair and the Goodyear Aircraft Corporation. By the time the partnership made its report, in December 1958, Project Mercury was already well underway. Although MER was undoubtedly the most ambitious of these early projects, its emphasis on new hardware and cutting-edge techniques led many observers to doubt its chances of approval, with or without the advance of Mercury. Indeed, of all of the military plans, MISS probably came closest to fruition, before the clear direction was taken to place manned spaceflight in the hands of a civilian organisation.

The urgency with which NASA addressed the need to launch an astronaut was heightened by the fact that, within four months of Glennan's announcement, the Mercury Seven were in place. It was already known that the Soviet lead on space achievement was strong and that they were surely planning their own man-in-space effort, with 1960 or shortly thereafter considered the most likely timeframe for a human launch. Indeed, Time magazine told its readers in September of that year that the long-awaited Soviet shot "could happen tomorrow", adding that "few of the world's scientists doubted ... that man at last was nearly ready to launch himself boldly and bodily into space". Eighteen bold bodies survived the punishing tests at Lovelace and Wright-Patterson and their names were forwarded to a NASA selection committee, with the original intent to choose six astronauts. However, after firmly picking five names, officials and physicians could not agree between two competing volunteers and ended up selecting both of them.

In many ways, the Mercury Seven were quite distinct from their counterparts in the Soviet cosmonaut team. For a start, their ages were somewhat higher. According to Neal Thompson, in his 2004 biography of Al Shepard, NASA had opted for "steely, technology-savvy test pilots", who were "mature ... who'd been around, been tested and stuck it out", rather than inexperienced, wet-behind-the-ears young bloods for whom the fascination might lose its lustre when faced with the prospect of long hours and extremely hard work. As a result, Glennan's agency stipulated that the astronauts had to be 25–40 years old at the time of selection, around 1.8 m tall and no heavier than 80 kg, to ensure that they could fit comfortably inside the tiny, conical Mercury capsule. They were also required to possess degrees in medicine, physical science or engineering, together with several years of professional expertise, including test piloting credentials, and at least 1,500 hours in their flight logbooks. Interestingly, this eliminated some of the most famous names in American experimental aviation – Chuck Yeager did not hold the required academic qualifications and Scott Crossfield, the first man to fly at twice the speed of sound, was a civilian – although the former at least would publicly ridicule Project Mercury, believing that it did not require the talents or merits of a test pilot.

The choice of combat and test pilots seemed logical, but had actually come about after lengthy debate: submariners, high-altitude balloonists and even mountaineers were considered in the early days and original plans advocated a public call for volunteers, after which perhaps 150 might be chosen for testing and around a dozen finally selected. In fact, a notice to this effect, with an annual salary of between $8,330 and $12,770, had appeared in the Federal Register on 9 December 1958. Nowadays, of course, astronauts are chosen from both the military and civilian sectors, but the sheer unknowns surrounding space travel at the close of the Fifties prompted the selection committee and, in particular, Navy psychologist Bob Voas, to favour test fliers from the armed forces. It also did not hurt, wrote Deke Slayton, that "you wouldn't have to be negotiating salaries with active-duty officers who volunteered". Moreover, none of the Mercury Seven was obliged to resign their military commissions in order to work for a civilian agency and, indeed, continued Slayton, given the state of NASA in late 1958, "you'd have had to be an idiot to give up your Air Force or Navy career to join them". For his part, President Eisenhower heartily endorsed the idea of selecting purely from the military, effectively ending the national call for volunteers.

"The astronaut training programme," Glennan told the Dolley Madison audience that April afternoon in 1959, "will last probably two years. During this time, our urgent goal is to subject these gentlemen to every stress, each unusual environment they will experience in that flight." That training programme had scarcely begun and according to Chris Kraft, a legendary NASA flight director from those early days, "we were inundated with the newness of everything". The astronauts expected their preparation to include many hours in the cockpits of jet aircraft – "we didn't know what else to train on," Gordo Cooper remarked – but their actual training for one of the most audacious feats in human history would encompass much more: physical and psychological conditioning, together with intense, PhD-level technical, scientific and medical instruction, to enable them to understand the intricacies of the spacecraft and rockets upon which their lives would depend. Spaceflight training had never been attempted and, in many ways, NASA and its first seven astronauts were forced to make it up as they went along. Indeed, Bob Gilruth, head of the Space Task Group, which included Project Mercury, stressed that they were not merely 'hired guns' and that, unlike the military, "where direction comes from the top", their direct input with respect to spacecraft design was expected and desired.

One particular training contraption, known as the Multiple-Axis Space Test Inertia Facility (MASTIF), was used to simulate the motions of the Mercury capsule in orbital flight conditions. Located at the Lewis Research Center in Cleveland, Ohio, it comprised a system of three interlocking concentric 'cages', one inside the next, with the innermost resembling the spacecraft itself. The cages could be

Described as one of the most sadistic trainers ever created, the Multiple-Axis Space Test Inertia Facility (MASTIF) comprised three interlocking cages to simulate motions about roll, pitch and yaw axes.

programmed to spin, sometimes simultaneously, about all three axes – roll, pitch and yaw – at up to 30 revolutions per minute. Nitrogen-gas jets attached to the cages created these motions, which were intended to precisely mimic the worse-than-worst-case scenario of a complete loss of control of the capsule whilst in space. As the simulator tumbled, the astronaut, with all but his arms held firmly in place, had to read eye-level instruments and actuate the jets by means of a control stick to somehow interpret the motions and correct and steady the capsule accordingly. For three weeks in February and March 1960, all seven Mercury astronauts were wrung through the MASTIF, which often left them nauseous and vomiting and which all would agree was one of the most sadistic trainers they had ever ridden.

Elsewhere, punishing centrifuge runs at the Naval Air Development Center in Johnsville, Pennsylvania, subjected their bodies to forces as high as 16 G – enough to smooth back the skin on their faces and break blood vessels in their backs – in recognition of the fact that so little was known about deceleration during descent. Nicknamed, rather innocuously, 'the wheel', the centrifuge "took every bit of strength and technique you could muster to retain consciousness," according to John Glenn. These exercises were perverted yet further by so-called 'eyeballs-in, eyeballs-out' testing, where the forces were extended by simulating another worse-than-worst-case eventuality that the Mercury capsule could splashdown in the sea on its nose, rather than its base; the astronauts were rotated 180 degrees and thrown violently against their restraining straps, which Al Shepard sarcastically called "a real pleasure". Indeed, one NASA physician who underwent the test could not properly catch his breath for some time afterwards. It later became clear that his heart had slammed into one of his lungs and deflated it ...

The eyeballs-in, eyeballs-out testing eventually led to recommendations for more durable shoulder harnesses inside the capsules and, after their first visit to prime contractor McDonnell Aircraft Corporation's St Louis plant in Missouri, the astronauts realised to their surprise that no window – only a blurry periscope – existed for them to see outside. Although their suggestion to include a viewing window was implemented, the first three Mercury spacecraft had already been built and outfitted, meaning that at least the first American in space would have to rely instead on two small portholes and the fisheye view transmitted through the periscope lens onto a circular screen in front of his face. The astronauts' ability to apply their technical prowess and implement practical changes proved quite at odds with the experiences of Yuri Gagarin and his comrades, who had little or no input into the Vostok design process. In fact, with each of the Mercury Seven assigned a responsibility – Carpenter focused on communications and navigation, Cooper on Redstone rockets and trajectories, Glenn on cockpit layout, Grissom on controls, Schirra on environmental systems and space suits, Shepard on recovery equipment and Slayton on the Atlas booster – questions of whether to include aircraft-like rudder pedals or a control stick, whether to use gauges or easier-to-read tape-line instruments, where to position certain switches or handles to make them easily reachable or how best to remove the capsule's hatch in an emergency were encountered on a regular basis.

In spite of the intense preparation, some psychologists remained fearful that the two-year wait for the first manned mission could lead to 'over-training' and staleness, although Shepard and others would strongly disagree and remark that the similarity of training with actual flight conditions was a key factor in making the real thing feel 'routine'. The choice of Shepard as the first American in space was delivered on 19 January 1961 – the eve of John Kennedy's presidential inauguration – when Bob Gilruth personally visited the astronauts at their headquarters at Langley Research Center in Hampton, Virginia. After 20 months of training, it did not come as a great surprise. Gilruth had already asked them, weeks earlier, to write the name of the astronaut, excepting themselves, that they would like to fly first. "We all intuitively felt that Bob had to make a decision as to who was going to make the first flight," Shepard said, "and when we received word that Bob wanted to see us at five o'clock in the afternoon in our office, we sort of felt that perhaps he had decided."

Gilruth wasted little time and got straight to the point. Revealing that it was the hardest decision he had ever made, he announced that Shepard would fly first, Gus Grissom would fly second and John Glenn would support both missions. Years later, strong suspicions would abound that the choice of naval aviator Shepard had much to do with President Kennedy's own nautical background and more than one member of the Mercury Seven would attribute the decision purely to politics. Indeed, even the Shepard-Grissom-Glenn trinity neatly represented the United States Navy, Air Force and Marine Corps. (Army pilots had not been selected, on the basis that they lacked the required expertise in high-performance jets.)

No further public decisions on subsequent missions would be made, Gilruth told them, and, in fact, the three men's names as mere 'candidates' for the first flight would not be announced to the world for another five weeks. Shepard's selection as America's first astronaut would not be revealed publicly until 2 May, as his first launch attempt was in the process of being scrubbed. Until then, the trio had to run through their own training façade of 'not knowing' which of them was to fly – "a ruse," wrote Neal Thompson, "that all the astronauts thought was ridiculous and annoying". Oblivious, the media's favourite had always been John Glenn, whose appearance, eloquence and warm demeanour typified the 'all-American' hero, and much surprise abounded when he, in fact, was not picked. Some newspapers even implied that inter-service rivalry was so strong that the Air Force may have deliberately leaked Glenn's name to embarrass NASA and reduce his chances. Others supported Grissom, since his parent service – the Air Force – was already in the process of developing its own winged spacecraft called 'Dyna-Soar', together with an Earth-circling space station.

Glenn was the first to offer his hand to Shepard in congratulation, but others, including Wally Schirra, were "really deflated" by the decision. Deke Slayton, although he had privately ranked Shepard as the best in terms of piloting skills and general 'smartness', felt humiliated and could scarcely believe that he had not even made the cut of the final three. Gilruth's choice did not surprise Scott Carpenter, though, who had been aware for two years that Shepard was "single-minded in his pursuit of the first flight". Moreover, according to Walt Williams, NASA's director of operations for Project Mercury, Glenn's image-consciousness, his untiring effort to perfect his 'boy-next-door' image and his currying of favour with top brass, led some officials to consider Shepard the best. His intense focus on the mission, his desire to know every aspect of the engineering and capsule design and his superb flying skills left him the obvious choice.

Admittedly, all seven knew that only one of them could make the coveted first flight, even though it would amount to little more than a 15-minute suborbital lob into the heavens atop a converted Redstone, but the disappointment was tangible. "I think Life magazine got into the act with some horseshit about the Gold Team (Glenn, Grissom and Shepard) and Red Team (the rest of us)," wrote Deke Slayton, "and I even had to have a press conference ... a couple of days after the announcement to reassure everybody that we weren't depressed." In a situation once described as seven pilots all trying to fly the same aircraft, each man had been chosen at the very pinnacle of his profession; each was hyper-competitive and in their time together each had set himself the personal goal of ensuring that 'the other guy' never got so much as half a step ahead in 'the game'. That game, and that intense competitiveness, ran from flying jets to mastering the MASTIF to winning a dispute over an aspect of Mercury design to racing the fastest in their flashy sports cars.

With the exception that all were military pilots and all would someday be flying into space, little commonality existed between them and their Soviet counterparts. The former were screened from the world and venerated only after their missions. The Mercury Seven, on the other hand, were placed on a pedestal of hero-worship from the day of their selection. They were, in a sense, 'premature heroes', with their personal stories sold by their lawyer Leo D'Orsey to Life magazine for $500,000 and a variety of perks – from sleek Corvettes on one-dollar-a-year leases from General Motors to the choicest picks of real estate – headed their way. They would battle the evil Soviet empire, take democracy to the new 'high ground' of space, and Alan Bartlett Shepard Jr, it was hoped, would be the first man to do it.

THE COMPETITOR

"Who let a Russian in here?" Louise Shepard joked on the evening of 19 January 1961, when her husband announced that she had her arms around the man who would be first to conquer space. Her light-hearted words hinted at the closeness of the race between the United States and the Soviet Union in achieving that goal, but would prove unfortunately prophetic when, in less than three months' time, Yuri Gagarin would rocket into orbit. Al Shepard would not be the first man in space, but would come close, missing out by barely three weeks. Privately and publicly, the gruff New Englander would fume over the lost opportunity to make history. "We had 'em by the short hairs," he would growl, "and we gave it away."

Shepard had been born in East Derry, New Hampshire, on 18 November 1923, the son of an Army colonel-turned-banker father and Christian Scientist mother and the progeny of a close-knit, fiercely loyal and wealthy family. His key qualities – bravery, a spirit of adventure and an absolute determination to be the best – emerged at a young age: as a boy, he did chores around the home and a paper round gave him enough money to buy a bicycle, which he rode to the local airport, cleaning hangars and checking out aircraft. At school, his boundless energy led teachers to advise that he skip ahead two grades, making him the youngest in each class he attended. After spending a year at Admiral Farragut Academy in New Jersey, Shepard entered the Naval Academy in Annapolis, Maryland, receiving his degree in 1944 and serving as an ensign aboard the destroyer Cogswell in the Pacific theatre during the closing months of the Second World War.

He subsequently trained as a naval aviator, taking additional flying lessons at a civilian school, and received his wings from Corpus Christi, Texas, and Pensacola, Florida, in 1947. Shepard served several tours aboard aircraft carriers in the Mediterranean and was chosen in 1950 to join the Navy's Test Pilot School at Patuxent River – the famed 'Pax River' – in Maryland; whilst there, he established a reputation as one of the most conscientious, meticulous and hard-working fliers. On more than one occasion, he was hand-picked to wring out the intricacies of a new aircraft, purely on the basis of his technical skill and precision. His test work included missions to obtain data on flight conditions at different altitudes, together with demonstrations of in-flight refuelling systems, suitability trials of the F-2H-3 Banshee jet and evaluations of the first angled carrier deck.

Later, as operations officer for the Banshee, attached to a fighter squadron at Moffett Field, California, Shepard made two tours of the western Pacific aboard the Oriskany. A return to Pax River brought further flight testing: this time of the F-3H Demon, F-8U Crusader, F-4D Skyray and F-11F Tiger jets, together with posts as a project officer for the F-5D Skylancer and as an instructor at the school. Graduation from the Naval War College in Rhode Island in 1957 led to assignment to the staff of the commander-in-chief of the Atlantic Fleet as an aircraft readiness officer. By the time he was selected as an astronaut candidate by NASA in April 1959, Shepard had accumulated 8,000 hours of flying time, almost half of it in high-performance jets. His flight-test experience surpassed that of the other members of the Mercury Seven, although he was alone among them in having never flown in combat.

His standoffish attitude also set him apart from the others. Since childhood, perhaps in light of his family's wealth, Shepard had been a loner and in his years at NASA many fellow astronauts would comment on his notorious dual personalities: warm and smiling one minute, icy and remote the next. "If you were a friend of Al's," said Deke Slayton's wife Bobbie, "and you needed something, you could call him and he'd break his neck trying to get it for you. If you were in, you were in. It was just tough to get in." During the mid-Sixties, when he was grounded from flying due to an inner-ear ailment and serving as NASA's chief astronaut, Shepard's secretary would put a picture of a smiling or scowling face on her desk each morning to pre-warn astronauts of which personality to expect from 'Big Al' that day.

Reputation-wise, though, he was quick-witted, a top-notch aviator and, as a leader, possessed all of the characteristics of a future admiral – a rank which, even whilst attached to NASA and never having commanded a ship, he attained in 1971. In fact, when he told his father of his selection as an astronaut candidate, the older Shepard expressed grave misgivings that he was abandoning a promising naval career for what was perceived by many as an ill-defined programme with limited prospects, run by a newly-established civilian agency. For Shepard, though, Project Mercury represented a logical extension to a life spent looking for the next challenge. His competitive nature had become the stuff of legend years before Mercury and had gotten him into hot water with superiors on more than one occasion: after several illicit, close-to-the-ground flying stunts, known as 'flat-hats' – one over a crowded naval parade ground, another looping under and over the half-built Chesapeake Bay Bridge in Maryland and a third blowing the bikini tops off sunbathing women on Ocean City beach – he had come dangerously close to court-martial.

Undoubtedly, Shepard's less-than-reputable exploits had come to the attention of the NASA selection board, but Neal Thompson speculated that it was viewed as an aspect of his fearless and competitive personality, rather than as an excuse to discard his application. He indulged in other hobbies, too. After taking up water-skiing, he progressed rapidly from two skis to one and, later, even experimented on the soles of his bare feet. His wife, Louise, whom he had married in 1945 whilst at Annapolis, would remark that it was "characteristic" of Shepard to always be restless for new challenges. His biggest feat – and, he would say later, his proudest professional accomplishment – was selection to fly the first American manned space mission. "That was competition at its best," he said, "not because of the fame or the recognition that went with it, but because of the fact that America's best test pilots went through this selection process, down to seven guys, and of those seven, I was the one to go. That will always be the most satisfying thing for me."

DIFFERENT WORLD

The day after Bob Gilruth picked Shepard for the first American manned space mission, another selection was being ratified in a cold and snowy Washington, DC, under the auspices of Chief Justice Earl Warren and accompanied by Robert Frost's poetry. At 12:51 pm on 20 January 1961, the man who would truly define the United States' space ambitions for the new decade officially became its 35th president. John Fitzgerald Kennedy, the first incumbent of the office to have been born in the 20th century, famously encouraged Americans in his 14-minute, 1,300-word inauguration speech to participate as active citizens: to "ask not what your country can do for you; ask what you can do for your country". Only months later, speaking before a joint session of Congress in the wake of Shepard's flight, Kennedy would rally hundreds of thousands of Americans from all corners of the nation to participate, actively, in the greatest scientific endeavour ever attempted: to land a man on the Moon.

Today, he holds a somewhat nostalgic, even mystical, place in the hearts of space aficionados, as the first major world leader to truly support a peaceful exploration programme with words, deeds and serious money. Indeed, the lunar landing effort, known as Project Apollo, would consume more than $25 billion in a little over a decade of operations. However, Kennedy's motivations for funding it were at least partly political. At the time of his appointment, American missile and space technology had fallen seriously behind that of the Soviet Union, opening up a much-publicised 'gap' between the two superpowers and creating an issue which had been a central component of his election campaign. It is interesting to speculate when one considers Kennedy's words – that "the world is very different now" – whether or not the lunar effort would have gone ahead if such issues with the Soviet Union and the steady march of communism into south-east Asia had not been present.

The son of a businessman-turned-ambassador, Kennedy's grandfathers had both been important political figures in Boston, Massachusetts. After a stint in command of a torpedo boat in the Solomon Islands in 1943 – during which he famously brought his crew ashore after being hit by a Japanese destroyer – Kennedy remained undecided for a time over whether to enter journalism or politics in civilian life. He eventually settled on the latter, winning a seat in the House of Representatives in 1946, supporting President Harry Truman and advocating policies of progressive taxation, the extension of social welfare and increasing the availability of low-cost housing. Election to the Senate in 1952 was followed by his sponsorship of bills to provide federal financial aid for education, liberalise immigration laws and implement measures to require full disclosure of all employees' pension and welfare funds. He also wrote the Pulitzer-winning Profiles in Courage in 1956, becoming the first president to achieve the coveted literary prize.

Kennedy officially declared his intent to run for the presidency on 2 January 1960, defeating opponents Hubert Humphrey and Wayne Morse in the Democratic primaries. Despite his staunch Roman Catholic beliefs – which caused suspicion and mistrust of him in several states, particularly the largely-Protestant West Virginia – he succeeded in winning solid support and cemented his credentials. In a speech delivered to the Greater Houston Ministerial Association, he revealed himself to be "the Democratic Party's candidate for President, who also happens to be a Catholic". Further, he attacked religious bigotry and explained his belief in the absolute separation of Church from State. By mid-July, the Democrats had nominated him as their candidate, with Lyndon Johnson joining him for the vice-presidency.

During the first televised debate in American political history, Kennedy appeared relaxed opposite his Republican rival (and then-Vice-President) Richard Nixon, further increasing the momentum of his campaign. On 8 November 1960, he won the election in one of the most closely-contested votes of the 20th century, leading Nixon by just two-tenths of a per cent – 49.7 against 49.5 – although it might have been higher, had not 14 electors from Mississippi and Alabama refused to back him on the basis of his support for the brewing civil rights movement. Nevertheless, and despite Nixon lambasting Kennedy's lack of experience in senior politics, the second-youngest man ever to win the presidency duly took office.

A little more than three years later, he would also become the youngest – and the last – to be assassinated.

"WE GAVE IT AWAY"

As Kennedy battled through the closing months of his election campaign, NASA battled with similar tenacity and vigour to launch the first man into space. Many in the United States, however, were already echoing Louise Shepard's sentiment that the Soviets remained in pole position to accomplish the historic feat. Project Mercury, Time magazine told its readers in September 1960, "is not far behind, but it will be at least nine months before a US astronaut will enter orbit". 'Orbit' would prove the pivotal point, for neither America's first man in space, nor even its second, would achieve orbit – they would experience little more than 15-minute suborbital arcs over the Atlantic Ocean, into space and back down – and the nation's first piloted circuit of the globe would not come until February 1962. Still, in the weeks after Kennedy's inauguration, Al Shepard and John Glenn were dividing their time between Langley Research Center in Virginia and the swamp-fringed Cape Canaveral launch site in Florida, familiarising themselves with 'Spacecraft No. 7': the vehicle which, since October of the previous year, had been earmarked for the first mission.

Unlike the huge spherical Vostok which had ferried Yuri Gagarin into space, the Mercury capsule was a cone-shaped machine, 1.9 m across the blunt, ablative heat shield at its base and 2.9 m tall, with a total habitable volume of just 1.6 m^3 and an approximate weight at launch of 1,930 kg. The idea that a blunt cone was the most suitable shape to prevent a rocket-carried warhead from burning up in the atmosphere had arisen in the early Fifties, thanks to the work of NACA engineers Julian Allen and Al Eggers. Attached to its nose was a cylindrical parachute compartment and at its base a cylindrical package of three retrorockets. Its cramped nature prompted the astronauts to smirk that, far from 'flying' the spacecraft they actually 'wore' it. "You get in with a shoehorn," added McDonnell's pad leader Guenter Wendt, "and get out with a can opener!" During the early stages of ascent, capsule and astronaut would be protected by a pylon-like, solid-fuelled Launch Escape System (LES), capable of whisking them away from an exploding or malfunctioning rocket. This measured 5.15 m tall and produced 23,580 kg of thrust. Under normal circumstances, however, it was intended that the LES would be jettisoned shortly after the burnout of the rocket, although many engineers doubted its effectiveness and felt that a catastrophic failure would give an astronaut little chance of survival.

The Mercury capsule was equipped with attitude-control thrusters to enable yaw, pitch and roll exercises, but was incapable of actually changing its orbit. The three solid-fuelled retrorockets provided an ability to return to Earth, firing in sequence at five-second staggered intervals, in a 'ripple' fashion, although one was sufficient to complete this task if the others failed. To guard against temperatures as high as 5,200°C at its base during re-entry, a heat shield composed of fibreglass, bonded with a modified phenolic resin, was employed. By charring, melting and peeling off, taking heat with it, this 'ablative' material would protect the structure of the spacecraft from the high heat flux of hypersonic re-entry into the atmosphere. It was first tested atop an Atlas rocket in September 1959, surviving re-entry in remarkably good condition. The heat shield was not, in fact, an integral part of the spacecraft, but was held in place by a series of hooks. Between it and the base of the capsule was a folded rubber-and-glass-resin 'landing bag', 1.2 m deep, which would unfold and fill with air shortly before splashdown in the ocean. This would act as an absorber,

The Mercury spacecraft. Note the parachute container at the top and the retrorocket package at the base of the capsule.

softening the shock of landing from 45 G to 15 G, before filling with water to provide a kind of 'sea-anchor'.

Mercury was the brainchild of NACA aerodynamicist Max Faget, adapted from Allen and Eggers' blunt-cone design, and received the go-ahead on 7 October 1958, only six days after NASA's birth. The name arose from that of the fleet-footed messenger of Roman mythology and, wrote Loyd Swenson in 'This New Ocean', a seminal 1966 work on Project Mercury, "seemed too rich in symbolic associations to be denied. The esteemed Theodore von Karman had chosen to speak of Mercury, as had Lucian of Samosata, in terms of the 're-entry' problem and the safe return of man to Earth". By mid-January 1958, McDonnell had been awarded the $18.3 million contract to build the spacecraft, beating Grumman, which was heavily loaded with conceptual naval projects at the time. Faget's original design for a ballistic capsule envisaged that it would re-enter the atmosphere at an attitude 180 degrees from that of launch, such that the G forces would be imposed on the front of the body under acceleration and deceleration; in effect, its 'tail' during launch would become its 'nose' during the journey back to Earth. Initial sketches from late 1957 revealed a squat, domed body with a nearly flat heat shield, the former slightly recessed from the perimeter of the latter, leaving a narrow 'lip' to deflect airflow and minimise heat transfer. However, this configuration proved dynamically unstable at subsonic speeds, so Faget's group lengthened the capsule and removed the heat shield lip.

By March of the following year, the design resembled an elongated cone, which provided dynamic stability, but hypersonic wind tunnel tests showed that too much heat would be transferred by turbulent convection. Further, engineers could not figure out how to incorporate parachutes into the upper part of the nosecone, prompting its redesign into a rounded shape with a short cylinder attached to the top. Heat-transfer concerns, however, remained, and it was not until the late summer that the design, incorporating maximum stability, relatively low heating and a suitable parachute compartment, had been finalised. Faget's team argued that by launching the capsule on a ballistic trajectory, its automatic stabilisation, guidance and control equipment could be minimised and the only manoeuvre it would be required to make would be to fire the retrorockets to decelerate and dip into the atmosphere for aerodynamic drag. In fact, added Faget, even that manoeuvre did not need to be too precise to accomplish a successful recovery.

In theory, Spacecraft No. 7 – the seventh of 20 Mercury capsules built by McDonnell – should have been capable of flying Shepard almost immediately, but after delivery to Cape Canaveral on 9 December 1960, it became necessary to implement 21 weeks' worth of unexpected tests, repairs and rework. Additionally, the landing bag, beneath the heat shield, which would cushion its splashdown in the Atlantic Ocean, had to be installed and communications hardware checked. Its reaction-control system needed attention, whilst damaged and corroded hydrogen peroxide fuel lines required replacement and a variety of other obstacles surrounded equipment, minor structural defects and even the need to install a manual bilge pump to remove seawater. The need for the latter had been compounded by the successful, though harrowing, flight of a chimpanzee named Ham. He had been launched atop a Redstone in late January, but his capsule had suffered a multitude of niggling malfunctions. Firstly, a faulty valve had fed too much fuel into the rocket's engine, causing Ham to fly too high and too far, whereupon the tanks ran dry, the spacecraft separated too early and re-entered the atmosphere too fast and at the wrong angle. Temperatures soared and a glitch 'rewarded' Ham not with banana pellets for pulling the right levers and pushing the right buttons, but with electric shocks. At the end of the mission, with the capsule filling with seawater and about to sink, "a very pissed-off chimp" was safely fished from the Atlantic by the recovery forces.

Wernher von Braun, whose team had designed and built the Redstone, feared that Shepard's mission, then scheduled for March, could be similarly affected and opted for one final unmanned launch. The astronaut, however, pushed NASA officials and even von Braun himself to go ahead with his mission, regardless of the risk, feeling that he could handle and overcome any Ham-type problems. The German stood firm, though, and a nervous NASA stood beside him.

"We were furious," remembered Chris Kraft. "We had timid doctors harping at us from the outside world and now we had a timid German fouling our plans from the inside." Furthermore, Jerome Wiesner, recently picked by President Kennedy as his science advisor, warned of the harm a dead astronaut could cause the new administration and pressed for another test flight. In addition, having inherited chairmanship of the President's Science Advisory Committee (PSAC), he convened a panel of experts to assess the situation and recommend whether or not to proceed with Shepard's launch. After viewing astronauts 'flying' in the simulators, whirling in the MASTIF and pulling up to 16 G in the centrifuge, the panel concluded that the manned mission should proceed. Their report, ironically, landed on Kennedy's desk on the afternoon of 12 April 1961.

By this point, Shepard's launch had already been postponed until the end of the month and, despite the crushing disappointment of Vostok 1, both he and Glenn continued to train feverishly, rehearsing every second of the 15-minute 'up-and-down' mission that would arc 188 km into space and back to Earth, splashing into the Atlantic some 200 km downrange of the Cape. It would be a suborbital 'hop': the Redstone, capable of accelerating to around 3,500 km/h, lacked the impulse to deliver Shepard into orbit – an Earth-girdling flight would have to await the Atlas – but the mission would prove to the world that the United States was in the game. Today, wrote Chris Kraft in his foreword to Neal Thompson's biography of Shepard, it is easy to dismiss it and, when placed alongside Vostok 1, it was insignificant, but in the spring of 1961 it captivated not only America, but the world. "Add to this the fact that the reliability of a rocket-propelled system in 1961 was not much better than 60 per cent," wrote Kraft, "and you may begin to have a feel for the anxiety all of us were experiencing."

The Redstone itself was a direct descendant of the infamous V-2, used by Nazi Germany with such devastating effect in the Second World War, and had been employed as a medium-range ballistic missile to conduct the United States' first live nuclear tests during Operation Hardtack in August 1958. It remained operational within the Army until 1964, gaining a reputation as the service's workhorse and, as a non-military launcher, as 'Old Reliable'. Initial production, under the auspices of prime contractor Chrysler, had gotten underway at the Michigan Ordnance Missile Plant in Warren, Michigan, in 1952. Meanwhile, the Rocketdyne division of North American Aviation built its Model A-7 engine, Ford Instrument Company supplied its guidance and control systems and Reynolds Metals Company fabricated its fuselage. As a weapon, it could be armed with an atomic warhead with a yield of 500 kilotons of TNT or a 3.75 megaton thermonuclear warhead and, indeed, batteries of Redstones were stationed in West Germany until as late as 1964.

A direct outgrowth of the Redstone was the Jupiter-C intermediate-range ballistic missile, which, some observers believe, could have beaten Sputnik 1 by orbiting an artificial satellite in August 1956, had the political will been there. President Eisenhower's administration, however, preferred to launch America's first satellite atop a civilian rocket named Vanguard, rather than with a modified military weapon, and the chance was lost. The Vanguard failed spectacularly in December 1957, exploding on the pad, but less than two months later a Jupiter-C successfully lofted the United States' first satellite, Explorer 1, into space.

A number of modifications were incorporated into the Redstone from 1959 onwards to complete the metamorphosis from a warhead-laden weapon to a man-rated launch vehicle; its reliability as a tactical missile, though high, was inadequate for an astronaut. Since redesigning it to provide the required assurances could have meant implementing a totally new, expensive and lengthy development programme, it was decided instead to adapt the existing model with only the changes needed for a manned flight. In January 1959, the Army Ballistic Missile Agency (ABMA) received the go-ahead to convert the rocket and, two months later, the Space Task Group requested the implementation of an effective abort system. By June, ABMA had submitted its response and, throughout the remainder of the year and into 1960, the design was finalised and implemented: an automatic system, capable of shutting down the Redstone's engine and transmitting separation abort signals to the Mercury capsule and its attached LES tower. Had the rocket veered off-course, a range safety officer at the Cape would have had little option but to remotely destroy it. However, a three-second delay existed between the transmission of the abort command and the actual destruction of the Redstone, offering a hair's breadth of time for the capsule to be pulled clear of the conflagration.

It had long been recognised that some emergencies could develop too rapidly for a mission to be manually aborted and, moreover, the astronaut's own performance under the dynamic conditions of a launch were not known. During their analysis of this problem, ABMA engineers studied 60 Redstone flights, identifying a huge number of components which could conceivably fail. It would be impractical to accommodate them all. However, the study did find that many malfunctions – loss of attitude control and velocity, a lack of proper combustion chamber pressure in the engine or perhaps power supply problems – led to similar results, thus permitting the inclusion of relatively few abort sensors.

Constructed from aluminium alloy, the single-stage Redstone measured 25.4 m long and weighed 3,720 kg. Ignition of its engine was initiated from the ground and liftoff occurred when approximately 85 per cent of its rated thrust had been achieved. During ascent, carbon jet vanes in the exhaust of its propellant unit, coupled with air rudders, served to control its attitude and stability. Its Model A-7 engine, fuelled by a mixture of ethyl alcohol and liquid oxygen, together with a hydrogen peroxide-fed turbopump, yielded 35,380 kg of thrust and was essentially the same as that used by the military Redstone, although a number of improvements had been implemented for efficiency and safety. The Jupiter-C's use of a highly-toxic propellant mixture called hydyne had been ruled out in favour of alcohol, although the use of the latter was more erosive of the jet vanes. Engine operations continued until the Redstone had reached a pre-determined velocity, at which stage an integrating accelerometer emitted a signal to initiate shutdown by closing off the hydrogen peroxide, liquid oxygen and fuel valves. As pressures in the thrust chamber decreased, a timer started in the Mercury capsule which triggered its separation from the tip of the Redstone.

Other modifications included lengthened tanks, the walls of which were thickened to handle the increased loads of the capsule and heavier propellant haul, and changes were made to increase the reliability of critical electronic components in the Redstone's instrument section. Indeed, the entire layout of this section was extensively revamped to accommodate new control and abort systems. The elongated propellant tanks and increased payload weight, however, meant that the rocket tended to become more unstable in the supersonic region of flight, around 90 seconds after liftoff, and necessitated the inclusion of 310 kg of steel ballast. Stringers were also added to the inner skin of the Redstone's aft section to support the weight of the Mercury capsule. The overall 'burn time' of the engine for suborbital launches was shortened by 20 seconds to 143.5 seconds in total, prompting the addition of heat-resistant stainless steel shields for the stabilising fins. Additionally, nitrogen-gas purging equipment was added to the tail to prevent an explosive mixture from accumulating in the engine compartment whilst the Redstone sat on the pad.

The first three unmanned test flights evaluated each of these modifications and the combined performance of both the rocket and the capsule under real mission conditions. The first, named Mercury-Redstone 1 (MR-1), was intended to put the abort system fully through its paces, in addition to achieving the kind of velocities – around Mach 6.0 – that the suborbital astronaut would experience and demonstrating the ability of the capsule to separate satisfactorily from the rocket. A launch attempt on 7 November 1960 was scrubbed due to low hydrogen peroxide pressures in the capsule's thrusters and was rescheduled for the 21st. At 8:59 that morning, ignition occurred on time, but as the Redstone made to leave the pad, a shutdown signal was initiated. The thrust buildup was sufficient for the rocket to rise 10 cm, before it settled back onto its pedestal. However, the shutdown signal had caused the LES tower to fire, producing huge clouds of smoke which momentarily hid the Redstone from view. Flight Director Chris Kraft, watching the proceedings, was astonished by the tremendous acceleration, thinking it to be the actual liftoff . . . "but then the smoke cleared and the missile was still there!" Wally Schirra described the fiasco as "a memorable day, especially for someone who likes sick jokes".

The rocket swayed slightly, but remained upright and did not explode. Worryingly, though, the LES – which shot 1.2 km high and landed 360 m from

In full view of the world's media, the Redstone carries its first human passenger into space.

the pad – had not pulled the Mercury capsule clear of the Redstone and, as the shocked flight controllers watched, the drogue parachute popped out of its nose, followed by the main canopy and lastly, accompanied by a green cloud of marker dye, an auxiliary chute. All three fluttered pathetically down onto the pad. The rocket, meanwhile, was left alone as its liquid oxygen and high-pressure nitrogen were drained, its fuel and hydrogen peroxide tanks emptied, its circuits deactivated and its destruct arming devices removed. (Initial suggestions to relieve the pressurised propellant tanks by shooting holes in them with a rifle, thankfully, were squashed.)

"The press had a field day," Kraft recalled later. "It wasn't just a funny scene on the pad. It was tragic and America's space programme took another beating in the newspapers and in Congress." Time magazine bemoaned 'Lead-Footed Mercury' and ridiculed Wernher von Braun for downplaying the MR-1 fiasco, although a New York Times journalist urged President-elect Kennedy to persevere.

Investigators would find that the shutdown had been triggered by a 'sneak' circuit, created when two electrical connectors in a two-pronged booster tail plug separated in the wrong order. And why did the capsule fail to separate along with the LES? According to NASA's investigation report, it was because the G load sensing requirements had not been met. Ordinarily, after an engine cutoff, a ten-second timer was initiated and, upon its expiration, was supposed to separate the capsule if acceleration was less than 0.25 G. However, MR-1 had settled back onto the pad before the timer expired and the G-switch, sensing 1 G of acceleration, blocked the separation signal. On-board barostats, meanwhile, properly sensed that the rocket's altitude was less than 3 km and therefore activated the parachutes. "Once we realised that the capsule had made the best of a confusing situation and had gone on to perform its duties just as it would have on a normal flight," John Glenn said later, "we were rather proud of it." However, to avoid a recurrence, a 'ground strap' was added to maintain grounding of the vehicle during all umbilical disconnections and changes to the electrical network distributor prevented a cutoff signal from jettisoning future LES towers prior to 130 seconds after liftoff.

The undamaged spacecraft would be recycled and reused on the MR-1A flight just four weeks later. Despite some difficulties with a leakage in the capsule's high-pressure nitrogen line and a faulty solenoid valve in its hydrogen peroxide system, the mission was launched successfully at 11:15 am on 19 December. Thankfully, the abort system performed as advertised, although a malfunction of the velocity integrator caused the Redstone's velocity cutoff to occur 78 m/sec higher than planned, thus boosting the capsule 9.6 km above its intended 205 km altitude. Accelerations during re-entry were correspondingly more severe and high tail winds during the final portion of the flight led to MR-1A splashing into the Atlantic 32 km further downrange than anticipated. The source of the velocity integrator problem was traced to excessive torque against the pivot of the accelerometer, caused by electrical wires; five of these were replaced and a softer wire material was implemented. This solved the problem, as the chimpanzee Ham's MR-2 flight at the end of the following month would demonstrate.

Ham was not the first animal to have been launched by the United States. A pair of Rhesus monkeys, nicknamed 'Sam' and 'Miss Sam', from the School of Aviation

Ham, the chimpanzee occupant of MR-2.

Medicine in San Antonio, Texas, had been launched atop Little Joe rockets in December 1959 and January 1960, respectively. Although neither of their Mercury capsules reached space (Sam achieved an altitude of 88 km, Miss Sam of 15 km), their flights demonstrated that living creatures could survive a launch and return alive. Unfortunately, the flights of the Rhesus monkeys and chimpanzees, though significant, would offer an excuse for some test pilots to heap further ridicule on the Mercury Seven. When asked if he was interested in riding a capsule into orbit, Chuck Yeager had laughed. "It doesn't really require a pilot," he said, "and, besides, you'd have to sweep the monkey shit off the seat before you could sit down!"

Ham – the name was an acronym for the Holloman Aerospace Medical Center, based at Holloman Air Force Base in New Mexico, which prepared him for his mission – was launched at 11:54 am on 31 January 1961. Chosen specifically because of their close approximation to human behaviour, a colony of six chimpanzees, four female and two male, accompanied by 20 medical specialists and handlers from Holloman, had arrived at Cape Canaveral's Hangar S a few weeks earlier. The chimps were split into two groups to prevent the spread of any contagion and were led through training exercises with the help of Mercury capsule mockups in their compounds. By the end of the month, each of the chimps was somewhat bored, but nevertheless an expert at pulling levels and pushing buttons in the right order, receiving either banana pellets or mild electric shocks for doing the right (or wrong) thing. The day before launch, James Henry of the Space Task Group and Holloman veterinarian John Mosely examined the six chimps and settled on a particularly frisky and good-humoured male as the prime candidate, with a female as his backup. Both were put on low-residue diets, instrumented with biosensors and, early on the 31st, outfitted in their space suits, placed in their contoured couches and taken to the launch pad. With 90 minutes to go, Ham, described as "still active and spirited", was inserted inside the MR-2 capsule.

His home for the 16-minute mission boasted a number of significant innovations, including an environmental control system, live retrorockets, a voice communications device and the accordion-like pneumatic landing bag. The latter was attached to the heat shield and shortly before splashdown, the pair would drop 1.2 m, filling with air to help cushion MR-2's impact. In the water, the deflated landing bag and heat shield were intended to serve as an anchor, keeping the spacecraft upright.

Ham's liftoff was successful, although his Mercury capsule, programmed to travel 183 km into space and 468 km downrange of the Cape, actually flew 67 km higher and 200 km further downrange than intended. The chimp experienced six and a half minutes of weightlessness and endured 14.7 times the force of normal terrestrial gravity at one point during his re-entry. He survived and seemed to be in good spirits, despite having to wait for several hours before being picked up by the dock landing ship Donner. After splashdown, his heat shield had skipped on the water, bounced against the capsule's base and punched two holes in the pressure bulkhead. As MR-2 capsized, the open cabin pressure relief valve let in yet more seawater. By the time he was rescued, it was estimated that there was around 360 kg of seawater inside the capsule. Ham, however, seemed in good cheer, gobbling down a pair of apples and half an orange on the recovery ship's deck.

Post-flight analysis would reveal that the Redstone's mixture ratio servo control valve failed in its full-open position, causing early depletion of the liquid oxygen supply; consequently, the propellant consumption rate increased, the turbopump ran faster and led to higher thrust, an earlier-than-scheduled engine shutdown and the inadvertent 'abort' of the MR-2 spacecraft. Nonetheless, the basic controllability and habitability of Mercury was deemed a success. In the wake of Ham's flight, the reliability of the booster-capsule combination was reassessed, culminating in an estimated probability of success at somewhere between 78 and 84 per cent. However, many components had been designed to parameters which exceeded those demanded by the Space Task Group and launch operations personnel had devised their own methods which were more conducive to flight success. Taking this into account, the overall reliability of the system was judged at 88 per cent for launch and 98 per cent for the survival of the astronaut. These assurances were confirmed by one final test prior to Shepard's mission – the Mercury-Redstone Booster Development (MR-BD) flight, launched at 12:30 pm on 24 March.

Although it was doubtful that any of the problems experienced on either MR-1A or MR-2 would have endangered Shepard, had he been aboard, the Space Task Group's scrupulous attention to reliability meant that all significant outstanding modifications to the Redstone had to be dealt with. Von Braun also invoked one of the original ground rules, which insisted that no manned flight would be attempted until all responsible parties felt assured that everything was ready. Shepard's mission was fatefully postponed until 25 April. The MR-BD test, meanwhile, was perfect: the Redstone flew flawlessly, with its thruster control servo valve's closed position adjusted to 25 per cent open and flight sequencer timer changes prevented a recurrence of the problems on Ham's flight. Control manoeuvres were executed to evaluate the effect of higher-than-normal angles of attack, confirming that the Redstone could withstand additional aerodynamic loads. No attempt was made to separate rocket and capsule and they splashed down together, some eight and a half minutes after launch, before sinking to the bottom of the Atlantic. The success of MR-BD had cleared the way for MR-3 – the first manned mission – to launch.

To help them prepare more effectively for the flight, Shepard and Glenn had, since February, been using a pair of McDonnell-built Mercury simulators for 55–60 hours per week. They went through flight plans together and, indeed, Shepard 'flew' more than 120 simulated Redstone launches during this period. As February wore into March, the training became yet more exacting: both men even went through the ritual of their pre-flight medical examinations, just as they would on launch morning, and were instrumented with biosensors and outfitted in their silver pressure suits. A week after Gagarin's mission, on 19 April, Shepard sat in the actual capsule, atop its Redstone, on Pad 5 at the Cape, with the hatch open, meticulously plodding through each of the procedures he would follow.

By this time, he had nicknamed his tiny spacecraft 'Freedom 7' – not, as some observers would hint, in honour of the seven Mercury astronauts, but rather to reflect its status as the seventh capsule off the McDonnell production line. According to assistant flight director Gene Kranz, the name was adopted during the final Freedom 7 training exercises. On later missions, each member of the Mercury Seven would suffix their own spacecraft with the number as something of a good-luck charm.

By now, the launch was officially scheduled for 7:00 am on 2 May and, in late April NASA timetabled a full dress rehearsal, with Gordo Cooper standing in for Shepard. He duly suited-up, rode the transport van out to the base of Pad 5 and jokingly bawled "I don't want to go! Please don't send me!" before being shoved into the elevator. The assembled journalists, apparently, did not appreciate Cooper's gallows humour and the following morning's newspapers even went so far as to criticise NASA for its astronaut's inappropriate horseplay at such a tense moment. Meanwhile, Shepard checked out of a Holiday Inn where he had been staying with his wife, dropped her at the airport and drove to the astronaut quarters in the three-story Hangar S at the Cape. Since they were still required to maintain the official 'secret' that the first American in space could be any one of them, Shepard, Grissom and Glenn shared the same air-conditioned quarters, which had been specially decorated for them by their nurse, Dee O'Hara.

The heavens opened to heavy rain and storms early on 2 May, as the trio arose and ate a breakfast of bacon-wrapped filet mignon and scrambled eggs, together with orange juice and coffee. Since defecation in the spacecraft was, at best, difficult, such 'low-residue' launch-morning diets had been enforced by NASA. (Indeed, the astronauts' lawyer and agent, Leo D'Orsey, when told about the diet, had exclaimed "No shit?" Shepard responded with a grin, "Exactly!")

The intention was that the public 'final choice' of who was to fly would be made that morning, with some officials even suggesting bringing all three men out of their quarters wearing hoods to keep the charade going until one of them boarded the Pad 5 elevator. Shepard, wrote Neal Thompson, opposed this lunacy and opted instead to emerge from Hangar S in his pressure suit and wade through the teeming journalists. It made little difference: the rain was so bad that the launch was scrubbed, although not before the identity of America's first astronaut became known to the newsmen. "An alert reporter standing by the hangar door," wrote Gene Kranz, "had seen him and broke the story. The secret was out."

Originally planned as a 48-hour postponement, it was soon realised that an attempt early on 4 May would be impossible, so foul was the weather. However, at 8:30 that night, the two-part, ten-hour-long countdown began for a launch the following morning. The stunted nature of this countdown owed itself to past experience, which showed that it was preferable to run it in two short segments to permit the launch crews responsible for both Freedom 7 and the Redstone to be adequately rested and ready. A built-in hold of some 15 hours was called when the clock hit T-6 hours and 30 minutes, during which time various pyrotechnics were installed into the capsule and the hydrogen peroxide system to feed Freedom 7's thrusters was serviced. The countdown resumed at 11:30 pm and proceeded smoothly until another hour-long built-in hold at T-2 hours and 20 minutes, intended to check that all preparations had been made before Shepard's departure for the launch pad.

"LIGHT THIS CANDLE!"

At 1:30 am on 5 May, the prime and backup astronauts, clad in bathrobes, met again at breakfast, before parting. Glenn headed out to the pad to check Freedom 7, while Shepard underwent his pre-flight examination, conducted by physician Bill Douglas. He was instrumented with biosensors – four electrocardiograph pads, glued to his chest, then a respirometer taped to his neck and a rectal thermometer to gauge deep body temperatures – before being helped into a set of long underwear with built-in spongy pads to aid air circulation. He would confess later to "some butterflies" in anticipation of the impending flight and began the 15-minute effort to squeeze into his silver pressure suit, securing zips and connectors and ensuring that the rubber and aluminium-coated nylon garment and his portable, briefcase-like air-conditioning unit were ready. The latter proved essential: by the time he had donned his suit, wrote Time magazine, Shepard was sweating profusely and breathing hard.

The suit had been designed and built by the B.F. Goodrich Company, under a $98,000 contract awarded by NASA in July 1959, and followed similar principles to the Mark IV pressure garments worn by Navy fighter pilots. The company already had a long history in the field. Indeed, Goodrich engineer Russell Colley, who led the Mercury suit effort, had designed the pressurised ensemble worn by the legendary Thirties aviation pioneer Wiley Post. However, unlike the military Mark IV, which was hampered by problems of weight and mobility, the Mercury suit employed elastic cording, which arrested its tendency to 'balloon'. Moreover, at just 9.97 kg, it was the lightest military pressure suit yet built. Its other key features included a 'closed-loop' system, which eliminated a rubber diaphragm around the pilot's face; oxygen instead entered the suit through a hose at its waist, circulated to provide cooling and exited either through a hose on the right-hand side of the helmet or through the visor, if it was open. A small bottle, connected by a hose next to the astronaut's left jaw, was used to pressurise a pneumatic seal when the Plexiglas visor was closed. During flight, the suit would provide and maintain a 0.38-bar atmosphere to keep Shepard alive in the event that Freedom 7 lost cabin pressure.

Elsewhere, the dark-grey nylon outer shell of the military Mark IV was replaced with one of silvery aluminium-coated nylon for improved thermal control – additionally, black boots were substituted for white-coloured leather ones (and, later, by aluminium-coated nylon leather) for the same purpose – and straps and zips provided a snug, though uncomfortable fit. However, the gloves on Shepard's suit were zipped onto the sleeves, which prevented him from easily rotating his wrists to use the hand controllers. Post-flight modifications, implemented in time for Gus Grissom's suborbital mission in July 1961, would incorporate wrist bearings and ring locks for greater dexterity. The fingers of the gloves were curved to permit the astronaut to grasp controls and a 'straight' middle finger allowed him to better push buttons and flip toggle switches. Each member of the Mercury Seven was supplied with three individually-tailored suits: one for training, another for flight and a third as a spare, costing some $20,000 overall. Body moulds were taken by dressing the men in long underwear, covering them with brown paper tape and cutting the

Clad in his silvery space suit, Al Shepard prepares to clamber aboard Freedom 7. With him is Gus Grissom and in the background, clad in white cap and clean room garb, is John Glenn.

resultant mould to remove it when dry. So complex was the suit, Wally Schirra told Life magazine, that it required "more alterations than a bridal gown".

Its intense discomfort was caused by the fact that, when inflated, it took only one shape. Any change in this shape, perhaps by the astronaut trying to walk or sit, reduced the suit's volume and forced its wearer to exert himself to overcome the increased pressure. Simply walking left Shepard rapidly out of breath and, indeed, not until the Apollo missions would suits be built with 'constant-volume' joints to permit movement in the legs and arms without changing the pressure.

At 3:55 am, Grissom accompanied the fully-suited Shepard in the white transport van to Pad 5, after which technician Joe Schmitt fitted his gloves and Gordo Cooper briefed him on the countdown status. Meanwhile, at the top of the gantry, clad in white overalls and cap, John Glenn had spent the last two hours checking that every switch and instrument inside Freedom 7 was ready. At around 5:15 am, Shepard ascended the elevator to a green-walled room at the 20 m level – nicknamed 'the greenhouse' – which surrounded the capsule's hatch. Glenn and Schmitt helped him inside, an effort made all the more difficult by his bulky parachute. America's first spaceman, though, had an unexpected opportunity for a chuckle when he saw a girlie pin-up and a placard, put there by Glenn, which read 'No Handball Playing In This Area'. A grinning Glenn, normally considered a straight-arrow and no prankster, quickly pulled it down. Presumably, wrote Neal Thompson, he had second thoughts and did not want to risk the automatic cameras inside Freedom 7, soon to begin rolling, accidentally recording his joke for posterity.

For more than an hour, Shepard lay in his custom-contoured couch and was secured by straps across his shoulders, chest, lap, knees – "the only time we used knee caps," remembered Joe Schmitt in a 1997 interview, "because we didn't know what was going to happen when he went up" – and even caps over his toes. Meanwhile, other personnel fitted sensors and adjusted straps before Glenn reached in, shook his gloved hand and wished him luck. The hatch clanged shut at 6:10 am, at which point Shepard's heart rate quickened. Less than half an hour later, he began a 'denitrogenation' procedure, breathing pure oxygen to prevent aeroembolism – 'decompression sickness'; a pilot's equivalent of the bends – before sitting tight for liftoff, set for 7:00 am. This, however, was repeatedly postponed, first as banks of clouds rolled over Florida's south-eastern seaboard, then when one of the 400 hz power inverters to the Redstone experienced regulation difficulties.

The countdown was recycled to the T-35 minute mark and picked up again 86 minutes later, after the inverter had been removed and replaced. Next, an error surfaced in one of the IBM 7090 computers at NASA's Goddard Space Flight Center in Greenbelt, Maryland, responsible for processing Freedom 7's flight data. By the time engineers had solved this problem, in which they had to completely rerun the computers, Shepard had been on his back inside the capsule for over three hours. Adrenaline pumping, he checked and rechecked the switches and dials on his instrument panel, then peered through the periscope at the throng of spectators lining Cape Canaveral's beaches. Then, another, more personal, issue arose.

In addition to the discomfort brought about by lying in the cramped spacecraft for three hours, he found that he needed to urinate; the consequence, obviously, of too much orange juice and coffee consumed with Glenn earlier that morning. He eventually radioed "Man, I gotta pee" to Gordo Cooper, stationed in the nearby control blockhouse, and asked if Freedom 7's hatch could be opened.

Cooper, taken aback, passed Shepard's request up the chain of command, as far as Wernher von Braun, who emphatically declared that, no, "ze astronaut shall stay in ze nosecone". Exasperated, and in a tirade that was ultimately removed from the official transcript, Shepard warned that he would be forced to urinate in his suit if he was not allowed outside. Immediately, mission managers panicked – could the urine short-circuit the medical wiring attached to his body and the electrical thermometer in his rectum, they wondered – until the astronaut suggested that they switch off the spacecraft's power until he had 'been'. Eventually, agreement was received, Cooper confirmed "Power's off" and heard shortly thereafter a long, contented "Ahhhh" from Shepard over the radio. "I'm a wetback now," he told Cooper, as the warm fluid worked its way around his suit and pooled in the small of his back ...

"It caused some consternation," Shepard admitted in his tape-recorded debriefing of the flight an hour after landing. "My suit inlet temperature changed and it may possibly have affected the left lower chest sensor ... [but] my general comfort after this point seemed to be good." Still, the issue led to the inclusion of a proper, though hastily-engineered, urine-collection device in time for Gus Grissom's own suborbital mission 11 weeks later.

Fortunately, the urine was absorbed by his long cotton underwear and evaporated in the 100 per cent pure oxygen atmosphere of the capsule and Shepard, thankfully, received no electrical shocks. NASA was spared, wrote Neal Thompson, of the embarrassment of having to report that America's first spaceman had been electrocuted by his own piss. Humour aside, there remained a very real chance that 5 May 1961 might have ended with a dead astronaut. In such a dire eventuality, press spokesman John 'Shorty' Powers had readied statements to the effect that 'Astronaut Shepard has perished today in the service of his country', all adjusted slightly to take into account the point at which disaster struck: during launch, whilst in space or during re-entry. President Kennedy, too, was nervous, even though he had been assured by NASA Deputy Administrator Hugh Dryden back in March that no unwarranted risks would be taken. As late as a few weeks before launch, some senators, including Republican John J. Williams of Delaware and Democrat J.W. Fulbright of Arkansas, felt that it should be delayed and run in secret to avoid a much-publicised failure.

Such attempts were vetoed by most members of Congress. The Soviets, it was pointed out, had received much international criticism for staging Vostok 1 under such ridiculous secrecy – to such an extent that some observers doubted Gagarin had flown at all – and felt tradition dictated that the press should have free access to cover an event of such historic magnitude. Moreover, said Ed Welsh, secretary of the National Aeronautics and Space Council, "why postpone a success?" Still, even on 5 May, Kennedy questioned the need to televise the event live, even pressing NASA Administrator Jim Webb to play down the feverish publicity. Webb's assurance that the Redstone's LES tower would yank Freedom 7 and Shepard to safety in the event of a malfunction had little impact and Kennedy's fears continued throughout the mission. Only minutes before the television networks picked up the countdown, NASA public relations officer Paul Haney was forced to reassure Kennedy's press secretary Pierre Salinger that the escape tower would indeed save Shepard in an emergency. Salinger promised to pass this information to the president.

The decision to televise the launch in front of the media would also offer a poke in the eye for Soviet premier Nikita Khrushchev, whose regime had insisted on cloaking Gagarin's flight in secrecy until success could be confirmed. In many minds, America's 'bravery' at accomplishing the feat in front of the world placed it on a par with, or even above, the Vostok achievement. Khrushchev would publicly ridicule Freedom 7 as a "flea hop" in comparison to the Soviet triumph and, in truth, he was right, but on 5 May 1961 America won the moral high ground. A few days later, the Istanbul newspaper Millyet reported that Turkish journalists had asked the Soviet consul-general why his nation had not revealed the full story of Gagarin's mission. In response, a Tass correspondent was quoted as explaining, somewhat lamely, that Russia was "mainly interested in the people's excitement and reaction".

Shortly before 9:00 am, yet another halt was called when pressures inside the Redstone's liquid oxygen tanks climbed to unacceptable levels. Instead of resetting the pressure valves, which would have meant scrubbing the attempt for the day, it was decided to bleed off some of the pressure by remote control. After cycling the vent valves several times, pressures returned to normal. The decision to do this was aided, at least partly, by an irritable Shepard, who, after almost four hours on his back and now lying in his dried-up urine, snapped "I'm cooler than you are! Why don't you fix your little problem and light this candle?" The words, like Gagarin's "Poyekhali!" three weeks earlier, have since achieved immortality and truly epitomise the 'right stuff' which Shepard and everyone associated with Project Mercury had in shovelfuls. With two minutes to go, as the television networks started broadcasting live, the voice of Cooper in the blockhouse was replaced by that of Deke Slayton, serving as the first 'capsule communicator' – 'capcom' – from the Mercury Control Center at the Cape. Thirty seconds before launch, an umbilical cable, supplying electricity, communications and liquid oxygen, separated from the rocket as planned.

As the final seconds ticked away, Shepard's biosensors would testify that his pulse rose from 80 to 126 beats per minute; privately praying that he would not screw up, his hand tightened around the abort handle. Already described by the physicians as the calmest man on the Cape that morning, his pulse rate was comparable to that of a driver moving from a service road onto a freeway. In total, he had been lying supine for four hours and 14 minutes – the delays alone had cost more than three and a half hours, enough time for a dozen Freedom 7 missions – and the jolt he expected at the instant of liftoff was instead replaced by what he would describe as something "extremely smooth ... a subtle, gentle, gradual rise off the ground".

At 9:34 am, with 45 million Americans watching or listening in person, on television, on the radio or over loudspeakers, the Redstone roared to life, prompting Shepard, who had just activated the on-board timer, to confirm "Roger ... liftoff and the clock is started!" Slayton, with a nod to comedian Bill Dana's astronaut character José Jiménez, replied "You're on your way, José!" So significant was the next quarter of an hour that it brought much of the United States to a standstill. A Philadelphia appeals court judge interrupted proceedings to make an announcement, free champagne flowed in taverns, traffic slowed on Californian freeways and people danced and sang in Times Square. President Kennedy broke up a National Security Council meeting, walked into his secretary's office and stood, dumbstruck, hands in his pockets, watching Freedom 7's rise to the heavens.

"I remember hearing [the] firing command," Shepard recalled less than an hour later aboard the Lake Champlain recovery ship, "but it may very well be that, although Deke was giving me other sequences ... prior to main stage and liftoff, I did not hear them. I may have been just a little bit too excited. I do remember being fairly calm at T-0 and getting my hand up to start the watch when I received the liftoff from the control centre. I must say the liftoff was a whole lot smoother than I expected. I really expected to have to use full volume control to be able to receive ... [but] all my transmissions over UHF were immediately acknowledged, without any repeats being requested."

Shepard had little time to sit still and do nothing. He had already agreed with Project Mercury's operations director, Walt Williams, that he would talk as much as possible throughout the mission, to keep everyone updated on even the tiniest details. As the Redstone speared higher and higher, his voice crackled out the data – "This is Freedom 7," he exulted, "the fuel is go; 1.2 G; cabin is 14 psi; oxygen is go; all systems are go" – and, 16 seconds after launch, the rocket commenced a pitchover manoeuvre of two degrees per second, from 90 to 45 degrees, a procedure that was complete some 40 seconds into the flight. He had anticipated around 6 G during ascent, even though he had trained and endured more than twice as much in the centrifuge and the MASTIF. Although the liftoff was smooth, his ride turned bumpy and caught him off-guard when he reached the turbulent transition between the edge of the 'sensible' atmosphere and space. Eighty-eight seconds after launch, Freedom 7 began shuddering violently and Shepard's head, wrote Neal Thompson, was "jackhammering so hard against the headrest that he could no longer see the dials and gauges clearly enough to read the data". As a result, he waited until the vibrations had calmed before transmitting any more status updates to Slayton.

Lack of visibility, in fact, had been one of the fundamental medical effects that physicians had most feared: could the astronauts see properly in and around their capsules? Such questions seem trivial today, but aerospace scientists had for years written quite seriously about the possible impact of weightlessness on the muscle structure around and beneath the eyes and the possibility that it might change shape over several hours, permanently ruining their vision. John Glenn was assigned to investigate this possibility on his Friendship 7 mission, the first American orbital flight, in February 1962. "On the instrument panel," he told an interviewer in 1997, "is a little Snellen chart like the eye chart they use in doctors' offices, miniaturised for the distance from my eyes to the panel, and I was to read the smallest line I could read every 20 minutes during flight and report what that was, so if my eyes were changing shape or vision was changing, I would be able to report this."

Uncontrollable nausea and vertigo, triggered by the random movement of fluids in the inner ear, was another possibility and, although other astronauts have since been known to suffer 'space sickness', the irony of Project Mercury was that the capsules were too small and cramped to give their pilots an opportunity to move around and become disorientated. There were other concerns. "They didn't know whether you could swallow properly," explained Glenn and, added Gordo Cooper, "There were a lot of these medical experts who said that the cardiovascular system would not be able to function under zero gravity". The simple, yet vital, work done by the Mercury astronauts and their Vostok counterparts exemplified how little was known about how human beings could function in the weightless environment, high above their home planet, at the dawn of the Sixties. They were pioneers, taking the first steps into a strange new environment.

Even before he reached space, Shepard had satisfied one of the physicians' main worries: by proving that he could indeed survive the rigours of a rocket launch. Yet it was only after passing through 'Max Q' – a period of maximum aerodynamic turbulence; the phase at which Freedom 7, by now accelerating through the sound barrier and into the rarefied air of the high atmosphere, was subjected to massive loads – did he finally grunt to Slayton that the ride was "a lot smoother now". Then, 141.8 seconds after launch, the Redstone's engine finally burned out, followed, a second later, by the jettisoning of the LES tower. Although the latter should have been automatic, it was actually performed manually, but Shepard would not recall ever pulling the manual 'JETT TOWER' override ring. The Redstone burnout triggered the initiation of small explosive charges, which, 38 seconds later, severed the link with Freedom 7 and three posigrade rockets on the capsule pushed the pair apart at 4.6 m/sec. By now, Shepard's pulse had climbed to 132 beats per minute, but calmed dramatically when he saw and reported to Slayton that the green 'CAP SEP' indicator light confirmed the capsule had indeed successfully separated from the rocket.

Now flying free of the Redstone, Shepard's tasks – to prove that, unlike Gagarin, he was able to actively control his ship – got underway. Firstly, the attitude-control system moved the capsule into a heat shield-forward position for the rest of the flight, introducing momentary oscillations which were quickly damped out by a five-second firing of the automatic thrusters. He switched Freedom 7 from automatic to manual control about three minutes after launch and, using a MASTIF-like stick, tilted the capsule through pitch, yaw and roll exercises, 'controlling' his spacecraft for the first time, whilst travelling at 8,200 km/h – nearly eight times the speed of sound and almost three times faster than any American in history. Shepard found that he was able to exercise control crisply and Freedom 7 responded very much like the simulator, although his ability to hear the spurting hydrogen peroxide jets was virtually drowned out by the crackling of the radio. He actually operated his spacecraft by three means – fully automatic, manual and a 'fly-by-wire' combination of the two. He reported that the manual mode responded well, although the capsule tended to roll slightly clockwise. Post-flight inspections uncovered a piece of debris lodged in the hydrogen peroxide tubing, which probably caused their jets to leak a tiny increment of thrust.

At 9:38 am, four minutes after launch, Shepard experienced weightlessness for the first time, as his body gently floated from his couch and against his shoulder harnesses. Flecks of dust drifted past his face, together with a stray washer, which quickly vanished from view. As he neared the apex of his arc into space, he made an attempt to observe the world beneath him through Freedom 7's periscope. Unfortunately, during the morning's lengthy delays, to minimise the blinding sunlight, he had flicked a switch that covered the lens with a grey filter and had forgotten to remove it before launch. Now he was greeted only with a grey-coloured blob on the screen before him. When he tried to reach across the cabin to flick off the filter, his wrist hit the abort handle and he thought it best to leave well alone.

Shepard's dramatic description of Earth – "What a beautiful view!" – was surely sincere, but was certainly not accompanied by glorious colour. Still, he later told NASA officials, the vista was "remarkable" and he was able to see Lake Okeechobee, on the northern edge of Florida's Everglades, together with Andros Island, shoals off Bimini and some cloud cover over the Bahamas. Although he would later tell a Life journalist of the "brilliantly clear" colours around Bimini, he would privately admit that the grey filter "obliterated most of the colours". When questioned by Wally Schirra, his response was "shit, I had to say something for the people!"

At the top of the long arc over the Atlantic – rising, at apogee, to almost 188 km – the periscope automatically retracted and Shepard was obliged to strain to look for stars and planets through the two small, awkwardly-placed portholes, one to his upper-left side and the other to his lower-right. He wanted the chance to see what Yuri Gagarin had claimed to have seen, but in fact could see nothing, no matter which way he twisted or turned. Looking for stars and planets placed him slightly behind schedule. Although the entire mission would span only 15 minutes, and barely a fraction of that would be spent 'in space', NASA had overloaded him with tasks, lasting a minute here or two minutes there. To catch up, Shepard feverishly put Freedom 7 through its paces, before eventually radioing to Slayton that the three retrorockets had successfully fired at their prescribed five-second intervals and his spacecraft was positioned in its proper re-entry attitude.

Six minutes and 13 seconds after launch, as the tiny capsule began its plummet towards the ocean, the now-spent retrorocket package automatically separated. Even though his time in weightlessness had been so brief, he would later remark aboard the recovery ship Lake Champlain, "there is no question about it: when those retros go, your transition from zero-G to essentially 0.05 G is noticeable". Flying with its base facing in the direction of travel, such that it could properly absorb the intense heat caused by friction with the upper atmosphere, Shepard was shoved into his couch with 11 times the force of normal terrestrial gravity. This part of the ride, he knew, was among the most physically demanding and "not one most people would want to try in an amusement park". In less than 30 seconds, Freedom 7 slowed from its 8,200 km/h suborbital velocity to around 800 km/h. During this time, the astronaut could scarcely speak, so high were the G forces, and could barely manage a series of grunted "okays" to Slayton. Then, from an altitude of 24 km down to 12 km, the frictional heating raised temperatures at the base of the capsule to a blistering 1,200°C; within, however, conditions held steady at 38°C and, inside his pressure suit, Shepard experienced a balmy, but relatively comfortable, 28°C. It was, he said, "like being in a closed van on a warm summer day".

As the descent continued, the automatic stabilisation and control system detected the onset of re-entry and initiated a roll of ten degrees per second to keep Freedom 7 on track. Six and a half kilometres above the Atlantic, and still barely nine minutes after launch, Shepard felt relief as the drogue parachute popped from the spacecraft's nose, followed, seconds later, by the jettisoning of the antenna capsule and deployment of the 19.2 m-diameter orange and white main canopy. This blossomed open, arresting the capsule with "a reassuring kick in the butt". A snorkel valve opened to equalise cabin pressure with the outside air, after which the heat shield dropped 1.2 m and the landing bag was extended. Shepard would later describe the deployment of the main chute, not surprisingly, as the most beautiful sight of the whole mission. It slowed the capsule to a stately 30 km/h and even the splashdown itself, some 490 km east of Cape Canaveral and 160 km north of the Bahamas, felt no worse than the shove he used to experience from the catapults aboard naval aircraft carriers.

His precise landing co-ordinates were 75 degrees 53 minutes West longitude and 27 degrees 13.7 minutes North latitude. Freedom 7 initially listed over to its right side, about 60 degrees from an upright position, but righted itself within a minute or so. The parachutes cast loose to prevent dragging the capsule and a patch of fluorescent green marker dye spread across the water, although recovery forces had already been monitoring Shepard's descent for several minutes and were closing in. America's first manned spaceflight had lasted 15 minutes and 28 seconds.

Shortly thereafter, Wayne Koons, the pilot of one of the five Marine Air Group 26 rescue helicopters despatched from the Lake Champlain, was hovering overhead and his co-pilot George Cox had snagged Freedom 7 with hook and line (though not before the spacecraft's high-frequency antenna had pronged upwards and dented the base of the chopper). Shepard, still midway through removing his helmet and releasing his restraints, asked the impatient Koons to lift the capsule slightly above the waterline. Eventually, America's first astronaut popped open the hatch and leaned out to grab the 'horse's collar' – a padded harness that Cox had lowered – then pulled it towards him and looped it over his head and under his arm. He gave a thumbs-up and was pulled to the helicopter. Koons' crew had trained for more than a year for this moment and had established themselves as experts at hovering above Mercury capsules, hooking them and getting astronauts out. Indeed, Cox had successfully fished Ham out of the drink just a few months earlier.

Even now, it was a nervous time for Shepard. Only hours before, he had read a disturbing report of the harrowing experience of fellow naval aviators Malcolm Ross and Victor Prather. On 4 May, only hours after the Freedom 7 countdown had begun, the pair had ascended 34.6 km above the Gulf of Mexico in a balloon gondola, part of the Navy's Stratolab high-altitude research effort. During their nine-hour ascent, the two pressure-suited men had been subjected to temperatures as low as –70°C and, since their weights were doubled by their equipment, they had found it virtually impossible to move within the gondola. Their mission to the very edge of space was successful and satisfactorily evaluated the performance of their pressure suits, but, after landing, Prather, mistakenly thinking himself to be out of danger, opened his helmet visor. As he clambered up the ladder to the rescue helicopter, he slipped, fell and drowned when his suit filled with water. Prather's tragic death would surely have reinforced to Shepard that, even home from space, he would not be truly 'safe' until he was standing on the deck of the recovery ship. Indeed, his weight on the end of the winch actually caused Koons' helicopter to drop slightly and the astronaut's splayed legs splashed briefly back into the Atlantic, before finally being pulled clear.

The efforts to ensure his safety had been nothing short of extraordinary. Fire trucks had been stationed close to Pad 5, ready to offer support in the event of a launch accident, whilst helicopters stood by with technicians, physicians and frogmen to recover Shepard if he landed unhappily. Waiting out at sea were naval speedboats, whilst other craft were prepared to fish Freedom 7 from the Banana River, a lagoon between Cape Canaveral and Merritt Island. Meanwhile, near the prime recovery zone in the Atlantic, the Lake Champlain bristled with its recovery helicopters and a flotilla of six destroyers was strung out along the tracking range.

An exhausted Shepard is welcomed aboard the Lake Champlain.

Elsewhere, at the Cape itself, radars and high-flying aircraft monitored the skies for virtually every second that the astronaut was aloft.

Declaring that it was truly "a beautiful day", Shepard was flown back to the recovery ship, where 1,200 sailors covered the decks, cheering his success. Koons and Cox lowered Freedom 7 – soon to be exhibited at 1961's Paris Air Show – onto a specially-made stack of mattresses, disconnected it and touched down in what Shepard would call "the most emotional carrier landing I ever made". Barely 11 minutes after splashdown, he set foot on the carrier deck.

Similar emotions were being played out across the nation: Floridian crowds cheered, John Glenn jovially asked the recovery ships to remain in the Atlantic in the hope that NASA might set up another Redstone for him, New Hampshire's governor visited East Derry, schools closed and military aircraft dropped confetti as Shepard's proud parents and sister rode in an open-topped convertible. For the astronaut's wife, Louise, the calm after the storm came when she received word from NASA that her husband was safely aboard the Lake Champlain. As she chatted to journalists outside her Virginia Beach home, a Navy jet spelled out the letter 'S' in the sky to honour the United States' newest hero.

The hero himself, after guzzling orange juice in the quarters of the ship's captain, was handed a tape recorder and asked to record his initial thoughts. He was then grilled by the physicians, as he relived every detail of the 15-minute flight – and the hours on the pad beforehand: no, he did not sleep, no, he did not defecate, yes, there was a noticeable odour in the cabin (urine) and so on. Midway through this debriefing, he received the first of many calls from his commander-in-chief, President Kennedy, who congratulated him. Privately, the administration could breathe a sigh of relief now that, 23 days after Gagarin's mission and still smarting from the Bay of Pigs, the United States finally had something in which to take pride. That afternoon, the president announced that "this is an historic milestone in our own exploration of space". Added journalist Julian Scheer, later to become NASA's public affairs officer: "Shepard bailed out the ego of the American people. As a nation, we desperately wanted a success and we got not only a success, but an instant hero".

An hour after his arrival aboard the Lake Champlain, the astronaut set off aboard a two-engined C-1 transport aircraft, which took him to Grand Bahama Island for three days of tests. He was greeted by Wally Schirra, who had watched his launch from the front seat of an F-106 aircraft that morning, together with Gus Grissom and capcom Deke Slayton. Thirty-two specialists debriefed him, with Carmault Jackson questioning his medical health, Bob Voas probing his performance as Freedom 7's pilot and Harold Johnson and Sigurd Sjoberg focusing on the operation of the capsule's systems. After downing a huge shrimp cocktail, a roast beef sandwich and iced tea, Shepard learned that he had lost 1.3 kg in weight since breakfast. Nonetheless, the doctors proclaimed him in good shape and jubilant spirits. His $400 million mission had cost each American taxpayer $2.25 and the astronaut himself was surely overjoyed to receive an extra $14.38 in naval flight pay.

TO GO TO THE MOON

In the days that followed Freedom 7, John Kennedy wanted to talk of nothing but space. Years later, commentators would speculate cynically that his motivations for supporting the space programme and, in particular, a manned expedition to the Moon, were based purely on political concerns: the need to beat the Soviets, overcome the Bay of Pigs embarrassment and prove America's technological mettle. However, when Shepard met the president shortly after the flight, he saw a true statesman and, as the two became friends, he believed that the attraction of space exploration was very real for Kennedy. "He was really, really a space cadet," Shepard said later, "and it's too bad he could not have lived to see his promise." Three weeks after Freedom 7, Kennedy nailed his colours to the space mast in one of the most rousing and inspiring addresses ever given in United States political history.

First came the adoration. On 8 May, each of the Mercury Seven, together with their wives and NASA Administrator Jim Webb, arrived in Washington, DC, to participate in a ceremony at the White House. There, in the Rose Garden, Kennedy pinned NASA's Distinguished Service Medal onto Shepard's chest. So nervous was the president that he dropped the decoration, joking that it had "gone from the ground up". Later, as they rode in an open-topped limousine along Pennsylvania Avenue, Vice-President Lyndon Johnson stated the obvious: "They love you!" he said of the crowd's applause. "You're a famous man, Shepard!"

Others keen to be seen with the famous man included New York Mayor Robert F. Wagner, who hosted Shepard in a tickertape parade to rival that of Charles Lindbergh, first to fly solo across the Atlantic in 1927. Elsewhere, in Los Angeles, Mayor Norris Poulson tried to outdo Wagner with his city's own celebration, whilst the people of East Derry – the astronaut's hometown – organised their biggest-ever parade in honour of their famous son. Schools were named for him, including a newly-built one in Deerfield, Illinois, and everyone from senators and congressmen to foreign dignitaries and journalists to members of the public sought his autograph, a chance to meet him and the opportunity to shake his hand.

Shepard already knew of Kennedy's intention to support NASA's steadily-growing lunar landing project and he and the other Mercury Seven astronauts expressed their desire to participate. The gigantic rocket needed for the momentous journey, called 'Saturn', had already been under development for a couple of years and would undertake its maiden test launch in October 1961. Indeed, one of Wernher von Braun's conditions for joining NASA had been that the agency should continue its work on the Saturn. Alongside the rocket was the spacecraft which would actually transport three-man crews to the Moon and this, too, had received a name: 'Apollo', after the ancient Greek sun god. However, many of the technical facets of how astronauts would reach their target remained unanswered.

Still, on 25 May, Kennedy stood before a joint session of Congress and publicly declared his vision. He knew, after a joint recommendation from Jim Webb and Secretary of Defense Robert McNamara, that the presence of men in space, rather than just machines, would truly capture the imagination of the world. The first part of his speech focused upon ways in which the United States could exploit its economic and social progress against the menace of communism, called for increased funding to protect Americans from a possible nuclear strike … and lastly hit Congress with his lunar bombshell. "I believe," he told them, "that this nation should commit itself to achieving the goal, before this decade is out, of landing a man on the Moon and returning him safely to the Earth. No single space project in this period would be more impressive to mankind or more important for the long-range exploration of space … and none would be so difficult or expensive to accomplish."

Expense, indeed, was a major stumbling block. By the end of 1961, NASA's budget would have grown to more than $5 billion, ten times as much as had been spent on space research in the past eight years combined and roughly equivalent to 50 cents for each taxpayer. Kennedy acknowledged that it was "a staggering sum", but reinforced that most Americans spent more each week on cigars and cigarettes. Still, he would have to face off harsh criticism from many quarters by placing the lunar goal ahead of education projects and other social welfare efforts for which he had campaigned so hard during his years in the Senate. He would admit in his speech that he "came to this conclusion with some reluctance" and that the nation would have to "bear the burdens" of the dream.

Some did not wish to accept such burdens. Immediately after the speech, a Gallup poll revealed that barely 42 per cent of Americans supported Kennedy's push for the Moon. Yet he also gained immense support, both as a risk-taker and a bold statesman. The decision, said his science advisor Jerome Wiesner, was one that he made "cold-bloodedly". It was also a decision that he firmly stood by. Only hours after Gus Grissom flew America's second suborbital mission on 21 July 1961, Kennedy signed into law an approximately $1.7 billion appropriation act for Project Apollo. In a subsequent address, given at Rice University in Houston, Texas, in September 1962, he admitted that Apollo and Saturn would contain some components still awaiting invention, but remained fixed in his determination to "set sail on this new sea, because there is new knowledge to be gained and new rights to be won".

Kennedy's visit to Rice coincided with his dedication of NASA's new Manned Spacecraft Center (MSC), on the outskirts of Houston, and a year later he would again visit the growing space establishment. On 18 November 1963, he toured the Cape Canaveral launch facilities in Florida, then flew to Texas and was shown around MSC, which, by this point, had supported its last Mercury mission and was gearing up for the two-man Gemini flights. During this project, ten teams of astronauts would rehearse many techniques needed to get to the Moon – from rendezvous and docking to spacewalks to changing their orbits. In late 1963, America's chance of achieving its lunar goal seemed to be growing brighter and the gap in space achievement with the Soviets showed signs of closing. Kennedy knew this and, around the same time, approached Nikita Khrushchev for the second time to propose a joint mission. Khrushchev agreed and it is possible that, had the two leaders remained in office, a co-operative venture sometime in the Sixties might have borne fruit.

An earlier attempt by Kennedy to interest Khrushchev in a joint project had fallen on deaf ears in June 1961, which some historians have seen as evidence of the Soviet leader's reluctance to open up the real limitations of his space infrastructure. Or, as one analyst has observed: "The USSR literally had nothing to hide: if they didn't hide it, everyone would know they had nothing".

The road to the Moon, however, remained fraught with risk. Reminding his audience that it was still "a time for pathfinders and pioneers", Kennedy recounted in San Antonio on 21 November the story of a group of Irish boys who came to an orchard wall, seemingly too high to climb. Throwing their caps over the wall, they now had no choice but to force themselves to find a way to scale it. "This nation has tossed its cap over the wall of space and we have no choice but to follow it," he said. "Whatever the difficulties, they will be overcome."

The following day, whilst travelling through the streets of Dallas in an open-topped car, Kennedy was shot dead.

GRUFF GUS

Had Gus Grissom lived longer, wrote Deke Slayton, he would have been the first man on the Moon. Slayton, who was grounded from flying throughout most of the Sixties, found himself in 1962 in charge of the selection and training of astronauts for the two-man Gemini and Moon-bound Apollo missions. It was Slayton who, after Grissom's death in a spacecraft fire on the launch pad, chose Neil Armstrong to command the first manned lunar landing. Yet, he wrote, "had Gus been alive, as a Mercury astronaut, he would have taken that step ... my first choice would have been Gus, which both Chris Kraft and Bob Gilruth seconded".

The man to whom Slayton, Kraft and Gilruth would have offered this honour was a small, tough fighter pilot, the first person to fly into space twice, the first astronaut to eat a corned beef sandwich whilst gazing down on Earth and a man who fiercely guarded his family's privacy. "Betty and I run our lives as we please," he

Gus Grissom inspects his spacecraft's periscope.

once said. "We don't care about fads or frills. We don't give a damn about the Joneses."

Virgil Ivan Grissom, America's second man in space, was born in the Midwestern town of Mitchell, Indiana, on 3 April 1926. Small for his age, he was nicknamed 'Greasy Grissom' as a child and grew up with a determination to "prove I could do things as well as the big boys". His father worked for almost half a century on the Baltimore and Ohio Railroad and Grissom, too small to participate in school sports, eventually established himself as a Boy Scout, where he led the Honour Guard. Like Al Shepard, he delivered newspapers and, in the summer months, picked peaches and cherries for the local growers to earn enough money to date his school sweetheart, Betty Moore, whom he married in July 1945. By this time, he had left school – described by his principal as "an average, solid citizen, who studied just about enough to get a diploma" – and served a year as an aviation cadet. His hopes of joining the theatre of war evaporated a month later, when Japan surrendered.

Unwilling to fly a desk, Grissom left the Air Force and took a job installing doors on school buses, before deciding to study mechanical engineering at Purdue University. Whilst there, his wife worked as a long-distance operator and Grissom himself flipped burgers at a local diner. He received his degree in 1950, crediting Betty for making it possible, and eventually re-enlisted in the Air Force, finished cadet training and won his wings the following year. His completion of training coincided with the outbreak of war in Korea and Grissom soon found himself in the thick of the conflict for six months, flying a hundred combat missions in sleek F-86 Sabre jets as part of the 334th Fighter-Interceptor Squadron. An interesting tale surrounds his early days in Korea. Each morning, the pilots would ride an old school bus from the hangar to the flight line and only those who had been involved in air-to-air combat were permitted to sit. The uninitiated had to stand. Grissom stood only once; testament, perhaps, to the continued determination that had dogged him since boyhood, only now he planned to be even better than the big boys.

His first taste of war came as something of a surprise – "For a moment, I couldn't figure out what those little red things were going by," he said later, "then I realised I was being shot at!" – and he returned to the United States to be awarded both the Air Medal with a cluster and the Distinguished Flying Cross. Subsequent assignments, which he actually considered more dangerous than combat flying, included instructing new cadets at Bryan Air Force Base in Texas, studying aeronautical engineering at the Institute of Technology at Wright-Patterson Air Force Base and being chosen in October 1956 for the famous Test Pilot School at Edwards Air Force Base in California. By this time, Grissom had gained a reputation as one of the best 'jet jockies' in the service, with more than 3,000 hours of flight time, and was also father to two young boys, Scott and Mark. When the Soviets launched Sputnik, Grissom took notice, but was far too preoccupied with his job of wringing out new jets at Wright-Patterson to give much consideration to space travel. Then, a little over a year later, he received a teletype message, labelled 'Top Secret', which instructed him to go to Washington, DC, in civilian clothes for a classified briefing.

Mystified, Grissom found that he had been picked as one of 110 candidates for

Project Mercury and the events of 1959 would truly change his life. "I did not think my chances were very big when I saw some of the other men who were competing for the team," he said later. "They were a good group and I had a lot of respect for them, but I decided to give it the old school try and take some of NASA's tests." Whilst at Wright-Patterson Aeromedical Laboratory, undergoing test after test, his run on the treadmill had to be stopped abruptly when his heart soared to almost 200 beats per minute. On the other hand, he endured the heat chamber perfectly, keeping cool by reading a dog-eared copy of Reader's Digest, "to keep from getting bored".

He nearly flunked, however, when the physicians discovered he was a hay fever sufferer, but Grissom, without missing a beat, convinced them that the absence of ragweed pollen in space would not pose a problem. He viewed the psychological tests, too, as illogical. "I tried not to give the headshrinkers anything more than they were actually looking for," he said. "I played it cool and tried not to talk myself into a hole." Fortunately for Grissom, 'talking' was not one of his strengths. Astronauts and managers alike would recall that he rarely spoke unless he had something to say and during a visit to the Convair Corporation in San Diego, prime contractor for the Atlas rocket, he told the workers to "do good work". Ironically, those three words turned into a motto of incalculable value for the Convair workforce.

Even after his selection as one of the Mercury Seven, Grissom would privately question why he had volunteered to fly a bomb-carrying missile into space. The answer came instantly: "I happened to be a career officer in the military and, I think, a deeply patriotic one. If my country decided that I was one of the better-qualified people for this new mission, then I was proud and happy to help out." However proud he might have been, one thing that Grissom despised was the moniker 'astronaut'. In his mind, it had an irritating PR undertone. One day, he even announced: "I'm not 'ass' anything. I'm a pilot. Isn't that good enough?"

ROOM WITH A (BETTER) VIEW

Grissom's efforts during his first two years as an astronaut proved pivotal in securing him his seat on Mercury-Redstone 4, planned as a virtual duplication of Freedom 7, in July 1961. It, too, would last barely 15 minutes, arcing high above the Atlantic Ocean and splashing down a few hundred kilometres east of Cape Canaveral. However, Grissom's spacecraft – which he had nicknamed 'Liberty Bell 7' – would differ, visibly, from that of Shepard in an important way: it was the first to boast a large trapezoidal window, instead of two, 25.4 cm portholes. This window provided a field of view of 30 degrees in the horizontal plane and 33 degrees in the vertical, allowing him to look 'upward' and see directly outside, and had actually been one of the Mercury Seven's earliest recommendations when they visited the capsules in production at McDonnell's St Louis plant in Missouri. Manufactured by the Corning Glass Works of Corning, New York, the window comprised an outer pane of 8.9 mm-thick Vycor glass and a dual-layered inner pane. Its strength closely paralleled that of the capsule's hull and during re-entry it was capable of withstanding temperatures as high as 980°C. To reduce glare, it was also treated with a magnesium fluoride coating.

Additional improvements implemented since Shepard's flight included a make-shift urine collector, pieced together by the astronauts themselves the day before Liberty Bell 7 launched, which Grissom would later nonchalantly remark "worked as advertised". More importantly, however, the second suborbital mission would feature an explosively-actuated side hatch. Early plans had called for the astronauts to exit their capsules through an antenna compartment in the nose, but the awkwardness of this exercise, coupled with the need to remove a small pressure bulkhead to do it, led the Space Task Group and McDonnell engineers to develop two hatches: one activated manually, the other explosively. The hatch itself was held in place by 70 titanium bolts, each measuring 6.35 mm in diameter, and the mechanical version was employed on both Ham's and Shepard's missions. However, it weighed 31 kg – three times as much as its explosive counterpart – and was deemed too heavy for the orbital Mercury flights. Thus, to put it through its paces, an explosively-actuated hatch was installed on Grissom's capsule.

Built by Honeywell's ordnance division of Hopkins, Minnesota, it contained a mild detonating fuse, installed in a channel between an inner and outer seal around the periphery of the hatch. When fired, the gas pressure between the two seals fractured each of the 70 screws and blew off the hatch. Small holes, drilled into each bolt, provided weak points and aided the fracturing process. The fuse could be triggered manually by the astronaut himself, using a knobbed plunger close to his right arm, or from outside the capsule by means of an external lanyard. The performance of the hatch on the Liberty Bell 7 mission would, in some eyes, tarnish Grissom's reputation for the rest of his life and even today continues to arouse fierce debate.

The manual controls he would use during his 15-minute flight, too, had been extensively modified. A new rate stabilisation control system enabled him to manage the spacecraft's attitude movements by small twists of the hand controller, rather than by simply jockeying the device into the desired position. It provided a kind of 'power steering', offering finer control and easier handling qualities. The instrument panel, moreover, had been upgraded with a new Earth-path indicator, showing Liberty Bell 7's precise position. A redesigned fairing for the spacecraft adaptor clamp ring, a rearrangement of the instruments and more foam padding to the headrest of Grissom's couch, it was hoped, would prevent a recurrence of the vibrations and blurred vision experienced by Shepard.

Other changes to the flight plan included adjusting the sequencing of retrofire to ensure that the Redstone and capsule would be at least 1.2 km apart on their suborbital paths as they neared the end of their mission. A new life raft, weighing less than 2 kg, had been provided by NASA's Langley Research Center and the Space Task Group, and Grissom's pressure suit featured a better wrist fitting for improved movement and a convex, chest-mounted mirror – the 'Hero's Medal' – which would allow the spacecraft's camera to record both the astronaut and the instrument panel readings.

During training and throughout the production of Liberty Bell 7, Grissom established his reputation as a 'hands-on' pilot, attending meetings, supervising some of the engineering work and, he said, "fretting a little over whether all of the critical parts would arrive from the subcontractors on time and get put together". Among his concerns were mistakenly-switched instruments, which caused the spacecraft to yaw to the left instead of the right, and the failure of the attitude controls, which did not properly centre themselves after manoeuvres. The capsule itself arrived at Cape Canaveral on 7 March 1961, followed by its Redstone three months later. On 22 June, the rocket was erected on Pad 5 and Liberty Bell 7 installed a few days thereafter. Throughout June, however, nagging problems were encountered whilst testing Grissom's pressure suit and the need to replace the spacecraft's rusted on-board clock. As for Grissom, to ensure that he did not inadvertently take himself out of the running for the flight, he gave up water-skiing and even calmed down some of his raucous exploits in his General Motors-provided Corvette. As souvenirs for the flight, he took two rolls of Mercury dimes – a hundred in all – which he stuffed into the pocket of his pressure suit. He would regret this decision later.

The Liberty Bell 7 mission would, in many ways, be substantially different from Freedom 7. Al Shepard's flight had been literally overloaded with activities during barely five minutes of weightlessness. Grissom's plan would 'weed out' a number of communications obstacles and allow him more time to use the new trapezoid window to learn about an astronaut's visual abilities in space. Shepard had controlled his spacecraft by one axis at a time; Grissom would assume full manual control, taking over all three axes simultaneously. "I also planned to fire the retrorockets manually," he said, "instead of automatically, as they had been fired on Freedom 7."

Grissom's selection as the prime pilot for Liberty Bell 7, with John Glenn again serving as backup, was ratified by Bob Gilruth on 15 July 1961 and launch was scheduled for three days later. Part of what was becoming traditional was naming the capsule. Grissom's choice honoured the famous Liberty Bell, today housed in Philadelphia, Pennsylvania, and considered one of the most important symbols of the American War of Independence, epitomising freedom, nationhood and the abolition of slavery. Cast in 1751, its most famous ringing, supposedly, occurred a quarter of a century later to summon Philadelphia's citizenry for a reading of the Declaration of Independence. When questioned about the name, Grissom explained that the bell's message of freedom and, indeed, its similarity of shape with the spacecraft had influenced his decision. To further honour the original bell – which had cracked during its first ringing, was repaired, cracked again in 1846 and was eventually rendered unusable – it was decided to paint a large white fracture along the side of the spacegoing Liberty Bell. "No one seemed quite sure what the crack looked like," said Grissom, "so we copied it from the tails side of a 50-cent piece."

On 17 July, the day before the planned launch, Grissom and Glenn relaxed in the crew quarters at Hangar S. During this time, they set to work on their urine-collection device. They took a pair of condoms, snipped off the ends and secured some rubber tubing which ran to a plastic bag taped to Grissom's leg. Flight surgeon Bill Douglas even asked the astronauts' nurse, Dee O'Hara, to buy a panty girdle for Grissom to wear in order to hold the hastily-made contraption in place.

Late that same night, the launch was scrubbed by low cloud cover. Fortunately, the decision was taken before the lengthy procedure of loading the Redstone with liquid oxygen had begun. This meant that only a 24-hour delay, as opposed to a 48-hour stand-down, would be necessary. Bill Douglas duly woke Grissom at 1:10 am on 19 July and informed him that the scheduled launch time had been moved up by an hour, to 7:00 am, in the hope that the mission could be underway before an anticipated spell of bad weather settled over the Cape. The plan was for Grissom to eat breakfast and undergo his pre-flight physical examination. Unfortunately, the astronaut recalled later, "someone forgot to pass the word about the earlier launch time, because breakfast was not ready at 1:45 as it was supposed to be". As a result, Douglas and psychologist George Ruff checked Grissom and, for breakfast, he was joined by Glenn, Scott Carpenter and Walt Williams.

After the now-customary procedure of gluing sensors to his body, Grissom was helped into his pressure suit and, at 4:15 am, clambered into the white transfer van. Inside, technicians had helpfully stencilled a sign which read 'Shepard and Grissom Express'. Despite the presence of cloudy skies along the entire Atlantic coast, "from Canaveral on north", the astronaut was given the go-ahead at 5:00 am to board his spacecraft. To alleviate any boredom, should a lengthy, Shepard-like situation materialise, Douglas handed Grissom a crossword book. The countdown, though, proceeded normally, albeit with a keen eye on the weather, until T-10 minutes and 30 seconds, when the clock was stopped in the hope that conditions might improve. They didn't. The launch was scrubbed and, since the Redstone had been fully loaded with liquid oxygen, a 48-hour turnaround was now unavoidable.

Early on 21 July, Grissom again suited-up and headed out to Pad 5. Delay after delay hit the countdown: firstly, one of the 70 titanium bolts around the rim of the hatch became cross-threaded, then the pad's searchlights had to be switched off to prevent them from affecting telemetry from the rocket and, all the while, cloud cover was given the opportunity to move away from the launch area. The astronaut, meanwhile, spoke briefly to his wife, did deep-breathing exercises and flexed his arms and legs so as not to become too stiff. At length, at 7:20:36 am, Liberty Bell 7 lifted-off. Again paying tribute to Bill Dana's reluctant Mexican spaceman character, José Jiménez, Capcom Al Shepard radioed "Loud and clear, José. Don't cry too much!" as Grissom headed for space. Passing through the sound barrier at an altitude of 11 km, he experienced none of the vibrations that had affected Freedom 7 and the Redstone's engine shut down, as planned, 142 seconds after liftoff. The astronaut felt a "brief tumbling sensation" at this stage and would later describe the clear sound of the LES tower jettisoning. "Actually," he wrote in his post-flight report, "I think I was still watching the tower at the time the posigrade rockets [which separated Liberty Bell 7 from the Redstone] fired ... the tower was still definable as a long, slender object against the sky."

Two minutes after launch, at an altitude of 30 km, Grissom noticed the sky turn rapidly from dark blue to black. He also noticed what he believed to be a faint star, roughly equivalent in brightness to Polaris, but which actually turned out to be the planet Venus; this won him a steak dinner from John Glenn, who had bet him that he would not be able to see any stars or planets. Observing Earth proved somewhat more problematic. Cloud cover over the Gulf of Mexico coastline between Apalachiocola, Florida, and Mobile, Alabama, made it virtually impossible for him to discern any land masses. Still, unlike Shepard's grey-tinted view, Grissom was granted a fascinating glimpse through the trapezoid window. "I could make out brilliant gradations of colour," he remembered later, "the blue of the water, the white of the beaches and the brown of the land."

His attempts to manoeuvre Liberty Bell 7 through all three axes were hampered when the yaw and pitch controls overshot their marks; overall, he judged the system as "sticky and sluggish". These problems, together with Grissom's observations through the window, put the mission behind schedule and the planned roll manoeuvres had to be abandoned. He did, however, successfully execute a manual retrofire five minutes into the flight and, as the retrorocket pack fell away, Cape Canaveral came clearly into view. "The Banana and Indian Rivers were easy to distinguish," he said later, "and the white beach all along the coast was quite prominent. I could see the building area on Cape Canaveral. I do not recall being able to distinguish individual buildings, but it was obvious that it was an area where buildings and structures had been erected."

Re-entry posed no significant problems, with the exception that it gave Grissom the peculiar sensation that he had reversed his backward flight through space and was actually moving face-forward. Plummeting towards the Atlantic, he saw what appeared to be the spent retrorockets passing the periscope view. Nine minutes and 41 seconds after launch, the drogue chute deployed, slowing Liberty Bell 7, before the descent was arrested by the jolt of the main canopy. "The capsule started to rotate and swing slowly under the chute as it descended," he said later. "I could feel a slight jar as the landing bag dropped down to take up some of the shock." In spite of a small, L-shaped tear in the main canopy, it did its job and the spacecraft impacted the Atlantic at 7:35 am, completing a mission of 15 minutes and 37 seconds – barely nine seconds longer than Shepard's flight – with what Grissom described as "a good bump". After splashdown, it nosed underwater, with the astronaut lying on his left side with his head down, but slowly righted itself as the landing bag filled with water and acted as a sea anchor.

Shortly thereafter, he disconnected his oxygen inlet hose, unfastened his helmet from his suit, released the chest strap, lap belt and shoulder harness and detached his biosensors. At first, he considered not bothering to unroll a rubber neck dam to keep air in and water out of his suit. "It's a chore to secure the dam," he said of the device, which had been designed by fellow astronaut Wally Schirra, "and I didn't think I'd need it. Fortunately, I reconsidered."

"THE FIRST THING I HAD EVER LOST"

As Grissom moved smartly through his post-landing checks, a quartet of Sikovsky UH-34D helicopters, despatched from the recovery ship Randolph, were already on the scene. One of their crews, Jim Lewis and John Rinehard, had been tasked with raising Liberty Bell 7 from the water, after which the astronaut would explosively blow the hatch, exit the capsule and be winched aboard the chopper. Seconds after splashdown, Grissom radioed Lewis, callsigned 'Hunt Club 1', to ask for a few minutes to finish marking switch positions. Finally, after confirming that he was ready to be picked up, he lay back in his couch and waited. All at once, he recounted, "I heard the hatch blow – the noise was a dull thud – and looked up to see blue sky . . . and water start to spill over the doorsill". The ocean was calm, but Mercury capsules were not designed for their seaworthiness, particularly with an open hatch, and Liberty Bell 7 started to wobble and flood. Grissom, who later admitted that he had "never moved faster" in his life, dropped his helmet, grabbed the right side of the instrument panel, jumped into the water and swam furiously. "The next thing I knew," he said, "I was floating high in my suit with the water up to my armpits."

Although both cap and safety pin were off the detonator, Grissom would later explain that he did not believe he had hit the button to manually blow the hatch. "The capsule was rocking around a little, but there weren't any loose items . . . so I don't know how I could have hit it, but possibly I did," he told a debriefing that morning aboard the Randolph. Lewis, meanwhile, had to dip his helicopter's three wheels into the water to allow Rinehard to hook a cable onto the now-sinking Liberty Bell 7. "Fortunately," Lewis recounted, "the first time John tried, he managed to hook-up while the capsule was totally submerged." Grissom, by now in the water, was puzzled, anxious and then angry when the helicopter did not lower a horse collar to hoist him aboard. Lewis, whose own training had shown him that Mercury pressure suits "floated very well" and had seen the astronauts apparently enjoying their time in the water, had no idea that Grissom was actually close to drowning. The astronaut had inadvertently left open an oxygen inlet connection, which allowed water to seep into his suit and air to leak out, thus reducing his buoyancy. Although he closed the inlet, some air also seeped from the neck dam, causing him to sink lower and regret the weight of souvenirs in his pockets.

Grissom did not know that Lewis was himself struggling with the spacecraft: in addition to the waterlogged capsule, the landing bag had filled with seawater and it now weighed in excess of 2,000 kg – some 500 kg more than the helicopter was designed to lift. Although Lewis felt he could generate sufficient lift to raise Liberty Bell 7 and take it back to the Randolph, every time he pulled it clear of the water and it drained, a swell would rise and fill the capsule again. Lewis' instruments told him that the strain on the engine would allow him only five minutes in the air before it cut out. He therefore released the $2 million capsule to sink in 5,400 m of water and requested that another chopper fish Grissom from the water while he nursed his own aircraft back to the ship.

Unaware of the difficulties the astronaut was having – they assumed that his frantic waving was to assure them that he was fine – it was several more minutes before the second helicopter, with the familiar face of George Cox aboard, dropped him a horse collar, which he looped around his neck and arms (albeit backwards) and was lifted to safety. Grissom was so exhausted that he could not even remember the helicopter had dragged him across the water before he finally started ascending. He had been in the water for only four or five minutes, "although it seemed like an eternity to me," he said later. His first request upon arrival on the Randolph's deck

The unsuccessful attempt to hoist Liberty Bell 7 from the ocean.

was for something to blow his nose, as his head was full of seawater. A congratulatory call from President Kennedy fell on deaf ears as, for the first time, "my aircraft and I had not come back together. In my entire career as a pilot, Liberty Bell 7 was the first thing I had ever lost". Worse was to come. At his first post-mission press conference, and in the years to follow, Grissom would be grilled by journalists, not over the success of his mission, but over the festering question of whether he had contributed to the loss of his spacecraft by blowing the hatch. It was an accusation that Grissom would refute until the day he died.

Not surprisingly, his temperature and heart rate were both high when he arrived aboard the Randolph. Physicians described him as "tired and ... breathing rapidly; his skin was warm and moist". Years later, although it was known that Grissom had an abnormally high heart rate, Tom Wolfe, in his bestselling book 'The Right Stuff', would point to his physiological state as 'evidence' that he had panicked inside Liberty Bell 7 and possibly blown the hatch. Even Grissom, at his first post-flight press conference in Cocoa Beach's Starlight Motel, admitted that he was "scared" during liftoff, an admission later jumped upon by the media as proof that America's second spaceman had displayed a chink of weakness. Test pilots at Edwards Air Force Base in California scornfully mocked that Grissom had "screwed the pooch" – had made a terrible mistake – and even the astronaut's two sons were lambasted by their schoolmates for the loss of the capsule. In his own summing-up for Life magazine, Grissom admitted that "if a guy isn't a little frightened by a trip into space, he's abnormal". Chris Kraft agreed, pointing out that "if you weren't nervous, you didn't know what the hell the story was all about".

A subsequent investigation from August to October 1961, which included Wally Schirra on its panel, would determine that the astronaut did not contribute in any way to the mysterious detonation of the hatch. Indeed, said Schirra, whose design of the neck dam had helped save Grissom's life, "there was only a very remote possibility that the plunger could have been actuated inadvertently by the pilot". During the inquiry, Schirra, fully-suited, even wriggled into a Mercury simulator himself and, no matter how hard he tried, could not 'accidentally' trigger the hatch's detonator. One of the conclusions reached by the Space Task Group was that a 76 cm-diameter balloon would be installed in future capsules to allow recovery ships to pick up the spacecraft if the helicopters were forced to drop them.

Many other engineers and managers shared the astronauts' conviction that Grissom was blameless. However, even though confidence in him remained high and he went on to command the first two-man Gemini mission, the stigma refused to go away. Some engineers continued to mutter of "a transient malfunction", but had no ability to identify it because the evidence lay on the floor of the Atlantic. Not until 1999 would Liberty Bell 7 be salvaged and raised to the surface.

Grissom himself participated exhaustively in the investigation. "I even crawled into capsules and tried to duplicate all of my movements," he said, "to see if I could make the whole thing happen again. It was impossible. The plunger that detonates the bolts is so far out of the way that I would have had to reach for it on purpose to hit it ... and this I did not do. Even when I thrashed about with my elbows, I could not bump against it accidentally." Moreover, to hit the plunger manually would have required sufficient force to produce a nasty bruise, which Grissom did not have. Possibilities explored over the years have included the omission of the ring seal on the detonator's plunger, static electricity from the helicopter, a change of temperature of the exterior lanyard after splashdown or – a hypothesis that Grissom supported – the entanglement of the lanyard with the straps of the landing bag. Walt Williams, writing in Deke Slayton's autobiography, considered the astronaut to be blameless, but thought it "very possible" that he had bumped the plunger accidentally with his helmet.

Launch pad leader Guenter Wendt, speaking in 2000, fiercely discounted all theories but one: the entanglement of the exterior lanyard. "It is the most logical explanation," he said, but acquiesced "Can we prove it? No." It is a pity that the mishap – however it happened – should have, in the eyes of the public, marred what had otherwise been a hugely successful mission and which cleared the way for John Glenn's historic orbital flight in February 1962. Was the unfortunate, twice-cracked Liberty Bell to blame for its spacegoing namesake's watery demise? All Grissom would say was that Liberty Bell 7 "was the last capsule we would ever launch with a crack in it!"

'GAS BAG'

Despite the hatch malfunction, Grissom's flight validated the Mercury capsule sufficiently to encourage the Space Task Group to do away with plans for two more suborbital Redstone missions; in fact, MR-6 had already been discarded from consideration since early June. Although Bob Gilruth was happy with this plan, NASA's head of spaceflight programmes, Abe Silverstein, felt that the Liberty Bell 7 data should be fully appraised before abandoning MR-5. Moreover, the public knew that three astronauts – Shepard, Grissom and Glenn – were in training for Redstone missions and fully expected each to fly. On the other hand, by expediting Project Mercury and accomplishing a five-hour, three-orbit mission, the achievement of Yuri Gagarin would be eclipsed. To do this, the abilities of the Redstone were simply insufficient; the larger Atlas missile, a rocket with a long history of development problems, would be needed. Then, on 7 August 1961, less than three weeks after Grissom's mission, all hopes of beating Gagarin were quashed when Gherman Titov completed 17 circuits of the globe in the day-long Vostok 2.

After the analysis of the Liberty Bell 7 data, it became clear that little more could be accomplished with the Redstone and, on 14 August, the Space Task Group's Paul Purser drafted a termination recommendation for Gilruth to submit to Silverstein. In it, Purser argued that the Redstone had successfully qualified the Mercury spacecraft, had validated NASA's training hardware and, despite problems, had not presented anything that could hinder a manned orbital flight. Four days later, NASA Headquarters announced the effective termination of Mercury-Redstone and the decision was made that the next manned mission would orbit Earth three times. However, although the capsule was ready, the Atlas rocket had a very bad habit of exploding either on the pad or shortly after liftoff and would require further qualification before it could be entrusted with a human pilot.

Unlike the Redstone, which owed its genesis to the Army and Wernher von Braun, the Atlas was an Air Force effort, inaugurated in 1946 to develop the United States' first intercontinental ballistic missile. Initial studies were awarded to the Convair Corporation of San Diego and led to Project MX-774 or what was described as "a sort of Americanised V-2". Its novel design would control the rocket by swivelling its engines, using hydraulic actuators which responded to commands issued by gyroscopes and an autopilot. Unfortunately, President Harry Truman's administration offered the Air Force the choice of having funding cut for either its intercontinental manned bombers and interceptors or its advanced weapons designs; the latter option was taken and, as the first MX-774 test vehicle neared completion, it was abruptly cancelled. This left the United States with no intercontinental ballistic missile, a problem made all the more worrisome when the Soviets detonated their first live nuclear device in 1949. A dramatic turnaround followed, with Truman ordering the development of hydrogen-fusion warheads on a priority basis and the outbreak of the Korean War boosting military budgets overnight. The Army began planning the Redstone and the Air Force, at last, was able to resume efforts to build an intercontinental missile which, in 1951, assumed the name 'Atlas'.

An initially cautious approach to its development was altered dramatically late the following year, when the Atomic Energy Commission conducted the world's first thermonuclear explosion on Eniwetok Atoll in the Pacific Ocean and increased emphasis was imposed on the Air Force to give its highest consideration to work on long-range ballistic missiles. By 1955, Convair's rocket gained a new lease of life with a long-term contract for its fabrication: Atlas truly became a high-profile, 'crash' project. During its development, another Air Force missile, the intermediate-range Thor, was designed by Douglas Aircraft Company as a stopgap nuclear deterrent, while the Army and (at first) the Navy assumed joint responsibility for a rocket dubbed 'Jupiter'.

Convair, meanwhile, was busy tackling several fundamental problem areas with the Atlas, one of which led to an entirely different airframe. The principle of this airframe, nicknamed 'the gas bag', utilised stainless steel sections thinner than paper, which were rigidised through helium pressurisation at between 1.7–4.2 bars. This led to a huge reduction in the ratio between the Atlas' structure and total weight – its 'empty' weight was less than two per cent that of its propellant weight – and yet the airframe remained capable of withstanding heavy aerodynamic loads. Meanwhile, a three-engine design for the missile, employing two boosters and one sustainer, producing a total thrust of 163,000 kg, together with small vernier jets, was devised by the Rocketdyne Division of North American Aviation.

The technique of igniting the boosters and sustainer on the ground provided an advantage of avoiding the need to start the Atlas' second stage in the high atmosphere. Firing the sustainer at liftoff also meant that smaller engines could be used. These would be fuelled by a combination of liquid oxygen and a hydrocarbon mixture known as Rocket Propellant-1 (RP-1) – a highly-refined form of kerosene – brought together by an intricate system of turbopumps, lines and valves, which fed them into the Atlas' combustion chambers at a rate of 680 kg per second. Appearance-wise, it also made the Atlas 'fatter' than the Redstone or Thor. Its original length was nearly 23 m, its diameter at the fuel tank section was 3.3 m and its fully-loaded weight was around 118,000 kg. At burnout, it was capable of a speed of some 25,750 km/h and a range of 14,480 km. Later Atlas variants, including those used for Project Mercury's orbital missions (the Atlas-D), were thicker-skinned and employed radio-inertial guidance systems to detect aerodynamic forces and calculate and adjust position, speed and direction.

A key stumbling block, though, involved preventing the warhead inside the Atlas nosecone from burning up as it entered the denser atmosphere at several times the speed of sound; this also had important ramifications from a man-in-space standpoint. During the development of Project Mercury, discussion flared over whether to include a beryllium heat sink or an ablative shield, with both concepts being developed in tandem for a time, until the latter option was finally selected. The Atlas' role as an orbital rocket became more acute when the Soviets launched the first Sputniks in 1957, by which time it was only partway through its verification programme, plagued by turbopump and fuel-sloshing problems. Nonetheless, on 8 December 1958, the Space Task Group formally approved it as the launch vehicle for its orbital missions and ordered nine flight units.

Since the Mercury-Atlas combo was taller than the weapons-carrying version of the missile, the gyroscopes had to be installed higher in the airframe, in order to more precisely gauge attitude changes during flight. The Mercury spacecraft would use its own posigrade rockets to separate from the Atlas, but because there was a chance that they could burn through the thin-skinned liquid oxygen dome, a fibreglass shield was affixed to the capsule-to-rocket mating ring. Also, the two small vernier jets were adjusted to reduce weight and complexity and increased aerodynamic loads and buffeting problems with the attached capsule forced engineers to thicken the skin of the Atlas' forebody. New instrumentation was installed to carefully monitor liquid oxygen and differential tank pressures, attitude rates about all three axes, engine manifold pressures and primary electrical power, all of which had the potential to lead to catastrophe.

Its maiden suborbital flight with a capsule, dubbed Mercury-Atlas 1 (MA-1), got underway on the morning of 29 July 1960. Despite holds for heavy rainfall, the cloud ceiling rose high enough to be considered acceptable and, after other delays caused by problems topping-up the Atlas' liquid oxygen tanks and ensuring telemetry was sound, the rocket lifted-off at 9:13 am. The early part of ascent went like clockwork, but, around a minute into the flight, the pressure difference between the liquid oxygen and fuel tanks went to zero and all contact with the Atlas was lost. Unfortunately, cloud cover over the Cape was so thick that visual and photographic evidence was virtually impossible, although it subsequently became apparent that the Atlas' walls had ruptured due to vibrations set up by mechanical resonance in the capsule-to-rocket adaptor. The rocket and spacecraft reached a peak altitude of 13 km, before descending to impact the Atlantic. One of the few saving graces was that the Mercury capsule maintained its structural integrity until it hit the ocean.

A stainless steel reinforcing 'belly band', strapped around the upper part of the

The crumpled MA-1 spacecraft after its ill-fated Atlas launch.

rocket, was implemented and the capsule-to-Atlas adaptor was stiffened. The MA-2 suborbital mission was duly launched at 9:12 am on 21 February 1961, passing successfully through 'Max Q' a minute later, and after reaching an apogee of 183 km, the now-separated capsule commenced its ballistic fall towards the South Atlantic. Splashdown occurred 18 minutes after launch and a proud NASA described the mission as "nominal in nearly every respect", with MA-2 recovered in good condition. When asked at a press conference later that day if an astronaut could have survived the test, Bob Gilruth beamed with a resounding "Yes".

Nine weeks later, at 11:15 am on 25 April, MA-3 lifted-off on what should have been an orbital flight, carrying an electronic mannequin capable of 'inhaling' and 'exhaling' man-like quantities of gas, heat and water vapour. This time, however, the Atlas failed to properly follow its roll and pitch manoeuvres due to a transient voltage. "The roll and pitch program normally changed the initial vertical trajectory of the launch into a more horizontal one that would take the Atlas out over the Atlantic," wrote Gene Kranz. "This Atlas was still inexplicably flying straight up, threatening the Cape and the surrounding communities." It was remotely destroyed after just 43 seconds, but, fortunately, the LES tower saved the Mercury capsule by pulling it free as planned. It impacted the Atlantic seven minutes after launch and was in such good condition that it was used on the very next Atlas flight. That flight was itself repeatedly postponed, firstly by delays in the delivery of its rocket to the Cape and also by the need to extensively overhaul the old MA-3 capsule back at McDonnell's St Louis plant. During its time in Missouri, the spacecraft was meticulously cleaned, its heat shield replaced and other repairs implemented.

At length, at 9:04 am on 13 September, only weeks after Gherman Titov's 17-orbit mission, MA-4 succeeded where its predecessor had failed, splashing down safely 280 km east of Bermuda. Finally, on 29 November, the chimpanzee Enos was blasted aloft in MA-5 to evaluate the capsule's life-support systems and the Atlas' performance with a living passenger. NASA Administrator Jim Webb's office had questioned the Manned Spacecraft Center (MSC) about the need for this and, indeed, Washington newspapers suggested that another such mission would invite Soviet ridicule. However, the decision was taken for a "necessary preliminary checkout" of the hardware before committing a human pilot. Enos, one of four chimps shortlisted for the flight, owed his name to the Hebrew word for 'man', hopefully indicative that the next Mercury-Atlas would be flown by a somewhat less hairy hominid. He underwent 1,250 hours of training – more than Ham, because Enos would be exposed to a much longer period of weightlessness and higher G loads – which included psychomotor preparations and aircraft flights. President Kennedy drew laughs from the Senate when he announced that the just-launched Enos "reports that everything is perfect and working well".

The Atlas successfully placed MA-5 into an orbit of 159 x 237 km. Originally intended to fly three orbits – the same as was planned for John Glenn on MA-6 – the capsule encountered difficulties with its attitude-control system when a metal chip in a fuel line caused one of its roll thrusters to fail. This allowed the spacecraft to drift from its normal attitude and, although the automatic system worked to correct this, some 4 kg of fuel was wasted trying to keep it properly aligned during its second orbit. Coupled with this problem, the environmental system experienced glitches and the temperature of Enos' pressure suit rose to 38.1°C. The problem later resolved itself, but engineers' concerns over fuel consumption prompted them to request a re-entry at the end of the second orbit. As Enos hurtled over Point Arguello in California, Flight Director Chris Kraft decided to bring MA-5 home early and the retrofire command was transmitted to the spacecraft.

Three hours after launch, Enos' capsule was bobbing in the Atlantic, just off the coast of Puerto Rico. It was hauled aboard the destroyer Stormes and its hatch explosively detonated. Enos, who, like Ham, had been 'rewarded' with electric shocks for operating the correct controls, thanks to an equipment malfunction, was bloodied and "excitable", but nonetheless alive and happy to see his rescuers.

Significant though it was, the flight of Enos – who would die of dysentery less than a year later – quickly faded as public attention became riveted on the impending mission, tentatively scheduled for 19 December. At around the same time, some members of the media speculated that Glenn, Shepard and Scott Carpenter had been selected as candidates for the first orbital mission. Glenn, however, having served as backup for the last two missions, had already been picked by Bob Gilruth to fly. Barely a day after Enos splashed down, his launch vehicle, designated 'Atlas 109D', arrived at the Cape and Mercury's operations director Walt Williams told journalists that three shifts were now working around-the-clock, seven days a week, in a bid to get an American into orbit before the end of the year. That plan evaporated on 7 December, when it was announced that "minor problems dealing with the cooling system and positioning devices in the Mercury capsule" had obliged a postponement until January. Admittedly, many senior managers had known since October that the timeframe for a December launch was tight. Said the agency's deputy administrator, Hugh Dryden: "You like to have a man go with everything just as near-perfect as possible. This business is risky. You can't avoid this, but you can take all the precautions you know about."

With the completion of MA-5, NASA felt confident and ready for the manned orbital mission. By the end of February 1962, a somewhat different hominid – a Marine pilot, transcontinental record-holder and 'Name That Tune' winner named John Herschel Glenn Jr – would ride the temperamental Atlas not only into space, but into orbit and into the history books. Yet the risks were pervasive and enormous. The success of Enos' mission did not detract from the reality that the rocket had exploded on a number of occasions. "John Glenn is going to ride on that contraption?" asked the Redstone's designer, Wernher von Braun. "He should be getting a medal just for sitting on top of it before he takes off!"

THE ALL-AMERICAN

Before his selection as an astronaut, Glenn's two claims to fame in the public eye were setting a supersonic cross-country record for flying from America's west to east coasts in just three hours, 23 minutes and 8.4 seconds as part of the Navy's Project

Bullet ... and appearing alongside ten-year-old Eddie Hodges to win the CBS television show 'Name That Tune'.

In the wake of his astronaut career, Glenn's fame skyrocketed. Not only did he become the first American to orbit the Earth, which, in many minds, actually eclipsed the achievement of Al Shepard, but he enjoyed a highly successful career in politics, serving with distinction for a quarter of a century as a senator and running unsuccessfully for the presidency in the 1984 election. During his final days in the Senate, I wrote to Glenn's office and was pleasantly surprised to receive a personally-signed black-and-white photograph of the great man, clad in the silver pressure suit that he wore for his five-hour Mercury mission more than three decades earlier. His fame resurfaced in October 1998, when he hurtled into orbit aboard Space Shuttle Discovery, securing yet another record which still stands: at 77, he remains the oldest person ever to journey into space.

To the media, Glenn was a hero from the very day that he and the other members of the Mercury Seven were introduced in April 1959. Freckle-faced, witty, articulate and charismatic, he was described by some journalists as epitomising 'all-American' qualities and many were surprised when he did not secure the first American suborbital flight. Others did not buy Glenn's 'boy-next-door' act, including Chris Kraft, who had dealings with him years earlier when working on the Vought Crusader naval fighter. Kraft considered Glenn's head to be "up and locked" – a tongue-in-cheek reference to the retraction of an aircraft's landing gear – and, indeed, the astronaut's politicking to secure MR-3 for himself had even led to a stern reprimand from Bob Gilruth to stop backbiting. After Gilruth's decision to fly Shepard first, Glenn had complained bitterly that the December 1960 peer vote had turned the selection process into a popularity contest. Glenn felt strongly that his advice, given to his Mercury Seven colleagues about the damaging effect of womanising on their public image, had cost him what he perceived to be his rightful seat on MR-3. "I didn't think being an astronaut was a popularity contest," Glenn wrote years later. "I was wrong about that."

By the time he finally flew, in February 1962, Glenn was 40 years old, making him by far the oldest among the Mercury Seven. He had been born in Cambridge, Ohio, on 18 July 1921, although he grew up and received his education in the town of New Concord, studying engineering. Already, as a youth, he had undergone flight training and took the Army Air Corps' physical examination and passed. However, when no orders materialised, he took the Navy's physical, which he also passed and was sworn into the Naval Aviation Cadet Program. Initial training at the University of Iowa was followed by preparation at Olathe, Kansas, and finally at Corpus Christi, Texas. It was whilst stationed at the latter base that Glenn learned of his eligibility to apply for the Marine Corps, which he did, winning his wings and lieutenant's bars in 1943. That same April, he married Annie Castor.

After a year of training, Glenn joined Marine Fighter Squadron 155, flying F-4U combat missions in the Marshall Islands of the South Pacific during the Second World War. He returned to the United States shortly before the end of the conflict to begin test pilot work at Patuxent River, Maryland, evaluating new aircraft. Subsequently, he served as an instructor in advanced flight training at Corpus Christi from 1948–50, completed marine amphibious warfare training and flew 63 combat missions during the Korean conflict. It was whilst in south-east Asia that Glenn shot down three MiGs along the Yalu River, earning himself the nickname 'MiG Mad Marine'. (He was also tagged 'Magnet Ass' for his ability to attract flak.)

Overall, in the Second World War and Korea, he flew 149 combat missions and his chestful of medals proved it: six Distinguished Flying Crosses and an Air Medal with 18 clusters. After Korea, he joined the Navy's Test Pilot School at Pax River, later serving as project officer for a number of advanced fighters. Whilst serving in this capacity for the F-8U Crusader, in July 1957, he set the transcontinental speed record by flying non-stop from Los Alamitos Naval Air Station in California to Floyd Bennett Field on Long Island. The attempt arose from Glenn's desire to kill two birds with one stone: running the Crusader's engines in afterburner at full combat power whilst at high altitude and seizing the Air Force-held transcontinental speed record, which then stood at three hours and 45 minutes.

"We could do the test," he wrote in his memoir, "and also call attention to the fine plane the Navy had purchased as its frontline fighter." Under the name 'Project Bullet' – so-called because the Crusader flew faster than the muzzle velocity of a bullet from a .45-calibre pistol – Glenn volunteered himself as pilot for the attempt. "The plane flew beautifully," he wrote of the epic flight on 16 July 1957. Glenn, who beat the previous record by 21 minutes, was awarded another Distinguished Flying Cross. It was in the wake of the flight that he was approached with an invitation to appear on the television quiz show 'Name That Tune', alongside Eddie Hodges from Mississippi. The pair won the $25,000 first prize.

Consequently, in the eyes of the public, Glenn was probably the most recognisable when the Mercury Seven were introduced at the Dolley Madison House less than two years later. "He was already first among equals," wrote Scott Carpenter in his autobiography, co-authored with daughter Kris Stoever, "the oldest of the seven, with the most military and combat experience, a television celebrity and holder of a transcontinental speed record. He wore old clothes, old cowboy hats and lived next to his dearest friend, Tom Miller, his roommate and wingman from World War II." His purchase, at the height of the sports-car craze, of a tiny $1,400 Prinz, barely big enough for two passengers, looked somewhat comical parked alongside Al Shepard's brand-new Corvette. By his own admission, Glenn bought it for its great mileage, which got him from his home in Arlington to Langley for less than a dollar. Yet Glenn turned the humour around, one day writing on a classroom blackboard a quote that he had seen in Reader's Digest: "Definition of a sports car: a hedge against male menopause!"

Even during the initial press conference, when each astronaut candidate spoke about their lives, their wives, their families and their dreams, Glenn was by far the most eloquent: speaking at length about love of God, family and his desire to serve his country. The others rolled their eyes, but, in Glenn's mind, it was part and parcel of achieving the goal of being the best of the seven and becoming the first man in space. Indeed, he was so adored by the public and the media that questions were asked of NASA press spokesman Shorty Powers as to why Shepard, not Glenn, had received the coveted Redstone mission. Powers' response was that, out of the three final candidates, Shepard was the best, but Glenn remained furious. Only when Gilruth firmly set him in line did he grudgingly abandon his effort to undermine Shepard and train as his backup. For the Marine Corps, however, the reason behind the decision was obvious: NASA was saving 'the best' flight, the first orbital flight, for a Marine.

SPY SWAP

A few days before Glenn's historic mission, another historic event was underway on the Glienicke Bridge, linking Potsdam to West Berlin, as the Soviet intelligence officer Colonel Vilyam Fisher was exchanged for the American U-2 pilot Francis Gary Powers. Two years earlier, in May 1960, Powers had been shot down near Degtyarsk in the Urals by a salvo of S-75 Dvina surface-to-air missiles. He had been despatched from an American communications facility at Badaber, close to Peshawar in Pakistan, to photograph Soviet ballistic missile sites and was scheduled to land at Bodø in Norway. The incident came two weeks before the opening of a major East-West summit in Paris – a summit which Soviet premier Nikita Khrushchev would leave in disgust when Dwight Eisenhower refused to apologise – and proved hugely embarrassing for the United States.

Powers had succeeded in ejecting from his stricken aircraft and parachuted to the ground, whereupon he was captured and placed on trial in Moscow. Khrushchev, meanwhile, announced to the world that a "spyplane" had been shot down, but deliberately omitted to detail the fate of its pilot. The Eisenhower administration, assuming that Powers had been killed, set to work creating a cover story that he had actually been flying a "weather research aircraft", which accidentally strayed into Soviet airspace after the pilot had reported "difficulties with his oxygen equipment" over Turkey. No attempt, continued Eisenhower, was made to deliberately violate Soviet territory. On 7 May, Khrushchev proved this to be a lie, revealing that the pilot was indeed alive and the remains of his largely-intact spyplane were displayed at the Central Museum of Armed Forces in Moscow. Powers' survival pack, hardly representative of a weather research pilot, included 7,500 roubles in cash and jewellery for women and was also placed on display.

He was convicted that August of espionage and sentenced to three years' imprisonment and seven years' hard labour, but on 10 February 1962 was exchanged for Fisher. The latter, born in Newcastle-upon-Tyne of Russian-German parentage, had moved from England to the Soviet Union with his Bolshevik-sympathising parents in the early Twenties, where he became a translator and, after military duty, trained for the secret services. Sent to Canada and, later, the United States, to recruit and supervise intelligence agents, he was captured by the FBI in June 1957 and sentenced to 30 years in prison. His exchange for Powers and an American economics student named Frederic Pryor was followed by continued service with the KGB until his death in 1971.

Powers, meanwhile, was criticised upon his return to the United States for having failed to activate the U-2's self-destruct charge, which would have eliminated the camera, photographic film and other classified components. He had also not used a CIA-provided suicide pin, secreted inside a hollowed-out silver dollar, to avoid capture and the possibility of torture. Three weeks after his release, as John Glenn paraded in triumph through the streets of Washington and New York, Powers testified before the Senate Armed Services Select Committee and was found to have followed orders appropriately and praised "as a fine young man under dangerous circumstances". He subsequently worked for the U-2's contractor, Lockheed, as a test pilot and was later hired by the Los Angeles television station KNBC to fly its new telecopter. In August 1977, returning from an assignment to cover brush fires in Santa Barbara, his telecopter ran out of fuel and crashed, killing both Powers and KNBC cameraman George Spears.

THE BEST ... AT LAST

John Glenn's flight – dubbed Mercury-Atlas 6 or, in keeping with Shepard and Grissom's spacecraft-naming tradition, 'Friendship 7' – was eagerly awaited by the United States, although it proved a long time coming. The choice of name, Glenn recalled in his memoir, had been made by his children, Dave and Lyn. "They pored over a thesaurus and wrote dozens of names in a notebook," he wrote. "Then they worked them down to several possibilities, names and words, including Columbia, Endeavour, America, Magellan, We, Hope, Harmony and Kindness. At the top of their list was their first choice: Friendship." Although the name would be kept quiet until the morning of launch, Glenn had privately asked Cecelia Bibby, the artist at Hangar S, to inscribe the name on his capsule in script-like characters, adding more individuality than the block lettering employed to stencil Freedom and Liberty Bell onto Shepard and Grissom's spacecraft.

"From what John Glenn told me later, [he] had decided that he wanted the name of his spacecraft applied in script and applied by hand," Bibby said, "because Al Shepard's and Gus Grissom's names had been applied by some mechanic who went into town, got a can of spray paint, a stencil-cut of the names and then spray-painted them onto the capsule." Apparently, added Bibby, Glenn felt that men had such poor handwriting that a female artist would be preferable. When she painted the name on the capsule, Bibby, clad in white clean-room garb, became the only woman to ascend the gantry to Pad 14 at Cape Canaveral and was even told by a disgruntled Guenter Wendt that she did not belong there. So pleased was Glenn with the design that Gus Grissom dared Bibby to secretly paint naked women on the spacecraft as well.

She rose to the challenge, not by painting on the exterior of Friendship 7, but by drawing a naked woman on the inside of a cap used to cover the periscope. Although the cap would be jettisoned before launch, it would be seen by Glenn as he boarded the capsule and hopefully might give him a laugh. Reading 'It's just you and me against the world, John Baby', the drawing was placed there by Bibby's friend, launch pad engineer Sam Beddingfield. The launch itself was scrubbed, but Bibby got into work the following morning to find a note from Glenn, "telling me he had

With artist Cece Bibby proudly looking on, John Glenn displays the Friendship 7 logo
on the side of his spacecraft.

gotten a big kick out of the drawing". Bibby was almost fired for her practical joke,
although fortunately both Grissom and Glenn intervened on her behalf and saved
her. Later in the launch preparations, she sent Glenn another gift: this time a
drawing of a frumpy old woman in a house dress, bearing mop and bucket and the
legend 'You were expecting maybe someone else, John Baby?' Not long afterwards,
Glenn's backup, Scott Carpenter, requested a naked woman for his own capsule,
Aurora 7, which would fly the second American orbital mission in May 1962 ...

Sadly, the news at the beginning of the year was nowhere near as light-hearted: a
launch attempt on 16 January was postponed by at least a week, due to technical
problems with the Atlas rocket's fuel tanks. With each successive delay, more
criticism was voiced from journalists and congressmen, who questioned whether
Project Mercury – already a year behind the Soviets – would ever succeed in placing
a man into orbit. Even President Kennedy, at a news conference on 14 February,
expressed disappointment, although he felt that the final decision on when to launch
should be left to the Mercury team. Others, however, commended NASA's frankness
in conveying the reasons for each delay. It was stressed that the orbital mission had
been planned for over three years and a few more weeks' delay was of little

consequence, a sentiment shared by Glenn himself, who described being not "particularly shook-up" by the postponements.

Indeed, according to planning charts issued by NASA in April 1960, the orbital mission was originally scheduled for May 1961, then July, October and ultimately December. A variety of manufacturing changes to Glenn's capsule – Spacecraft No. 13 off the McDonnell production line – had contributed to delays in its progress: a shortage of environmental control components had virtually stalled work in October 1960 and then extensive re-planning ordered after the MA-3 failure in April 1961 prompted NASA to assign No. 13 to the first manned orbital mission. By the end of August, the capsule had been delivered to Cape Canaveral and early in January 1962 was mated to its Atlas launch vehicle on Pad 14.

Following the 23 January postponement, caused by poor weather, another attempt was scheduled for the cloudy morning of the 27th. Glenn rose early for his low-residue breakfast of filet mignon, scrambled eggs, orange juice and toast with jelly, before undergoing the laborious process of having biosensors glued onto his body and his pressure suit fitted. That day, he lay inside Friendship 7 for more than five hours, hoping for a break in the overcast skies. It never came and, at T-20 minutes, Walt Williams scrubbed the launch. "It was one of those days," Williams remembered later, "when nothing was wrong, but nothing was just right either." Another event, back in Arlington, Virginia, which did not go right, at least for Vice-President Lyndon Johnson, was his plan to visit Glenn's home ... complete with a television crew and a horde of the media.

Johnson also asked for Life journalist Loudon Wainwright, who was in attendance at Glenn's house as part of the Mercury Seven magazine deal, to leave. Annie Glenn, who wanted nothing less than to have television lights in her home and wanted Wainwright to stay, flatly rejected Johnson's request. "I understand the vice-president was pretty pissed off," wrote Deke Slayton, "and that he wasn't too happy with Jim Webb or Webb's astronauts at that point." In the weeks that followed, there were theories, Slayton added, "that Webb had gotten ticked-off at John Glenn" as a result of the episode and had begun searching for a way in which to better 'control' his astronauts. Some observers would speculate that Slayton, assigned to fly the next Mercury-Atlas orbital mission, would be an unfortunate victim of Webb's politicking. Although it was not a theory that Slayton himself supported, he remained convinced, years later, that the decision made about his career just weeks after Glenn's flight "was political".

After the 27 January postponement, Glenn's launch was initially targeted for 1 February, necessitating the emptying, purging and refilling of the Atlas' propellant tanks. Then, two days before launch, on the 30th, as the ground support team began refuelling, a mechanic discovered, by routinely opening a drain plug, that there was fuel in the cavity between the structural bulkhead and an insulation bulkhead which separated the propellant tanks. Initial estimates suggested at least a ten-day delay to correct the problem and recheck the rocket's systems. The 600 accredited members of the media at the Cape could do little but groan as John Glenn's launch was postponed yet again, this time until no earlier than 13 February.

Most of the journalists quickly dispersed, together with Glenn himself, who spent a few days with his family at home in Arlington, before travelling to the White House for a brief visit with President Kennedy. For the astronaut, it was time of peaks and troughs. "I think people normally build up to a peak when they are getting ready for an event as complicated as this," he said later, "and here we had a situation where we kept building up psychologically and nothing happened. It was like crying 'wolf' over and over again. But I needn't have worried at all. These people kept working and preparing and lost none of their sharpness." Some psychologists were concerned that he would suffer emotionally under the strain. In Glenn's mind, the delays simply gave him extra time to run each day, to study, to read and respond to mail (one of which told him that it was God's way of letting him know that he shouldn't tamper with the heavens) and to work in the simulators.

"If I was suffering," he said, "I wasn't aware of it and neither were the psychiatrists whose job it was to keep track of my emotions. The nearest I came to getting upset was after I visited a friend's house for a home-cooked meal and a quiet evening with his family. A couple of days later, it turned out that the friend's children had the mumps. As far as I could remember, I'd never had them. Delays for weather and for technical difficulties were facts I could accept, but a postponement or a possible replacement while the astronaut recovered from a childhood disease seemed a bit silly. It would make quite a headline!"

On 13 February, although weather conditions remained foul, NASA personnel began to move back into position to attempt a launch. The media's pessimism was reflected in their turnout: by that evening, only 200 had checked in at the nearby Cocoa Beach motels. Their doubt was well-placed and the launch gradually slipped towards the end of the month. By the 19th, with liftoff rescheduled for the following morning, the Weather Bureau predicted only a 50 per cent chance of a launch: conditions in the recovery zones were fine, but the Cape was poor. A frontal system had been observed moving across central Florida, which, it was surmised, could cause broken cloud over the Cape in the early hours of the next day.

Glenn rose early on the morning of 20 February, to be greeted by physician Bill Douglas at 2:00 am, who told him that the weather still offered little more than a 50–50 chance of a successful launch. After breakfast, he underwent the now-customary pre-flight examination and was outfitted with biosensors and helped into his silver pressure suit. Technician Joe Schmitt tested the suit and Bill Douglas ran a hose into a fish tank to check the purity of the air supply – dead fish meaning bad air – which offered Glenn the chance for some humour. "Bill, did you know a couple of those fish are floating belly-up?" Douglas' shocked reaction as he rushed over to the tank was soon arrested by a broad grin on Glenn's face.

Out at Pad 14, clouds rolled overhead by the time the astronaut arrived outside the capsule at 6:00 am. However, forecasters were predicting possible breaks by mid-morning, producing a different atmosphere on the gantry, with less casual chatter, as if everyone sensed, said Glenn, "that we were going for real this time". Weather caused the original launch time to be missed and a broken microphone bracket inside Glenn's helmet required repair before Friendship 7's hatch could be finally closed and bolted at 7:10 am. One of the bolts sheared, necessitating the removal of the hatch while it was replaced. (Several months earlier, Gus Grissom

"Godspeed, John Glenn!" The Atlas takes flight with a man aboard.

flew with a defective hatch bolt, but this time Walt Williams was taking no chances.)

Forty minutes later, the countdown resumed. By the time the pad crew moved clear of the Atlas, Glenn – whose pulse varied from 60–80 beats per minute – was granted his first view of blue skies as the realisation took hold that 20 February might be 'The Day'. He was also assailed by the peculiar, eerie sense of being atop the silvery rocket. "I could hear the sound of pipes whining below me as the liquid oxygen flowed into the tanks and heard a vibrant hissing noise," he said later. "The Atlas is so tall that it sways slightly in heavy gusts of wind and, in fact, I could set the whole structure to rocking a bit by moving back and forth in the couch!" Thirty-five minutes before launch, the rocket's liquid oxygen supply was topped off and, despite another brief hold caused by a stuck fuel pump outlet valve and a last-minute electrical power failure at the Bermuda tracking station, the clock resumed ticking.

With 18 seconds to go, the countdown switched to automatic control and, at four seconds, Glenn "felt, rather than heard" the engines roaring to life far below. At 9:47:39 am, with a thunder that overwhelmed Scott Carpenter's "Godspeed, John Glenn" send-off, the Atlas' hold-down posts separated and the enormous rocket began to climb. The 'gas bag' was on its way.

AN AMERICAN IN ORBIT

"Liftoff was slow," Glenn recounted in his 1999 memoir. "The Atlas' thrust was barely enough to overcome its weight. I wasn't really off until the umbilical cord that took electrical communications to the base of the rocket pulled loose. That was my last connection with Earth. It took the two boosters and the sustainer engine three seconds of fire and thunder to lift the thing that far. From where I sat, the rise seemed ponderous and stately, as if the rocket were an elephant trying to become a ballerina." For the first few seconds, the Atlas climbed straight up, before its automatic guidance system placed it carefully onto a north-easterly heading; a transition which Glenn found noticeably "bumpy". However, his checklist took priority, requiring him to tick off cabin pressure levels, oxygen and fuel limits and ampere readings on Friendship 7's batteries. Forty-five seconds into the climb, the rocket entered 'Max Q', the highly dynamic phase of the flight in which MA-1 had faltered two years earlier.

"It lasted about 30 seconds," said Glenn. "The vibrations were more pronounced at this point. I did not expect any trouble, but we knew there were certain limits beyond which the Atlas and capsule should not be allowed to go … Since it is difficult for the human body to judge the exact frequency and amplitude of vibrations like this, I was not sure whether we were approaching the limits or not. I saw what looked like a contrail float by the window and I went on reporting fuel and oxygen and amperes. The G forces were building up now. I strained against them, just to make sure I was in good shape." A successful liftoff and passage through Max Q completed two of the four major hurdles needed to achieve his mission: the third, shutting down the Atlas' outboard boosters, followed perfectly, leaving only the

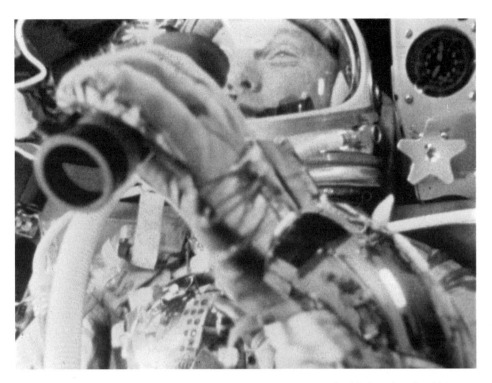

Demonstrating the cramped nature of the Mercury spacecraft, this interior view shows John Glenn hard at work during his five-hour mission.

central sustainer engine to continue the push into orbit. "There was no sensation of speed," recalled Glenn, "because there was nothing outside to look at as a reference point."

Shortly before 9:50 am, two and a half minutes into the ascent, and by now outside the 'sensible' atmosphere, the Atlas' LES was released, "accelerating [away] at a tremendous clip". The fourth hurdle – insertion into orbit – followed with the sustainer engine's cutoff, separation of the Atlas from the capsule and the firing of Friendship 7's posigrade rockets to push it clear. By 9:52 am, according to the operations team in their flight logs, the mission was "through the gates" and Glenn was in orbit.

Although he was able to describe the magnificent views he was seeing, it was on this mission that the rest of the world was also able to capture something of the grandeur of Earth – thanks to a camera. Glenn had discussed this idea with Bob Gilruth some months earlier and the two men had begun searching for an appropriate device: small enough to operate with one hand, yet adaptable, so that he could advance the film with his thumb and snap the shutter with his forefinger whilst encased in a pressure suit. Their search achieved little success. Then, one day, whilst getting a haircut in Cocoa Beach, Glenn saw "a little Minolta camera in a display case. It was called a Hi-Matic." The camera, he noted, had automatic exposure; he would have no need to fiddle with light meters and f-stops. He bought it on the spot for $45. NASA technicians adapted it for the spacecraft and, in tests, Glenn found that it was the easiest camera to use, even wearing his pressure suit gloves. It would yield some of the most amazing images of the entire mission.

Five minutes after launch, as planned, Glenn achieved orbital speed of 28,200 km/ h, an altitude of some 160 km and – for the first time – experienced the strange state of weightlessness. "Zero-G and I feel fine," he exulted; words that he would repeat more than three decades later after reaching orbit aboard the Space Shuttle Discovery. "Capsule is turning around," he added as Friendship 7 slowly swung around into the re-entry attitude. "Oh, that view is tremendous!" he exclaimed as the horizon appeared in the window and he caught his first sight of the curvature of the Earth and the fragile atmosphere.

At length, Friendship 7 oriented itself into its 'normal' operating position, with its blunt end facing into the direction of travel, flying eastwards. Looking back, Glenn could clearly see the spent Atlas making slow pirouettes as it tumbled away. At first, the parameters of his trajectory were so good that he was given a go-ahead by Capcom Al Shepard for "at least seven orbits". With all systems running as expected, Friendship 7 crossed the Atlantic and passed over the Canary Islands.

Glenn's first tasks involved checking the spacecraft's roll, pitch and yaw attitude controls in case an emergency re-entry became necessary. At 9:59 am, he crossed the coast of western Africa – "a fast transatlantic flight," he later wrote – and felt one of his earliest sensations of weightlessness: a grey-felt toy mouse, pink-eared, with a long tail, drifted from an equipment pouch. It was one of Shepard's jokes, a reference to Bill Dana's astronaut character, who had always felt sorry for the experimental mice sent aloft in rocket nosecones. It also offered a subtle 'gotcha', following Glenn's own placement of a girlie pin-up in Shepard's Freedom 7 nine months earlier. Next, out came the Minolta, which floated comically before Glenn's eyes. "I found that I had adapted to weightlessness immediately," he recalled in his memoir. "When I needed both hands, I just let go of the camera and it floated there in front of me. I didn't have to think about it. It felt natural."

After quickly checking his blood pressure for the Canaries ground station, Glenn turned his attention to photographing selected spots on Earth's surface. One of the first was a patch of cloud covering the Canaries, followed, at 10:08 am, by shots of enormous dust storms brewing over the Sahara Desert. Next came further checks of his capsule's attitude controls, which he reported were "well within limits", followed by exertion tests with a bungee cord attached beneath the instrument panel and then reading the vision chart at eye-level. Glenn's vision, it seemed, was not changing. Head movements, too, did not cause him any disorientation, "indicating," he wrote later, "that zero-G didn't attack the balance mechanism of the inner ear".

Forty minutes into the mission, Friendship 7 drifted into darkness as Glenn approached his 240 km apogee. A sunset from space was one of the features of the flight about which he was most excited. He was already a lover of the beauty of terrestrial sunsets, but seeing one from orbit, "was even more spectacular than I imagined and different in that sunlight coming through the prism of Earth's atmosphere seemed to break out the whole spectrum, not just the colours at the red end, but the greens, blues, indigos and violets at the other. It made 'spectacular' an understatement for the few seconds' view. From my orbital front porch, the setting Sun that would have lingered during a long earthly twilight sank 18 times as fast. The Sun was fully round and as white as a brilliant arc light and then it swiftly disappeared and seemed to melt into a long thin line of rainbow-brilliant radiance along the curve of the horizon. I added my first sunset from space to my collection''.

As he hurtled onwards, Glenn was also able to describe the absolute blackness of the sky above him, although he admitted that he could see stars, successfully identifying the Pleiades cluster. When astronaut Gordo Cooper, the capcom at the Muchea tracking station, just north of Perth in Australia, came within communications range, Glenn related the shortness of the passage from orbital daytime into darkness and proceeded through the humdrum blood pressure readings. Then, almost 55 minutes after launch, as Friendship 7 passed directly over Perth and Rockingham, Cooper asked him if he could see lights. "I can see the outline of a town," he replied, "and a very bright light just to the south of it." Glenn thanked the residents of Perth for turning on their lights to greet him.

Travelling over the South Pacific, close to the tiny coral atoll of Canton Island, midway between Fiji and Hawaii, the astronaut lifted his visor and ate: squeezing some apple sauce from a toothpaste-like tube into his mouth, gobbling some small malt tablets and confirming that weightlessness posed no obstacles. As he approached orbital sunrise, Glenn was surprised to see around the capsule a huge field of particles, like thousands of swirling fireflies. "They were greenish-yellow in colour," he said later, "and they appeared to be about six to ten feet apart. I seemed to be passing through them at a speed of three to five miles an hour. They were all around me and those nearest the capsule would occasionally move across the window, as if I had slightly interrupted their flow. On the next pass, I turned the capsule around so that I was looking right into the flow and though I could see far fewer of them in the light of the rising Sun, they were still there."

The particles diminished in number as he flew eastwards into brighter sunlight. Scott Carpenter, who duplicated Glenn's mission in May 1962, would also see them and they would be attributed to little more than ice crystals venting from Friendship 7's heat exchanger.

"LEAVE THE RETRO PACKAGE ON"

Meanwhile, Glenn continued putting his capsule through its paces, evaluating its abilities in manual, automatic and fly-by-wire modes. It was shortly after passing the two-hour mark of the mission, however, that he received an unusual request from Mission Control: to keep the switch for Friendship 7's landing bag in the 'off' position. He confirmed that the switch was indeed off and pressed on with his work. Later, as he flew over Muchea, Gordo Cooper asked him to confirm it again. Then, during another pass over Canton Island, Glenn overheard an indication from a flight controller that his landing bag – located between the base of the spacecraft and the heat shield – might have accidentally deployed. He queried Mission Control and was assured that the ground was merely monitoring the situation. Glenn began to suspect that the fireflies could be related, perhaps, to some shifting of his heat shield and landing bag.

Already, as early as the second orbit, an engineer at the telemetry console named William Saunders, had noted that 'Segment 51' – an instrument providing data on the landing system – was generating unusual readings and Mercury Control instructed all tracking sites to monitor it carefully. A little over four hours into the flight, however, it became clear that the landing bag might be deployed or, at least, not securely locked into position. Whilst over Hawaii, the duty capcom informed Glenn that the signal was probably erroneous, but, to be sure, asked him to set the landing bag switch in its 'auto' position. "Now, for the first time, I knew why they had been asking about the landing bag," Glenn wrote. "They did think it might have been activated, meaning that the heat shield was unlatched. Nothing was flapping around. The package of retrorockets that would slow the capsule for re-entry was strapped over the heat shield. But it would jettison and what then? If the heat shield dropped out of place, I could be incinerated on re-entry."

If the green landing bag light came on, Glenn wrote, it would clarify that it had indeed accidentally deployed. However, "if it hadn't, and there was something wrong with the circuits, flipping the switch to automatic might create the disaster we had feared". He flipped the switch. No light. This suggested that the landing bag was secure. As retrofire approached, Capcom Wally Schirra, based at Point Arguello in California, told Glenn not to jettison his retrorocket package at least throughout his passage across Texas. It marked the first of several efforts to ensure that, if the heat shield had been loosened, the retrorocket package might hold it in place just long enough to survive the hottest part of re-entry.

Glenn, meanwhile, had worries of his own. From the end of his first orbit, he had experienced problems with Friendship 7's automatic control system, as the capsule started to swing over to one side along its yaw axis, corrected itself at considerable expense of hydrogen peroxide fuel, before doing the same again. "It became necessary for me to control the capsule's movements by hand," he said later. "For most of the rest of the trip, I controlled the capsule myself. This did cut down on the other activities we'd planned. It meant that I had to cancel several of the experiments and observations I wanted to make on the second and third orbits." These would have included observations of the solar corona, terrestrial cloud structures, his ability to adapt his eyes to orbital darkness and further studies of the effects of weightlessness. "I was able to take far fewer pictures than I'd intended," Glenn continued, "and I had to pass up my plan to have two meals during the flight to test my ability to get food down under various conditions."

Throughout most of his second orbit, as the ground pondered the Segment 51 situation, Glenn persevered with his efforts to determine the problem with the automatic control system. "I could hear the large fuel thrusters outside the capsule as they popped off their bursts of hydrogen peroxide in first one direction and then the other," he said. "I could feel the slight throb of the smaller nozzles when I cut them in. The manual system had been a little mushy. It did not respond quite as crisply as I thought it should have, but I still had good control. It worked best when I switched to the fly-by-wire mode, which combines the manual control stick and the fuel nozzles which are operated by the automatic system. This meant that I could work the automatic system by hand and conserve its fuel. Though this routine took most of my attention and kept me rather busy for the next three hours, I thoroughly enjoyed it. The idea that I was flying this thing myself and proving on our first orbital test that a man's capabilities are needed in space was one of the high spots of the day. The value of this outweighed the loss of some of the things I did not get to do." Coupled with the landing bag situation, however, these malfunctions contributed to shortening the mission from a possible seven orbits to the original three.

Six minutes before retrofire, Glenn duly manoeuvred Friendship 7 into a 14-degree, nose-up attitude. At 2:20 pm, the first retrorocket fired, causing a dramatic braking effect on the capsule and making him feel momentarily as if he was flying backwards, towards Hawaii. The second and third retrorocket firings came at five-second intervals, slowing the capsule sufficiently to drop it out of orbit. Once again, Schirra repeated: "Keep your retro pack on until you pass Texas". The tension at Mission Control was palpable. "I looked around the room," wrote Gene Kranz, "and saw faces drained of blood. John Glenn's life was in peril."

Throughout the early stages of re-entry, Glenn and Schirra chattered like a pair of tourists exchanging travel notes, as Friendship 7 came within sight of El Centro and the Imperial Valley, followed by southern California's Salton Sea. Then, as he passed over Corpus Christi, Glenn was told again, with evident urgency, to "leave the retro package on through the entire re-entry". This meant that he would have to override the 0.05 G switch – which sensed atmospheric resistance and started the capsule's re-entry program – and manually retract the periscope. Glancing at the on-board clock, which had ticked off four hours, 38 minutes and 47 seconds since launch, his suspicions resurfaced that the reason for keeping the retro package on throughout re-entry was because the heat shield was loose.

He would later admit to irritation at being kept, officially, in the dark about the potential disaster looming ahead. Information about his spacecraft was his lifeblood, he wrote in his post-flight report, and it was knowledge, not absence of knowledge, which informed each of his decisions. Despite asking for Mission Control's reasons for wanting the package kept in place, he was told only that the decision was "the judgement of Cape Flight". In Florida, Al Shepard gave Glenn as much information as he had available. "We're not sure whether or not your landing bag has deployed," he said. "We feel it is possible to re-enter with the retro package on. We see no difficulty at this time with this type of re-entry."

Although the package and its three metal straps would eventually burn up as Friendship 7 plunged deeper into the atmosphere, Glenn assumed that, by keeping it on, the capsule might be protected for long enough until the thickening air could firmly hold the heat shield against Friendship 7's base. Max Faget, the father of the Mercury spacecraft, agreed with the plan of using the retrorocket package to hold the heat shield in place, but admitted that it posed its own risks: any unused solid fuel could explode when re-entry temperatures grew too hot, destroying the capsule. Flight Director Chris Kraft chimed in with additional concern that the attached retrorocket package could cause Friendship 7 to tumble, incinerating both it and

Glenn. Kraft, although eventually persuaded by Faget and Walt Williams, strongly felt that the 'cure' could be worse than the 'disease'.

As the main phase of re-entry got underway, Glenn switched Friendship 7 from manual to fly-by-wire control, placing the capsule in a slow spin to hold it onto its correct flight path through the atmosphere. A little more than a metre behind his back, the base of the spacecraft steadily heated to a maximum of 5,200°C. Its ionised envelope of heat also blacked out communications, as Al Shepard's voice in Glenn's headset faded to nothing. A thud reverberated through Friendship 7 as the retrorocket package's three metal straps melted, a fragment of which "fell against the window, clung for a moment and burned away". Glenn would recount years later that he expected, as each second passed, to feel the heat on his back and along his spine, but continued working his procedures, as flaming bits of – something – streamed past the window. He had no idea if they were pieces of the retrorocket package or, indeed, of the heat shield.

He was preoccupied for much of the re-entry with damping out the capsule's oscillations with the hand controller, but had the brief opportunity to glance through the window and see the sky turn to a bright orange, which he described to the automatic tape recorder as "a real fireball outside". Every few seconds throughout the communications blackout, he attempted to contact Shepard, all the while working to keep Friendship 7 on the straight and narrow. At length, and by now through the worst of the re-entry heating, he heard the crackle of Shepard's voice over his headset. The relief on the ground was audible.

Descending at subsonic speeds, by the time he reached an altitude of 13.7 km, Glenn felt the capsule rocking wildly and, as he looked upwards, could see "the twisting corkscrew contrail of my path". Eight and a half kilometres above the Atlantic, the stabilising drogue chute deployed automatically, followed by the main canopy a few seconds later. Shortly before splashdown, Glenn flipped the landing bag deployment switch. As expected, it lit up. Green. Subsequent investigation would discover that the rotary switch to be actuated by the heat shield deployment had a loose stem, which caused the electrical contact to break when the stem was moved up and down. This was believed to account for the false landing bag deployment signal. However, thinking back over the decision to keep the retro package attached throughout re-entry, Chris Kraft would resolve never to agree to such a dangerous exercise again.

Glenn splashed down with what he later described as "a good solid thump" at 2:43 pm. His landing co-ordinates were later given as 21 degrees 20 minutes North and 68 degrees 40 minutes West, some 320 km north-west of Puerto Rico. He was 60 km off-target, a discrepancy caused by retrofire calculations which had not taken into account Friendship 7's weight loss in consumables. Seventeen minutes later, the destroyer Noa, codenamed 'Steelhead', drew alongside the capsule, followed, shortly afterwards, by helicopters from the Randolph, which had recovered Gus Grissom a few months earlier.

"I could hear gurgling sounds almost immediately," Glenn said of his first few seconds back on Earth. "After it listed over to the right and then to the left, the capsule righted itself and I could find no traces of any leaks." He released his straps and shoulder harness, removed his helmet and put up his neck dam. He was sweating profusely, as physicians would later determine, and despite the open snorkels in Friendship 7's hull, the humid air offered little respite. When the Noa arrived, he glanced through the window – coated with a smoky film from re-entry – and saw a deck full of sailors, so high in number that he asked the destroyer's captain if anyone was actually running the ship! Within minutes, Friendship 7 had been winched aboard and, after obtaining clearance from the bridge, Glenn detonated the hatch. He received two skinned knuckles, through his pressure suit gloves, as the plunger snapped back. (It would be his only injury and, later, the realisation that manually activating the plunger caused such an injury would work in Gus Grissom's favour.)

According to physicians Robert Mulin and Gene McIver, Glenn was clearly fatigued, sweating and dehydrated; his only water intake had been from the apple sauce pouch, although his urine collector was full. After drinking water and showering, he became more talkative. He debriefed into a tape recorder aboard the Noa, before being flown to the Randolph for X-rays and an electrocardiogram and, later that evening, to Grand Turk Island for a welcoming committee of astronauts, physicians and NASA officials. When asked by psychiatrist George Ruff if there had been any unusual activity during his mission, Glenn replied "No ... just a normal day in space!"

EUPHORIA, DISAPPOINTMENT ... AND A BOSS

The weeks after Friendship 7 would be filled with euphoria over John Glenn's achievement, tempered with disappointment over the fate of Deke Slayton, the man meant to follow him into orbit on the next Mercury-Atlas mission.

Relief at Glenn's safe return was evident on many faces, including that of lawyer Leo D'Orsey, who had personally endorsed a $100,000 cheque for Annie to cover her husband's life insurance in the event of a disaster. (Despite D'Orsey's attempts to purchase million-dollar policies, the astronauts were uninsurable.) Now, thankfully, it did not need to be cashed. "Boy, am I glad to see you!" D'Orsey told Glenn later. Elsewhere, Henri Landworth, former manager of Cocoa Beach's Starlight Motel and later of the Holiday Inn frequented by the astronauts, had made a 400 kg cake, shaped like Friendship 7. He even rigged up an air-conditioned truck to prevent it from spoiling. After Glenn sampled a slice, Landwirth told him that it had been made in time for the original January launch date and was a month old!

At Cape Canaveral, President Kennedy awarded the astronaut and Bob Gilruth with NASA's Distinguished Service Medal, then flew America's newest hero to Washington, DC, aboard Air Force One to address Congress and, later, enjoy a tickertape parade through the streets of New York. Whilst airborne, Kennedy introduced his four-year-old daughter Caroline to Glenn. Clearly accustomed to the flights of Ham and Enos, the little girl's disappointed reaction was: "But where's the monkey?"

Like Shepard, Glenn was impressed by the president's enthusiasm and passion for the space programme. "He believed ... that it was not just a scientific journey," the

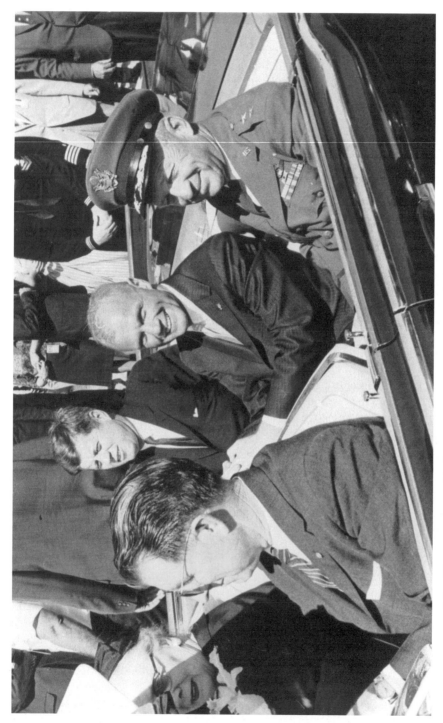

John Glenn experiences his first taste of public adoration, with President Kennedy at his side.

astronaut wrote in his memoir, "but a source of inspiration that could motivate Americans to pursue great achievements in all fields." Glenn had been accorded the rare privilege of addressing a joint session of Congress, which he did on 26 February, before the tickertape parade through New York. Four million people reportedly lined the streets to greet him and his speech before Congress led some columnists, including Arthur Knock and James Reston of the New York Times, to remark on his suitability for politics and his embodiment of "the noblest human qualities". One group of Republicans from Nevada even called upon him to run for the presidency. Glenn would admit at the time that his interests were apolitical, but a cultivated friendship with, among others, Attorney-General Bobby Kennedy would gradually direct him towards a career in the Senate.

His diplomatic abilities were also put to the test in May 1962, when he was invited, along with Vostok 2 cosmonaut Gherman Titov, to address the Committee on Space Research, part of the third International Space Symposium, in Washington, DC. Glenn described Titov as "cordial but forceful and thoroughly indoctrinated … charts and photographs had supplemented my presentation, but not his; he followed the Soviet line that disarmament would have to precede full sharing of information". Nonetheless, he hosted the cosmonaut during his visit to Washington, even debating the existence of God in front of journalists.

Titov's standpoint was that he did not see anyone in space – offering clear "proof for the communist position" that such a deity did not exist – but Glenn responded that 'his' God was not so small that he expected to "run into Him a little bit above the atmosphere". Diplomacy was extended still further when the Glenns invited the Titovs to a barbecue at their house; after initially refusing, the cosmonaut's delegation changed its mind at the last moment. The Glenns, unprepared, were forced to ask their neighbours for spare steaks and send their police escorts to buy vegetables. Meanwhile, Al Shepard picked up the Titov delegation and bought the Glenns some time by taking a few wrong turns on his way to Arlington …

Not only was Glenn deemed valuable to the nation, but so too was Friendship 7, which, despite some discolouration, was in remarkable condition and today resides in the National Air and Space Museum in Washington, DC. In 1962, however, it undertook what was popularly nicknamed its 'fourth orbit': a global tour of 17 countries before eventually being installed in the Smithsonian, close to the Wright Brothers' original aircraft and Charles Lindbergh's 'Spirit of St Louis'.

As Glenn basked in his new-found fame, 38-year-old Donald Kent 'Deke' Slayton fought a losing battle to fly his own Mercury-Atlas mission, tentatively scheduled for April 1962. He had already picked a name for his capsule – 'Delta 7', the fourth letter in the Greek alphabet for the fourth manned Mercury mission – which he would describe in his autobiography as "a nice engineering term that described the change in velocity". Slayton's own velocity, both in spacecraft and high-performance aircraft, would decline markedly, thanks to a minor, yet persistent, heart condition, known as 'idiopathic atrial fibrillation'. This took the form of occasional irregularities of a muscle at the top of the heart, caused by unknown factors and extremely rare in highly-fit thirtysomethings like Slayton. It had first arisen during a centrifuge run at Johnsville in August 1959, when physicians noticed traces of sinus arrhythmia, which Bill Douglas later wrote "wasn't uncommon in healthy young men and ... the kind of thing that often went away with exertion". After the run, however, it was still present, prompting Douglas and his team to undertake a clinical electrocardiogram at the Philadelphia Navy Hospital.

They concluded that Slayton had a 'flutter' in his heartbeat, although, in 1959, the astronaut himself "had no idea how much of a problem it was". Further tests at the Air Force's School of Aviation Medicine in San Antonio, Texas, verified that the condition was of no consequence and should not influence Slayton's eligibility for a spaceflight. Douglas informed Bob Gilruth, who briefed NASA Headquarters on the issue late in 1959, as well as the Air Force Surgeon-General, who advised that no further action was necessary. The 'Slayton File', for a time, lay dormant.

The problem resurfaced a little over two years later, a week before Friendship 7's launch, when speculation arose that John Glenn had a heart murmur. Apparently, wrote Slayton, the call to Bill Douglas came from Air Force physician George Knauf, attached to NASA Headquarters, and had originated from "a source higher than the Department of Defense". Douglas denied that Glenn had a problem, but effectively opened a can of worms. Knauf asked next if Glenn's backup, Scott Carpenter, had a heart murmur: again, the response was negative. Then Douglas, to reinforce the point that the matter was of little consequence, revealed that Slayton had long been known to have a minor condition. He expected this to be the end of the matter. It wasn't.

Back in 1959, flight surgeon Larry Lamb had examined Slayton at Brooks Air Force Base in San Antonio and had become convinced the heart fibrillation should disqualify him from the selection process. "He hadn't said so in 1959," wrote Slayton, "but he said so now. I don't think it was anything personal – this was just his medical opinion." Lamb's judgement was very much a voice in the wilderness, but unfortunately he also happened to be Lyndon Johnson's cardiologist, and in the spring of 1962 began to question the astronaut's suitability for flight. Three weeks after Friendship 7, Jim Webb reopened Slayton's medical files and the astronaut and Douglas were summoned to the office of the surgeon-general of the Air Force in Washington, DC. A panel of military physicians signed him off as fit to fly, a decision endorsed by the Air Force's chief of staff, General Curtis LeMay.

For Webb, though, it was not enough. The Secretary of the Air Force, Eugene Zuckert, suggested that a civilian panel of physicians should also examine Slayton at NASA Headquarters. On 15 March, less than two months before launch, Slayton was poked and prodded and had his heart monitored by Proctor Harvey of Georgetown University, Thomas Mattingley of the Washington Hospital Center and Eugene Braunwell of the National Institutes of Health. In a turnaround which, in Slayton's own words, would leave him "devastated", Deputy Administrator Hugh Dryden entered the room and told him, point-blank, that he was off the flight. None of the physicians had found a specific medical reason to keep him off Delta 7, but their consensus was that if NASA had pilots 'without' his condition, one of them should fly the mission instead.

Years later, Slayton would refute theories that Lyndon Johnson's annoyance over the Annie Glenn incident had anything to do with the decision, but certainly felt it was a political move. "NASA knew it would have to publicly disclose my heart condition prior to my flight," he wrote. "There would be medical monitors at tracking stations all over the world who wouldn't know how to react otherwise. Everybody expected this to be a big deal. NASA would be opening itself up to a lot of medical second-guessing." Bill Douglas felt that problems could arise if Slayton started fibrillating on the pad – "do you scrub the launch or go ahead?" – but he, Bob Gilruth and Walt Williams had confidence that he was the best person to follow John Glenn. All three men were prepared, personally, to take the heat, but Jim Webb's fear that it could trigger adverse headlines for the agency drew a line in the sand. "It didn't matter that a whole lot of doctors thought I didn't have a problem," Slayton wrote of Webb's actions. "He was only going to listen to the few who did."

In Webb's mind, an Atlas abort could subject the astronaut to acceleration loads as high as 21 G and conjured the very real possibility that Slayton, dehydrated and perhaps fibrillating, could die as a result. The impact on NASA, on President Kennedy's promise to land a man on the Moon before the end of the decade and on the ongoing contest with the Soviets could be profound.

The next day, 16 March, the now-grounded, and furious, astronaut was forced to sit through a lengthy press conference, in which the minutiae of the case were examined. Hugh Dryden remarked that, despite the decision, Slayton might remain eligible for future flights. One journalist asked if the problem had been caused by stress, to which Slayton responded no, and further that he did not even know about it until he had been hooked up to an electrocardiogram in 1959. The most stressful part of the space business, he explained caustically and with more than a hint of sarcasm, was "the press conference after the flight". Bill Douglas' own departure from NASA within days of the announcement, to return to the Air Force, was leapt upon by some journalists as 'evidence' of his bitterness over Slayton's treatment. In truth, Douglas was already at the end of a three-year detachment to NASA and his transfer had been in the works since mid-1961.

Despite his grounding, Slayton did not give up on flying. "I made some changes in my lifestyle," he wrote, "gave up drinking, started working out more regularly – quit doing everything that was fun, I guess!" Thanks to Bill Douglas, he also secured an examination by Dwight Eisenhower's cardiologist, Paul Dudley White, in June 1962. White advised him that two-thirds of people with his condition would die young, whilst the remainder would probably never know they had it and might never be affected. The verdict: "Young man, you're going to live a long time." (Slayton lived to be 69.) However, White's report, which highlighted that Slayton did not appear to have a problem, also advised that if astronauts were present without the condition, it would be preferable to assign them in his stead. As it became clear that he would not draw any of the remaining Mercury-Atlas flights, Slayton turned his gaze to the two-man Gemini project, only to be told by Bob Gilruth that his condition would make him a "hard sell" to senior management. Shortly thereafter, the Air Force decreed that he no longer met the qualifications for a Class I pilot's licence – effectively, he could no longer fly solo – and, at the end of November 1963, Slayton tendered his resignation from the service.

Although he would eventually get his ride into space, it would not come until July

1975, when he was 51 years old. A lesser man might have thrown in the towel and departed NASA for pastures new, but not Slayton. With no guarantee that he would ever fly, he decided to stay and in the summer of 1962, as the agency prepared to expand its astronaut corps by picking nine new pilots, he was appointed as Co-ordinator of Astronaut Activities. Initial plans to bring in a manager from the outside to oversee the corps were quashed by the astronauts themselves. "What we wanted the least," wrote Wally Schirra, "was somebody who would outrank us and issue orders in a military way. We wanted someone who knew us, who trained with us. Deke was the one and only choice." During the next few years, as America pushed for the Moon, Slayton, though non-flying, would be all-powerful within the astronaut corps, deciding the career paths of the men who would someday walk on the lunar surface ... and those who would not.

'BLACK-SHOE' CARPENTER

In theory, with Slayton's removal from Mercury-Atlas 7, the new pilot should have been his backup, Wally Schirra. However, since Scott Carpenter had only recently been primed as John Glenn's reserve and was considered the best-prepared, his name was announced instead at the 16 March press conference. "I figured," wrote Walt Williams years later, "that MA-7 was likely to be more a repeat of John's flight than anything groundbreaking, so why not give it to Scott, since he had already trained for something pretty similar. We were thinking about a seven-orbit flight later in the year, and that would be perfect for Wally." Carpenter had trained since October 1961 in Glenn's shadow and had accrued nearly 80 hours of 'pre-flight checkout and training' time, considerably more than Schirra or even Slayton had accumulated during their preparations for MA-7.

Yet Schirra would learn of the assignment during an impromptu gathering at the Carpenters' home. Moreover, what should have been the most exhilarating moment of his career turned into an ordeal for Carpenter and his wife, Rene. Slayton's anger at having lost Delta 7, coupled with Schirra's annoyance at having been dropped in favour of Glenn's backup, led Carpenter to spend more time apologising than training. One evening, he got home and declared to Rene: "Damn it! I'm tired of apologising. This is my flight now." Although he felt no bitterness towards Carpenter, Schirra would comment in his autobiography that he "felt the system was rotten". As far as Schirra was concerned, although Carpenter had been through test pilot school, he was a multi-engine aviator and had been a communications officer aboard an aircraft carrier ... not a fighter pilot. In Schirra's mind, Carpenter represented 'black-shoe Navy', a seagoing fleet officer, and despite his impressive flying credentials, was not truly a 'brown-shoe' naval aviator.

"To make it worse," Schirra wrote, "I was designated Scott's backup! I did my best and worked my tail off on Scott's mission. I don't think anyone knew how angry I was." Even Schirra, though, had to admit that his disappointment was nothing compared to the devastating news Deke Slayton had just received. The man who would effectively replace them both had, in the December 1960 peer vote, actually been John Glenn's personal choice for the first American in space. Malcolm Scott Carpenter had been born in Boulder, Colorado, on 1 May 1925, the son of chemist Dr Marion Scott Carpenter and Florence 'Toye' Noxon. Both parents had met as undergraduates at the University of Colorado, but separated soon after their son's birth and divorced in 1945. His mother was hospitalised with tuberculosis for several years in Carpenter's infancy and the boy – nicknamed 'Buddy' – attended school in Boulder, graduating in 1943 and entering the Navy's V-12a wartime officer flight training programme at the University of Colorado.

A year later, he moved to St Mary's Pre-flight School in Moraga, California, undergoing six months of training, followed by another four months at Ottumwa in Iowa. On his personal website, Carpenter would write that, despite his relief when the Second World War ended, as a fledgling naval aviator, he "was deeply dejected that I had not taken part in what I assumed was the greatest aeronautical contest of the century". In fact, he and his classmates had logged barely a few hours in the Stearman N2S 'Yellow Peril' training aircraft when atomic bombs were dropped on Hiroshima and Nagasaki; three months later, he was demobilised. He did, however, win a regimental wrestling contest whilst a member of V-12a.

At war's end, Carpenter enrolled in the University of Colorado to read mechanical engineering, with an aeronautical option. "CU did not then offer a degree in aeronautical engineering," he wrote on his website, www.scottcarpenter.com. A near-fatal car accident in September 1946 severely disrupted his studies, but he returned to university early the following year. However, he missed his final examination in thermodynamics, leaving him one requirement shy of a complete undergraduate degree. He would make up for this on his one and only spaceflight. Carpenter married Rene Price in September 1948 and would father five children – four of whom survived – from the first of his four marriages.

He then joined the Navy, receiving flight training at Pensacola, Florida, and Corpus Christi, Texas, before working in the Fleet Airborne Electronics Training School in San Diego, California, volunteering for transitional training for Lockheed's P2V Neptune patrol bomber. His decision to fly patrol planes was, he wrote in his autobiography, a difficult one. "His boyhood dream, held all through high school and beyond was to be a fighter pilot," Carpenter and Stoever wrote. "But he was now ... a husband and a father ... His ego demanded he be a fighter pilot, but he remembered as a boy how he had hated being fatherless." The decision, he wrote, continues to haunt him. Still, in November 1951, he was assigned to Patrol Squadron Six at Barbers Point in Hawaii and, throughout the Korean conflict, engaged in anti-submarine patrols, shipping surveillance and aerial mining activities in the Yellow Sea, South China Sea and Formosa Straits.

After the war ended, Carpenter entered the Navy's Test Pilot School at Pax River, graduating in the top third of his class, and subsequently conducted flight testing of the A-3D Skywarrior strategic bomber and the F-11F and F-9F fighters. He also tested numerous other naval aircraft – single- and multi-engine and propellor-driven fighters, attack planes, patrol bombers and seaplanes – before attending Naval General Line School at Monterey, California, in 1957 and the Naval Air Intelligence School in Washington, DC, the following year. His next assignment, in August 1958, placed him on the Hornet anti-submarine aircraft carrier, and he was serving as an air intelligence officer when he received cryptic orders from the Pentagon to report to Washington for a classified briefing. It was whilst on their way to the airport, after discussing the endless possibilities, that Rene, reading her copy of Time magazine, spotted a report about Project Mercury. "Their excitement mounted," Carpenter wrote in his autobiography, "as they went through a list that described, well, Lieutenant M. Scott Carpenter."

Whilst still assigned to the Hornet, he was invited to attend the second stage of testing, but met with the resistance of his skipper, Captain Marshall White, who emphatically declared that the young lieutenant was about to embark on an important training cruise. Carpenter's entreaties fell on deaf ears, it seemed, and he was obliged to call NASA's Dr Allen Gamble, who personally contacted the Chief of Naval Operations, Admiral Arleigh Burke. The admiral, betwixt some "real sailor language", agreed to deal with the matter. He promptly spoke to White and, wrote Carpenter in his autobiography, the skipper "went on that training cruise without his air intelligence officer".

Interestingly, as the selection process for Project Mercury got underway, Carpenter and Deke Slayton were members of the same group at the punishing Lovelace Clinic in Albuquerque. Slayton would later write that he had been particularly impressed to see Carpenter, during one test, blow into a tube of mercury "for about three minutes, twice as long as anybody else!" Carpenter's remarkable endurance was also demonstrated after selection, when, during a centrifuge run at Johnsville, he devised a breathing technique, akin to explosive grunting, which allowed him to withstand 18 G with few ill-effects.

SCIENCE FLIGHT

When Carpenter was named to pilot MA-7 in March 1962, he decided on the moniker 'Aurora 7' for his capsule. "I think of Project Mercury and the open manner in which we are conducting it for the benefit of all as a light in the sky," he wrote later. "Aurora also means 'dawn' – and, in this case, the dawn of a new age. The Seven, of course, stands for the original seven astronauts." By now, the suffix had become commonplace and, coincidentally, 'Aurora' also happened to be the name of one of two streets bordering Carpenter's boyhood home in Boulder. The capsule which bore this name was Spacecraft No. 18 off the McDonnell production line and arrived at Cape Canaveral on 15 November 1961, followed by its Atlas booster just two weeks after John Glenn's mission.

Owing to the 'experimental' nature of Friendship 7 – "for all its first-time danger," wrote Carpenter and Stoever, "MA-6 had been designed to answer the simple question: Could it be done?" – the next mission was intended to encompass far more engineering and scientific tasks, including observations, photography and extensive manoeuvres. Deke Slayton, when the flight was still his to fly, had expressed consternation at the sheer volume of tests and experiments. "Everybody and his brother came out of the woodwork," Slayton wrote. "One guy wanted me to

Scott Carpenter prepares for his flight.

release a balloon to measure air drag. Another guy had some ground observations I was supposed to make. One damn thing after another. I had my hands full trying to resist it." From 16 March 1962, with barely ten weeks to go, Scott Carpenter found that the scientific demands of MA-7 were his to handle: they included combined yaw-roll manoeuvres to study orbital sunrises, using terrestrial landmarks and stars for navigational reference and flying in an inverted attitude to determine the effect of 'Earth-up/sky-down' orientation on the pilot's abilities.

Furthermore, Homer Newell, head of NASA's Office of Space Sciences, had established a formal panel to outline experiments and objectives for future flights. Astronomer Jocelyn Gill of NASA Headquarters was appointed to run this 'Ad-Hoc Committee on Scientific Tasks and Training for Man-in-Space' and her enthusiasm for Carpenter, who had an impressive background in navigational astronomy following his experience aboard the P2V with Patrol Squadron Six, was evident. Gill's committee considered a number of possible experiments and Kenny Kleinknecht, now in charge of the Mercury Projects Office, appointed Lewis Fisher to lead a newly-established Mercury Scientific Experiments Panel. With the Fisher group overseeing the Gill committee "from an engineering feasibility standpoint" and on the basis of their "scientific value, relative priority and suitability for orbital flight", a consensus was reached on 24 April to propose five major experiments for Aurora 7. In his autobiography, Carpenter wrote that he "liked and admired scientists" and "liked being a champion of embattled groups with high purpose ... And in 1962, the scientists at NASA were already a beleaguered group".

Deke Slayton's perspective had always been that scientific tasks should be kept to a minimum, particularly in light of John Glenn's problems with the flight controls. "Scott had a different perspective," Slayton wrote. "He was always at home with the doctors and scientists – I think he was genuinely curious about the things that interested them. But it bit him in the ass during his flight." Within NASA, added Carpenter, scientific experiments were viewed "with a mixture of suspicion and ridicule, the butt of jokes when the reporters weren't around" and the astronaut found himself at loggerheads with Flight Director Chris Kraft. Although their relationship would never turn antagonistic, many observers have commented over the years that Carpenter's performance during Aurora 7 would lead Kraft to declare openly that he would never fly in space again.

The five experiments recommended by the Fisher panel required Carpenter to observe, measure, analyse and photograph (1) a tethered, multi-coloured balloon, (2) the behaviour of liquids inside a sealed flask, (3) different visual phenomena – both celestial and terrestrial – using a modified photometer, built by psychologist Bob Voas and nicknamed 'The Voasmeter', (4) weather patterns and land masses and (5) the 'airglow' layer of the upper atmosphere. Of these, the balloon was the most visible. Measuring 76.2 cm in diameter and weighing 900 g, it was an inflatable Mylar sphere, divided into five equal sections painted an uncoloured aluminium, Day-Glo yellow, Day-Glo orange, white and a phosphorescent coating which appeared 'white' by day and 'blue' by night. The intention was for it to be inflated with a small nitrogen bottle immediately after release from Aurora 7's antenna canister at the end of the first orbit. Carpenter would then observe and photograph the effects of space and sunlight on the different colours at different times, perhaps aiding in the design of future lunar spacecraft and their docking systems, which would require exceptionally good visibility. To better understand atmospheric density at orbital altitudes, it would be fitted with a 'tensiometer' – a strain gauge – to measure tension on the 30 m tether. Carpenter would add his own observations, monitoring the amount of atmospheric drag and turbulence in the balloon's slipstream by carefully watching its oscillations and general behaviour as it trailed behind Aurora 7. During launch, the balloon would be folded, packaged and housed with its nitrogen bottle in the antenna canister on the spacecraft's nose.

Meanwhile, the fluid flask – just behind Carpenter's right ear in the cabin – was designed to build on theoretical and experimental work at NASA's Lewis Research Center in Cleveland, Ohio, where it was already known that liquids behave differently under microgravity conditions. Its inclusion was intended to provide preliminary answers to questions of how fuels and other spacecraft fluids could be transferred from one storage tank to another during the long-duration Gemini and Moon-bound Apollo missions. Terrestrial aircraft flights and drop-tower tests conducted at Holloman Air Force Base and the service's School of Aviation Medicine in Texas had been too short. One of the leading suggestions was that surface tension could be used to pump fluids, using capillary action, between tanks. The flask on Carpenter's mission contained a small capillary, or meniscus, tube and by observing the behaviour of the fluid it would be possible to determine how effectively surface tension translated into pumping action, simply by measuring how far the liquid was drawn up into the tube. It was 20-per cent-filled with some 60 ml of a mixture of distilled water, green dye, an aerosol solution and silicone.

Observations of the constellations were also planned and considered important for future navigational purposes. Moreover, the Massachusetts Institute of Technology had requested photographs of the 'daylight' horizon through blue and red filters to define more precisely the Earth's limb as seen from above the atmosphere. John O'Keefe of NASA's Goddard Space Flight Center in Greenbelt, Maryland, sought a distance measurement of the airglow above the atmosphere, together with its angular width and a description of its characteristics. For this study, Carpenter would use the Voasmeter. (It was lucky that Voas lent his name to the device, for its formal title was the 'extinctospectrophotopolariscopeoculogyrogra-vokinetometer'; a name requiring 20 syllables!) The astronaut would also have a German-made 35 mm SLR camera, called a 'robot recorder', capable of exposing two frames per second from a 250-frame magazine, which would provide images of the daylit horizon, considered valuable for the design of Apollo's navigation system. Another Goddard scientist, Paul Lowman, requested images of North America and Africa.

In addition to analysing events beyond his spacecraft, Carpenter was also charged with monitoring himself: by performing numerous exercises at specified intervals, followed by blood pressure readings. Aurora 7 would be the most science-heavy Mercury mission so far and the numerous problems encountered by Carpenter would be at least partly attributed to an overloaded work schedule.

RISE AND FALL OF THE AURORA

Preparations for the MA-7 mission had begun long before Deke Slayton's grounding. In mid-November 1961, Spacecraft No. 18 arrived in Hangar S at Cape Canaveral, followed, shortly after the Friendship 7 flight, by its Atlas. Checkout problems with capsule and rocket delayed an original mid-April launch to mid-May. The landing bag switches which had caused problems for John Glenn were rewired so that both had to be closed in order to activate a 'deployed' signal. Engineers also determined that the cause of the flight control system glitches lay in the fuel line filters, which were replaced with platinum screens and new, stainless steel fuel lines. Finally, on 28 April 1962, Aurora 7 was attached to its rocket at Pad 14. A simulated flight proved satisfactory, although decisions were made to install an extra barostat in the capsule's parachute circuitry, fit temperature survey instrumentation and replace flight-control canisters in the launch vehicle. Additional delays were caused by Atlantic Fleet tactical exercises which required participation by the recovery ships and aircraft for several weeks. Other concerns arose following the failure of an Atlas-F missile a few weeks earlier. However, the different engine start-up sequences of the Atlas-F and Carpenter's own Atlas-D eliminated any doubts over its reliability.

The installation of the extra barostat postponed the mid-May launch attempt and, on the 19th, an effort to get Aurora 7 into space proved fruitless when irregularities were detected in a temperature control device on the Atlas' flight control system heater. Five days later, however, Carpenter was awakened at 1:15 am and proceeded through the usual pre-flight breakfast ritual, was examined by Bill Douglas, suited-up by Joe Schmitt's team and departed for Pad 14. He was aboard the capsule by 5:00 am, to enjoy one of the smoothest countdowns in Project Mercury, with only persistent ground fog and cloud and camera-coverage issues complicating matters. During a 45-minute delay past the original 7:00 am launch time, Carpenter sipped cold tea from his squeeze bottle and chatted to his family over the radio. His wife Rene and their four children represented the first astronaut family to journey to the Cape and watch the launch. To avoid media attention, a neighbour provided a private flight to Florida and a car, which Rene drove to the astronauts' hideaway – nicknamed the Life House – near Pad 14, wearing huge sunglasses, a kerchief over her conspicuous blonde coif and her two daughters hidden under a blanket. The media, anticipating the arrival of a blonde mother of four, instead saw only a well-disguised mother of two ...

Sixteen seconds after 7:45 am, the Atlas' engines ignited, prompting all four Carpenter children to abandon the television set and rush onto the beach to watch their father hurtle spaceward. Elsewhere, at the Cape and across the nation, an estimated 40 million viewers watched as America launched its second man into orbit. Carpenter himself would later describe "surprisingly little vibration, although the engines made a big racket" and the swaying of the rocket during the early stages of ascent was definitely noticeable. In his autobiography, he would express surprise, after so many years of flying aircraft and 'levelling-out' after an initial climb, to see the capsule's altimeter climbing continuously as the Atlas shot straight up.

Already, however, the first glitches of what would become a troubled mission

Aurora 7, atop its Atlas, is readied for launch.

were rearing their heads. Aurora 7's pitch horizon scanner, responsible for monitoring the horizon to maintain the pitch attitude of the spacecraft, immediately began feeding incorrect data into the automatic control system. When this 'wrong' information was analysed by the autopilot, it responded, as designed, by firing the pitch thruster to correct the perceived error – in effect, wasting precious fuel. Forty seconds after the separation of the LES tower, the scanner was 18 degrees in error, indicating a plus-17-degree nose-up attitude, whilst the Atlas' gyros recorded an actual pitch of minus 0.5 degrees. It had reached 20 degrees in error by the time Carpenter achieved orbit. As the flight wore on, the error persisted and, wrote Carpenter and Stoever, produced "near-calamitous effects" as Aurora 7 neared re-entry.

Sustainer engine cutoff came as a gentle drop in acceleration, with a pair of bangs providing cues that explosive bolts had fired and posigrade rockets had pushed Aurora 7 away from the spent Atlas. The astronaut reported, with clear elation in his voice, "I am weightless! Starting the fly-by-wire turnaround." Deciding not to rely on the automatic controls, his use of fly-by-wire smartly turned the capsule around at a fuel expense of just 725 g, as compared to 2.3 kg on Friendship 7. Carpenter would later describe that he felt no angular motion during the turnaround and, in fact, his instruments provided the only evidence that a manoeuvre was being executed. No sensation of speed was apparent, although he was travelling at 28,240 km/h and was soon presented with his first "arresting" view of Earth. Carpenter watched the Atlas' sustainer tumble into the distance, trailing a stream of ice crystals two or three times longer than the rocket stage itself. As he flew high above the Canaries, he could still see its silvery bulk, tagging along with Aurora 7.

Five and a half minutes into the mission, Capcom Gus Grissom radioed the good news: Carpenter's orbit was good enough for seven circuits of the globe. The astronaut got to work. "With the completion of the turnaround manoeuvre," he wrote, "I pitched the capsule nose down, 34 degrees, to retro attitude, and reported what to me was an astounding sight. From Earth orbit altitude, I had the Moon in the centre of my window, a spent booster tumbling slowly away and looming beneath me the African continent." He pulled his flight plan index cards from beneath Aurora 7's instrument panel and Velcroed them into place; these would provide him with timing cues for communications with ground stations, when and for how long to use control systems, when to begin and end manoeuvres, what observations to make and when to perform experiments. Minute-by-minute, they mapped out his entire flight. First, he took out the camera, adapted with strips of Velcro – "the great zero-gravity tamer" – to begin photographing Sun-glint on the Atlas sustainer. Next came filters to measure the frequency of light emissions from Earth's atmospheric airglow, followed by star navigation cards, worldwide orbital and weather charts and bags of food.

Orienting the capsule such that the sustainer was dead-centre in his window, Carpenter reported to the Canaries ground station that he could see "west of your station, many whirls and vortices of cloud patterns". His view of the heavens was somewhat less clear, with the stars too dim to make out against the black sky, although the Moon and terrestrial weather patterns were obvious. Then, 16 minutes after launch, the astronaut noted that his spacecraft's actual attitude did not seem to be in agreement with what the instruments were telling him. Aware of problems that John Glenn had experienced with his gyro reference system, and cognisant of the fact that he had other work to do, Carpenter dismissed it.

"A thorough check, early in the flight, could have identified the [pitch horizon scanner] malfunction," he later wrote. "Ground control could have insisted on it, when the first anomalous readings were reported. Such a check would have required anywhere from two to six minutes of intense and continuous attention on the part of the pilot. A simple enough matter, but a prodigious block of time in a science flight – and in fact the very reason [such] checks weren't included in the flight plan." With so much to do, it would not be until his second orbit that Carpenter would again report problems with the autopilot.

Passing over the ground station at Kano, in north-central Nigeria, Carpenter successfully photographed the Sun for physicists at the Massachusetts Institute of Technology, then, over the Indian Ocean, acquired initial readings for John O'Keefe's airglow study. However, conditions aboard the spacecraft were becoming uncomfortable, as the cabin temperature increased. Years later, in 'The Right Stuff', Tom Wolfe would describe Aurora 7 as "a picnic" and that its astronaut had "a grand time"; Carpenter, however, countered that his lengthy training as Glenn's backup and shorter-than-normal preparation for his own mission made it anything but a walk in the park. "To the extent that training creates certain comfort levels with high-performance duties like spaceflight," he wrote, "then, yes, I was prepared for, and at times may have even enjoyed, some of my duties aboard Aurora 7. But I was deadly earnest about the success of the mission, intent on observing as much as humanly possible, and committed to conducting all the experiments entrusted to me. I made strenuous efforts to adhere to a very crowded flight plan."

Admirably, for the first 90 minutes of his mission, Carpenter focused on his Earth-observation tasks, photographing rapid changes in light levels as the spacecraft crossed the 'terminator' – the dividing line between the darkened and sunlit sides of Earth – and expressing sheer astonishment as the Sun disappeared below the horizon. "It's now nearly dark," he remarked in the flight transcript, "and I can't believe where I am!" Passing over Muchea in Australia, Carpenter discussed with Capcom Deke Slayton possible ways of establishing attitude control on the dark side of Earth with no moonlight and relayed what reliable visual references he had through the window or the periscope. Pitch attitude was not a problem, thanks to scribe reference marks on Aurora 7's window, but accomplishing the correct yaw angle was much more difficult and time-consuming.

"At night, when geographic features are less visible, you can establish a zero-yaw attitude by using the star navigation charts, a simplified form of a slide rule," Carpenter wrote. "The charts show exactly what star should be in the centre of the window at any point in the orbit – by keeping that star at the very centre of your window, you know you're maintaining zero yaw. But there are troubles even here, for the pilot requires good 'dark adaption' to see the stars and dark adaption was difficult during the early flights because of the many light leaks in the cabin." Among the most annoying of these leaks were Aurora 7's instrument panel lights and, particularly, the glowing rim around the spacecraft's on-board clock. Carpenter told Slayton that his pressure suit's temperature was higher than normal, before crossing over the Woomera tracking station, with the intention of observing four flares of a combined one million candlepower, fired from the Great Victoria Desert as a visibility check. To see the flares, Carpenter was required to undertake "a whopping plus-80 degrees yaw manoeuvre and a pitch attitude of minus-80 degrees", but, unfortunately, cloud cover was too dense. "No joy on your flares," he told Woomera.

Another aspect of the mission about which no joy was forthcoming was the multi-coloured balloon, which he released an hour and 38 minutes after launch. For a few seconds, the expected 'confetti spray' signalled a successful deployment, but it soon became clear that the balloon had not inflated properly. Due to a ruptured seam in its skin, it deployed to a third of its expected size and only two of its five colours – Day-Glo orange and dull aluminium – were visible. Two small, ear-like appendages, each about 15–20 cm long and described as "sausages", emerged on the edges of the partially-inflated sphere. Its movements turned out to be erratic and, although Carpenter succeeded in acquiring a few drag-resistance measurements, the 30 m tether quickly wrapped itself around Aurora 7's nose. Consequently, the aerodynamic data was of limited use. Carpenter attempted to release the balloon during his third orbit, whilst flying over Cape Canaveral, but it remained close to the spacecraft. There it stayed until retrofire and eventually burned up during re-entry.

By this time, Mission Control was keeping a close eye on Aurora 7's fuel usage, which, by two hours into the flight, was at the 69-per cent capacity for both its manual and automatic supplies. As Carpenter passed over Nigeria early in his second orbital pass, the manual supply had dropped still further to just 51 per cent. He told the Kano capcom that he felt he had expended additional fuel trying to orient the spacecraft whilst on the dark side and blamed "conflicting requirements of the flight plan". During each fly-by-wire manoeuvre, very slight movements of the control stick would activate the small thrusters, whereas bigger movements would initiate larger thrusters. For every accidental flick of his wrist, Carpenter could activate the larger thrusters and would then have to correct them, thus wasting valuable fuel. "The design problem with the three-axis control stick, as of May 1962," he wrote later, "meant the pilot had no way of disabling, or locking-out, these high-power thrusters." Subsequent Mercury flights would employ an on-off switch for just that purpose.

The still-unknown problem with the pitch horizon scanner, though, remained. A little over two hours into the mission, the Zanzibar capcom informed Carpenter that, according to the flight plan, he should now be transitioning Aurora 7 from automatic to fly-by-wire controls. The astronaut opposed this, preferring to remain in automatic mode, which was supposedly more economical with fuel consumption. Unfortunately, this proved not to be the case, because the malfunctioning pitch horizon scanner was feeding incorrect information into the autopilot, which, in turn, was guzzling far more fuel than it should. A few minutes later, in communication with a tracking ship in the Indian Ocean, Carpenter reported difficulties with the automatic control mode and switched to fly-by-wire in an effort to diagnose the problem.

Although a malfunctioning automated navigational system in orbit was tolerable, its satisfactory performance was essential for retrofire to ensure that the spacecraft was properly aligned along the pitch and yaw axes to begin its fiery descent through the atmosphere. "Pitch attitude ... must be 34 degrees, nose-down," wrote Carpenter. "Yaw, the left-right attitude, must be steady at zero degrees, or pointing directly back along flight path. The [autopilot] performs this manoeuvre automatically, and better than any pilot, when the on-board navigational instruments are working properly." Sadly, on Aurora 7, they were not. The astronaut could align his capsule manually, but with difficulty: by either pointing the nose in a direction that he thought was a zero-degree yaw angle and then watching the terrain pass beneath him (considered near-impossible over featureless terrain or ocean) or use a certain geographical feature or cloud pattern for reference.

"Manual control of the spacecraft yaw attitude using external references," he wrote later, "has proven to be more difficult and time-consuming than pitch and roll alignment, particularly as external lighting diminishes ... Ground terrain drift provided the best daylight reference in yaw. However, a terrestrial reference at night was useful in controlling yaw attitudes only when sufficiently illuminated by moonlight. In the absence of moonlight, the pilot reported that the only satisfactory yaw reference was a known star complex nearer the orbital plane."

Carpenter had other worries, too. His cabin and pressure suit temperatures were climbing to uncomfortable levels; the former, in fact, peaked at 42°C during his third orbit, while the latter rose to 23.3°C and a "miserable" 71 degrees of humidity. The capcom's query as to whether the astronaut felt comfortable, having fiddled with his suit's controls, was greeted with a non-committal "I don't know". After the flight, the high cabin temperatures were attributed to the difficulty of achieving high airflow rates and good circulation, as well as the vulnerability of the spacecraft's heat exchanger to freezing blockage when high rates of water flow were used. Meanwhile, Carpenter was also required to take frequent blood pressure readings, pop xylose pills for post-flight urinalysis and monitor each of his scientific experiments. He did also manage to eat solid food during the mission: the Pillsbury Company had prepared chocolate, figs and dates with high-protein cereals, whilst Nestlé provided some 'bonbons', composed of orange peel with almonds, high-protein cereals with almonds and cereals with raisins. These had been processed into particles a couple of centimetres square and were coated with edible glazes. The astronaut sampled them, but found them to crumble badly, leaving pieces floating around the cabin.

He succeeded in shooting photographs of the Sun for the Massachusetts researchers, acquired photometric readings on the star Phecda (more formally, Gamma Ursae Majoris) and his work on the liquid-behaviour experiment showed that capillary action could indeed pump fluids in space. However, he also reported worrying decreases in his fuel, which had hit just 45 per cent in the case of the manual supply. Indeed, Flight Director Chris Kraft, writing in his post-flight report on Aurora 7, would comment that the mission had run smoothly thus far, with the exception of the "over-expenditure of hydrogen peroxide fuel". At this point, Kraft felt that sufficient fuel remained to achieve the retrofire attitude, hold it steady and re-enter the atmosphere with either the automatic or manual control systems.

In his autobiography, Carpenter suggested that Kraft's frustration with him began to emerge at this point, the flight director having apparently concluded that the astronaut had deliberately ignored a request to conduct an attitude check over Hawaii. Kraft also voiced serious concerns to California capcom Al Shepard that Carpenter was to tightly curb his automatic fuel use prior to retrofire. By this time, Aurora 7 was restricted to long periods of drifting flight, with both automatic and manual fuel quantities now dropping to less than 50 per cent. Years later, Gene Kranz would blame ground controllers for waiting too long before addressing the fuel status and felt that they should have been more dogged and forceful in getting on with the checklists. "A thoroughgoing attitude check, during the first orbit," added Carpenter, "would probably have helped to diagnose the persistent, intermittent and constantly varying malfunction of the pitch horizon scanner. By the third orbit, it was all too late." Whilst drifting, Carpenter beheld one of the most spectacular sights of the mission: his final orbital sunrise, witnessed four hours and 19 minutes after launch, shortly before retrofire. "Stretching away for hundreds of miles to the north and the south," he wrote, sunrise presented "a glittering, iridescent arc" of colours, which faded into a purplish-blue and blended into the blackness of space.

This blackness, he would write in his post-flight report, together with brilliant shades of blue and green from the sunlit Earth, were "colours hard to imagine or duplicate because of their wonderful purity. Everywhere the Earth is flecked with white clouds". The South Atlantic, he recounted, was 90 per cent cloud-covered, but western Africa was completely clear and Carpenter was granted a stunning view of Lake Chad. He spotted patchy clouds over the Indian Ocean, a fairly clear Pacific and an obscured western half of Baja California. He described the atmospheric airglow layer in detail to Slayton when he came within range of Muchea. "The haze layer is very bright," he reported. "I would say about eight to ten degrees above the real horizon … and I would say that the haze layer is about twice as high above the horizon as the bright blue band at sunset is."

His long period of drifting flight also meant that he had the opportunity to witness the 'fireflies' seen by John Glenn three months earlier. By rapping his knuckles on the inside walls of the spacecraft, he could raise a cloud of them and determined that they came from Aurora 7 itself. "I can rap the hatch and stir off hundreds of them," he reported. To Carpenter, they appeared more like snowflakes and did not seem to be 'luminous', actually varying in size, brightness and colour. Some were grey, some white and one in particular, he said, looked like a helical shaving from a lathe. Carpenter then decided, with only minutes remaining before retrofire, to yaw the spacecraft in order to get a better view with the photometer. Shortly thereafter, he passed over Hawaii, whose capcom told him to reorient Aurora 7, go to autopilot and begin stowing equipment and running through pre-retrofire checklists.

More problems arose. Four hours and 26 minutes after launch, with retrofire barely six minutes away, Carpenter reported that the automatic system did not appear to be working properly and confirmed that the "emergency retro-sequence is armed and retro manual is armed". In his autobiography, he would recount that the autopilot was not holding the spacecraft steady and, indeed, that achieving the correct pitch and yaw attitudes were critical to ensuring that he would descend along a pre-determined re-entry flight path and plop into the waters of the Atlantic, just south-east of Florida. Carpenter promptly switched to the fly-by-wire controls, but forgot to shut off the manual system, which wasted even more fuel. At around the same time, two fuses overheated and the astronaut noticed smoke drifting through the cabin.

Concerned that the critically-timed retrofire would now be delayed by the autopilot malfunction, Carpenter initiated it manually. He fired the rockets three seconds late, but admitted later "at that speed, a lapse of three seconds would make me at least 15 miles 'long' in the recovery area". Although he radioed to Shepard that he felt his spacecraft attitudes were good, privately, Carpenter was not sure and added, almost as an afterthought, that "the gyros are not quite right". Years later, he would describe the difficulty in dividing his attention between two attitude reference systems and attempting to accomplish a perfect retrofire. "It appears I pretty much nailed the pitch attitude," he wrote, "but the nose of Aurora 7, while pitched close to the desirable negative 34 degrees, was canted about 25 degrees off to the right, in yaw, at the moment of retrofire. By the end of the retrofire event, I had essentially corrected the error in yaw, which limited the overshoot. But the damage was already done."

The 25-degree cant alone would have caused Aurora 7 to miss its planned splashdown point by around 280 km; however, the three-second delay in firing the retrorockets and a thrust decrement – some three per cent below normal – contributed an additional 120 km to the overshoot. On the other hand, if Carpenter had not bypassed the autopilot and manually fired the retrorockets, he could have splashed down even further off-target. At this stage, his fuel supplies were holding at barely 20 per cent for manual and just five per cent for automatic. Carpenter survived re-entry, but experienced a wild ride back through the atmosphere, with Aurora 7 oscillating between plus and minus 30 degrees in pitch and yaw. The astronaut was able to damp out many of these oscillations with the fly-by-wire controls and the post-flight report would commend him as having "demonstrated an ability to orient the vehicle so as to effect a successful re-entry". It provided clear evidence that a human pilot could overcome malfunctioning automatic systems.

Carpenter's descent and the trapezoidal window offered a spectacular view of Earth, zooming towards him. "I can make out very small farmland, pastureland below," he reported, some four hours and 37 minutes after launch. "I see individual fields, rivers, lakes, roads, I think." Five minutes later, Gus Grissom, the Florida capcom, informed him that weather conditions in the anticipated recovery zone were good. By this time, shortly before ionised air surrounding the capsule caused a communications blackout at an altitude of around 22 km, Carpenter began to see the first hints of an intense orange glow as particles from the ablative heat shield formed an enormous 'wake' behind him. Then came distinct green flashes, which the astronaut assumed were the ionising beryllium shingles on Aurora 7's hull. As the re-entry G forces peaked at 11 times their normal terrestrial load, telemetred cardiac readings at Mission Control revealed the substantial physical effort needed by

Carpenter to speak, announce observations and make status reports. His breathing technique, perfected in the Johnsville centrifuge, would come in useful.

Five minutes before splashdown, at an altitude of 7.6 km, he manually deployed the drogue parachute, which steadied the capsule and damped out what he had earlier described as "some pretty good oscillations". The drogue was soon followed by the main chute, again manually deployed, although Carpenter's announcements of each milestone over the radio to Grissom fell on deaf ears. The capcom could not hear his transmissions and was forced to broadcast 'in the blind' to inform him that his splashdown point would be some 400 km 'long' and advise that pararescue forces would arrive on the scene within the hour. A minute before splashdown, Carpenter acknowledged Grissom's call. The impact with the water, 215 km north-east of Puerto Rico, was not hard, but Aurora 7 was totally submerged for a few seconds. It popped back up and listed sharply, 60 degrees over to one side, before the landing bag filled and began to act as a sea-anchor.

Keen to get out as soon as possible and probably thinking back to Grissom's own misfortune, Carpenter opted to exit the capsule through the nose, becoming the first and only Mercury astronaut to do so. It took him four minutes and required him to remove the instrument panel from the bulkhead, exposing a narrow egress passage up through the spacecraft's nose, where the two parachutes had resided. As he squirmed his way through the cramped space, Carpenter decided, in defiance of standard egress procedures, not to deploy his pressure suit's neck dam. He was already overheating and felt the gently swelling seas would make it unnecessary. Next, still perched in the nose of the capsule, he dropped his life raft into the water, where it quickly inflated and a Search and Rescue and Homing (SARAH) beacon came on automatically. The latter would allow recovery forces to home in on his position.

Preparing himself for a long wait, Carpenter tied the raft to the side of the capsule, deployed his neck dam, said a brief prayer and relaxed. He stretched out on his raft and was joined, he wrote later, by "a curious, 18-inch-long black fish who wanted nothing more than to visit". It was his first physical contact with another living being and his first moment of calm in four hours and 56 minutes since launch.

For those watching the mission from afar, however, there was no relaxation. At Cape Canaveral, CBS anchorman Walter Cronkite played up the drama by describing for his audience Mission Control's repeated attempts to contact Aurora 7, then highlighted that Carpenter had endured "half a ton" of pressure during re-entry and finally recapped that flight controllers were still "standing by" after losing voice contact with the astronaut. "While thousands watch and pray," Cronkite told his audience, "certainly here at Cape Canaveral, the silence is almost intolerable." In Manhattan's Grand Central Terminal, a hush fell over the crowd gathered before a huge CBS screen, while in the White House a direct telephone link with the Cape had been set up to provide President Kennedy with news. In fact, the SARAH beacon had already given Carpenter's co-ordinates and his telemetred heartbeat had been clearly heard in Mission Control throughout re-entry. Moreover, his splashdown point was almost exactly where the IBM computers at NASA's Goddard Space Flight Center in Greenbelt, Maryland, had predicted him to be after factoring-in radar tracking data and the yaw error at the point of retrofire.

Aboard the destroyer John R. Pierce – nicknamed the 'Fierce Pierce' – the attitude was quite different thanks to the reception of the strong SARAH signal. "Believe you me," reported CBS journalist Bill Evenson from aboard the destroyer, "this bucket of bolts is really rolling now and what a happy crew we've got!"

It was a Lockheed P2V Neptune, one of the same breed of patrol aircraft that Carpenter had flown a decade earlier in Korea, which finally greeted him. The astronaut signalled the pilot with a hand mirror and was acknowledged when the Neptune began circling his position. Shorty Powers, upon hearing the news, announced that "a gentleman by the name of Carpenter was seen seated comfortably in his life raft". One hour and seven minutes after splashdown, at 1:48 pm Eastern Standard Time, Airman First Class John Heitsch and Sergeant Ray McClure from an SC-54 transport aircraft joined the astronaut in the water, opened their rafts and tethered them together. Carpenter offered them some of his food rations, which were politely declined. Eventually, the astronaut was picked up by the Intrepid, originally earmarked as the prime recovery ship, but delayed in its arrival by Aurora 7's 400 km overshoot. The Fierce Pierce, meanwhile, successfully recovered the spacecraft itself and delivered it, on 28 May, to Roosevelt Roads in Puerto Rico.

Carpenter, meanwhile, was hot and wet after almost an hour on his back on the launch pad, followed by five hours in space and more than an hour in the Atlantic. Soon after boarding the rescue helicopter, he borrowed a pocket knife, cut a hole in the sock of his pressure suit and let his sweat and seawater drain out of the makeshift toe hole. Army physician Richard Rink asked Carpenter how he felt. In true Mercury Seven fashion, perfectly demonstrative of the 'right stuff' about which Tom Wolfe would later write, America's fourth man in space replied simply: "Fine".

PUSHING THE ENVELOPE

The situation within Mission Control, Deke Slayton recounted, was far from fine. Slayton had been stationed throughout Aurora 7's flight at the capcom's mike at the Muchea tracking site in Australia, which he described as "a good place to be, all things considered". Flight Director Chris Kraft and many other mission controllers were furious, accusing Carpenter of having recklessly endangered himself during a botched re-entry. Their anger was exacerbated when, aboard the recovery ship, the astronaut off-handedly remarked that "I didn't know where I was ... and they didn't know where I was, either". Retrofire controller John Llewelyn is said to have retorted: "Bullshit! That son-of-a-bitch is damned lucky to be alive!"

Kraft, apparently, was considerably more caustic. In his autobiography, he wrote of Carpenter's "cavalier dismissal of a life-threatening problem" – the failure of the spacecraft's navigational instruments – and troublesome re-entry and swore that the astronaut would never fly again. Carpenter was never assigned another mission, not even in a backup role. After a month-long tour in the Navy's Sealab-II underwater habitat, off the coast of La Jolla, California, he would resign from NASA early in 1967. Some have seen Carpenter's mistakes and omissions and his forgetting to do certain critical tasks as evidence that the early Mercury flights were simply too overloaded with experiments and manoeuvres and, further, that Mission Control was at least partly to blame for failing to identify the pitch horizon scanner malfunction for what it was. Tom Wolfe, for his part, later wrote that any speculation that Carpenter had panicked made no sense "in light of the telemetred data concerning his heart rate and his respiratory rate".

Psychologist Bob Voas weighed in with his own judgement: "The astronaut's eye on the horizon was the only adequate check of the automated gyro system," he told Carpenter and Kris Stoever. "With its malfunctioning gyros, the spacecraft could not have maintained adequate control during retrofire. Mercury Control may have viewed the manually controlled re-entry as sloppy, but the spacecraft came back in one piece and the world accepted the flight for what it was: another success."

Aurora 7, though harrowing, was certainly viewed as a success by Carpenter's family and hundreds of thousands of residents of his home state, Colorado. In Denver, a 300,000-strong crowd cheered the nation's newest astronaut son in their own ticker-tape parade. The city of Boulder declared 29 May as 'Scott Carpenter Day', sponsored its biggest-ever celebration and the University of Colorado named the astronaut its most accomplished graduate. Years earlier, Carpenter's own father, a research chemist, had achieved the same accolade from the same institution. In the case of the younger Carpenter, however, it also came with the formal conferring of his engineering degree, which he completed in 1949, save for a final examination in thermodynamics. The university granted the degree on the grounds that his "subsequent training as an astronaut has more than made up for the deficiency in the subject of heat transfer".

Carpenter's flight brought Project Mercury to another crossroads. In August 1961, the question had been whether to eliminate further Redstone missions in favour of moving towards the Atlas. Now, nine months later, discussion within NASA centred on whether enough had been learned from the three-orbit flights of Friendship 7 and Aurora 7 to justify a still-longer venture. Speaking before the Exchange Club in Hampton, Virginia, NASA engineer Joe Dodson pointed out that the lessons derived from Glenn and Carpenter were pleasing and speculation arose that a day-long mission, to rival that of Gherman Titov in Vostok 2, could be attempted as early as 1963. Indeed, many congressional observers supported a flight to surpass Titov. The debate ended on 27 June, barely five weeks after Carpenter's re-entry, when NASA Headquarters announced that Wally Schirra would fly Mercury-Atlas 8, possibly as early as September, and attempt up to six circuits of the globe.

Perhaps in reference to the same engineering influence with which Slayton's Delta 7 had been named, Schirra chose to call his capsule 'Sigma 7'. "Sigma, a Greek symbol for the sum of the element of an equation," wrote Schirra in his 1988 autobiography, "stands for engineering excellence. That was my goal – engineering excellence. I would not settle for less." Nor, indeed, would the ground team, who prepared Sigma 7 for launch with such tenacity and engineering precision ... and even humorously placed a car key on the capsule's control stick and stowed a carefully-wrapped steak sandwich in Schirra's ditty bag. The astronaut sought to honour them, too, during his mission. "All these little things do really help to make you realise that there are a lot of other people interested in what you're doing," he said later. "We know this inherently, but these visible examples of it do mean a lot."

The mission would double the number of orbits achieved by Glenn and Carpenter, lasting around nine hours, and as a consequence the Sigma 7 capsule required 20 modifications to provide more consumables. "I think probably the best part of my Mercury mission," Schirra wrote later, "was naming it Sigma 7. Naming it the sum of engineering effort, I wanted to prove that it was a team of people working together to make this vehicle go. That's why I talk so wildly about knowing the engineers, how they were brothers and buddies ... and all of them were! That's what I saw as the ultimate on that mission, was that [it was] an engineering test flight, where we weren't going to look around for fireflies. We weren't going to look for the lights of Perth. We weren't going to give prayers to the peasants below. We were going to make this thing work like a vehicle!"

THE JOKER

Right from the start, Walter Marty Schirra Jr was known for his 'gotchas'.

Born in Hackensack, New Jersey, on 12 March 1923, he was once described as having aviation in his blood; his parents having both engaged in 'barnstorming' and 'wing-walking' during the Twenties. As his father Walter, a veteran First World War fighter pilot and engineer, handled the controls of a Curtiss Jenny biplane, his mother Florence danced on the lowermost wing, using the struts for support. Spectators in Oradell, New Jersey, paid up to five dollars a time to witness the Schirras' stunts. Fortunately, wrote the astronaut, his mother "gave up wing-walking when I was in the hangar!" Nevertheless, he would use his mother's experience to his advantage: unlike the celebrated, faster-than-sound test pilots Chuck Yeager and Scott Crossfield, Schirra was flying even before he was born ...

He graduated from Dwight Morrow High School in Englewood, New Jersey, in 1940 and attended the Newark College of Engineering, before being appointed to the Naval Academy in Annapolis, Maryland. His disappointed father had wanted him to attend West Point as an Army officer, but in his autobiography Schirra would recall seeing a naval aviator during his boyhood, clad in "green uniform, the sharp gold wings above his left pocket and his polished brown shoes shiny. From that day on, I always wanted to go to Navy". Schirra underwent an abbreviated class, "a five-year programme ... crammed into three", received his degree in 1945 and served two years at sea in the Pacific. Not only was his education abbreviated, but so too was his whirlwind romance with Jo Fraser, whom he met, courted for seven days and finally married whilst on leave in February 1946.

A tour of duty in China, attached to the staff of the commander of the Seventh Fleet as a briefing officer, meant that Schirra was a witness to the communist revolution sweeping across the most populous nation on Earth. "A high crime rate in the neighbourhood in which Jo and I lived," he wrote, "practically a robbery a night, was an expression of revolutionary contempt for the American 'imperialists' ... I knew when we left that China would never be the same again." Shortly thereafter,

Wally Schirra (right) and Wernher von Braun.

and proving that aviation was truly in his blood, Schirra became the first member of his academy to be detailed for flight training, transferring to Pensacola Naval Air Station in Florida and receiving his wings in June 1948. Like Scott Carpenter, he started off by soloing in a Yellow Peril biplane and then flew naval fighters for three years. Upon the outbreak of war in Korea, Schirra volunteered for active service as an exchange aviator with an Arkansas-based Air National Guard unit. He spent eight months in south-east Asia, flew 90 combat missions in the F-84 Thunderjet fighter-bomber and shot down two MiG-15s – "a tough little adversary" – for which he was awarded the Distinguished Flying Cross.

Following Korea, Schirra served at the Naval Weapons Center in China Lake, California, during which time he participated in the initial development of the Sidewinder air-to-air missile and later served as chief test pilot for the F-7U Cutlass and FJ-3 Furyjet fighters at Miramar Naval Air Station in San Diego. Although he praised the usefulness of the Cutlass in better understanding the aerodynamics of a delta-winged aircraft, Schirra would reject it on the basis that if it stalled with its leading-edge slats 'in', its motions became wild and random, with ejection the pilot's sole option. The Cutlass would later be declared operational, much to the chagrin of Schirra and the other members of his flight-test group, who had seen a number of fatalities. Indeed, over a quarter of all Cutlasses built would be lost in accidents. Years later, he would refer to it darkly as a "widow-maker".

Schirra completed the Naval Air Safety School and a tour of the Far East aboard the Lexington, before being selected for and reporting to the Naval Test Pilot School at Patuxent River, Maryland, in January 1958. It was at this time, he wrote, that he learned to communicate effectively with engineers, "the most valuable asset that I took from test pilot school to the space programme". For each test, Schirra was required to report in depth on tactical manoeuvres, power settings and data points. Within the next two years, he would be applying this expertise to the development of Project Mercury. Whilst at Pax River, Schirra met three other hotshot naval aviators, Pete Conrad, Dick Gordon and Jim Lovell, who would themselves later become astronauts. Graduating in late 1958, he assumed duties as a fully-fledged test pilot, transferring to the famed Edwards Air Force Base in California to help evaluate the F-4H Phantom-II long-range supersonic fighter-bomber.

Then, like more than a hundred other Navy, Air Force and Marine fliers across the United States, in the spring of 1959 Schirra received classified orders to attend a briefing in Washington, DC. Initially, he was reluctant to undergo the months of training for the much-lambasted 'man-in-a-can' effort to send an American into space. "I wanted to be cycled back to the fleet with the F-4H, get credit or take blame for its performance and put it through its paces as a tactical fighter," he wrote. "I saw myself as the first commander of an F-4H squadron. The space programme to me was a career interruption." Despite his reluctance, Schirra underwent the gruelling tests at Lovelace and Wright-Patterson, realising that as other hotshot pilots fell by the wayside, he was on the cusp of joining the most elite flying fraternity of all.

Undoubtedly, Schirra's experience had placed him in mortal danger on many occasions, yet his lifetime motto remained: "Levity is the lubricant of a crisis". In fact, on the day the Mercury Seven were introduced to journalists in April 1959, Deke Slayton remembered Schirra telling a joke. It was not his first, nor would it be his last, and he would be in good company in the dark-humour stakes. All seven of them would dream up their own fiendish practical jokes – called 'gotchas' – which they inflicted on each other, on flight surgeons, on their nurse, Dee O'Hara, on Henri Landwirth and on an unfortunate Life photographer named Ralph Morse. The latter, who had succeeded in his own 'gotcha' by tracking them down on a desert-survival training exercise in Reno, Nevada, received his comeuppance at the hands of Schirra and Al Shepard. The pair planted a smoke flare in the exhaust pipe of Morse's jeep and told him to move the vehicle, whereupon the unsuspecting photographer hit the starter and – boom! – was engulfed in a cloud of green smoke and dust. "The jeep had to be towed back to Reno," Schirra recounted with glee, "and sold for scrap."

On other occasions, the Mercury Seven would coat the bottoms of each other's metal ashtrays with thin films of gasoline, causing a flash fire whenever one of them inadvertently flicked hot ash. "Fiendish, but fun," Schirra wrote. Henri Landwirth, who frequently played host to the astronauts at his Holiday Inn, once found a live alligator in his pool, cunningly secreted there by Al Shepard. Flight surgeon Stan White, who had purchased a new sports car, liked to brag about its high efficiency. "So we plotted his comeuppance," wrote Schirra. "For a week, we added gasoline to his tank, a pint a day, and he raved about the great mileage he was getting. The following week, we siphoned off a pint a day and he went berserk. White never did figure it out."

Schirra's humour would form an essential part of astronaut morale over the coming years, under the respective shadows of both triumph and tragedy. Only weeks before his first spaceflight aboard Sigma 7, Dee O'Hara became his latest victim ...

THE SIX CIRCUITS OF SIGMA 7

Throughout training, Schirra and the other Mercury astronauts would allow only O'Hara to draw blood samples, fearing that the physicians – highly educated, but lacking practical skills – would collapse a vein. She came to their quarters every morning during training, was among the last to see them off before a flight and, as a devout Catholic, "counted beads when we went on missions". One day in September 1962, Schirra and his backup, Gordo Cooper, decided to give O'Hara their daily urine sample ... with a difference. They filled a huge bottle with warm water, added iodine to colour it and laundry soap to make it foamy, then dumped it onto O'Hara's desk. The two astronauts, for good measure, tagged the bottle with the time of delivery in Greenwich Mean Time and added a handful of lollipops, then disappeared to await their nurse's arrival. O'Hara took one horrified look at Schirra's enormous specimen bottle and burst into tears of laughter. Schirra would later dedicate a photograph of O'Hara, clutching the bottle, with the inscription: "Gotcha!"

By this time, both he and Cooper had been preparing for Sigma 7 for almost two months. After NASA managers decided to proceed with a six-orbit mission, their names had been announced on 27 July and last-minute adjustments to the flight plan were being made as late as September. Unlike the test flight of Friendship 7 and the exclusively scientific nature of Aurora 7, the mission internally dubbed 'Mercury-Atlas 8' would be devoted to engineering objectives, with Schirra expected to focus on the management and operation of his spacecraft's systems, including the hydrogen peroxide attitude-control fuel and electrical power. In fact, the only 'scientific' experiments on the agenda would be an attempt to observe a ground-based xenon floodlight at Durban in South Africa, which would blaze with 140 million candlepower, and a one-million-candlepower quartet of flares at Woomera, which had eluded Carpenter. A few terrestrial and weather photography tasks were timelined and eight ablation panels were fused onto Sigma 7's hull to evaluate passive thermal effects on a variety of materials.

Technical planning for such a lengthy mission had begun in February, when it was recognised that oxygen supplies, reaction-control system reserves and power considerations had to be taken into account. A three-orbit Mercury mission, with all systems operating, consumed about 7,080 watt-hours of battery power from an available 13,500 watt-hours, but a six-circuit plan was predicted to leave a reserve supply of only 6.7 per cent. Engineers insisted that at least a ten-per cent post-landing reserve should be available as a safety factor and made recommendations as to how to achieve this: either that some unneeded systems be turned off during a substantial portion of the flight or that telemetry-transmission and radar-beacon operations be transferred to ground command. This, it was felt, could raise the reserve power levels as high as 15 per cent and provide a healthy safety margin. Oxygen posed another problem: some 1.9 kg was expected to be consumed during a six-orbit flight, leaving an insufficient remainder for emergencies. Unwilling to relax safety rules, a vigorous effort was implemented to reduce cabin-leakage rates to 600 cm^3 per minute, two-thirds of the previous value, and a higher capacity of lithium hydroxide would be carried to remove carbon dioxide from the cabin atmosphere throughout the longer mission. (Eventually, the cabin oxygen-leakage rate was reduced to just 460 cm^3 per minute.)

Even as these efforts were ongoing, NASA and McDonnell engineers had their sights set on a much longer Mercury mission, up to a day in duration and completing as many as 18 orbits. However, this would very much depend on the ability of Sigma 7 and Schirra to validate the changes and verify that an astronaut could indeed tolerate the weightless environment over long periods. The Mercury Project Office also suggested alternating a combination of automatic and manual modes to provide safer fuel reserves at the end of a flight; then, in the case of a malfunction in one of the modes, Schirra would be assured of an adequate supply in the other. Recovery procedures also needed adjustment, for Sigma 7's flight path during its fourth, fifth, sixth and (conceivably) seventh orbits predicted a splashdown point in the northern Pacific Ocean, some 440 km north-east of Midway Island.

Mission rules dictated that a contingency recovery capability must be in place within 18 hours after splashdown; a capability easily met for flights lasting up to six orbits, but requiring additional recovery forces for seven orbits. Consequently, Sigma 7 would be restricted to six circuits of the globe. Two Mercury capsules – Spacecraft No. 16 and No. 19 – had been delivered to Cape Canaveral in January and March 1962, respectively, with the first of these selected for the mission. By April, work needed to validate the capsule for its lengthy sojourn was well underway: temperature surveys of its critical points were complete, its environmental system passed altitude-chamber tests and its reaction-control system was put successfully through its paces. However, niggling glitches with emergency oxygen rate valves, water coolant and a higher-than-allowable oxygen leakage rate conspired to delay Sigma 7 past August to at least 18 September and, ultimately, into the first week of October.

Aurora 7's fuel-thirsty mission also prompted engineers to slightly redesign the spacecraft's attitude-control thrusters. Despite Scott Carpenter's reservations, the heavy periscope – which he felt was useless on Earth's night side – would be retained. The post-flight inspection team had reported that Aurora 7's landing error had been caused by a faulty yaw attitude, mainly because the astronaut had used the trapezoidal window as his primary means of reference. However, it was speculated that the periscope might have assisted in correcting the attitude and reducing Carpenter's overshoot. As a result, Schirra would test both periscope and window as a spacecraft-attitude reference point on Earth's day and night sides. He would then check his visual judgement to gauge his attitude and compare his own abilities against those of the periscope and the spacecraft's instruments. Retaining the periscope, though, proved unfortunate for the Goddard Space Flight Center's Albert Boggess, who had hoped to fly his ultraviolet airglow spectrograph in its place.

Additional fuel-saving techniques, at Carpenter's prompting, included integrating the control-mode selector switch with the control system, thus 'sealing-off' the high thrusters until they were needed for fast-reaction manoeuvres. A radiation dosimeter was also installed in Sigma 7's hatch, the astronaut was provided with a hand-held unit and several more instruments were affixed to his pressure suit, thanks to the effects of the Atomic Energy Commission's Operation Dominic high-atmosphere nuclear test in July 1962. This test, conducted high above the Pacific Ocean, had created a new zone of radiation which was already thought to have affected the solar batteries of several satellites, including the Ariel 1 physics mission. Admittedly, by the end of August, the radiation hazard seemed negligible, but the doubts of Project Mercury managers remained.

Schirra's Atlas rocket, too, had experienced its own fair share of problems. Originally scheduled to be delivered to Cape Canaveral in July, it failed its initial composite test at Convair's San Diego facility and arrived the following month. Next, the Air Force revealed that its military Atlas programme had suffered four recent turbopump failures and advised NASA that the rocket assigned to Sigma 7 would be put through a flight-readiness static test-firing. This postponed the launch by another week, but, even before the test was made, a fuel leak was found in a seam weld, effectively pushing the mission back from 24 September to 3 October. Still, Schirra's Atlas promised to be the safest so far: its engines would utilise hypergolic fluids instead of pyrotechnics, thus eliminating a two-second 'hold-down' at ignition, which saved fuel and provided for smoother initial combustion. Additionally, baffled injectors prevented combustion instability.

Five relay stations aboard a quintet of Air Force C-130s, based in Florida, Puerto Rico and on Midway Island, would augment the tracking network by covering areas previously out of communications range. Nineteen ships stood by in the Atlantic and nine in the Pacific, while no less than 134 aircraft covered the primary and secondary splashdown areas. In total, around 17,000 personnel would support the Sigma 7 recovery effort.

Early on 3 October, Schirra was awakened in the quarters of Hangar S by Air Force flight surgeon Howie Minners, after which he showered, dressed and ate the traditional pre-launch breakfast of steak and eggs with Bob Gilruth, Walt Williams and the newly-appointed Co-ordinator of Astronaut Activities, Deke Slayton. Another item also took pride of place on the breakfast menu. The previous evening, Schirra and Slayton had gone fishing – Cape Canaveral being renowned for its excellent surf fishing, especially in the spring and autumn – and had hooked several bluefish. They were "in the five-pound range," wrote Schirra, "but they fought free by severing our leaders with their razor-sharp teeth. I managed to land one by slinging it on the beach and pouncing on it before it could wriggle back to the surf". The bluefish, it seemed, was not the only individual with a shock in store. The two astronauts were aware of a military Thor-Delta rocket on a nearby launch pad, but they did not know how close it was to launch!

"It wasn't until we heard a roar that we realised the Thor-Delta was lifting-off," wrote Schirra. "We were looking right up the tailpipe of its monster engine and we knew right away that we were in the danger zone. Had there been an abort, it would have been a bad day for Mercury, with the chief astronaut and the pilot of MA-8 incinerated like the legendary rattlesnakes." Fortunately, he added, the bluefish for breakfast the next morning was delicious. Another pleasant surprise at the breakfast table, literally hot off the press, was a copy of the New York Times. It had been flown from New York to Florida that very morning, at no small expense, and Schirra was so impressed that he kept it.

The drive to the launch pad was uneventful, with the exception that the astronaut fell asleep, to be awakened – for the second time that morning – by Howie Minners. Several months later, Gordo Cooper would actually doze off whilst atop the Atlas rocket, waiting out a hold in the countdown. It was testament, Schirra explained, "to our training, and it shows the confidence we had in the people who supported us, both from NASA and the contractors. We could ask questions of technicians at the pad or construction guys and we'd get straight answers. We could call the executives, like Mr Mac of McDonnell, and they too would level with us. That's one reason we completed Mercury with seven healthy astronauts".

By 4:40 am, Schirra, assisted by Cooper and Pad 14 leader Guenter Wendt, was aboard Sigma 7. The countdown proceeded with exceptional smoothness, the only minor problem being a radar malfunction at the Canary Islands tracking station, and the United States' third orbital mission set off at 7:15:11 am. In his post-flight debriefing aboard the destroyer Kearsarge, Schirra would describe the ascent as "disappointingly short", with all of his training to handle emergencies rewarded by a perfectly nominal climb into orbit. "I still believe that the amount of practice we had prior to [orbital] insertion is important," he debriefed, "in that you must be prepared for reaction to an emergency, rather than thinking one out."

Sigma 7's rise was not, however, entirely nominal. Ten seconds after liftoff, it became clear that the clockwise roll rate of the Atlas was greater than planned, giving flight controllers cause for concern. "My course was being plotted against an overlay grid called a 'harp', since it's shaped like the musical instrument," Schirra recalled in his autobiography. "Green lines in the middle of the grid designate the 'safe' zone and on the outer limits the lines go from yellow to red. I was headed into the yellow area. If I had reached the red, there was a likelihood that the Atlas would impact on land, possibly in a populated area." Such a dire eventuality would have forced the Cape's range safety officer to abort the mission, ejecting Sigma 7 and destroying the rocket. Indeed, the primary and secondary sensors within the Atlas had registered a 'rifling' roll only 20 per cent short of an abort condition ...

Thirty seconds into the climb, Schirra reported that the Atlas was "getting noisy", after which he briefly lost contact with the control centre. "You had your transmitter keyed," Capcom Deke Slayton told him two minutes after launch, when communications were restored. "That's why we couldn't read." The remainder of the ascent proceeded normally and, at 7:18 am, Slayton radioed Schirra with the cryptic, but loaded, question: "Are you a turtle today?"

The roots of the strange question were explained by Schirra on his website, www.wallyschirra.com. He tells the tale of a good and noble man, sickened by the vulgar minds of those around him and gradually driven to despair over whether he will meet someone of his own intellect. At length, he retreated, turtle-like, into a protective shell and found his only respite in the consumption of alcohol ("for purely medicinal purposes, of course"). Over time, he sought out other like-minded individuals to join his 'Ancient and Honourable Order of Turtles' drinking fraternity. However, he needed money and gambled his most prized possession – a donkey, which he had raised from birth – on a horse running at long odds at the local track. Fortunately, he won the bet and kept his donkey. To commemorate his triumph, all future members of the fraternity, when asked "Are you a turtle?" were expected to reply, without blinking or hesitating, "You bet your sweet ass I am!" Failure to do so would consign the victim to buy drinks for everyone close enough to have overheard the question.

Unfortunately for Schirra, the question had been asked over a 'hot-mike' and, if he did not answer correctly, the number of people 'within earshot' demanding alcoholic beverages could have run into hundreds! On the other hand, even though 'ass' in this context referred innocently to the donkey, responding correctly to Slayton's question over the open mike could have led to misunderstanding, embarrassment and a reprimand from NASA. Schirra was in a quandary. Then, he told Slayton "Going to VOX record only" and announced the correct response, privately, into Sigma 7's on-board voice recorder. Schirra would not need to buy drinks, but Slayton had scored his own launch-day 'gotcha' and put his friend briefly in the hot seat.

Schirra's membership of the Turtles would arise again during his Apollo 7 mission in October 1968, with the exception that this time he would gain his revenge on Slayton, by asking him – through the capcom – if he was a turtle. Even President Kennedy, also a Turtle, was asked at a press conference about his membership of the order, to which he responded that he would buy the questioner a drink later. Aside from the time-honoured tale of its formation, it can supposedly trace its origins back to the Second World War, when pilots used it as a means of amusement whilst relaxing between combat missions. "It was not meant to be serious," founder Hugh McGowan once told his son, "it had no constitution or by-laws and was a relief from the horrors and dangers we saw every day on our missions." To attain membership, candidates had to answer correctly at least four out of 25 questions; each of which, although suggestive of an obscene answer, was actually quite innocent. Examples included: What does a woman do sitting down, that a dog does on three legs and a man does standing up? (Shake hands), What is a four-letter word, ending with 'k', that means 'intercourse'? (Talk), What is long, hard and filled with seamen? (A submarine) and so on.

Schirra's hot seat rapidly turned into a weightless one, when, a little more than five minutes into the flight, the Atlas' sustainer shut down and Sigma 7 cleanly separated from the rocket. "I have SECO," the astronaut announced as sustainer cutoff occurred, then continued "Cap sep and in aux damp and it's very pleasant. Going to fly-by-wire." By now at an altitude of some 280 km and an orbital speed of 28,250 km/h – higher than any other Mercury astronaut – Schirra set to work evaluating his spacecraft's systems. "With my eyes fixed on the control panel, studiously ignoring the view," he wrote later, "I began a slow – four degrees per second – cartwheel. Once in the correct orbital position, I checked my fuel. I had used less than half a pound of hydrogen peroxide. The thruster jets worked perfectly. They responded crisply to my touch and shut off without any residual motion. I was able to make tiny, single-pulse spurts, the micromouse farts, to assume an exact position." During this time, Schirra oriented his capsule to gain a better view of the sustainer as it tumbled away. He would conclude in his autobiography that his efforts demonstrated that rendezvous in space was possible. On his next mission, in barely three years' time, he would achieve just that.

The manual system, however, seemed somewhat "sloppy", with a tendency to 'overshoot', and Schirra switched next to the third control mode: the autopilot, which he referred to as "chimp mode". At around this time, his pressure suit began overheating. The suit had been one of Schirra's areas of responsibility during training and he would comment later on the seriousness of the situation; in fact, Flight Director Chris Kraft even considered terminating the mission after just one orbit. Before launch, Schirra had developed his own technique should overheating occur: he would inject cool water very slowly into the system, advance the temperature knob by half a mark at a time and then wait for ten minutes. He did not want to rush the water, lest the heat exchanger freeze. It worked. By the end of his first orbit, the temperature had dropped to 32°C, which Schirra considered "hot but not unbearable". Flight surgeon Chuck Berry advised Kraft to press on with a second orbit and the happy news was relayed to Schirra by Capcom Scott Carpenter, based in Guaymas in Mexico.

Schirra was pleased, not only with the prospect of a full-length mission, but with the realisation that the problem was "solved through no great amount of ingenuity, but the point was it was solved by a human". The necessity of flying people into space, in his mind, had been vindicated. After the mission, Schirra would receive a plaque, signed by Frank Samonski of the spacecraft's environmental control system team and emblazoned with the legend that they had 'sweated more than you did during the first orbit of MA-8'. Attached to the plaque was the valve used by Schirra to control the water flow to his suit.

Later, as Sigma 7 began its third orbit, Schirra entered a period of drifting flight, during which time he undertook a psychomotor experiment: closing his eyes and touching certain dials on the control panel. "I missed only three out of nine," he wrote later, "concluding that my sense of direction and distance had not been impaired by weightlessness." Repowering Sigma 7 over the Indian Ocean, Schirra switched back to fly-by-wire and successfully fixed his attitude using the Moon in the window as a reference point. Aware that flight controllers had been closely monitoring his fuel consumption, he expressed a hearty "Hallelujah!" when Capcom Gus Grissom, situated at Kauai in Hawaii, radioed approval to fly a full six orbits.

After another hour of drifting, during which he focused his attention on Earth observations and photography, Schirra told Capcom Al Shepard, based on a tracking ship in the Pacific Ocean, that his pressure suit temperature had dropped to a more comfortable 20°C and his fuel in both the manual and automatic tanks stood at around 80 per cent; good news, it seemed. On the fifth orbit, busily working through his pre-retrofire checks, Schirra was infuriated to hear a voice interrupt from Quito in Ecuador. "We had what we called a 'mini-track' there," he wrote, "a miniature tracking station, and it wasn't supposed to come on the air except in an emergency." The speaker asked the astronaut if he had any words for the people of South America. Schirra, irritated, wished them "Buenos dias, you all" After splashdown he would receive a handful of telegrams, complaining of his brusqueness. "But there was one," he wrote, "I treasured from a US diplomat in Ecuador. He said in effect that Schirra had proved his devotion to the people of Latin America by wishing them a good day." By addressing them as 'you all', the diplomat continued, Schirra was simply noting that he would soon become a resident of Texas …

Scientific experiments on the mission, thankfully, had been trimmed down, lessons having been learned from the grossly overloaded work schedule on Scott Carpenter's flight. Indeed, more time was granted to Schirra to complete each of his tasks. One of the earliest, a little under an hour after launch, was another attempt to see the million-candlepower flares from Woomera in Australia. Initially, Schirra thought that he had seen them – only to discover that he had actually witnessed enormous lightning flashes – and, when notified that the flare firings were imminent, was hampered by heavy cloud cover. A similar test over Durban in South Africa was ruined by rain showers. The astronaut also had the opportunity to examine the fireflies observed by Carpenter and John Glenn, reporting them to be far too small to photograph and moving, in clouds, at a slower velocity than Sigma 7 itself.

Another experiment was conducted on both Schirra and his capsule. Since the Operation Dominic detonation in July 1962, concern had remained over radiation intensities in low-Earth orbit and the astronaut was outfitted with five thermo-luminescent dosimeters over his eyes, on his chest and on his interior thigh. Sigma 7 was instrumented with two nuclear emulsion packs, on either side of the control panel, to monitor its internal proton dosage. No major issues were raised, with upward of 65 per cent of the electrons from the artificial 'belt' of radiation created by the Dominic test having energies less than 1.3 MeV. This, it was noted, was insufficient even to penetrate the capsule's hatch cover, its least-protected region. Post-flight analysis concluded that the radiation had a negligible impact on both Schirra and his spacecraft.

Preparations for re-entry commenced over Africa, when Schirra transitioned to fly-by-wire control to orient Sigma 7 using celestial reference points, then switched to the autopilot to ensure that it operated satisfactorily in this mode. "The flight plan called for me to position the spacecraft using manual controls," he wrote in his autobiography, "then switch to automatic for retrofire, because the power of the big thrusters offsets the unbalancing force of the retrorockets. The retrorockets exert their force through the centre of the spacecraft, so they don't kick it off course." Manual control during retrofire and throughout re-entry had always remained a reserve option for the astronauts. Indeed, on Gordo Cooper's Mercury mission in May 1963, he would experience a failure of his electronics systems, forcing him to initiate and guide retrofire entirely by hand.

In Schirra's case, however, the retrorockets of his "sweet little bird" fired perfectly and in sequence under automatic conditions at 4:07 pm. Shortly thereafter, he switched the capsule to fly-by-wire, although Sigma 7 was "steady as a rock", with so much fuel remaining in its automatic and manual supplies that he had to 'dump' it during re-entry. Minor adjustments were needed to damp out wobbles after the jettisoning of the retrorocket package, but the descent proved to be as 'textbook' as the mission itself. Schirra manually deployed the drogue and main chutes, his elation evident in his words to Capcom Gus Grissom, and splashed down just 7.2 km from the Kearsarge at 4:28:22 pm. Within minutes, wrote Schirra, "a whaleboat was alongside and the underwater team had the flotation collar in place". Stepping aboard the Kearsarge, he noted that it was barely ten hours – and six full orbits – since his launch from Cape Canaveral. Another surprise awaited him in the admiral's quarters: an oversized urine-collection device, sent specially by Dee O'Hara. Almost identical to the one he and Cooper had dumped on her desk a few weeks earlier, it concluded the perfect mission with the perfect Gotcha.

FIRE HOSES AND POLICE DOGS

For Birmingham, Alabama, the early months of 1963 were a time of turmoil. Its 350,000-strong population was two-thirds white and a third black and it had earned itself a reputation as one of the United States' most racially segregated cities. Indeed, this segregation affected, as a legal requirement, all public and commercial facilities, covering every aspect of everyday life and was rigidly policed. Barely ten per cent of the city's blacks were registered to vote and their average income was less than half that of the whites. No black police officers, firefighters, bus drivers or shop assistants could be found anywhere in the city and manual labour in steel mills or work in black neighbourhoods were among the few available options. Racial violence was rife: since 1945, some 50 bombings had earned Birmingham the nickname 'Bombingham', one area even being dubbed 'Dynamite Hill'.

Efforts to implement change, spearheaded by Fred Shuttlesworth and Martin Luther King, initially focused on challenging the city's segregationist policies through partially-successful legal action and protests. These began with a selective buying campaign to impose pressure on business leaders to open retail jobs to blacks and end segregated facilities in stores; when this failed, King and his followers instituted sit-ins at libraries and white churches and marches to provoke arrest, even irresponsibly recruiting children to their campaign at one stage. The demonstrations were denounced by white religious leaders, who argued that their case should be pressed in the courts and not on the streets, and the infamous public safety commissioner Eugene 'Bull' Connor obtained an injunction to bar future protests. The injunction was ignored as a denial of constitutional rights but on Good Friday, 12 April 1963, King was among 50 Birmingham residents, aged between 15 and 81, who were arrested. King's four-day incarceration in 'Birmingham Jail' inevitably focused the eyes of the nation on the city. The Children's Crusade, although it attracted criticism from Attorney-General Bobby Kennedy, served to maintain this focus.

By the first week of May, Birmingham's jails were overflowing and Connor's efforts to keep protestors out of the downtown business area had degenerated into using high-pressure fire hoses and police dogs. Witnessing the traumatic events was photographer Charles Moore, then working for Life magazine, who was hit by a brick meant for the police and one of whose "era-defining" images showed three teenagers being hit by a water jet powerful enough to rip bark off a tree. When published, it was labelled 'They Fight A Fire That Won't Go Out'. Television cameras broadcast harrowing pictures of unprotected protestors being attacked by police dogs. A photograph of student Walter Gadsden being charged by one of the dogs ended up on the front page of the New York Times, which President Kennedy called "sick" and the scenes in Birmingham "shameful". Moore would later reflect that the events of those violent days were "likely to obliterate in the national psyche any notion of a 'good Southerner'".

The tactics employed by King in Birmingham certainly focused the entire nation on the troubles. New York Senator Jacob Javits declared that America would refuse to tolerate such scenes, pressing Congress to pass a civil rights bill, while Oregon Senator Wayne Morse compared the situation to South Africa's loathsome apartheid regime. President Kennedy sent Assistant Attorney-General Burke Marshall to Birmingham to negotiate a truce. As the crisis deepened and the city's infrastructure virtually collapsed, it aroused international debate and condemnation. The Soviet Union devoted a quarter of its news broadcasts to the demonstrations, accusing Kennedy's administration of neglect and "inactivity".

By 8 May, white business leaders had agreed to most of the protesters' demands and two days later Shuttlesworth and King told journalists that they had received an agreement from the City of Birmingham to desegregate lunch counters, restrooms, drinking fountains and fitting rooms within three months and to hire black workers as salesmen and clerks. President Kennedy urged a number of workers' unions to raise bail money to free the demonstrators from the city's jail – a move condemned by Connor and Birmingham's outgoing mayor, Art Hanes – and 3,000 federal troops were deployed to restore order. A new mayor, Albert Boutwell, took office and Bull Connor ended his time as Commissioner.

Desegregation took longer to achieve. Indeed, at around the same time, bombs destroyed the hotel where King had stayed and damaged the house of his brother, Reverend A.D. King. Many observers felt that the protestors had settled for "a lot less than even moderate demands"; it was feared that the three-month desegregation agreement meant, in effect, that a single black clerk hired by mid-August would suffice. Nonetheless, by the summer of 1963, some lunch counters in department stores had complied with the new rules, while parks and golf courses opened to blacks and whites and Mayor Boutwell established a committee to discuss additional changes. At the same time, however, no black policemen, firefighters or clerks were hired and no black lawyers were admitted to the city's bar.

The reputation of King, though, had soared. Later that year, he would deliver his famous "I have a dream" speech in Washington, DC, and receive the Nobel Peace Prize in 1964. Longer-term consequences of the 'Birmingham Campaign' and other protests included the drawing up a Civil Rights bill by President Kennedy to prohibit racial discrimination in employment and access to public places, eventually passed into law and signed by his successor, Lyndon Johnson.

Kennedy's civil rights drive extended, indirectly, to the space programme, too. By mid-1963, NASA had two groups of astronauts in training – a new nine-man team having been picked under Deke Slayton's auspices the previous September – and a third was expected before the year's end. Bobby Kennedy, in particular, felt that a black pilot should be a member of Group Three. Unfortunately, wrote Slayton, "the Navy didn't have anyone remotely qualified, but in the Air Force there was a black bomber pilot, Captain Edward Dwight". With a multi-engine flying background, no engineering degree and no test piloting credentials, Dwight was underqualified, but, in what Aerospace Research Pilot School (ARPS) commandant Chuck Yeager felt was a case of 'reverse racism', the Attorney-General pressurised Air Force Chief of Staff Curtis LeMay to accept the black pilot.

A deal was struck, whereby Dwight was admitted to ARPS, on condition that better-qualified white pilots ahead of him were also accepted. In a laughable chain of events, which even obliged Yeager to appoint a tutor for the pilot, Dwight graduated from ARPS "and did okay", Slayton wrote, "but okay wasn't really enough: had he been white, he wouldn't even have been a serious candidate". NASA was not simply looking for test pilots, but for Navy, Marine and other Air Force fliers, together with civilians and research scientists, many of whom had far better qualifications than Dwight. Some of those candidates – including the Air Force's Mike Collins and the Navy's Dick Gordon – were actually among the best pilots in their respective services and had been passed over in the September 1962 astronaut intake to gain additional experience before being picked the following year.

"As I heard it," wrote Slayton, "Dwight himself wasn't particularly driven to become an astronaut: he wanted to move up in the Air Force." Dwight himself, it seemed, was being used as little more than a pawn in the Kennedys' civil rights crusade; a crusade which would win them both enemies and admirers in the coming years.

THE END ...

Gus Grissom had two reasons to be grateful to Wally Schirra. The first came immediately after his ill-fated splashdown on 21 July 1961, when he owed his life to the neck dam designed by his Mercury colleague. The second, however, came only minutes after Schirra's splashdown in the Pacific at the close of Sigma 7. Grissom's misfortune had prompted both John Glenn and Schirra to refrain from opening their capsules' hatches in the water and choosing instead to explosively blow them when on the deck of the recovery ship.

"I blew the hatch on purpose," Schirra wrote in his autobiography, "and the recall of the plunger injured my hand – it actually caused a cut through a glove that was reinforced by metal. Gus was one of those who flew out to the ship and I showed him my hand. 'How did you cut it?' he asked. 'I blew the hatch,' I replied. Gus smiled, vindicated. It proved he hadn't blown the hatch with a hand, foot, knee or whatever, for he hadn't suffered even a minor bruise." Already close from their three years training together, the two men were also neighbours in Houston and Schirra had agreed to act as the executor of Grissom's will. Little did he know that he would be called upon to do just that a little over four years' time.

The euphoria which surrounded Schirra's return from one of the most productive Mercury missions to date was evident. After a greeting at Pearl Harbour and a day-long stay in VIP quarters in Hawaii, the nation's latest astronaut hero found himself surrounded by the state's governor, a senator and military top brass. He was also pleased to be able to verify that he had, in fact, responded correctly to Deke Slayton's 'turtle' question, uttered during the launch. Whilst still aboard the Kearsarge, he asked the communications officer for a copy of the transcript of the first few minutes of his flight – and there it was, on his microphone's voice recorder, the correct answer: "You bet your sweet ass I am!"

Some voices within NASA opted to end Project Mercury immediately, its brief of placing a man into orbit for a lengthy period having been met. The next step on the road to the Moon, the two-piloted Gemini series, was just around the corner, with an inaugural unmanned venture scheduled for sometime in 1964. However, another of Mercury's original goals had been to fly a mission lasting at least one full day and, although erased in October 1959 due to the growth of the capsule's weight and power requirements and the limitations of the tracking network, this option returned to the fore shortly after Gus Grissom's flight. Among officials at the burgeoning Manned Spacecraft Center (MSC) in Houston, Texas, the decision was easy, particularly since long-duration experience prior to Gemini was highly desirable. Early in January 1962, Project Gemini was publicly named as the nation's interim stepping stone to Apollo and both NASA and McDonnell were hard at work planning a 'Manned One-Day Mission' (MODM) to round out Project Mercury in style.

By this time, MSC had effectively replaced the Langley Space Task Group and NASA Headquarters had reorganised Abe Silverstein's Office of Space Flight Programs into an Office of Manned Space Flight, directed by D. Brainerd Holmes. Silverstein himself had been made director of the Lewis Research Center. By September 1962, days before Schirra's launch, negotiations with McDonnell settled on a number of configuration changes needed for the MODM flight, which Bob Gilruth hoped to launch as early as April of the following year, using Spacecraft No. 20. The success of Sigma 7 prompted an emboldened NASA, in November 1962, to extend the MODM from 18 to 22 orbits, which would require a stay in orbit of around 34 hours. Such an ambitious venture – surpassing Vostok 2, though little more than a quarter as long as Andrian Nikolayev's Vostok 3 – was anticipated to cost in the region of $17.8 million and require truly enormous tracking support, since its orbital path would carry it over virtually all of Earth's surface between latitudes 33 degrees north and south of the equator.

Twenty-eight ships, 171 aircraft and around 18,000 military personnel would be needed to support the mission. Its duration also meant that, for the first time, round-the-clock control operations were required, with a Red Shift flight director (Chris Kraft) and a Blue Shift counterpart (a Canadian engineer named John Hodge). On 14 November 1962, Schirra's backup, Gordo Cooper, was assigned as the MODM pilot, with Freedom 7 veteran Al Shepard backing him up.

Other problems surrounded the Atlas rocket, whose 'F-series' military variant had suffered two inexplicable failures. When Cooper's Atlas-D was rolled out of its Convair factory in San Diego in late January 1963, it failed to pass its initial inspections and was returned for rewiring of its flight control system. This led NASA, on 12 February, to officially postpone the originally scheduled mid-April launch until mid-May. Meanwhile, the MODM capsule itself – at the centre of the mission designated 'Mercury-Atlas 9' – was being outfitted with more than 180 engineering changes: heavier and larger-capacity batteries for more electrical power, an additional oxygen bottle, extra cooling and drinking water, more hydrogen peroxide manoeuvring fuel, a full load of consumables for the life-support system, various other modified components and, of course, an expanded scientific payload. Providing partial compensation for the added weight, the periscope, which Schirra had considered virtually useless on Sigma 7, was deleted, together with UHF and telemetry transmitters and a rate stabilisation control system. Other plans included removing Cooper's fibreglass couch and replacing it with a lighter hammock, but fears that its material might stretch and the astronaut might 'bounce' meant that this proposal never materialised.

Still, the increasing weight of the later Mercury missions prompted an extensive requalification of the spacecraft's parachute and landing systems. Other changes included the installation of a slow-scan television unit to monitor both the astronaut and his instruments. In fact, at a press conference on 8 February, Cooper had referred to his mission as "practically ... a flying camera", in recognition not only of the television unit, but of a 70 mm Hasselblad, a special zodiacal-light 35 mm camera and a 16 mm all-purpose moving-picture camera. Cooper himself would wear a pressure suit which sported a mechanical seal for its helmet, new gloves with an improved inner liner and link netting between the fabrics at the wrist and a torso which afforded greater mobility. His lightweight boots were integrated, providing better comfort and reducing the time it took to put them on. All in all, the suit was much less bulky than its predecessors.

The mission appeared to be back on track by mid-March 1963, when the Atlas passed its acceptance inspection, this time without even a single minor discrepancy. In fact, having defined an offset of the engines to counteract the threatening roll rate experienced by Schirra during his liftoff, the rocket's contractor confidently believed that they had produced their best bird to date. The delays had, however, pushed the MODM into mid-May and on 22 April the Atlas and its Mercury capsule were mated. As launch drew nearer, Cooper and NASA had their hands full with other problems. Four years after the selection of the Mercury Seven, attitudes towards manned spaceflight had already begun to change, with Philip Abelson, editor of the journal Science, Warren Weaver of the Alfred P. Sloan Foundation and Senator J.W. Fulbright of Arkansas arguing that the high cost of President Kennedy's Moon project neglected urgent social and political problems at home.

Therefore, in spite of Schirra's success, Project Mercury and the manned space effort still had much to prove as the days ticked down towards Cooper's launch. After much consideration, he had named his 'spacecraft' – no longer called a 'capsule' – as 'Faith 7', to symbolise, he said, "my trust in God, my country and my teammates". Within the higher echelons of the space agency, concerns were

Gordo Cooper trains for the last Mercury mission.

expressed over the name: a mission failure, the Washington Post told its readers, could produce unfortunate headlines, such as 'The United States today lost Faith'. Much consideration was also given to a 'Mercury-Atlas 10' mission, flown by Al Shepard for up to three days, thereby further closing the space-endurance gap with the Soviets. Tests had already shown, as part of NASA's Project Orbit in February 1963, that a Mercury spacecraft could theoretically endure a four-day mission, although the effects of freezing or sluggishness in its hydrogen peroxide thrusters remained a lingering worry. Shepard himself, naturally, was in favour of a three-day flight, whose allocated spacecraft he had already nicknamed 'Freedom 7-II'.

Had it gone ahead, it would have been launched sometime in October 1963. Shepard, for his part, even went so far as to lobby John Kennedy to support the extended-duration flight, although the president deferred the final decision, rightly, to NASA Administrator Jim Webb. "After Cooper finished his day-and-a-half orbital mission," Shepard reflected in a February 1998 interview, "there was another spacecraft ready to go. My thought was to put me up there and just let me stay until something ran out – until the batteries ran down, until the oxygen ran out or until we lost a control or something; just an open-ended kind of a mission."

Even before Cooper's flight, however, on 11 May, NASA's newly-appointed deputy assistant administrator for public affairs, Julian Scheer, had emphatically declared that MA-10 would not fly. Webb himself killed off the plan a few weeks later, arguing that Gemini was already planned for long-duration missions – why prove something, only once, with an obsolete system, he asked – and that an accident on MA-10 could postpone subsequent ventures. In mid-June, the mission was officially removed from consideration, its spacecraft placed into storage and the shift to Project Gemini began in-earnest.

For Gordo Cooper, therefore, his own launch, set for 14 May, would be the end of the beginning.

HOTSHOT

Cooper almost missed out on flying in Project Mercury entirely. Since his selection in April 1959, he had steadily gained a reputation for himself, firstly as a hotshot pilot with a passion for fast cars, but also as a complainer who pulled dangerous stunts, including one in an F-106 jet which screamed right outside, and below, Walt Williams' office window. Moreover, Deke Slayton wrote of his personal surprise that Cooper had been chosen as an astronaut at all. "My first reaction was, something's wrong," he noted. "Either he's on the wrong list or I am. Gordo was an engineer at Edwards. As far as I was concerned, he wasn't even a test pilot."

Test pilot or not, if Schirra had been flying before he was born, then Leroy Gordon Cooper Jr was all but born in a pilot's seat. His father, an Air Force lawyer, county judge and pilot from Shawnee, Oklahoma, frequently plopped his young son onto his lap in the cockpit of an old Command-Aire biplane, even allowing the boy to take the controls at the age of six. Later, in his teens, Cooper would hang around at the airport in Shawnee to pay for lessons in a J-3 Piper Cub trainer; inspired to fly, it seems, from his own experiences and from his father's tales of the famed aviation aces Amelia Earhart and Wiley Post. He soloed, 'officially' at least, at the age of 16. It would garner a lifelong fascination with aviation which Cooper would retain for the rest of his life. Even in his seventies, he once told an interviewer that "I get cranky if I don't fly at least three times a month!"

His love of fast cars also became legendary during his astronaut days, as Gene Kranz, arriving at Cape Canaveral for his first day at work, related in his book 'Failure Is Not An Option'. "After the plane rolled to a stop," Kranz wrote, "a shiny new Chevrolet convertible wheeled to a halt just beyond the wing tip. An Air Force enlisted man popped out, saluted and held open the car's door for a curly-haired guy in civilian clothes, a fellow passenger who deplaned ahead of me." The curly-haired man offered Kranz a lift to the Cape, which he accepted, then "peeled into a 180-degree turn and raced along the ramp for a hundred yards, my neck snapping back as he floored the Chevy. I had never driven this fast on a military base in my life." For a while, Kranz wondered if he had a madman behind the wheel as the driver seemingly broke every rule in the book and apparently cared nothing for being pulled over by the Air Police. "Hitting the highway," Kranz continued, "he made a wide turn and a hard left, burning rubber. In no time, he had the needle quivering between 80 and 90 miles an hour. After a joyful cry of 'Eeeee-hah', he turned and offered his hand, saying 'Hi, I'm Gordo Cooper'. I'd just met my first Mercury astronaut!"

Born on 6 March 1927 in Shawnee, Cooper attended primary and secondary schools in his hometown and in Murray, Kentucky, and enlisted in the Marine Corps after graduation. The Army and Navy flying schools, he found, were not taking any new candidates that year. He promptly left for Parris Island, South Carolina, but the Second World War ended before he had an opportunity to see combat and he was assigned to the Naval Academy's Preparatory School and was an alternate for Annapolis; Cooper was given Marine guard duty in Washington, DC, and was serving there with the Presidential Honour Guard when he and other reservists were released from service. After his discharge, he moved to Hawaii to live with his parents – his father, at the time, was assigned to Hickam Air Force Base in Honolulu – and it was whilst there that he met his future wife, Trudy. A drum majorette at the University of Hawaii, she owned a third interest in a Piper Cub and taught flying. She would be the only Mercury Seven wife to hold a pilot's licence in her own right. In fact, when Cooper joined the astronaut corps, he and Trudy were the only members of the Mercury Seven to own an aircraft: a Beechcraft Bonanza.

The couple married in Honolulu in August 1947 and lived there for two years as Cooper pursued his degree at the University of Hawaii. Whilst studying, he received a commission from the Army's Reserve Office Training Corps, transferred to the Air Force and was called to active duty for flight training at Perrin Air Force Base, Texas, and Williams Air Force Base, Arizona. Cooper received his pilot's wings in 1950 and was attached to the 86th Fighter-Bomber Group at Landstuhl, West Germany, flying F-84 and F-8 jets and later commanding the 525th Fighter-Bomber Squadron. Whilst in Europe, he attended an extension of the University of Maryland's night school, returning to the United States in 1954 for detachment to the Air Force Institute of Technology at Wright-Patterson Air Force Base in Dayton, Ohio. From here, he received a degree in aeronautical engineering in August 1956 and was sent to Edwards Air Force Base in California for a year at test pilot school.

It was at around this time, in Denver, that he first flew with another Air Force pilot named Gus Grissom; the pair crashed a T-33 jet off the end of the runway at Lowry Air Force Base, though thankfully both were unhurt. Graduation from Edwards brought rapid reassignment to the fighter section of the famed base's Flight Test Engineering Division as a project engineer and test pilot. Whilst there, Cooper worked on the F-102A and F-106B development efforts. Then, early in 1959, he read an announcement that McDonnell had been awarded the prime contract to build a space capsule. Shortly afterwards, he received mysterious orders to attend a classified briefing in Washington. After undergoing the Lovelace and Wright-Patterson tests, he was so confident that he would be picked by NASA that he told his boss to start looking for a replacement and took two weeks' leave to move his family to Langley, Virginia. When NASA called him to ask how soon he could get to Langley, Cooper replied "How about now?"

Despite his flying credentials and engineering talent – he designed a personal survival knife and chaired the Emergency Egress Committee for Project Mercury – Cooper's early days within the astronaut corps were somewhat less than illustrious and would lead several senior managers to consider bypassing him entirely for a spaceflight. He was, some said, a complainer, unpredictable, with a seemingly indifferent stance towards the public image that NASA wanted each of its astronauts to display. Cooper protested, for example, about the lengthy periods away from his family, about the lack of opportunities to fly jets and collect flight pay and, in fact, when Deke Slayton was grounded from Delta 7, he even threatened to leave the programme. Flying a chase plane over Cape Canaveral during Gus Grissom's July 1961 ascent, Cooper buzzed the launch site, momentarily disrupting communications and earning him a severe ticking-off from superiors. On another occasion, he flew to Huntsville in Alabama, landed on a runway that was too short and asked to be refuelled. When ground crews objected that it was too dangerous for him to take off again, Cooper shrugged, took off regardless and made it to a nearby air base with fumes in his tanks ...

Even in the weeks leading up to Cooper's own mission, Faith 7, there were persistent rumours in the press that he might be dropped in favour of his backup, Al Shepard. In fact, so shaky was operations director Walt Williams' 'faith' in Cooper that he had approached Shepard several months earlier and strongly hinted that the Freedom 7 pilot might be tipped to fly instead. Believing the mission to be his, Shepard continued training feverishly, but Deke Slayton – removed from his own flight – felt that Faith 7 belonged to Cooper. Others agreed that it would reflect badly on NASA if the astronauts were switched so soon before launch.

A timely intervention by Wally Schirra, who threatened to raise the roof if his friend was overlooked, eventually contributed towards securing Cooper his seat on the very last Mercury mission. Shepard was livid and Williams admitted that the Freedom 7 flier could have done a better job, but that the decision had been made and it was now his job to ensure that Cooper was as prepared as possible. As partial compensation, Williams half-promised Shepard the three-day MA-10 mission. This never transpired. (Shepard later gained his revenge by lending Williams his Corvette for the day ... then, as the operations director drove off, phoned security to inform them that 'someone' had just stolen his car.)

Perhaps reacting to these frustrations, two days before the scheduled launch, Cooper took a flight in an F-106 and, to the great surprise of Walt Williams and Chris Kraft, made a very low pass over Cape Canaveral. "We were talking," Kraft recalled of that Sunday afternoon in Williams' office, "and a sudden roar came upon us. The roar was a jet airplane diving onto the Cape at a very high rate of speed, which was forbidden. We looked out the window to see none other than Gordo." Cooper flew beneath the second-floor office window and the astonished managers were actually able to look down on the screaming jet. The Cape, of course, was restricted airspace and its switchboard quickly lit up with frantic calls. Williams went berserk, according to onlookers, and threatened to have Cooper's "ass on a plate!"

He called Deke Slayton, who had to shout down the phone to be heard over the F-106's roar, and Williams argued that Cooper should lose Faith 7. He even contacted Al Shepard, asking him if he and his pressure suit were ready to go. Slayton, however, refused to pull Cooper off the mission, but expressed serious reservations about the astronaut's judgement. Both he and Williams allowed Cooper to sweat about his flight status for a day to put some fear into him. Not until late on the evening before launch did the operations director finally relent and agree to let him fly. Although many would come to regard him as a daredevil, Cooper's supporters described him as a good, smart pilot, a man with a mission "to go a little bit higher and a little bit faster". In May 1963, he would fly his highest and fastest mission so far.

LAST MERCURY OUT

Cooper's hotshot characteristics were balanced by a misleadingly quiet voice and laid-back personality, to such an extent that he frequently fell asleep during the lengthy physical checks ... and, famously, dozed off aboard his Faith 7 spacecraft, atop the fully-fuelled Atlas, on launch morning. Al Shepard, for his part, had lost his last chance to fly the final Mercury mission. Despite having himself engaged in flat-hatting as a naval aviator, he told Walt Williams that he felt Cooper had shown "unusually bad judgement". However, wrote Neal Thompson, "it wasn't the height Shepard thought was dumb; it was buzzing the administration building".

Four hours after a still-enraged Williams had given his consent to let Cooper fly, early on 14 May, the prime and backup astronauts ate breakfast ... and Shepard got more revenge for his 'lost' mission through another tension-relieving, though somewhat mean-spirited, gotcha. Press spokesman Shorty Powers arrived early that morning with a pair of cameramen to shoot some behind-the-scenes footage of Cooper as he prepared for launch. However, they found, to their shock, that none of

Cooper's Atlas 130D booster is prepared for launch.

the overhead lights were working, nor, indeed, were any of the electrical sockets. Someone, it seemed, had cut the wires, removed every light bulb, inserted thick tape into the sockets and replaced the bulbs. No one pointed any fingers, but Powers recognised the grin on Shepard's face "that is typical of him when he has a mouse under his hat".

Another gift from Shepard awaited Cooper when he boarded Faith 7 at 6:36 am: a small suction-cup pump on the seat, labelled with the legend 'Remove before launch', in honour of the new urine-collection device aboard the spacecraft. Cooper would become the first Mercury astronaut who would be able to urinate in a manner other than 'in his suit'. At this stage, the only expression of doubt over whether Faith 7 would fly came from meteorologist Ernest Amman. His fears were soon realised, not because of the weather, but due to a malfunctioning C-band radar at the mission's secondary control centre in Bermuda. Shortly after this had been rectified, at 8:00 am, with an hour remaining before the scheduled launch, a simple 275-horsepower diesel engine, responsible for moving the gantry away from the Atlas, stubbornly refused to work. More than two hours were wasted in efforts to repair a fouled fuel injection pump on the engine and the count resumed around noon. The gantry was successfully retracted, but the failure of a computer in Bermuda – crucial for a 'go/no-go' launch decision to be made – caused the attempt to be scrubbed.

Cooper, after six hours on his back inside Faith 7, remained upbeat and summoned a forced grin. "I was just getting to the real fun part," he said. "It was a very real simulation." He spent part of the afternoon fishing, while checkout crews prepared the Atlas for launch the following morning. Arriving at Pad 14 for the second time, he greeted Guenter Wendt, with mock formality, reporting as "Private Fifth Class Cooper", to which the pad fuehrer responded in kind. The roots of their joke came two years earlier, when Cooper had stood in for Al Shepard in a launch-day practice run prior to Freedom 7. Upon arriving at the pad, Cooper had expressed mock terror, begging Wendt not to make him go, in true José Jiménez fashion. Some of the assembled media were amused, but NASA's public affairs people were not and one even suggested that Cooper be "busted to Private Fifth Class". Ironically, the astronaut and Wendt liked the idea and ran with it.

This time, his wait inside the spacecraft lasted barely two and a half hours. The countdown ran smoothly until T-11 minutes and 30 seconds, when a problem developed in the rocket's guidance equipment and a brief hold was called until it was resolved. In fact, so smooth was the countdown that flight surgeons were astonished to note that Cooper's heart rate had fallen to just 12 beats per minute: he had dozed off. It took Wally Schirra, the capcom at Cape Canaveral, to bellow his name over the communications link to awaken him. Agonisingly, another halt came just 19 seconds before liftoff to allow launch controllers to ascertain that the Atlas' systems had assumed their automatic sequence as planned.

Thirteen seconds after 8:00 am on the morning of 15 May, the Atlas rumbled off its launch pad in what Cooper would later describe as a smooth but definite push. A minute into the climb, the silvery rocket initiated its pitch program and the astronaut felt the vibrations of Max Q, after which the flight smoothed out and he heard a loud clang and the sharp, crisp 'thud' of staging as the first-stage boosters cut off and separated. Unneeded, the LES tower was jettisoned and, at 8:03 am, Faith 7's cabin pressure sealed and held, as intended. Two minutes later, the sustainer completed its own push, shutting down and inserting the spacecraft perfectly into orbit. It was so good, in fact, that the heading was 0.0002 degrees from perfect, Cooper's velocity was right on the money at 28,240 km/h and his trajectory set him up for at least 20 circuits of the globe. Said Wally Schirra as America's sixth spaceman entered orbit: "Smack-dab in the middle of the plot!"

Cooper watched for about eight minutes as the sustainer tumbled away and then moved to his checklists, running through temperature readings, contingency recovery areas and began the process of adjustment to weightlessness. So rapid was Faith 7's passage across the Atlantic – accomplished in a matter of minutes – that he expressed surprise when called by the capcoms in the Canaries and Kano in Nigeria. Sigma 7 had been near-perfect and it seemed that Cooper's mission would match or excel it; he dozed off for a few minutes during his second orbit, as the spacecraft passed over a lonely stretch of the Pacific, between Hawaii and California. Flight surgeons would note that his heart rate surged momentarily from 60 to 100 beats per minute, suspecting that he was having an exciting, though somewhat brief dream. At one stage, things were running so well that Capcom Al Shepard had nothing to say, except to offer Cooper some quiet time. Not until the following day, 16 May, would serious problems arise and allow him to demonstrate his skills as a pilot.

He was by no means inactive. His tasks including eating – brownies, fruit cake and some bacon – as well as Earth observations, photography, collection of urine samples and monitoring Faith 7's health. His efficient use of the cabin oxygen even prompted Shepard to tell him to "stop holding your breath and use some oxygen if you like". Cooper's response was that, as the only non-smoker among the Mercury Seven, his lungs were in better shape than his colleagues. Not only was his oxygen expenditure economical, but so too was his fuel usage, prompting mission managers to nickname him, good-naturedly, a "miser". As Faith 7 embarked on its second orbital pass, Shepard reiterated that the flight was proceeding beautifully and "all of our monitors down here are overjoyed". In fact, Cooper's only complaint during this period was of a thin, oily film on the outside pane of his trapezoidal window.

Beginning with the third orbit, the astronaut set to work on the first of 11 scientific experiments assigned to his mission. One of these was a 15 cm sphere, instrumented with two xenon strobe lights, part of efforts to track a flashing beacon in space. Three hours and 25 minutes after launch, he clicked a squib switch and heard and felt the experiment separate successfully. However, despite repeated efforts, he could not see the flashing beacon in orbital darkness. He would later catch a glimpse of it pulsing at sunset, during his fourth circuit of the globe, telling Capcom Scott Carpenter with jubilation: "I was with the little rascal all night!" Cooper reported seeing the beacon flickering during his fifth and sixth orbits, too. Another major experiment, the deployment of a 76 cm Mylar balloon, painted fluorescent orange for visibility, was less than successful. Nine hours into the mission, he set cameras, attitude and spacecraft switches to release the balloon from Faith 7's nose, but it refused to move. Another attempt also proved fruitless. The intention of the balloon – similar to that flown on Carpenter's mission – was for it to inflate with nitrogen and extend out on a 30 m nylon tether, after which a strain gauge would measure the differences in 'pull' at Faith 7's apogee of 270 km and perigee of 160 km. Sadly, the cause of the failure was never determined.

Cooper was, however, able to observe not only a flashing beacon in space, but also a xenon ground light of three million candlepower, situated at Bloemfontein in South Africa. He would also make detailed mental notes throughout the flight as he flew over cities, large oil refineries near Perth in Australia, roads, rivers, small villages and even saw smoke from Himalayan houses. Although he pointed out that the finer details could only be seen if lighting and background conditions were right, his sightings were disputed after the mission, but Gemini astronauts would later confirm them. Further theoretical confirmation came from visibility researchers S.Q. Dunt and John H. Taylor of the University of California at San Diego. In a paper published in October 1963, they highlighted Cooper's observation of a dust cloud, presumably kicked up by a vehicle travelling along a dirt road near El Centro, on the border between Mexico and the United States.

"Calculation shows that the vehicle, plus the dust cloud behind it, is more visible than the road itself," agreed Dunt and Taylor in their report. "It is possible, moreover, that the appearance of the dust cloud would create the impression of having a lighter tip at its eastern end. There is reason to believe, therefore, that the presence of a moving Border Patrol vehicle on the dirt road near El Centro could have been seen from orbital altitude under the atmospheric and lighting conditions which we believe to have prevailed at the time of Major Cooper's observation."

Several other scientific experiments, in fact, encompassed photography. Before the mission, Cooper spent time with University of Minnesota researchers on an investigation into the mysterious phenomena of the zodiacal light and the nighttime airglow layer, as part of efforts to better understand the origin, continuity, intensity and reflectivity of visible electromagnetic spectra along the basic reference plane of the celestial sphere. His work would also help to answer questions about solar energy conversion in Earth's upper atmosphere. Many of the zodiacal light photographs turned out to be underexposed and the airglow shots overexposed, but they were nonetheless of usable quality and complemented Carpenter's images from Aurora 7. Flying over Mexico, Cooper photographed horizon-definition imprints in each quadrant around his local vertical position, part of a Massachusetts Institute of Technology project to design a guidance and navigation system for Apollo. Light-heartedly complaining that all he seemed to be doing was taking pictures, Cooper acquired some excellent imagery, including infrared weather photographs.

Surpassing Wally Schirra's nine-hour endurance record for the United States, Cooper settled down to a battery of radiation experiments to ascertain that the effects of the Operation Dominic artificial aurora were indeed diminishing. He also undertook the hydraulic tasks of transferring urine samples and condensate water between storage tanks. Physicians had expressed particular interest in urine checks and the Soviets had already highlighted significant accumulations of calcium in their cosmonauts' urine, suggesting that extended spaceflights could adversely affect human bones. Cooper found the hypodermic-type syringes used to pump liquid manually from bag to bag to be unwieldy and exasperatingly leaky, even telling his on-board tape recorder that "this pumping under zero-G is not good. [Liquid] tends to stand in the pipes and you have to actually forcibly force it through".

Ten hours into the mission, the Zanzibar capcom officially informed Cooper that his flight parameters – circling the globe every 88 minutes and 45 seconds – were good enough for 17 orbits. Shortly before retiring for a scheduled sleep period on his ninth revolution, Cooper ate a supper of powdered roast beef mush, drank some water and checked Faith 7's systems to ensure that they could be powered down for the next few hours. His orbital speed was truly phenomenal: after speaking to Capcom John Glenn, based on the Coastal Sentry Quebec tracking ship, near Kyushu, Japan, he swept south-eastwards over the Pacific and gave a full report to the telemetry command vessel Rose Knot Victor, positioned near Pitcairn Island . . . just ten minutes later!

The Pitcairn communicator told Cooper to get some rest, but that proved almost impossible. Passage over South America, then Africa, northern India and Tibet, during daylight, offered wonderful viewing and photographic opportunities. The Tibetan highlands, with their thin air and visibility seldom obscured by haze, allowed him to make rudimentary estimates of his speed and ground winds from the direction of chimney smoke. In their paper, Dunt and Taylor suggested that ground-reflectance modelling made it not impossible for Cooper to have seen such fine details. Thirteen and a half hours into the flight, Glenn told him that the communicators would leave him alone and Cooper pulled a shade across Faith 7's window to get some sleep. The astronaut dozed intermittently for around eight hours, anchoring his thumbs at one stage inside his helmet restraint strap to keep his arms from floating freely. He woke briefly when his pressure suit's temperature climbed too high and over the next several hours he napped, took photographs, taped status reports and cursed to himself as his body-heat exchanger crept either too high or too low.

Faith 7 swept silently over the Muchea tracking site on its 14th orbit and Cooper, by now fully alert, again checked its systems, finding his oxygen supply to be plentiful and around 65 per cent and 95 per cent of hydrogen peroxide fuel, respectively, in his automatic and manual tanks. At around this time, he said a brief prayer to offer thanks for an uneventful mission: "Father, we thank you, especially for letting me fly this flight. Thank you for the privilege of being able to be in this position, to be in this wondrous place, seeing all these many startling, wonderful things that you have created." Slow-scan television images of Cooper, the first ever transmitted by an American astronaut, were broadcast during his 17th orbit and he even sang one revolution later. The prayers and light moments, it seemed, actually marked the beginning of Faith 7's troubles.

Early on his 19th circuit of Earth, some 30 hours after liftoff, the first of several serious problems reared its head. Cooper was flying over the western Pacific, out of radio communications with the ground, when he dimmed his instrument panel lights . . . and noticed the small '0.05 G' indicator glow green. This should normally have illuminated only after retrofire, as Faith 7 commenced its manoeuvre out of orbit, and should also have been quickly followed by the autopilot placing the spacecraft into a slow roll. Initial worries that Cooper had inadvertently slipped out of orbit were refuted a few minutes later by the Hawaii capcom, who told him his orbital parameters held steady, suggesting either that the indicator was faulty or that the autopilot's re-entry circuitry had been triggered out of its normal sequence.

An orbit later, the astronaut was advised to switch to autopilot and Faith 7 began to roll. This had its own implications. For proper flight, Time magazine told its readers a week later, there were other functions for the autopilot to perform prior to retrofire. Since each function was sequentially linked to the next, Mercury Control knew that several earlier steps had not been performed. Cooper would have to control them by hand, a situation not entirely unpalatable, since Scott Carpenter had flown part of his re-entry in a similar manner. Still, at Cape Canaveral, a training mockup of the spacecraft in Hangar S was set up to practice various scenarios and provided an assurance that all would be well. Then, on its 20th orbit, Faith 7 lost all attitude readings and, a revolution later, one of its three inverters, needed to convert battery power to alternating current and operate the autopilot, went dead. Cooper tried to activate a second inverter, but could not. (The third inverter was needed to run cooling equipment inside the cabin throughout re-entry.) His autopilot, in effect, was devoid of all electrical power.

A Mercury capsule after splashdown.

As flight controllers scrambled to relay questions, corrections and instructions and practice procedures on the ground – including the possibility of bringing Cooper back to Earth on battery power alone – the astronaut himself remained calm, though he watched in dismay as carbon dioxide levels rose both inside the cabin and within his pressure suit. The lack of electrical power meant that he could not rely on his gyroscopic system to properly orient Faith 7 for re-entry; it would have to be lined up manually. He could not even rely on the spacecraft's clock. "Things are beginning to stack up a little," he told Capcom Scott Carpenter in a cool and typically understated manner, but acquiesced that he still had fly-by-wire and manual controls as a backup. "We would have found some way to fire the retros," Mercury engineer John Yardley said later, "if it meant telling him what wires to twist together."

Guided by John Glenn, aboard the Coastal Sentry, Cooper ran smoothly through his pre-retrofire checklist, steadying Faith 7 with the hand controller and lining up a horizontal mark on his window with Earth's horizon; this brought the spacecraft's nose down to the desired 34-degree angle. Next, he lined up a vertical mark with pre-determined stars to gain the correct yaw angle. Glenn counted him down to retrofire and Cooper hit the button on time, receiving no light signals, because of his electrical system problems, but he confirmed that he could feel the three small engines igniting. Re-entry was uneventful, with Cooper damping out unwanted motions and manually deploying his drogue and main parachutes. The spacecraft broke through mildly overcast skies and splashed into the Pacific, some 130 km south-east of Midway Island, only 6.4 km from the recovery ship Kearsarge. Floundering briefly, Faith 7 quickly righted itself and Cooper requested permission, as an Air Force officer, to be allowed aboard a naval carrier.

Forty minutes later, permission having been granted, the hatch was blown and America's sixth astronaut set foot on the deck of the Kearsarge. His mission had lasted 34 hours, 19 minutes and 49 seconds – nowhere close to the four days chalked-up by Andrian Nikolayev a year earlier, but a significant leap as NASA prepared for its ambitious series of long-duration Gemini flights. Even more significantly, Cooper had returned to Earth as all the astronauts had wanted: as a pilot in full control. It also offered a jab at the test pilot community, some of whom had ridiculed Project Mercury as little more than 'a man in a can' or, even more deridingly, as 'spam in a can'. Walt Williams, who only days earlier had tried to have Cooper removed from the flight, now warmly shook the astronaut's hand. "Gordo," he told him, "you were the right guy for the mission!"

The future seemed bright. Ahead, in a year's time, lay Gemini ... and then the Moon.

3

Space Spectaculars

JOURNEY TO SPACE

The frigid cold of space held no fear for Alexei Arkhipovich Leonov. Growing up in central Siberia at the height of Stalin's purges, he had seen far worse times. One of the most traumatic was in January 1938. He was just four months shy of his fourth birthday when neighbours in the tiny village of Listvyanka, near the point where the Angara River leaves Lake Baikal, arrived in the bitterness of midwinter to strip bare his family's belongings, their clothes, their meagre furniture and even their food. The young boy, one of a dozen Leonov children, was even told to remove his trousers. It was the punishment meted out by the harsh Soviet regime on Arkhip Leonov, a staunch Bolshevik falsely accused of being an 'enemy of the people'. He had, it was said, deliberately allowed seeds for the next year's harvest to dry out.

"My father was thrown into jail without trial," Alexei Leonov wrote decades later in a joint autobiography, co-authored with American astronaut Dave Scott. "As we were then regarded as the family of an 'enemy of the people', we were branded subversives. Our neighbours were encouraged to come and take from us whatever they wanted." His elder siblings were removed from school and the family was forced to leave Listvyanka. Ultimately, remembered Leonov, his father was absolved from blame, thanks to glowing testimony from a former commanding officer, received compensation and was offered the headship of his local collective farm. Arkhip Leonov declined, however, and chose to work instead at a power plant in Kemerovo, on the Tom River, to the north-east of Novosibirsk, with his sister and brother-in-law. It was here that the man who would one day become the first to walk in space first experienced his life's two passions: art and aviation.

In his autobiography, he recounted drawing pictures on the whitewashed stoves of his neighbours' rooms, earning extra bread and receiving pencils and paints for his efforts. "I loved to draw," Leonov wrote, and his interest was indulged by his parents, who stretched bed sheets over wooden frames to provide rough canvasses for his work in oils. His ambition to become a professional painter was, he added, eclipsed one day in 1940 by the desire to fly. "It happened the first time I set eyes on a Soviet pilot," he explained, "who had come to stay with one of our neighbours. I remember how dashing he looked in his dark-blue uniform with a snow-white shirt, navy tie and crossed leather belts spanning his broad chest. I was so impressed. I used to follow him everywhere, admiring him from a distance."

The pilot, after a time, noticed the young boy's interest and demanded to know why he was being followed. When Leonov told him that he, too, desired to be a pilot, the aviator smiled and explained that he would need to grow physically strong and study hard at school. Equally importantly, Leonov would have to wash his face and hands each morning with soap. "Like most little boys," wrote Leonov, "I was not too keen on soap and water," but he followed the pilot's advice, running to him every morning to proudly display his clean face and hands.

Acquittal from false accusation was by no means the end of the family's troubles, as German forces rolled into the Soviet Union in a sweeping advance on a very broad front. Leonov recalled seeing truckloads of wounded Russian soldiers arriving in Kemerovo and the mass construction of chemical plants, two of which were blown up by Nazi sympathisers. In the autumn of 1943, when Leonov started school, the Red Army had repulsed the Germans at Stalingrad, although times remained grim. Years later, he would remember chanting thanks to Stalin in school for his 'happy childhood' and, indeed, would grow up believing the despotic Soviet system to be the best in the world. Not until his mid-teens would he begin reading and learning of other, happier worlds beyond the borders of Russia. When Nikita Khrushchev came to power, Leonov took the black armband he had worn in mourning of Stalin's passing and burned it.

His ambition to become a professional artist culminated, in early 1953, with a journey in the back of an open lorry to Latvia to apply for a college place in Riga. This, sadly, came to nothing. Despite being accepted by the principal, Leonov's realisation that the expense of living in the Latvian capital was simply too high pushed him in another direction, toward his second love: aviation. In the autumn, he was offered a place at the Kremenchug Pilots' College in the Ukraine and for two years learned to fly propellor-driven aircraft, then moved on to the higher military academy in Chuguyev to train on MiG-15 jets. Shortly afterwards, the effects of the 1956 Hungarian Uprising placed Leonov and other young Soviet Air Force fighter pilots on full combat alert. Graduation from Chuguyev coincided broadly with the launch of Sputnik, at which time he was flying MiG-15s, specially modified to take off and land on soil airstrips at night and during the daytime.

One harrowing incident in particular brought him to the attention of a mysterious recruiting team ... and paved the road to the cosmonaut corps. Whilst flying in heavy cloud, a pipe in his jet's hydraulic system snapped. Alerted by his on-board instruments to a fire, Leonov shut off the fuel supply to the engine and performed an emergency landing. He was too low, he wrote, to parachute to safety. His actions drew the recruiters to ask "if I would be willing to join a school of test pilots". In October 1959, two years after Sputnik 1, he was one of 40 semi-finalists selected from a pool of thousands of highly-qualified MiG-15 and MiG-17 fighter pilots. Interestingly, the same questions had arisen in Russia as the United States over what kind of individuals were best suited to space travel: and pilots were considered the best option, in light of their ability to work under extreme conditions, react with lightning speed and demonstrate a range of complex engineering skills.

For a month, he and the other candidates underwent gruelling physical and mental evaluations. "We were put in a silent chamber," Leonov wrote, "and set a series of complex tasks while blinking lights, music and noise were played to distract us. We were given mathematical problems to solve while a voice was piped into the chamber giving us the wrong answers. We were put in a pressure chamber with very little oxygen in extreme temperatures to see how long we could withstand it." By now, it was becoming clear to the 25-year-old pilot that this evaluation was for something more serious than test flying and suspicions abounded that missions into space were on the agenda.

At length, the candidates returned to their air bases – Leonov himself was shortly to be posted to East Germany, barely 20 km from the border with the West – to await further orders. Before leaving, he married his girlfriend, Svetlana. Within months, however, in March 1960, he was recalled to Moscow to commence cosmonaut training. It was during this period that he first met Sergei Korolev, the man whom Leonov would credit with masterminding virtually the entire early Soviet space programme. "Our training was intensive," he wrote, "a punishing regime which pushed us beyond what we thought we were physically capable of. Every day started with a 5 km run, followed by a swim, before we even began our individual programmes. Every aspect of our daily routine was carefully monitored by a team of doctors and nutritionists." The trainees were also enrolled in the Zhukovsky Higher Military Academy for academic accreditation.

Five of Leonov's colleagues rocketed into orbit between 1961 and 1963, together with a woman, and at around the time of Valentina Tereshkova's flight, he received his first introduction to a new type of spacecraft: the 'Voskhod' (translated variously as 'Dawn', 'Sunrise' or 'Ascent'). During a visit to Korolev's OKB-1 bureau, he was captivated by the "more interesting design" of one capsule in particular. It had, he wrote, "a transparent airlock attached, with a movie camera installed". Korolev explained that all sailors on ocean liners were required to swim and, by extension, all cosmonauts should learn how to 'swim' in open space. The airlock, which extended like a large blister from the Voskhod, would be used for just such an exercise. Leonov was told to don a training space suit and evaluate the airlock. At that moment, he recalled years later, Yuri Gagarin, the first man in space, clapped him on the back and whispered that Korolev had just chosen Leonov for the assignment.

That assignment, which would reach fruition with a launch from Tyuratam on 18 March 1965, would make him not only the first man to walk in space, but also half of the world's first-ever two-man cosmonaut crew. The Voskhod vehicle that Leonov and fellow pilot Pavel Ivanovich Belyayev were to fly would look physically similar to Vostok and had actually been adapted to carry crews of two or three men for, conceivably, up to two weeks at a time. It was, admittedly, another of Nikita Khrushchev's cynical and short-sighted ploys to outdo American plans to launch their two-man Gemini capsules, three-man Apollo missions and conduct spacewalks. Although only two manned Voskhods ever flew – the first with a crew of three, including a scientist and physician, the second with Leonov and Belyayev – they

Voskhod 2 is readied for launch. Note the 'blister' in the side of the nose shroud, caused by the projecting airlock.

would prove themselves to be among the most dangerous and reckless space missions ever attempted.

SARDINES

Plans for Voskhod had arisen in the wake of Tereshkova's flight, when it became increasingly unlikely that ten more Vostok capsules would be built. This was later downsized to four additional spacecraft and, in July 1963, Korolev laid plans to use them to fly a dog to high altitude for ten days, followed by an eight-day solo mission and a dual-spacecraft joint endeavour lasting around ten days. By December, the manned missions remained more or less unchanged, although the canine flight had been extended to 30 days, to stretch the spacecraft's life-support and other resources to their limits. All four were intended purely as stopgap measures as Korolev's bureau struggled to prepare its next-generation spacecraft, Soyuz, for a maiden flight sometime late in 1964.

Unfortunately, the four additional Vostoks would not be available until the middle of that year, and in February it would appear that an order was received 'from above' to attempt a three-man mission to upstage the Americans and cloak the reality that the Soviets were falling behind in what was now being coined 'the space race'. Certainly, Nikolai Kamanin hinted that the three-man stunt originated in 'discussions' between Korolev, government officials Leonid Smirnov and Dmitri Ustinov and the chair of the Soviet Academy of Sciences, Mstislav Keldysh, and a schedule for the mission was established by the Military-Industrial Commission on 13 March 1964. This resulted in the delivery of four rockets and four spacecraft to Tyuratam in June and July, with plans to launch one of them as early as August. Meanwhile, on 13 April, a government resolution officially declared the Soviet Union's intent to conduct the three-man mission, together with the mention of an 'extravehicular' – spacewalking – flight, dubbed 'Vykhod' ('Exit').

In his diaries, Kamanin revealed that neither he, nor Korolev, were happy with the notion of cramming three men into a one-seater Vostok. "It was the first time that I had seen Korolev in complete bewilderment," wrote Kamanin. "He was very distressed at the refusal to continue construction of the Vostok and could not see a clear path on how to re-equip the ship for three in such a short time." The crafty Chief Designer, however, ultimately turned the situation to his advantage, apparently agreeing to build a three-man Vostok in exchange for Nikita Khrushchev firmly committing the Soviet Union to a lunar landing project. Others, including Khrushchev's son, have countered more recently that it was Korolev himself who originally proposed the idea of modifying the craft for three-man crews.

"It is easy to forget," acquiesced Asif Siddiqi, "that Korolev himself had an almost pathological desire to be first – to beat the Americans at all cost. It would not have been contradictory to his personality to pursue the three-cosmonauts-in-a-Vostok plan simply to upstage the early Gemini missions." Siddiqi added that Korolev was, after all, firmly committed in 1963 to flying four more Vostoks right up to the limits of their survivability in space. At the same time, "the proposal to usurp

Gemini ... completely ignored the natural progression of space vehicles and inserted a diversionary programme that would ultimately result in little qualitative gain for Korolev's grand vision of an expansive space programme".

Plans for a trio of unmanned precursor missions of this new machine were reduced to just one and, in August, its launch was scheduled for 15–20 September 1964. The name 'Voskhod' was devised to convince western observers that it was actually a totally new spacecraft, whereas in reality it was little more than a slightly-modified and somewhat heavier Vostok. Its launch vehicle, too, had been upgraded with additional lift capacity. However, it is said, Korolev opposed the idea of sending three men aloft in a converted Vostok as being unsafe, an assertion supported by his deputy, Vasili Mishin. "Fitting a crew of three people, and in space suits, in the cabin of the Voskhod was impossible," Mishin said later. "So, down with the space suits! The cosmonauts went up without them! It was also impossible to make three hatches for ejection. So, down with the ejection devices! Was it risky? Of course it was. It was as if there was a sort of three-seated craft and, at the same time, there wasn't. In fact, it was only a circus act for three people who couldn't do any useful work in space. They were cramped, just sitting."

Other engineers and managers, too, were sceptical of the new spacecraft's safety. Konstantin Feoktistov, who had played a crucial role in the development of Vostok and who would actually fly aboard Voskhod 1, recounted years later that "we argued that it would be unsafe, that it would be better to be patient and wait for the Soyuz to be built ... In the end, of course, [Korolev] got his way". And how? Korolev offered one of the seats on the three-man Voskhod to an engineer from the OKB-1 design bureau. "Well," continued Feoktistov, "that was a very seductive offer and a few days later we produced some rough sketches. Our first ideas were accepted. We unveiled our plans for this new ship in March or April [1964]." According to Siddiqi, it was Feoktistov himself who proposed omitting ejection seats and space suits from the cabin; the only means possible of fitting three men inside.

The remarkable achievement of sending three men into orbit at the same time thus hid the reality that they had no protection in the event of a depressurisation, no means of emergency escape and, unlike Vostok fliers, had no option but to remain in their spacecraft until landing. Further, the sheer volume available to house consumables meant that Voskhod 1 could not easily remain aloft for much longer than 24 hours. Saving graces came in the form of a backup retrorocket atop the spherical crew cabin to reduce the risk of stranding in space, together with two parachutes, instead of one, to bring Voskhod and its cosmonauts safely to the ground. The eliminated ejection seats, which made a soft-landing capability essential, were replaced with a trio of couches, fitted at a 90-degree angle to the Vostok position.

This soft-landing system, known as 'Elburs', consisted of probes attached to the parachute lines, whose contact with the ground triggered a solid-propellant braking rocket to effect a zero-velocity touchdown. So successful was this mechanism that the Voskhod 1 cosmonauts would recall that they did not notice the instant of contact. Still, the whole effort would later be seen as something akin to a 'Potemkin village': a false façade built over a shabby building. Unlike the 'spam-in-a-can' of

Project Mercury, Voskhod 1 represented something worse: sardines in a can!

Indeed, wrote Alexei Leonov, the capsule very much reminded him of Vostok. "Some of the control panels I was familiar with from Vostok had been shifted to different positions," he explained. "The optical orientation system had been moved 90 degrees to the left ... My first impression was that the cabin was very cramped. I later found that in zero-gravity, Voskhod took on a more spacious feel and could even become a quite comfortable and reliable temporary home." Of course, on Leonov's own mission, Voskhod 2, he and Belyayev would have more room available to them than the three-man Voskhod 1 crew, but this advantage was balanced by the reality that they had to both wear space suits. Although only Leonov would perform the world's first spacewalk, Belyayev needed to be equally attired in case of depressurisation.

By March 1964, plans were laid to begin training the Voskhod 1 crew. The pilot, Vladimir Mikhailovich Komarov, selected as a cosmonaut four years earlier, would be joined by physician Boris Borisovich Yegorov and – interestingly – a scientist named Konstantin Petrovich Feoktistov, one of the physicists who had helped design Voskhod ... and who had advocated the elimination of space suits and ejection seats. Nikolai Kamanin, for his part, felt that launching such 'untrained' men into space was insane and highly dangerous, suggesting that Korolev, Keldysh and Smirnov had gone too far in the ludicrous bid to beat the Americans. It has also been said that Nikita Khrushchev himself, when advised of the risks, opted to pursue the project regardless. Herein lay one of the key obstacles in the way of a truly competitive and guided Soviet space effort in the Sixties: it was governed, funded and operated on an ad-hoc and very much whimsical basis by a fickle Russian leadership. With Korolev's death in 1966, its focus would drift yet further.

Vladimir Komarov, in command of Voskhod 1, would become one of the first physical victims of this faltering effort. Not only would the Soviet Air Force lieutenant-colonel lead one of the most dangerous missions to date, but in the spring of 1967 he would acquire the unenviable record of becoming the first man to die during the course of a spaceflight. Born in Moscow on 16 March 1927, he was among the oldest of the cosmonauts, serving as Pavel Popovich's backup on Vostok 4 and probably headed for a later solo mission had the plans for ten more capsules not been scrapped. He had been raised in an old house in a district typical of 'Old Moscow', excelling in mathematics and working on a collective farm during the Second World War, later proudly declaring that he could saddle a horse equally as well as fly a jet.

Graduation from aviation school coincided with Victory Day in 1945 and, despite his mother's admonitions to refrain from 'dangerous' flying, the young Komarov was determined to pursue high-speed, high-altitude test piloting. The Borisoglebsk and Bataisk schools introduced him to the skills and requirements of combat aviation and a period in Moscow's Air Force Academy imbued him with the engineering knowledge essential to test flying. Completion of his work at the academy in 1959 was soon followed by the same mysterious telegram received by numerous other Soviet pilots, summoning Komarov to Moscow for weeks of medical and psychological testing. His selection, at the age of 32, made him one of the oldest cosmonauts and Asif Siddiqi has suggested that his experience and education carried him through. Although none of the 1960 cosmonaut selectees were test pilots, Komarov, as an aircraft test engineer, came closest.

Aboard Voskhod 1, four years later, he would sit shoulder-to-shoulder with two men from very different backgrounds. Neither Konstantin Feoktistov nor Boris Yegorov possessed test-piloting credentials, but had established themselves as experts in the fields of physical science and medicine. Feoktistov, indeed, was lucky to be alive at all. Born on 7 February 1926 in the south-western Russian city of Voronezh, close to Ukraine, he was caught up in the Great Patriotic War shortly after the defeat at the Battle of Moscow. Amidst the retreating remnants of the Red Army, his mother gathered her belongings and, with the young Feoktistov, joined the steady stream of refugees fleeing eastwards.

At a village where they stopped to rest, Feoktistov met a group of Red Army soldiers, one of whom remembered him trying to enlist a short time earlier and offered to make him a scout. In early July 1942, Feoktistov provided his first information to his superiors; information which earned him a commendation from his commanding officer. Then, walking the streets of Nazi-occupied Voronezh, he was stopped by a patrol, marched around the city and ordered to stop near a pit. Shortly afterwards, Feoktistov felt a sharp pain close to his chin, as a bullet grazed his throat, after which his legs caved in and he toppled face-first into the pit. The Nazis, thinking him dead, left. Feoktistov waited until nightfall, crawled out of the pit and returned home. In later life, a scar on his neck and the proudly-worn medal 'For Victory Over Nazi Germany' would be his mementoes of the day – and night – when 'Kostya's' luck held out.

After the war, Feoktistov, who had nurtured a fascination with space exploration since childhood, graduated from the Bauman Moscow Higher Technical School as an engineer and would subsequently complete a doctorate in physics. In 1955, as a member of Mikhail Tikhonravov's design bureau, he was part of the team which eventually placed Sputnik into orbit and worked on the design of ion-powered spacecraft for flights to Mars. When 'cosmonauts' were sought for Earth-orbital missions, Feoktistov volunteered, but was overlooked.

For his part, Sergei Korolev had long desired civilian engineers from his own OKB-1 design bureau to fly aboard space missions, but had thus far been thwarted by the Soviet Air Force. When Voskhod appeared on the horizon, he succeeded in persuading Mstislav Keldysh to approve this, although the latter seemed more interested in flying a qualified scientist than an engineer. Others, including Deputy Minister of Health Avetik Burnazyan, added their weight behind putting a physician on the crew. In March 1964, the decision was made to fly a pilot, a scientist and a physician on Voskhod 1 and by the end of the following month, the names of Komarov, Leonov, Yevgeni Khrunov and Boris Volynov had been thrown into the pot as candidates for the position of command. Then, on 26 May, a pool of physicians and scientists – Yegorov, Vladimir Benderov, Georgi Katys, Boris Polyakov, Vasili Lazarev and Alexei Sorokin – were selected for consideration. Two weeks later, as the sole member of 'Civilian Specialist Group One', Feoktistov was picked.

Early in July, Nikolai Kamanin selected Volynov, Katys and Yegorov to fill command, scientist and physician posts on Voskhod 1, with Komarov, Feoktistov and Sorokin backing them up and Lazarev in 'reserve'. The second mission, meanwhile, dubbed 'Vykhod' and intended to make the first spacewalk, was assigned a team of pilots: Belyayev, Leonov, Khrunov and Viktor Gorbatko. Years later, Leonov would recall their intense work schedule. "Every week," he wrote, "I returned to the spacecraft as its design was modified, to familiarise myself with every inch of the vessel. I knew every nut and bolt in that spacecraft. I used to sit in the cabin regularly in my space suit without turning the ventilation system on, to test my stamina." Of the Vykhod candidates, Kamanin favoured Leonov and Khrunov, considering them both to have sharp, analytical minds. Pavel Belyayev and Viktor Gorbatko were assigned as candidates for the post of Voskhod 2's commander.

Before Vykhod, however, would be the three-man stunt. On 12 August 1964, Volynov, Katys and Yegorov were confirmed as the prime crew, although Korolev expressed his desire for Feoktistov, a man with unrivalled technical knowledge of the Vostok and Soyuz spacecraft, to fly instead. Kamanin opposed the idea, considering the 38-year-old engineer to be in poor medical condition, "suffering from ulcers, near-sightedness, deformation of the spine, gastritis and even has missing fingers on his left hand". To Kamanin and many physicians, Feoktistov was uncertifiable. After heated debates, the Voskhod 1 backup crew was redefined as consisting of Komarov with physicians Lazarev and Sorokin. Circumstances changed quickly. In late August, Marshal Sergei Rudenko, the Soviet Air Force's deputy commander-in-chief, objected to the selection and recommended the inclusion of an engineer on the crew, rather than two physicians. Meanwhile, a hastily-convened panel under Avetik Burnazyan cleared Feoktistov to fly, infuriating Kamanin, who felt that a fair and rational selection process was now being derailed by a hand-picking leadership.

The shift of Yegorov and Feoktistov from the backup to the prime Voskhod 1 crew came swiftly. At the end of August, it was discovered that Georgi Katys had a brother and sister living in Paris, a fact that he apparently did not reveal during the selection process. Combined with the fact that his father had been executed by the Soviet state, his suitability as a cosmonaut was immediately thrown into question. Katys fought for his seat, arguing that he knew nothing of his Parisian siblings, who had been born before 1910, long before his own birth. Moreover, his father was executed in 1931, when Katys was barely five years old. Ironically, Katys had the unwavering support of Mstislav Keldysh and several other academicians. It made little difference, with Nikolai Kamanin writing that Katys' unfavourable background "spoils the candidate for flight". Towards the end of September, Sergei Korolev again pressed for Feoktistov to take Katys' place. Kamanin still opposed it, feeling that, with Yegorov, there would now be two 'invalids' aboard Voskhod 1.

On 24 September, Kamanin detailed his arrival at Tyuratam in his diary and recorded telling the prime and backup crews – and Feoktistov – that all seven must remain physically and psychologically prepared for the mission, since the final decision over who would fly would not come until a couple of days before launch. Nevertheless, when the cosmonauts flew to the site in October for final preparations, Kamanin was sure that they knew the State Commission had ratified Komarov,

Feoktistov and Yegorov for the mission. The third crew member, Yegorov, born in Moscow on 26 November 1937, was recognised – despite his youth – as an authority on the vestibular apparatus in the inner ear, responsible for controlling the sense of balance. He came from a distinguished medical family, his father having been a prominent heart surgeon and his mother an ophthalmologist.

Indeed, it has been said that Yegorov's mother, able to speak German, French and English, with a plethora of hobbies from drawing to singing and playing the piano to a love of mathematics, was the making of him, even though she died when he was a teenager. The young boy grew up with a wide range of interests, devising gadgets to switch on lights and radios from his bed and open drawers, even crafting an eight-valve television device when he realised that the factory-made set in the family's living room was not 'his own'. Yegorov graduated from the First Moscow Medical Institute in 1961 and it has been suggested that his father's influence within the Presidium assured him of his seat on Voskhod 1. Together with medicine, his interest in physics, cybernetics and radio electronics remained. Ironically, despite the label 'physician' which he held on Voskhod 1, Yegorov would not actually receive his doctorate until 1965, courtesy of Humboldt University in Berlin. He would also earn the degrees of a candidate and doctor of medical sciences some years later.

"THE WORLD'S FIRST PASSENGER SPACESHIP"

As the crew debate continued, the spacecraft itself was put rigorously through its paces in the summer of 1964. Air-drop tests to verify its new soft-landing system were conducted close to the Black Sea resort of Feodosia. These proved initially successful, but on 29 August problems materialised when jettisoning the parachute hatch: an error in the circuit design caused it to fail and the test capsule – which some sources suggest was Gherman Titov's old Vostok 2 – was destroyed. In retrospect, Korolev claimed the test capsule's electrical system was not representative of a 'production' Voskhod and, at a State Commission on 18 September he declared that he was ready to certify the new spacecraft as ready to fly. The final air-drop on 3 October, with Korolev present, was successful. However, other glitches remained.

Firstly, the launch of a Vostok-based Zenit 4 reconnaissance satellite, employing a rocket identical to that planned for Voskhod 1, was aborted on the pad when one of its first-stage strap-on boosters failed to ignite. It was the first such failure in more than a hundred launches of Korolev's Little Seven. Rescheduled for mid-September, the Zenit liftoff was normal, quantifying the rocket's capabilities and clearing it for use with Voskhod 1. Then, as the launch neared, the spacecraft's Tral telemetry system exhibited discrepancies, requiring a week to fix. Finally, on 6 October, a full-duration, day-long unmanned dress rehearsal of the Voskhod 1 mission was flown under the cover name of 'Cosmos 47'.

Early plans called for dogs to be flown, although this was eventually set aside in favour of full-sized mannequins. The Cosmos 47 spacecraft duly entered a 177–413 km orbit, inclined 64.8 degrees to the equator, and flew for 24 hours before landing in Kustanai. It was returned to Tyuratam on the 8th for examination, which confirmed that both its interior, exterior and – perhaps most importantly – its soft-landing parachute were in good shape. Indeed, Cosmos 47 was described as having "zero velocity" on impact with the ground, penetrating merely 90 mm into the soil, and, although strong winds dragged the parachute and capsule some 160 m after touchdown, it was decided that a cosmonaut crew could jettison the canopy and endure this. The Little Seven, too, performed well, despite a slight depletion in thrust which was supplemented by the engine controller.

By 11 October, the Moscow rumour mill was billing Voskhod 1 as the 'Soviet Apollo'; a false illusion that would endure for many years. Wire services relayed news of a forthcoming flight with a cosmonaut known only as 'K', who was described as a violin player, a full Communist Party member and the bearer of a Ukrainian accent! Some observers already suspected Vladimir Komarov's involvement, particularly in light of an earlier rumour, in mid-August, when the Vostok 3 and 4 cosmonauts revealed that their backups were ready to fly. Those backups happened to be Komarov and Boris Volynov, which made some sense.

More problems with the Tral system on the evening before launch, which necessitated its last-minute replacement and, according to Nikolai Kamanin in his diary, caused Korolev to fly into a rage, did not conspire to delay the flight. The morning of 12 October dawned frosty, wrote Kamanin, although he considered it ideal: wind speeds were gentle and visibility extended to more than 20 km. The State Commission approved the launch at 3:00 am Moscow Time and, shortly thereafter, the Voskhod 1 crew – Komarov, Yegorov and Feoktistov – were awakened. The men washed, ate breakfast and were fitted with biosensors and dressed in blue flight garments. Since no pressurised suits would be worn, all three were fully outfitted by 7:00 am and ready to ride to the pad. During this quiet time, Kamanin advised them of secret code words to be used during the mission: "Outstanding" would mean just that, "Good" would denote the appearance of problems and "Satisfactory" would request an immediate emergency landing.

By 8:15 am, the cosmonauts had arrived at Gagarin's Start and Komarov rendered a smart salute and a declaration to the State Commission's chairman that he and his crew were ready to perform their mission. The two fliers whom Kamanin had labelled 'invalids' were first to enter the capsule; donning suede slippers, Yegorov boarded first, then Feoktistov moved to his middle seat and Komarov brought up the rear in his couch, closest to the hatch.

Launch itself came at 10:30 am Moscow Time – or 12:30 pm local time in Tyuratam – and the ride to orbit, thankfully, was uneventful. Indeed, Kamanin and Korolev had already discussed the crew's dire predicament in the event of a booster failure: a safe recovery was simply impossible for at least the first half-minute of the ascent and even an abort during the remainder of the climb to orbit only "should" have been achievable and survivable. (Korolev, apparently, was so nervous that he was visibly shaking during Voskhod 1's ascent.) With the benefit of hindsight, it is perhaps fortuitous that only two of these exceptionally high-risk ventures were ever attempted with men aboard.

Five hundred and twenty-three seconds later, Voskhod 1 entered orbit. Communications between Komarov and fellow cosmonaut Yuri Gagarin at

The crew of sardines: Feoktistov, Komarov and Yegorov.

Tyuratam had been clear and consistent throughout the ascent phase and the first few hours of the mission were characterised by only minor problems. A false reading was registered by Yegorov's biosensors, then there was a two-hour delay in confirming the correct operation of Voskhod 1's orientation system and the cabin temperature rose unexpectedly from 15°C to 21°C. However, by the seventh orbit, all cabin readings – pressure, humidity, gas composition and temperature – were normal and voice contact and televised images from the cabin proved crisp and clear.

At 11:46 am, a UPI wire revealed that "... the Soviet Union today launched the world's first passenger spaceship with three men aboard ..." and Radio Moscow's famous wartime announcer Yuri Levitan boomed out the news to an astonished world. Orbital parameters were given as 178–408 km, inclined 64.9 degrees to the equator, and radio hams in western Europe and North America picked up Morse transmissions from the spacecraft, identifying Komarov's callsign of 'Ruby'. The three cosmonauts extended greetings to the athletes of the 1964 Olympic Games, which had begun two days earlier in Tokyo, and spoke to both Nikita Khrushchev and his deputy, Anastas Mikoyan. At one stage, Khrushchev declared that Mikoyan "is standing next to me and is keen to take the telephone receiver from me". Within hours, not only the telephone receiver, but also the mantle of power as head of the Soviet Union, would have been taken from him. By the time Voskhod 1 parachuted to terra firma on the afternoon of 13 October 1964, Russia would be under new management.

FROM THAW TO STAGNATION

By the summer of 1964, opinions of Khrushchev within the Presidium were hardly complimentary. He had long been considered a boorish leader, which some blamed on his limited education. He had twice interrupted a speech by British Prime Minister Harold Macmillan, labelled Chairman Mao as an "old boot", famously pounded his fists – and shoe – on the desk during a United Nations General Assembly meeting and declared, in reference to capitalist nations, "We will bury you!" Not only did he prove hugely embarrassing for the Soviet Union's ruling elite, but many of his policies were ill-conceived and ill-considered. For example, in a bid to solve his country's agricultural woes, he had suggested the mass-planting of maize, on a scale akin to the United States, without realising that the inappropriate Russian soil and climate made this impractical.

Pivotally, the events of October 1962 humiliated Soviet hardliners, who perceived his removal of missiles and withdrawal from Cuba as a victory for the United States. Equally, the more liberal members of his government opposed his moves in Cuba as reckless 'adventurism'. Khrushchev's deposition came as the result of a conspiracy among a Communist Party leadership that could no longer disguise its irritation at his major political mistakes. Led by Leonid Brezhnev and Alexander Shelepin, together with KGB chief Vladimir Semichastny, the conspirators struck whilst their premier was on holiday at the resort of Pitsunda in Abkhazia, Georgia.

Writing in Time magazine two decades later, Sergei Khrushchev recalled the tension his father had felt when he called Leonid Smirnov on 12 October 1964 to demand news of the Voskhod 1 launch. Smirnov, barked the premier, should have kept him fully informed of the launch and circumstances pertaining to the mission. Little did he know that his most senior colleagues, including Smirnov – an aide of the past 30 years – were already deserting him. That evening, as Khrushchev and Anastas Mikoyan walked together along the Pitsunda beachfront, they were approached by a duty officer who told the premier that he had a telephone call from Presidium and Secretariat member Mikhail Suslov. 'Questions' needed to be asked, Suslov told him, which could not wait for the end of Khrushchev's two-week vacation, nor could 'discussion' of the Soviet Union's agricultural problems. Eventually, the premier agreed to return to Moscow the next day, 13 October.

After a three-hour flight to the capital aboard an Ilyushin-18 aircraft, during which time Khrushchev was uncharacteristically quiet and demanded to be left alone with Mikoyan, he was greeted by ... nobody. Not a soul from the Central Committee was waiting for him on the tarmac. His first contact was Vladimir Semichastny, who offered a polite welcome and informed the premier in a low voice that "everybody" had gathered at the Kremlin to meet him. Various charges were levelled against Khrushchev: unsatisfactory performance in dealing with Soviet agricultural problems, disrespectful treatment of members of the Presidium – including Brezhnev – and disdain for their opinions, the embarrassing withdrawal from Cuba, deteriorating relations with China, ongoing events at the Suez Canal and others.

Khrushchev had already decided, his son later wrote, to resign his powers without a struggle. He would stand down on the basis of ill-health and his son would recall him returning home on 14 October, thrusting a black briefcase into his hand and declaring "It's over ... retired". He was granted a dacha and city apartment for life, a pension of 500 roubles per month, together with his own security staff, car and chauffeur. Still, until his death in 1971, the events of those days would prove painful and he would complain bitterly of the spinelessness of his colleagues, including Brezhnev, who succeeded him as First Secretary and who had previously not given the slightest hint of wishing to oust him. Mikoyan, alone among the Central Committee to have supported Khrushchev, wanted to employ him as a 'consultant' to the Presidium, but, predictably, the request was turned down.

By 15 October 1964, two days after Voskhod 1 returned to Earth, the Presidium of the Supreme Soviet formally accepted Nikita Sergeyevich Khrushchev's resignation. It also ratified the appointment of Alexei Kosygin as the new premier, Brezhnev as First Secretary of the Communist Party and Mikoyan as chair of the Presidium. Kosygin would attempt to implement economic reforms to shift the paradigm of the economy from heavy to light industry and the production of consumer goods, although Brezhnev opposed this and as the Sixties wore on it was the latter who came to be seen as master of the state. Disturbingly, in a May 1965 speech, Brezhnev mentioned Stalin in a positive light for the first time in more than a decade – even adopting the dictator's old title of 'General Secretary' – and set about implementing increasingly conservative, regressive and repressive reforms.

The 'thaw' of de-Stalinisation which Khrushchev had overseen was steadily replaced with a period of socioeconomic stagnation under Brezhnev, which, ultimately, would pave the way for perestroika and the end of the Soviet Union. Notoriously, in February 1966, the writers Yuli Daniel and Andrei Sinyavsky would be tried and sentenced to hard labour for penning 'anti-Soviet' satirical texts, published under pseudonyms in western Europe. This prompted several historians to link the infamous episode with starting the movement to end communist rule. Under Brezhnev, the KGB would attain a similar level of power to that which it enjoyed under Stalin. In 1968, Czechoslovakia would be invaded at his direction, in response to Alexander Dubček's proposed reforms, and relations with China would decline to the extent of armed clashes along their borders on the Ussuri River.

In January 1969, as the blood-spattered decade drew to its close, an attempt would be made on Brezhnev's life. For cosmonauts Valentina Tereshkova, Georgi Beregovoi and Alexei Leonov, it almost signalled the end of their lives, too.

RISKY ENDEAVOUR

The Voskhod 1 fliers were still in orbit when the first of the calamitous events of October 1964 took place. Indeed, when Komarov requested a one-day extension to the mission, he was cryptically refused by Korolev, who quoted Hamlet with the words "There is more in heaven and earth, Horatio, than are dreamt of in your philosophy". This has been taken by some historians as a veiled hint at Khrushchev's removal from office and, perhaps, that Voskhod 1 was originally intended to fly a somewhat longer mission. After a full day aloft, during which time Yegorov took pulse and respiration measurements, Feoktistov checked Voskhod's equipment and atmosphere and Komarov evaluated the spacecraft's orientation system, the time came for the fiery plunge to Earth.

According to Asif Siddiqi, the extreme shortness of Yegorov and Feoktistov's training regime was reflected in their reactions to weightlessness. "Within two or three hours of the launch," he wrote, "both began to experience disorientation in space. Yegorov felt as if he was bent over face-downward, while Feoktistov actually felt he was upside down. Although the sensations apparently did not impair their ability to work, both suffered these feelings throughout the entire length of the mission ... an anomaly that had not been detected on any of the earlier Soviet space missions. Both cosmonauts also felt dizzy when they moved their heads sharply. It seems that Yegorov had been more afflicted, with his unpleasant sensations peaking about seven hours after launch ..."

Shortly after 9:00 am Moscow Time on 13 October, Komarov was advised of the landing instructions and completed the orientation of Voskhod and the firing of its retrorocket. Indications that everything did not go entirely smoothly were, however, relieved by an electrical signal from the spacecraft, which confirmed that the main capsule had separated from the instrument section. At 10:26 am, a tracking station in the Caucasus picked up the signal and followed Voskhod 1 as it hurtled over the Caspian Sea, sped high above the fishing port of Aralsk in south-western Kazakhstan and finally touched down 312 km north-east of Kustanai. Further electrical signals confirmed that the parachute hatch jettisoned properly, but for an anxious Sergei Korolev the key concern was whether the parachutes themselves had deployed. Airman Mikhailov, aboard an Ilyushin-14 some 40 km east of Marevka, confirmed that he saw two parachute canopies ... then, thankfully, the welcome sight of the capsule on the ground with the three men, alive and well, waving at him.

Landing occurred at 10:47 am, completing a mission of scarcely a few minutes more than 24 hours, and the three cosmonauts were helicoptered to Kokchetav and thence to Kustanai and finally Tyuratam, arriving late in the afternoon. Komarov, the only 'real' cosmonaut on the crew, was described as looking tired, but the two 'invalids', Feoktistov and Yegorov, were in good condition and high spirits. Plans for the men to speak to Khrushchev, mysteriously, were postponed when it became apparent that he had returned from his dacha in Pitsunda to Moscow. All attempts to call him at his office in the capital were unsuccessful; only later would it become apparent that as Voskhod plummeted Earthward, so Khrushchev's premiership had plummeted to its own ignominious end.

In their first conversations with the physicians, Feoktistov and Yegorov would describe their first – and only – experience of weightlessness: the former said that he found conditions not at all unpleasant and the latter, while admitting to feeling unwell during the first few orbits, recovered thereafter. Khrushchev remained uncontactable and, indeed, Marshal Sergei Rudenko was ordered to fly back to the capital immediately, delaying the cosmonauts' report to the State Commission until the next day. In his diary, Nikolai Kamanin noted his concern that "something unusual was happening in Moscow". The cosmonauts' post-flight visit to Red

Square was cancelled and their first meeting with the Soviet leader did not come until 19 October, nearly a full week after their landing. That leader was not Nikita Sergeyevich Khrushchev, but Leonid Ilyich Brezhnev.

The three-strong crew, Time magazine told its readers on 23 October, "was a sure promise of multi-man space stations", adding that "none of these feats have yet been accomplished by the lagging US space programme". Indeed, at the beginning of the second decade of the human exploration of space, following their loss of the Moon race, the Soviets would become the first to establish a true foothold in the heavens with a long-term orbital base called Salyut. It was a remarkable achievement that would, by the end of the 20th century, see cosmonauts spending more than a year apiece in space, and despite the harsh economic downturn after the collapse of the Soviet Union, Russian expertise continues to be drawn upon in today's International Space Station effort.

However, neither Time nor many of its readers were entirely fooled. The lack of transparency in the early Soviet missions meant that it was virtually impossible to determine what kind of spacecraft Komarov, Yegorov and Feoktistov had ridden into orbit. Rumours quickly abounded in the western press that longer multi-manned flights were planned. Other rumours, which suggested that Voskhod 1 had been brought home early due to communications difficulties, the illness of a cosmonaut or a malfunctioning rocket, were firmly refuted by the Soviets. In fact, the only light they publicly cast on the 'new' spacecraft was that it was lined with "a snowy-white, soft, sponge-like synthetic fabric", that its trio of cosmonaut seats were arranged in a row and that its instrumentation consisted of a navigation globe, telegraph key and a multitude of buttons and switches. It was, admitted Time, "little help in deciding whether the Sunrise was entirely new or merely an improved version of the standard one-man Vostok-type spaceships". Senator Clinton Anderson, chair of the Senate's space and aeronautics committee, speculated that it weighed some 6,800 kg – not too much higher than its real 5,680 kg – and, further, hinted that this would make it possible to fly aboard a similar rocket to that used by Vostok.

Within days, on 28 October, plans for subsequent missions were laid out. The Vykhod flight, to be flown by Pavel Belyayev and Alexei Leonov, was tentatively pencilled-in for the first quarter of 1965, after which five others would undertake long-duration sorties of up to 15 days, conduct scientific research and perform another spacewalk. These would decisively surpass American plans to fly their first two-man Gemini in the spring of 1965, conduct a spacewalk that summer and attempt a 14-day endurance run thereafter. Hopes were high, wrote Nikolai Kamanin at the end of December, that five or six cosmonauts could be flown in a pair of spacecraft which would rendezvous and dock in orbit as early as 1966. After this, perhaps, a flyby of the Moon could be attempted.

Key to the success of the Vykhod mission – Voskhod 2 – was the 250 kg inflatable airlock through which Leonov would have to squeeze his way, all the time encased in a cumbersome, pressurised suit which would almost claim his life in orbit. On Gemini missions, an astronaut would venture outside by depressurising the capsule and opening the hatch. This was feasible because the American miniaturised electronic systems were capable of operating in vacuum. The fact that the Soviet

Belyayev and Leonov, both clad in EVA suits, prepare for their audacious mission.

systems relied on air cooling meant that the capsule could not be depressurised and, hence, the airlock was born. Initial designs sketched out rigid structures, flexible contraptions and even one that could be rolled into a spiral before launch. The 'winning' design, codenamed 'Volga', envisaged a cylindrical device composed of 36 inflatable booms isolated in groups of 12. In this way, even if two groups of booms lost pressure, the airlock would retain its shape.

The airlock was not merely desirable, but necessary, since Voskhod's avionics were cooled by cabin air and would overheat if the entire cabin was depressurised. Physically, it comprised a metallic ring, 1.2 m wide, which fitted over the spacecraft's inward-opening hatch, and its length when fully deployed amounted to 2.5 m. Oxygen to inflate and pressurise the airlock booms was supplied by four spherical tanks and took around seven minutes to complete in orbit. The chamber boasted two lamps and three 16 mm cameras – two inside, one outside – and control of the inflation procedure would be performed from inside the cabin by Belyayev. However, a backup set of controls for Leonov's use were suspended on bungee cords inside the Volga airlock itself.

For two years, the cosmonauts and their backups – Dmitri Zaikin and Yevgeni Khrunov – trained to a point at which the spacecraft's cabin and airlock seemed like a second home, albeit a cramped one. In his autobiography, Leonov would recall his close friendship with Belyayev and express relief that attempts to remove 'Pasha' from Voskhod 2 on the basis of an old leg injury sustained in a parachute accident were not successful. "There were those," Leonov wrote, "who had wanted Yevgeni Khrunov to command the mission ... but I lobbied hard for Pasha, whom I thought more capable than Khrunov. I had worked with him more; I trusted him. In the end they agreed, though it caused some rancour with Khrunov." Instead of commanding the prime crew, Khrunov would instead shadow Leonov on the backup team. Other attempts to fly another cosmonaut, 43-year-old Georgi Beregovoi, were quashed by Nikolai Kamanin; not on the basis of his age, but in view of his height and weight, both of which were greater than Belyayev, Leonov, Zaikin or Khrunov.

Much of Leonov's preparation for the actual spacewalk, which would last between ten and 15 minutes, was undertaken in a modified Tupolev Tu-104 aircraft, flown in a series of parabolic arcs to simulate weightlessness for periods of up to 30 seconds at a time. It was time-consuming and imperfect, wrote Leonov, since "there was no way of simulating pure weightlessness in any laboratory on the ground ... such limited, disconnected periods meant that the hour and 15 minutes of exiting the spacecraft, performing the spacewalk and re-entering the airlock had to be practiced over the course of over 200 Tu-104 steep climbs". Astronaut Deke Slayton, who would fly with Leonov on the Apollo-Soyuz mission ten years later, remembered the world's first spacewalker recounting that he needed to build his upper-body muscles and indulged in diving "to test his equilibrium". Leonov also, according to Asif Siddiqi, "cycled about a thousand kilometres in less than a year, carried out more than 150 EVA training sessions and jumped by parachute 117 times". Elsewhere, pressure-chamber tests with Zaikin and Khrunov satisfactorily evaluated the performance of the airlock at conditions equivalent to an altitude of 37 km. At around the same time, early in February 1965, the mission was redesignated as Voskhod 2, rather than Vykhod ('Exit'), which it was felt would give away its true nature.

As their March 1965 launch date drew nearer, plans were afoot to despatch an unmanned precursor, Cosmos 57, to demonstrate the performance of the Volga airlock. Early on 22 February, three weeks before Belyayev and Leonov were due to fly, it was blasted into orbit ... and, shortly thereafter, it exploded! Telemetry signals intercepted by American intelligence hinted that its airlock had successfully deployed and tests of opening and closing its outermost hatch were in progress. Then, the hatch was automatically closed and further telemetry indicated that commands to repressurise the airlock had been issued and accepted by Cosmos 57. Television images transmitted from the spacecraft revealed that the airlock appeared to have fully inflated. "During the first orbit," wrote Nikolai Kamanin in his diary, "the craft was observed by special television circuit at Simferopol and Moscow [ground stations] ... Quite unexpectedly, a distinct image of the front part of the airlock appeared on the screen, causing an outburst of joy among all present."

Unfortunately, the Klyuchi and Yelisovo ground stations, both in Kamchatka, then issued simultaneous commands to deflate the airlock, which confused Cosmos 57 and which it interpreted as an instruction to commence descent. The spacecraft automatically set the retrofire process in motion, but the TDU-1 rocket fired improperly. Spaceflight analyst Sven Grahn has mentioned on his website that

Cosmos 57 apparently began to tumble; this was at least partly due to the unjettisoned airlock, whose presence would have displaced the spacecraft's centre of mass and induced the spinning. Twenty-nine minutes later, as programmed, the on-board destruct system – designed to prevent sensitive hardware from falling into enemy hands – was automatically activated to destroy Cosmos 57. A post-flight accident commission found that Klyuchi alone was permitted to transmit the airlock-deflation command and that Yelisovo should only have done so as a backup measure and only if directed by the Moscow control centre. In his diary, Kamanin would blame the poor security of commands being issued to orbiting spacecraft and felt that such weaknesses could be exploited by the United States.

Sergei Korolev faced a quandary. Only one other Voskhod capsule was ready to fly and the spacewalk could not be attempted until the airlock had been satisfactorily evaluated on a fully-successful mission. On 7 March, a Zenit reconnaissance satellite, under the cover name of 'Cosmos 59', was launched. A key concern was that, assuming the inflatable airlock jettisoned successfully after Leonov's spacewalk, its attachment ring would rise 27–40 mm above the surface of the capsule; however, if it only partially separated, the ring could be as much as 80 mm high. In such an eventuality there was a chance that the asymmetry of the ring on the upper heat shield could impart a rotation on the Voskhod, perhaps affecting the safe deployment of the drogue parachute during descent. Cosmos 59 was equipped with an identical airlock attachment ring to that planned for Voskhod 2 and its eight-day mission was entirely successful, proving, with just days to spare, that Belyayev and Leonov could survive re-entry and landing with this in place.

Nonetheless, in his autobiography, Leonov would recount the Chief Designer's concern that he and Belyayev should have the final decision over whether to fly their mission. Leonov also wrote of Korolev's visible exhaustion: he had suffered from high fever as a result of lung inflammation and, indeed, would die early the following year, after prophetically declaring Voskhod 2 as the "last major work of his life". Korolev told the men that they had two options: they could wait a year for another Voskhod to be built, fly one spacecraft unmanned to complete Cosmos 57's mission and then launch Belyayev and Leonov, or they could take a risk and fly the manned mission immediately. "Then, very cannily," wrote Leonov, "he added that he believed the Americans were preparing their astronaut Ed White to make a spacewalk in May. He knew how to get our competitive juices flowing. He must have known what we would say. We didn't want to lose a year."

A WALK OUTSIDE

Rollout of Voskhod 2, atop its R-7 booster, to Gagarin's Start occurred on 17 March, with an anticipated launch the following morning. That same day, data from the just-landed Cosmos 59 proved encouraging: during re-entry, despite the presence of the airlock attachment ring, the rate of rotation of the capsule never exceeded 40–100 degrees per second, well within design tolerances for both the crew and the parachute deployment mechanism. By the evening, intense rumours were circulating in Moscow of the impending flight of what was dubbed a 'space bus', but about which very little else was known.

Three cosmonauts suited up in the small hours of 18 March, one of them – Yevgeni Khrunov, who had trained for both Belyayev's and Leonov's positions – ready to take over from either prime crew member if needed. He was not needed on this occasion, but his spacewalking skills would be put to the test four years hence, when he embarked on a far riskier endeavour: to transfer between two orbiting Soyuz capsules in a pressurised suit. In his autobiography, Leonov recalled the customs he and Belyayev observed before launch: a breakfast of boiled eggs, a small sip of champagne, a brief moment of reflection with Yuri Gagarin and Korolev and, finally, at the pad, the time-honoured tradition of urinating on the wheel of the bus. Belyayev, the commander, was first aboard Voskhod 2, followed by Leonov. Also loaded into the capsule were a few personal items for the cosmonauts. Leonov's stash included a sketchpad and a set of coloured crayons. His childhood dream to become an artist would figuratively and literally reach new heights on this flight. He had already decided not to tell his wife of the spacewalk, instead informing her vaguely that he and Belyayev would be embarking on "a particularly complex and challenging mission". Indeed they would be.

Launch occurred at 10:00 am Moscow Time and the spacecraft entered orbit shortly afterwards. The entire ascent was flawless, although the first minute in particular proved stressful, with both men fully aware that they had no means of emergency escape. "As the engine of the rocket beneath us ignited, we felt a light vibration start to build," wrote Leonov. "Lifting away from the launch pad, we were pushed back into our seats. Now we felt the full force of the rocket propelling us upward through the Earth's atmosphere. It felt as if we were being lifted vertically by a speeding train. From this moment on we were required to report constantly on how we felt."

Leonov's first glimpse of Earth from the edge of space actually disappointed him, since it did not appear much different to the vistas he had seen from the MiG-15s he flew at Chuguyev almost a decade earlier. "I had expected to see the curvature of the horizon against a dark sky, but we were not yet high enough for that," he wrote. "Ten minutes into the flight, at an altitude of almost 500 km, our capsule separated from the rocket with a loud flap. We were flying far beyond the confines of the Earth's atmosphere." At this instant came weightlessness, which manifested itself in a flurry of loose objects drifting serenely through the cabin ... and by an absolute, ethereal silence. In fact, with the R-7's roar gone, it was now so quiet inside Voskhod 2 that the two men could even hear the clock on their instrument panel ticking.

"For two or three minutes," continued Leonov, "it was extremely uncomfortable. I had the feeling that I was suspended upside down, which is a well-documented phenomenon: once the force of gravity ceases, the senses become confused. But we quickly got used to it and started going through a complex series of checks to verify that all systems inside the capsule were operating normally." Shortly afterwards, Belyayev requested permission to extend the Volga airlock and promptly activated switches which pumped air into small rubber tubes running along the length of the chamber, inflating it from a coiled 74 cm to a fully-unfurled 2 m. Leonov, meanwhile, busily strapped bulky breathing apparatus, carrying sufficient oxygen for 90 minutes outside, onto his 'Berkut' ('Golden Eagle') space suit. When he was ready, Belyayev clapped him on the back and wished him luck.

In a similar manner to the ensemble already in the works for Gemini, the Berkut was of the purest white, "to reflect all possible sunlight," Time magazine told its readers, "for maintaining tolerable temperatures is one of the major problems in the design of space suits. Because sunlight in space is twice as strong as at the bottom of the atmosphere and contains ultraviolet rays that quickly weaken many materials, the outer layer of a space suit must not only ward off light and heat, but must be proof against ultraviolet". Equally hazardous was the intense cold, as Leonov passed from the direct sunlight of orbital daytime to the deepest black of frigid orbital nighttime, requiring the Berkut to perform adequately at both extremes. Not surprisingly, Time added, the Soviets had failed to describe the materials from which their suit had been manufactured, only admitting to its colour and general appearance. It would be many years before the Berkut was revealed in its entirety.

"Once inside the airlock," Leonov wrote, "I closed the hatch and waited for the nitrogen to be purged from my blood. To avoid suffering from what divers call 'the bends', I had to maintain the same partial pressure of oxygen in my blood once I emerged into space. With the pressure inside the airlock finally equal to zero pressure outside the spacecraft, I reported I was ready to exit." He was 'lying' on his back when the outermost airlock hatch opened, revealing the grandeur of Earth in its entirety for the first time. Years later, he would lucidly recall those first few, heart-racing moments as he pushed his upper body out of the airlock and, safely attached to Voskhod 2 by a 15 m tether, into open space. As the capsule neared orbital sunrise, he beheld the vast, deep blue panorama of the Mediterranean, together with the familiar shapes of Greece, Italy and, as Voskhod headed eastwards, the Crimea, the snow-capped Caucasus Mountains and the mighty Volga, largest river in Europe and national waterway of Russia.

Leonov brought his feet to the rim of the airlock and held tightly to a handrail for an instant, before letting go. The exhilarating feeling of being the first human being ever to do this – to actually leave the confines of a spacecraft – would remain with him with surreal clarity for decades; he felt both insignificant and overwhelmed by the importance of his achievement. His departure from the airlock came at 11:34:51 am Moscow Time, barely 94 minutes after launch, just before Voskhod 2 reached the radio horizon of the Yevpatoriya ground station in the Crimea. It was here that the ghostly images of humanity's first spacewalk were received.

"Dim and probably purposely fuzzy shots showed the round white top of a helmet poking slowly out of a hatch," Time magazine reported a week later on 26 March. "Then came the visored face of a man, followed by his shoulders and arms. He seemed to push something away with his left hand before he moved his left arm back and forth as if to test its freedom. He reached for a handrail and quickly his entire body came clear of the hatch. Now it could be seen that he was dressed in a bulky pressure suit, with cylinders strapped on his back and a thick cable twisting behind him ... " It was an image that would trigger dispute from some observers, who

Alexei Leonov during humanity's historic first spacewalk.

argued that the film had been faked in a terrestrial studio, that Sun-glint angles were not 'quite right' for it to be authentic. Rather than expressing disgust, however, Leonov acquiesced that "the race between our two countries for superiority in space was intense ... Personally, I did not believe in all this boasting about who did what first, the Soviet Union or the United States. If you did it, you did it".

Leonov may not have cared eitherway, but Leonid Brezhnev certainly did. In spite of the tumult surrounding the overthrow of his predecessor, Nikita Khrushchev, the new premier continued to support Soviet ambitions in space ... although his knowledge of its practicalities would later prove the butt of jokes when he suggested beating the Americans to the Moon by landing instead on the Sun. When advised that the cosmonauts would burn up, Brezhnev supposedly told them to land at night! On the morning of 18 March 1965, however, he was full of pride. "We, all members of the Politburo," the premier, surrounded by his stone-faced aides, began, "are here sitting and watching what you are doing. We are proud of you. We wish you success. Take care. We await your safe arrival on Earth." Brezhnev's pride was not, initially, shared by Leonov's young daughter, Vika, or his elderly father: the former hid her face in her hands and cried, while the latter, not understanding that the point of the mission was to venture outside, demanded that his son be punished for acting "like a juvenile delinquent" by abandoning his spacecraft in orbit.

Years later, Leonov would reveal more detail of the Berkut suit and his own activities outside Voskhod 2. He described the gold-plated filter across his visor, which, although satisfactory in cutting out nearly all ultraviolet sunlight, did not significantly improve his vision in the incessant glare. "It was like being somewhere in the south, Georgia maybe," he wrote, "without sunglasses on a summer's day." Every so often, he would ease open the filter to observe Earth through the clear faceplate of his helmet. The view was akin to a geography class, he said, with thousands of square kilometres laid out, map-like, beneath him. As an automatic television camera on the end of the airlock filmed his every move, the task of capturing the astonishing vista fell to Leonov himself. Mounted in the chest of his suit was a Swiss-built camera, together with a switch sewn into the Berkut's upper leg, which, unfortunately, turned out to be just beyond his reach!

At the time of writing, around 300 spacewalks and Moonwalks have been conducted since Leonov's excursion and some astronauts who did both would describe the sensation of floating high above the home planet as far more powerful than ambling across the surface of our closest celestial neighbour. The 'ethereal' nature of spacewalking, the almost godlike feeling of looking down from on high, was certainly not lost on its first practitioner. "I felt the power of the human intellect that had placed me there," Leonov wrote. "I felt like a representative of the human race ... I was overwhelmed by these feelings." He would also describe the profound tranquillity of floating in the void, the only sounds coming from his own breathing, the crackle of the radio and the noise of the life-support apparatus that kept him alive. Getting back inside Voskhod 2, however, would prove far from tranquil.

NEAR-DISASTER

The world's first spacewalk lasted barely 15 minutes, ending over eastern Siberia, when Belyayev radioed instructions for Leonov to begin preparations to re-enter the airlock. It would only become clear years later that the seemingly effortless 'swim' through space had actually required every ounce of physical exertion: the Berkut had ballooned, making bending extremely difficult, and Leonov noticed that his feet had pulled away from the boots and his fingers away from the tips of the gloves. It was, he wrote, "impossible to re-enter the airlock feet first" and his only option was to break mission rules and ease himself back into the Volga chamber head-first.

Reports hint that returning to the airlock in this manner caused him to get stuck sideways when he turned to close the outermost hatch. To relieve some of the pressure in his ballooned suit and move more easily, Leonov began bleeding off some of its oxygen by means of a valve in its lining, which placed him at severe risk of the bends. The Berkut, he found, behaved in a totally different manner in space to its performance on Earth. "The work became impossible. I tried to grab the handles [on the airlock] and my fingers wouldn't work – the gloves' fingers would just bend on me ... I decided I was breathing oxygen long enough to prevent boiling nitrogen in the blood. There was some risk, but I had nothing else to do, and once I did, everything started going normal."

However, even after bleeding off the oxygen pressure, the problem of how to turn himself around in the 1.2 m-wide airlock remained. "I literally had to fold myself to do this," he said later. "I spent tremendous effort trying to do this. I had a total of 60 litres [of air] for ventilation and breathing, which was not enough for this kind of action." Physicians would later discover that he almost suffered heatstroke – his core body temperature rising by 1.8°C during the 13-minute excursion – and the cosmonaut would later describe being up to his knees in sweat, to such an extent that it sloshed around in his suit as he moved. Similar problems of over-exertion were closely mirrored in the reports of American astronaut Gene Cernan following his own extravehicular outing in June 1966.

The world's first spacewalk, hazardous though it had been, ended at 11:47 am when Leonov re-entered the airlock. A minute and a half later, the outer hatch was finally closed and at 11:51 am he began repressurising the Volga. Shortly thereafter, with both pilots safely aboard the capsule, Belyayev fired pyrotechnic bolts to discard the airlock. Unfortunately, the explosive effect of the bolts placed Voskhod 2 into a 17-degree-per-second roll – ten times stronger than predicted – and, with only enough fuel for one orientation correction, the two men realised that they would be forced to live with it for the remaining 22 hours of their mission. Exhausted, and with no other option, they felt that they could bear it. Voskhod 2's real troubles, however, were only just beginning.

As Leonov worked his way through routine instrument checks, he noticed that the oxygen pressure in the cabin was steadily increasing from a normal level of 160 mm to 200 mm, then higher, eventually peaking at 460 mm, which – in the event of an electrical short – would be more than sufficient to cause an explosion. The cosmonauts were advised to lower Voskhod's temperature and humidity and, although it halted the upward climb of pressure, the situation remained highly dangerous. After a few hours of sleeplessness, they noticed to their relief that pressures had dropped below the critical level. Later, thankfully, the spacecraft's automatic landing system came into operation, stopped the rolling and, wrote Leonov, "we were able to enjoy a few delicious moments of tranquil flight".

Initiation of the automatic landing system, on Voskhod 2's 16th orbit, came from the Kamchatka ground station, but a solar orientation sensor fault meant that one command was not processed properly. It has been suggested that the effect of pyrotechnic gas from the jettisoned airlock led to the sensor failure. As a result, the rolling began again and, five minutes before the scheduled retrofire, Belyayev was forced to deactivate the automatic system. It was becoming apparent that the cosmonauts would have to perform a manual retrofire as Voskhod 2 passed over Africa on its 17th circuit, with the intention to land at around 52 degrees North latitude. At 10:16 am Moscow Time on 19 March, radio listeners overheard a ground station telling the cosmonauts – using their callsign 'Almaz' ('Diamond') – to perform a manual descent. The crew was asked, with more than a hint of urgency, to respond via Morse code.

Belyayev and Leonov would employ the Vzor optical device to orient their spacecraft, but this kept them out of their seats and delayed the retrofire by 46 seconds, which, coupled with an incorrect attitude, would ultimately conspire to bring them down in the wild Siberian taiga. They would land to the north of the industrial city of Perm, more than 2,000 km from their intended site. As Voskhod 2's navigator, Leonov felt that overshooting Perm should still bring them down in Soviet territory, but "we could not run the risk of overshooting so much that we came down in China; relations with the People's Republic were poor at the time". Nonetheless, in his autobiography, Leonov praised the superb skills of Belyayev, now charged with performing the Soviet Union's first-ever manual re-entry.

"In order to use the [Vzor] he had to lean horizontally across both seats inside the spacecraft," Leonov wrote, "while I held him steady in front of the orientation porthole. We then had to manoeuvre ourselves back into our correct positions in our seats very rapidly so that the spacecraft's centre of gravity was correct and we could start the retro-engines to complete the re-entry burn. As soon as Pasha turned on the engines we heard them roar and felt a strong jerk as they slowed our craft." The completion of retrofire was greeted with silence and should have been followed, ten seconds later, by the separation of Voskhod 2's instrument module. It did not happen. Leonov would recall the sight, also beheld by his comrade Yuri Gagarin four years earlier, of the useless section being dragged in the spacecraft's wake by the thread of a communications cable.

Not until an altitude of around 100 km, when the cable finally burned through, did the ride stabilise. In rapid succession, the cosmonauts felt a sharp jolt as, first, the drogue parachute and, next, the main canopy were automatically deployed. "Suddenly," wrote Leonov, "everything became dark. We had entered cloud cover. Then it grew even darker. I started to worry that we had dropped into a deep gorge. There was a roaring as our landing engine ignited just above the ground to break the speed of our descent. Finally we felt our spacecraft slumping to a halt." Voskhod 2 had landed in a couple of metres of snow, somewhere in the western Urals, at 59 degrees 34 minutes North latitude and 55 degrees 28 minutes East longitude. It was 12:02 pm Moscow Time and the mission had lasted a little over 26 hours, yet the cosmonauts had not even reached the halfway mark of their time aboard the capsule. A long, cold afternoon and an even colder night awaited them. Moreover, they would also have unwanted company.

For the outside world, everyone was bewildered by what might have happened to the two men. Official accounts gave away few details. Some media reports suggested that the cosmonauts were "resting" after their mission, while Radio Moscow suspended transmissions and played Mozart's 'Requiem' and Tchaikovsky's first piano concerto over and over in a sombre manner which hinted that Belyayev and Leonov had been killed.

Nikolai Kamanin would later write in his diary that tracking stations at Odessa and Saransk, both directly beneath Voskhod 2's re-entry flight path, had provided the first reports that the descent was underway, although he noted that no one knew about the cosmonauts' fate for at least four hours. Meanwhile, the Alma-Ata station in south-eastern Kazakhstan picked up a telegraph code via the high-frequency radio channel, which repeated 'VN' – 'Vsyo normalno' ('Everything normal') – over and over and the capsule's Krug radio beacon had provided a fix on its location, "but we wanted more convincing data as to the condition of the cosmonauts", wrote Kamanin. Voskhod 2 was eventually spotted, together with its red parachute and the two men, by the commander of one of the search-and-rescue helicopters, wedged between a pair of firs on the forest road between Sorokovaya and Shchuchino, some 30 km south-west of the town of Berezniki.

Immediately after impacting the snow, the first task for Belyayev and Leonov had been to release the spacecraft's hatch and get outside; unfortunately, upon flicking a switch, the explosive bolts activated and the sturdy plate of metal jerked, but refused to burst open. Only when they looked through one of the portholes did it become clear that the capsule was jammed between two firs. After much rocking backwards and forwards, Belyayev finally pushed the hatch away and the two men plopped out into the snow. Above them was the main canopy of their parachute, which had snagged the upper branches of firs and birches some 40 m high. Below, the base of the capsule, still simmering from the heat of re-entry, rapidly melted the snow and it thumped down onto solid ground.

As daylight faded and fresh snow began to fall, the two cosmonauts had reason to be grateful for their extensive experience in harsh climates. Leonov himself, of course, had been brought up in Siberia, whilst his older comrade, born on 26 June 1925 in Chelizshevo, in the Vologda region, north of Moscow, had spent much of his boyhood hunting in the forests near his home. As a youth, Belyayev dreamed of someday becoming a hunter and graduated from the Soviet Air Force Academy at Sarapul in 1944 and the Military Fighter Pilot School in Yeis the following year. He subsequently served in various Air Force units for more than a decade and became a squadron commander in naval aviation shortly before being selected as a cosmonaut candidate in 1960. Although he was the oldest member of the first group, Belyayev was hired for his experience, education and 900 hours of flying time. He would, wrote Asif Siddiqi, probably have flown in space sooner, but for an injury sustained during a parachute jump in August 1961.

As one of the older members of the corps, Belyayev's last claim to fame would occur on 10 January 1970, when he became the first flown spacefarer to die of natural causes: after several years overseeing the training of newer cosmonaut recruits, complications, including pneumonia, arose following an operation on a stomach ulcer. Colonel Pavel Ivanovich Belyayev, who had so expertly piloted Voskhod 2 through the Soviet Union's first manual re-entry, died at the age of just 44. Although fellow cosmonauts Yuri Gagarin and Vladimir Komarov preceded Belyayev to the grave, his state funeral honours would be somewhat less than theirs. His remains would not be interred in the Kremlin Wall, but rather in Moscow's Novodevich Cemetery, although pensions were paid to his wife and daughter and they were granted a seven-room apartment on Moscow.

With less than five years of life ahead of him, Belyayev, for now, felt that he could withstand anything. "Pasha and I both felt we had already been tested to our limits," wrote Leonov, "though we knew there was no way of telling how long we would have to fend for ourselves in this remote corner of our country." In the first few minutes after landing, they began transmitting their 'VN' code to confirm that they were alive and well. Interestingly, wrote Leonov, Moscow did not receive the signal, "because the vast expanse of forest in the northern Urals ... interfered with the radio waves", although listening posts as far afield as Kamchatka in the Soviet Far East and Bonn in West Germany did pick it up.

However, the area was so heavily wooded and so deeply coated in snow that the rescue helicopters could not hope to reach them until loggers had cleared a landing site. One civil helicopter, Leonov recalled, tried to extend a rope ladder, but in their bulky pressure suits the two cosmonauts had no chance of scaling it. As the afternoon wore on, other aircraft dropped supplies – two pairs of wolf-skin boots, thick trousers and jackets, a blunt axe and even a bottle of cognac – to keep the men alive through the night. The news of the safe landing was announced by Yuri Levitan at 4:44 pm Moscow Time, almost five hours after it had occurred, and, later that evening, a helicopter succeeded in touching down a few kilometres away, although its crew could not reach the cosmonauts.

As the last vestiges of daylight disappeared, the temperature in the taiga began to drop precipitously and the pool of sweat in Leonov's boots started to chill him. Fearing the onset of frostbite, both men stripped naked, wrung out their suits and underwear and separated the rigid sections from the softer linings, which they donned, together with boots and gloves. Their attempts to pull the snagged parachute from the trees for extra insulation proved fruitless and, as night approached, the snow started falling and temperatures plummeted still further to –30°C. Leonov would relate a cold and lonely night in the now-hatchless capsule, but stories would persist over the years that they were harassed by wolves which prevented them from disembarking and building a fire. Still others argued that mountain bears drew near Voskhod and others that the cosmonauts heard 'strange noises' outside.

Leonov mentioned nothing of this in his autobiography, although he admitted that when an Ilyushin-14 aircraft flew overhead at daybreak, the pilot revved his engine to scare away wolves in the vicinity. Later that morning, another helicopter reported seeing the cosmonauts chopping wood and setting a fire. At 7:30 am Moscow Time, an Mi-4 helicopter lowered a rescue team, including two physicians, to a point 1.5 km from the capsule and the first efforts began to fell trees and provide a suitable landing spot. Visibility was too poor to risk lifting them to a hovering helicopter and, as a result, the cosmonauts spent a second night in the dense taiga, together with their rescuers. "But this second night was a great deal more comfortable than the first," wrote Leonov. "The advance party chopped wood and built a small log cabin and an enormous fire. They heated water in a large tank flown in especially by helicopter from Perm ... And they laid out a supper of cheese, sausage and bread. It seemed like a feast after three days with little food."

It was a welcome relief to be among other human beings. At length, two landing spots were cleared, one of which lay just a few kilometres from the capsule, and at 8:00 am on 21 March the cosmonauts skiied there. They were then airlifted to Perm airport for a telephone call from Leonid Brezhnev and finally returned to Tyuratam at 2:30 pm, more than two full days after landing. Belyayev and Leonov would be rewarded and decorated for their efforts: each received a Hero of the Soviet Union award, together with 15,000 roubles, a Volga car and six weeks' leave. By the beginning of May, they had joined the circuit of official visits, international symposia and conferences and meetings with world leaders.

DECLINE

On the face of it, the Soviets remained in the lead in terms of space endeavours – even the first manned Gemini mission, launched a few days after Voskhod 2, ran for barely five hours and the United States' first spacewalk would not occur until June 1965. However, before the year's end, American astronauts would have not only surpassed Valeri Bykovsky's five-day endurance record, set on Vostok 5, but would have nearly tripled it. Moreover, they would have experimented with fuel cells for longer flights, demonstrated 'real' rendezvous techniques necessary for lunar sorties and their Apollo project was gearing up for its own missions from 1966 onwards. At the time, of course, many western observers would find it hard to fathom why the Soviets – once so far ahead – fell so far behind during this period. Their next manned mission, Soyuz 1, would not fly until April 1967 and would end with the death of its cosmonaut pilot, Vladimir Komarov.

Key to the Soviet slowdown was the death of Sergei Korolev, the famed Chief Designer, whose identity had been kept such a closely guarded secret that his importance would not become widely known until years later. In his autobiography, Alexei Leonov lamented that, even compared to Wernher von Braun, Korolev was both a giant and a genius. At a conference in Athens in August 1965, Leonov asked von Braun why America's supposed technological superiority had not enabled them to launch their own Sputnik, their own Gagarin, their own Voskhod 2, first. The man who designed the Saturn rocket which would win the Moon race in barely three years' time responded respectfully that the 'Chief Designer', his name still unknown in the west, was a far more determined man.

Determined, indeed, but by the middle of the Sixties, Korolev was also a sick man. Nikolai Kamanin had made numerous references in his diaries that Korolev had not been well and towards the end of 1965, as two American Gemini capsules rendezvoused in orbit, he was diagnosed as suffering from a bleeding polyp in his intestine, then admitted into hospital early in the new year. Released temporarily on 10 January to celebrate his birthday at home, he spent an evening with his closest friends, including Leonov and Yuri Gagarin, to whom he told the story of his remarkable life: from his early work in the field of rocketry to his incarceration in one of Stalin's gulags, near Magadan in the Kolyma region of the Soviet Far East, then his recall to Moscow to support Russia's war effort and, later, its space effort.

Only days later, on 14 January, after complications arose in what should have been a routine operation, Korolev died. The effect on the cosmonaut corps and upon the Soviet Union's direction in space was dramatic, with many recognising that the death of this previously-unknown man would severely affect future endeavours.

Pravda ran an obituary, Yuri Gagarin delivered a solemn eulogy – describing Korolev as "a name synonymous with one entire chapter of the history of mankind" – and Leonid Brezhnev, Alexei Kosygin and Mikhail Suslov took turns to carry his ashes for interment in the Kremlin Wall. The men who followed Korolev – his deputy, Vasili Mishin, who succeeded him, together with Georgi Babakin, Vladimir Chelomei and rocket engine designer Valentin Glushko – exhibited entirely different personalities which many cosmonauts felt damaged the Soviet Union's chances of beating America to the Moon.

In particular, Alexei Leonov has said, the lack of co-operation between Korolev and Glushko led to problems with the choice of propellants and the number of engines needed for the gigantic N-1 lunar rocket, while Mishin's apparent favoritism of newly-selected engineer-cosmonauts over the veteran pilots alienated many in the corps. Summing up, Leonov is not alone in having suggested that, had Korolev lived a little longer, "we would have been the first to circumnavigate the Moon". His optimism was far from misplaced. In fact, even under Mishin's leadership, early plans called for Leonov himself to command the first loop around the back of the Moon, scheduled, at one point, for mid-1967.

Judging from the ambitious Voskhod follow-on flights planned while Korolev was still alive, there is much reason to suppose that a Soviet man on the Moon was possible. Voskhod 3, notably, endured a lengthy and convoluted development and reared its head, drearily, on several occasions as yet another effort to upstage the Americans, this time by attempting a mission of almost three weeks in duration. Early plans from March 1965 envisaged a 15-day flight in October of that year, carrying a pair of cosmonauts – a pilot and a scientist – followed by the longer, 20-day Voskhod 4 in December, crewed by a pilot and a physician. By April, the first hints of crews appeared: Boris Volynov and Georgi Katys were favoured by Nikolai Kamanin for Voskhod 3, although some within the Soviet leadership contested this. Volynov, for example, was Jewish, whilst Katys' father, of course, had been executed by Stalin and the cosmonaut had half-siblings living in Paris. As the months wore on, Volynov was retained, paired firstly with Viktor Gotbatko and, finally, with Georgi Shonin. The next Voskhod to feature a spacewalk proved yet more controversial, with a crew of two female cosmonauts: Valentina Ponomaryova and Irina Solovyeva, backed-up, interestingly, by two men, Gotbatko and Yevgeni Khrunov.

In terms of space endurance, the United States seized the lead in August 1965, when Gemini V astronauts Gordo Cooper and Pete Conrad spent eight days in orbit, an endeavour which the Soviets, even with Korolev still alive, were powerless to prevent. Hopes of launching Voskhod 3 before the year's end to at least upstage Gemini V faded when it became clear that the challenges of modifying the spacecraft, its environmental system and controls to handle such a long flight were simply too great. The 14-day Gemini VII mission pressed the American lead still further in December. In the final months of his life, Korolev was overburdened with the development of the new Soyuz ('Union') spacecraft, the massive N-1 lunar rocket and plans to soft-land a probe on the Moon in early 1966. Privately, and with little direction from the government, he had already abandoned work on Voskhod 3.

The Soviet armed forces provided the impetus to jumpstart the proceedings when it became apparent that military activities had been conducted by the Gemini V astronauts. Ballistic missile detection experiments were duly added to Voskhod 3 and one of Korolev's projects, an artificial gravity investigation, which utilised a tether between the spacecraft and the final stage of the R-7 rocket, was also approved. Short-lived plans were even floated by the Soviet Air Force in August 1965 to stage a one-man Voskhod 4, lasting around 25 days, for exclusively military tasks. One of these would centre on a set of high-quality, Czech-built cameras known as 'Admira'. By the end of the year, Voskhod 3 had slipped into February 1966, much to the chagrin of many in the cosmonaut corps, who had already written to Leonid Brezhnev, complaining that the Soviet Union's lead in space was being hampered by its lack of focus and clear management.

Following Korolev's death, his successor Vasili Mishin pushed on with plans for a third Voskhod, pencilling it in for March 1966, although this date quickly became untenable due to nagging problems with ripped parachutes and an environmental control system which could not be qualified for missions longer than 18 days. On 22 February, the prime crew, Volynov and Shonin, passed their final examinations and were cleared to fly. In readiness for their launch, an unmanned Voskhod – under the cover name of 'Cosmos 110' – entered orbit on 28 February and completed a 21-day flight with two dogs, Veterok and Ugulyok. However, Voskhod 3 itself continued to drift further and further to the right. An R-7 failure provided the first postponement and Voskhod 3 was scheduled for May, but Leonid Smirnov, chairman of the Military-Industrial Commission, argued that the flight served no purpose for the Soviet government. Despite achieving a new record duration, it was not enough, Smirnov reasoned, to have a profound impact on the world.

For his part, Kamanin argued that valuable military experiments would be conducted by Voskhod 3 and Smirnov relented a little and the launch was rescheduled for sometime in late May. A state commission convened early that month and confirmed that problems with the R-7's engines, which had exhibited high-frequency oscillations in test-stand runs, would probably not occur under 'real' flight conditions. Later plans moved Voskhod 3 to July and even as late as October 1966, Mishin was ordered to prepare for its launch, but did so with little enthusiasm that it would actually go ahead as the new Soyuz project gained momentum. It has also been speculated that, just months after being appointed the new Chief Designer, Mishin simply did not want to begin his tenure under the cloud of a now-obsolete spacecraft which provided its cosmonauts with a limited margin of safety. In this way, as Mark Wade has pointed out on his website www.astronautix.com, Voskhod 3 was never really cancelled; it simply faded away.

Nikolai Kamanin had long since seen the writing on the wall: that Smirnov had killed the mission in favour of the more ambitious Soyuz project, which would demonstrate rendezvous and docking, long-duration flights, spacewalking, the potential to support an orbital station and whose crews would circumnavigate and land on the Moon. Placing their eggs in the Soyuz basket, it seemed, would give the Soviets a far better chance than Voskhod of decisively beating the American lead achieved by Gemini. The maiden voyage of the new spacecraft would suffer more than its own fair share of technical obstacles, but the loss of the Apollo 1 crew in a January 1967 flash fire offered increased hopes that the Soviets might yet beat the United States to the lunar surface. Then, just three months after the Apollo disaster, tragedy would strike the Russians in a manner that even their best propaganda apparatus could not fully conceal.

4

Pushing the Envelope

MEN WITH MISSIONS

On the evening before the first manned Gemini flight, astronauts Wally Schirra and Tom Stafford were men with a mission of their own. It involved a favour asked of them by Gemini 3 pilot John Young and a last-minute visit to a deli called Wolfie's on North Atlantic Avenue in Florida's Cocoa Beach. "Young mentioned to me a day or so before launch that there was not a meal scheduled on the flight," Schirra wrote in his autobiography, "since the flight was less than five hours. But add a two-and-a-half-hour countdown, he reasoned, and they were bound to get hungry." Schirra's solution: to buy Young and his command pilot, Virgil 'Gus' Grissom, a corned beef sandwich, "on rye with two dill pickle slices", which he kept in a refrigerator at the newly-renamed Cape Kennedy and passed word to Young that it was there. On Gemini 3's launch morning, 23 March 1965, the 34-year-old Young tucked the sandwich into one of his space suit pockets, ready to surprise Grissom when they reached orbit.

As Gemini 3's backup crew, Schirra and Stafford shadowed the training of Grissom and Young, to ensure that they would be capable of stepping in if the prime team was unable to fly. Of course, being Wally Schirra, the backup command pilot tasked himself with the additional responsibility of becoming the mission's prime prankster and the corned beef sandwich would be the latest in a long line of 'gotchas'. However, only months earlier, Schirra, by now NASA's second most experienced astronaut, with over nine hours in space to his name, had nothing to do with Gemini 3 at all. In fact, he had his sights set, with John Young as his pilot, on another mission which would have demonstrated rendezvous techniques in orbit. The original line-up for Gemini 3 would have seen Freedom 7 veteran Al Shepard in command, paired with Stafford, and backed-up by Grissom and Frank Borman. Then, with awful suddenness, in the summer of 1963, the hands of fate turned on Shepard and he was abruptly removed from active flight status.

By this time, Deke Slayton, himself grounded following a minor heart murmur, had taken charge of the astronaut office and was responsible for its day-to-day operations and for overseeing the selection and training of the first Gemini crews. It had already been recognised by NASA management that those crews would require a far larger pool of astronauts than the Mercury Seven could offer. Since the Gemini spacecraft, capable of supporting two-man teams, would be correspondingly larger than the Mercury capsule, Slayton devised his own set of selection criteria for the new astronauts. Specifically, the height limit was increased and the age limit changed. "One thing that got tougher," he wrote, "was that we dropped the maximum age from 40 to 35; in Mercury, we were looking at a programme that would conclude in three years. We knew that Apollo would be going until 1970 at the very least." Further, Slayton insisted on letters of recommendation from each candidate's previous employer.

One new selection of around ten candidates would be barely sufficient for Gemini and Slayton knew that a third group would be needed about a year later to provide long-term support for Apollo. In April 1962, he issued a formal announcement of NASA's intent to select a second group of astronauts and by the June deadline had received 253 applications. A series of medical tests at Brooks Air Force Base in San Antonio, Texas, winnowed these down to 33 finalists, who were interviewed by Slayton, Shepard and test pilot Warren North at the burgeoning Manned Spacecraft Center (MSC) in Houston. In his autobiography, Slayton wrote that he could have returned to the finalists from the 1959 Lovelace and Wright-Patterson tests – which included 1962 selectees Jim Lovell and Charles 'Pete' Conrad – but was glad in retrospect that he did not do so. "That second group," he explained, "is probably the best all-around group ever put together."

Among the nine candidates were Tom Stafford and John Young, who would train together for Gemini 3 and subsequently fly to the Moon together. "The most important change for me," Stafford wrote in his autobiography, co-authored with Mike Cassutt, "was that they had raised the height limit from five feet 11 inches to six feet even." By April 1962, he had just been accepted through his parent service, the Air Force, as a Harvard Business School student, at which point he noticed NASA's advertisement and realised that he fulfilled all of their requirements – he was a test pilot, had the necessary flying experience in high-performance jets, met the age criteria and had the right level of technical education. In July, after passing the Air Force's initial screening board, Stafford was summoned to Brooks Air Force Base "where we had the expected blood tests and EKG stuff but no centrifuge testing ... we were all pulling Gs on a regular basis in high-performance aircraft. You didn't need to be some kind of physical superman to fly in space".

Some tests, though, were pointless. Stafford recalled looking into an ocular device for long enough to see a sudden flash of light, part of evaluations of how an astronaut's eyes might respond to a thermonuclear explosion. "It wasn't enough to damage your eye – at least, I don't think it was – but you couldn't see for several minutes after the test," he wrote. The Brooks tests were followed by hour-long technical interviews in Houston in August 1962 and, although Stafford felt confident that he had a good chance of being selected, his attention was primarily focused on the impending start of classes at Harvard. In fact, his interview in Houston came partway through his family's move eastwards from his old station at Edwards Air

The Mercury Seven and the 'New Nine'. Front row (left to right) are Gordo Cooper, Gus Grissom, Scott Carpenter, Wally Schirra, John Glenn, Al Shepard and Deke Slayton. Back row (left to right) are Ed White, Jim McDivitt, John Young, Elliot See, Pete Conrad, Frank Borman, Neil Armstrong, Tom Stafford and Jim Lovell.

Force Base in California to Boston, Massachusetts. Arriving in early September, they unpacked enough belongings to temporarily live on and Stafford worked through three days of inaugural classes.

Upon returning home on 14 September, he was greeted by his next-door neighbour and the news that someone called Deke Slayton from NASA had called. The gruff Slayton told Stafford that, if he was still interested in the astronaut programme, he had been officially selected. Three days later, on his 32nd birthday, Stafford sat alongside fellow selectees Neil Armstrong, Frank Borman, Pete Conrad, Jim Lovell, Jim McDivitt, Elliot See, Ed White and John Young in their first press conference at Ellington Air Force Base near Houston.

In response to complaints from some sections of the press over the exclusive rights of Life magazine over the Mercury Seven's personal stories, NASA had issued a news release on 16 September to assure "equal access by all news media" and revealed that "specific guidelines were spelled out covering the sale by the astronauts of stories of their personal experiences ... [with] sharp prohibitions against such stories containing ... official information concerning the astronauts' training or flight activities not previously available to the public". Future missions, the release added, would benefit from a post-flight press conference, in which all accredited members of the media would have the opportunity to question the astronauts in depth. Privately, and in response to the business deals made by several of the Mercury Seven with the proceeds from their lucrative Life contracts, Slayton told the astronauts that, with regard to gratuities, they should follow the old test pilot's creed: "Anything you can eat, drink or screw within 24 hours is perfectly acceptable!" Many of the new astronauts, however, were far more interested in which of them would be the first to walk on the Moon.

The group, whose members dubbed themselves 'The New Nine', included four Air Force officers (Borman, McDivitt, Stafford and White), three naval aviators (Conrad, Lovell and Young) and two civilians: Armstrong, who had flown the X-15 rocket-propelled aircraft for NASA, and See, a General Electric test pilot. Several of them might well have made the cut for Project Mercury, but for minor discrepancies: Lovell, for instance, was dropped from the 1959 selection because of a minor liver ailment, whereas the wise-cracking Conrad had, said Slayton, shown "a little too much independence when it came to some of the medical tests". One of these, famously, had been when a psychologist asked him to comment on a blank piece of paper. Conrad replied that it was upside down! Young, who would ultimately become chief astronaut, walk on the Moon and command the first Space Shuttle mission, had missed Project Mercury because he was still at the Navy's test pilot school at Patuxent River in Maryland, whilst Stafford had been too tall.

One of the New Nine's first activities was to travel to Cape Canaveral and witness the launch of Wally Schirra and Sigma 7 on 3 October 1962. They were also pounced upon hungrily by the media, who knew that one of them would most likely be the first to set foot on the lunar surface, and were frequently forced onto the circuit of cocktail parties, signed many autographs and met countless officials and dignitaries. Indeed, in January 1963, under the tutelage of planetary scientist Gene Shoemaker, they visited a meteor crater in the desert near Flagstaff, Arizona, observed the Moon and examined lava flows.

After the completion of their basic science training, the New Nine was integrated with the Mercury Seven to form a 16-man unit, which, in June 1963, spent a week at the Caribbean Air Command Tropic Survival School at Albrook Air Force Base in the Panama Canal Zone. In addition to jungle-survival training, they focused on the identification and toxicity of tropical plants, their methods of preparation, local fauna and even interaction with indigenous people, all of which could someday prove essential in the event of an unhappy landing from a space mission. Three months later, at the Naval School of Pre-Flight at Naval Air Station Pensacola in Florida, they underwent water-survival training, including underwater egress, escaping from a dragging parachute, boarding a life raft and learning flotation techniques in a Gemini-specification space suit.

The suit – or, rather, 'suits', for each astronaut was granted three – provided them with an ensemble in which to train, in which to fly into space and, lastly, with a backup. "Since these suits were frightfully expensive," wrote astronaut Mike Collins, who worked on their development, "around $30,000 for a Gemini suit and more for an Apollo suit, three suits each may seem too many, but the argument was that some malfunction like a broken zipper on launch morning could not be allowed to delay or cancel a multi-million-dollar flight."

The New Nine also received their own technical assignments: Borman monitoring the development of the Titan II rocket, McDivitt handling spacecraft guidance and control, Young overseeing the Gemini pressure suits, Armstrong the simulators, Conrad the cockpit displays, See the electrical systems, White the flight controls, Stafford the range safety and communications and Lovell the re-entry and recovery techniques. Deke Slayton assigned Gus Grissom, already working on Gemini since shortly after his Liberty Bell 7 flight, to supervise their work. "They're all talented," Grissom admitted. "In fact, when one of them comes up with an answer for some problem, I think they are a lot smarter than our original group of seven." Indeed, several of the New Nine – Armstrong, Borman, See and White – had already secured advanced master's degrees.

Additionally, the Nine kept up their flying proficiency in high-performance aircraft, thanks to NASA's fleet of T-33s and F-102s, although plans were in the pipeline to upgrade to either the Air Force's T-38 Talon or the Navy's F-4 Phantom. Despite the objections of some in the corps, who felt the Mach 2-capable F-4 was the better choice and a 'hotter' jet, its complexity and expense of maintenance ultimately led NASA to opt for the T-38. It is a training aircraft still used by astronauts today.

Scarcely had the Nine completed their initial training than another, third group of astronauts was selected in October 1963. On this occasion, wrote Slayton, "we dropped the test-pilot requirement, figuring that we had just about drained that pool". Applications would also be accepted from candidates with operational flying backgrounds or advanced degrees in related fields of engineering or the physical or biological sciences. It was during the screening process for this group, which began in June 1963, that NASA became embroiled in the Edward Dwight affair – which Slayton called his "first, last and only political battle over astronaut selection" – and the reverse-racism demands of Bobby Kennedy to hire the black bomber pilot into the corps. In his autobiography, Slayton related his creation of a 'points system' to assess the suitability of candidates: up to ten points apiece would be awarded for their academic excellence, their flying performance and their character and motivation, with each man scored out of 30.

On 14 October 1963, the third group was finally chosen. Fourteen strong, they comprised six Air Force officers (Edwin 'Buzz' Aldrin, Bill Anders, Charlie Bassett, Mike Collins, Donn Eisele, Ted Freeman and Dave Scott), four naval aviators (Al Bean, Gene Cernan, Roger Chaffee and Dick Gordon), a Marine Corps pilot (Clifton 'C.C.' Williams) and two civilians (Walt Cunningham and Rusty Schweickart). Among this cadre was the first astronaut with a doctorate – Aldrin, whose ScD thesis had been in orbital rendezvous theory, a useful tool for lunar mission design – together with nuclear engineer Anders and two former military aviators in scientific fields, Cunningham and Schweickart. Within months of their selection, Slayton had already assessed each man's talents and weaknesses and began inserting them into his plan for flight assignments. In his autobiography, he wrote that, as long as he kept them, each man was eligible for any mission, although those with command, management or test-piloting experience would be his preferred choices for the complex rendezvous flights.

As new astronauts came aboard, however, other fliers' chances of returning to space diminished dramatically. John Glenn, for example, had already begun making noises of his intent to run for political office and his growing friendship with Bobby Kennedy seemed to be pointing him in the direction of a senatorial seat and, ultimately, in 1984, a run for the presidency. Moreover, there have been persistent suggestions over the years that President John Kennedy himself vetoed the idea of risking Glenn – the national hero, first American to circle the globe – on another mission. Scott Carpenter's performance on Aurora 7 had made him unacceptable to NASA management and Gordo Cooper, wrote Slayton, "was a question mark". It is interesting – ironic, even – that Al Shepard, the man who had beaten them all into space, was kept on flight status and would be 'risked' again.

In the summer of 1963, he was assigned to command the first manned Gemini, scheduled for sometime in October of the following year, with Tom Stafford as his pilot. Their backups would be Gus Grissom and Frank Borman. Gemini 3, as it was called, would come on the heels of two previous unmanned missions and would be the first of ten crewed flights, launched roughly every two to three months, with the project due to end sometime in 1966. The first four would be essential in demonstrating Gemini's capabilities: the initial manned shakedown, flown by Shepard and Stafford for five hours, after which would come the seven-day Gemini IV, a rendezvous on Gemini V and the 14-day Gemini VI. (Missions would be renamed with Roman numerals from Gemini IV onwards.) "I figured to handpick the crews for the first four missions," Slayton wrote, "since they had unique requirements, and treat the last six as more or less identical."

Gus Grissom's involvement with the development of Gemini would assure him a place in command of the 14-day mission, while Schirra's disinterest in a long-duration flight prompted Slayton to assign him the rendezvous. Moreover, Borman "was tenacious enough" to endure the 14-day flight, so was teamed with Grissom to backup Gemini 3 and fly Gemini VI. John Young would then join Schirra on the

Members of the Fourteen undergo desert survival training.

Gemini V rendezvous. Meanwhile, Gemini IV, at the time, was intended as a seven-day test of the spacecraft's capabilities, eclipsing Valeri Bykovsky's Vostok 5 record and perhaps involving a spacewalk or some limited rendezvous practice. Slayton assigned Jim McDivitt and Ed White as the prime crew, with Pete Conrad and Jim Lovell backing them up. It was at around this time that the 'three-flight rule' emerged, whereby each crew of four – two prime, two backups – would work as a single unit and the backups would become the prime crew for a mission three flights down the line. The last six Geminis would be three-day rendezvous, docking and spacewalking flights.

Plans, however, changed rapidly. The intention was for Gemini V to rendezvous with an unmanned target vehicle known as an Atlas-Agena; but this would not be ready in time and this mission was moved to Gemini VI. The result: Schirra and Young were reassigned as backups to Shepard and Stafford, while Grissom and Borman now became the Gemini V prime crew. Very soon after Gordo Cooper's Faith 7 mission, circumstances would also change dramatically for Al Shepard.

GROUNDED

By the middle of 1963, shortly after Faith 7, Shepard's chances of commanding the first Gemini looked bright. Then his career and health, figuratively and literally, started spinning. Years earlier, just after being selected as one of the Mercury Seven, he had complained about feeling light-headed during a game of golf; every time he attempted to swing the club, he felt that he was about to fall over. It was an isolated, peculiar incident, which did not resurface again until the summer of 1963. It came with a vengeance, usually striking him in the mornings and taking the form of a loud metallic ringing in his ears, coupled with feelings of intense dizziness and nausea. At first, Shepard dealt with the problem himself: he saw a private physician, who prescribed diuretics and vitamins such as niacin, which had little effect. It did not stop Slayton from assigning him to command Gemini 3 and, indeed, Shepard and Stafford completed the first six weeks of their training, visiting McDonnell's St Louis plant in Missouri to watch their spacecraft being built.

He told no one in the astronaut corps of the problem. However, very soon, it became impossible to conceal. An episode of dizziness whilst delivering a lecture in Houston forced him to admit his concerns to Slayton, who sent him to the astronauts' physician, Chuck Berry, for tests. In May 1963, unknown to everyone else in the corps, Shepard was temporarily grounded. The diagnosis was that fluids were regularly building up in the semicircular canals of his inner ear, affecting his sense of balance and causing vertigo, nausea, hearing loss and intense aural ringing. Although the incidents were intermittent, they proved sufficiently unpredictable and severe to render him ineligible to fly Gemini 3.

Known as Ménière's Disease, the ailment was a recognised but somewhat vague condition. Indeed, formal criteria to define it would not be established by the American Academy of Otolaryngology-Head and Neck Surgery until 1972. The academy's criteria would describe exactly the conditions suffered by Shepard:

fluctuating, progressive deafness – he would be virtually deaf in one ear by 1968 – together with episodic spells of vertigo, tinnitus and periodic swings of remission and exacerbation. Nowadays, it can be treated through vestibular training, stress reduction, hearing aids, low-sodium diets and medication for the nausea, vertigo and inner-ear pressure: such as antihistamines, anticholinergics, steroids and diuretics. In mid-1963, however, the physicians who examined Shepard had next to no idea what caused it, some speculating that it was a 'psychosomatic' affliction. Moreover, there was no cure.

His removal from flight status was temporarily revoked in August, with the prescription of diuretics and pills to increase blood circulation, in the 20 per cent hope that the condition would clear up on its own. This allowed Shepard to be internally assigned to Gemini 3, but when the early diagnosis was confirmed and no sign of improvement was forthcoming, he was formally grounded in October, after only six weeks of training with Stafford. During those weeks, the men had spent some time in the Gemini simulator, but little more. Not only was Shepard barred from spaceflight, but, like Deke Slayton, he also could not fly NASA jets unless accompanied by another pilot. Subsequent examinations revealed that he also suffered from mild glaucoma – a symptom of chronic hyperactivity – and a small lump was discovered on his thyroid. It was surgically removed in January 1964 and, the press announced, "would have no impact on his status in the space programme". In reality, Shepard had been effectively grounded for months by that point.

Ironically, at the same time, John Glenn, who had resigned from NASA after being told that his chances of flying again were slim, suffered damage to his own vestibular system. Glenn's friendship with Attorney-General Bobby Kennedy had led to the first inkling of a political career and, after leaving the astronaut corps in January 1964, he announced his candidacy to run for the Senate in his home state of Ohio. A few weeks later, he slipped and cracked his head on the bathtub in his apartment, resulting in mild concussion and, more seriously, swelling in his inner ear, which produced similar symptoms to Ménière's Disease. Glenn spent several weeks in a San Antonio hospital, virtually immobile, and was forced to withdraw from the Senate race in March.

Elsewhere, at the Rice Hotel in downtown Houston, Shepard pulled Stafford aside one evening that same March and asked him if Slayton had mentioned anything about the Gemini 3 assignment. No, Stafford replied, and could only listen open-mouthed as his former crewmate told him about the dizziness, the vertigo, the Ménière's diagnosis ... and the bombshell that Shepard was grounded. In his autobiography, Stafford recalled fearing for his own place on Gemini 3, and, indeed, in mid-April, the crew changes were announced. Slayton moved Gus Grissom up from the command slot on Gemini V to lead Shepard's old mission and replaced him with Gordo Cooper, who had established himself as capable of enduring a long-duration flight on Faith 7. Unluckily for Stafford, however, Slayton felt that John Young was a better personality match with Grissom and designated him as Gemini 3's new pilot. He had nothing against Stafford, of course, simply revealing in his autobiography that "Tom was probably our strongest guy in rendezvous, so it made sense to point him at [Gemini VI], the first rendezvous mission".

Stafford learned of his removal from the Gemini 3 prime crew from one of the flight surgeons, Duane Ross, who told him that he was now on Gus Grissom's backup team, paired with Wally Schirra. Grissom's original pilot, Frank Borman, would be "held for later" and another mission. In his biography of Grissom, Ray Boomhower cited fellow astronaut Gene Cernan as remarking that Grissom's and Borman's egos – both of them were strong-headed leaders – were too large to fit one mission. Indeed, in an April 1999 oral history for NASA, Borman hinted that he "went over to [Grissom's] house to talk to him about it ... and after that I was scrubbed from the flight". Borman, eventually, would command his own Gemini. Meanwhile, on 13 April 1964, the four-man unit for Gemini 3 set to work. Only days earlier, the first unmanned test to assess the compatibility of the spacecraft and its launch vehicle had proven a remarkable success. It came after almost three years of technical and managerial difficulties and a development programme laced with problems.

A NEW SPACECRAFT

The vehicle which Grissom and Young would fly, and which would demonstrate many of the techniques needed for missions to the Moon, represented a stopgap effort to bridge the gulf between Projects Mercury and Apollo. On 7 December 1961, Bob Gilruth announced in Houston the approval of a $530 million project to use a large Mercury capsule for a series of two-man flights, launched atop the Air Force's Titan II rocket, to practice rendezvous and spacewalking. It was originally dubbed 'Mercury Mark II' or 'Advanced Mercury', but the project name 'Gemini', with its nod toward 'the twins' of classical Greek lore, was suggested by Alex Nagy of NASA Headquarters. Nagy won a bottle of scotch for his trouble and the name was officially announced on 3 January 1962.

Until that time, Project Mercury remained the United States' only approved manned spacecraft, although plans were afoot to develop it further, and two concepts in particular emerged: one for a temporary orbital station, housing two or more men for several weeks or months, and another for a manoeuvrable vehicle with sufficient aerodynamic 'lift' to adjust its flight in the atmosphere. In its 1960 budget request to Congress, NASA asked for $300,000 to study ways of transforming Mercury into a long-duration laboratory, a million dollars to explore methods of making it manoeuvrable and a further three million to investigate rendezvous techniques. "These modest sums," wrote Barton Hacker and James Grimwood in their 1977 tome 'On the Shoulders of Titans', "signalled no great commitment. When NASA ran into budget problems, this effort was simply shelved and the money diverted to more pressing needs."

Still, the Space Task Group was interested in novel ways of controlling the landing of manned capsules and their attention was drawn to a technique devised by NACA engineer Francis Rogallo more than a decade earlier. He had worked on a flexible 'kite', with a lifting surface draped from an inflated fabric frame, which had the effect of producing more lift than drag; not as much, admittedly, as a conventional rigid wing, but it had the benefit of being foldable and lightweight. Early in 1959, Rogallo explained his concept to Gilruth, who was sufficiently impressed to implement further study of a follow-on, manoeuvrable Mercury spacecraft which could touch down precisely on land, thereby saving the cost of a naval recovery force. Other suggestions included a two-man Mercury capable of remaining aloft for three days, the addition of a 3 m cylinder at the back of the capsule to support two-week missions or even installing cabling to link the spacecraft to the booster, rotating them and providing experimental artificial gravity.

Unfortunately, with initial steps to develop Project Apollo, plans for advanced Mercury capsules were turned down by NASA Administrator Keith Glennan's budget analysis team in May 1960. Plans at the time amounted to achieving manned suborbital flight before the year's end and an orbital mission thereafter. These would be followed by an unmanned flight to Venus or Mars by 1962, a controlled robotic landing on the Moon, an unmanned circumlunar mission around 1964 and eventually crewed space stations and circumlunar flights by 1967. Manned landings on the Moon, it was expected, would be a longer-term goal for the Seventies. Of course, this plan would change dramatically with John Kennedy's speech to Congress in May 1961, but was limited at the time by the weight-lifting capacity of existing rockets and the widely-held assumption that lunar missions would be launched directly from Earth atop very large boosters.

The shift in climate from flying circumlunar missions to actual Moon landings began early in 1961, when George Low, head of manned spaceflight in NASA's Office of Space Flight Programs, advocated Earth-orbit and lunar-orbit rendezvous techniques at the quarterly meeting of the Space Exploration Program Council. In February, he submitted a report to Bob Seamans, NASA's newly-appointed associate administrator, stressing that orbital operations and large boosters would be needed, but that developing rendezvous techniques "could allow us to develop a capability for the manned lunar mission in less time than by any other means". By the end of that month, NASA Headquarters had taken greater notice and the possibilities of orbital rendezvous assumed centre stage in congressional hearings for the agency's budget. The House Committee on Science and Astronautics also expressed an interest, scheduling a special hearing on the subject for May and recommending that NASA be awarded the full $8 million it had requested for rendezvous research. The Bureau of the Budget had previously cut this rendezvous spending down to just $2 million, but the committee's inputs eventually led to NASA getting the funding it needed.

Elsewhere, Jim Chamberlin, an aeronautical engineer working for Toronto-based AVRO Aircraft Inc., joined the Space Task Group and was assigned by Bob Gilruth to oversee the development of an advanced Mercury capsule. Chamberlin seized the opportunity to effectively design a completely new spacecraft, retaining only the proven aerodynamic bell-like shape. In March 1961, at a weekend retreat in Wallops Island, Virginia, he described his plans to Gilruth and NASA's head of spaceflight programs, Abe Silverstein, sketching out an ambitious machine with its equipment located outside the crew compartment in a self-contained module that would be far easier to install and test. One of Chamberlin's suggestions was that the advanced

Mercury could be enlisted for circumlunar missions. Although Silverstein dismissed this lunar possibility, he and Gilruth expressed interest in the design itself and on 14 April the Space Task Group and McDonnell signed an amendment to the original Mercury contract, which provided for the procurement of long-lead-time items for six additional capsules. These items would then be used in support of what was now being dubbed 'Mercury Mark II'.

McDonnell's early efforts involved making no alterations to the shape of the spacecraft or its thermal protection system, but simply moving retrorockets and recovery equipment into modular subassemblies and, in Chamberlin's words, creating "a more reliable, more workable, more practical capsule". It would transform, effectively, from an experimental machine into an operational one. By June 1961, when Chamberlin revealed his Mercury Mark II design, some members of the Space Task Group were taken aback: not only did it fulfil the key requirements of extending the spacecraft's orbital lifetime and making it easier to test, but it essentially involved the repackaging and relocation of virtually every subsystem. This was needed, Chamberlin reasoned, because most of Mercury's components were inside the cabin, meaning that equipment had to be disturbed in order to reach and fix one malfunctioning device. As it stood, Mercury could do its job, but was far from being a convenient and serviceable spacecraft. Chamberlin's design allowed for any malfunctioning unit to be removed and tended, without the need to tamper with anything else. "If one system goes haywire," said Gus Grissom, "you take it out and plug in a new one."

It also tackled the problem of Mercury's sequencing system, in which many of its operations were automated for safety, by relying for the first time on pilot control; this, too, contributed to a far simpler machine. Chamberlin also advocated the inclusion of an ejection seat and eliminated the need for a Mercury-type escape tower, which he felt contributed hundreds of kilograms of weight to the capsule and argued that its extreme complexity made it inherently dangerous. Moreover, Mercury abort modes were automated, which could terminate a mission in some circumstances where such action may not be necessary. Flying an advanced Mercury with an ejection seat eliminated the option of using the Atlas – the seat could not push the pilots to safety quickly enough in the event that the rocket's volatile liquid oxygen and RP-1 hydrocarbon mixture exploded. In its stead, Chamberlin suggested the Titan II, which the Martin Company had been developing for the Air Force as an intercontinental ballistic missile.

Martin had already proposed the Titan II as a candidate for lunar missions and, although both Seamans and Silverstein doubted its usefulness, they were sufficiently interested to ask Gilruth to explore ways in which it could be used for other manned projects. Two and a half times more powerful than the Atlas, the Titan seemed, to Chamberlin, perfect for lofting a correspondingly heavier Mercury capsule. The rocket was fed by hydrazine and unsymmetrical dimethyl hydrazine, together with an oxidiser of nitrogen tetroxide. In a catastrophic failure, an ejection seat would be able to outrun the fireball of these less-explosive chemicals. This combination of 'hypergolics', capable of spontaneously igniting upon contact, meant that the Titan needed no ignition system and, since they could be held at normal temperatures,

A Gemini-Titan launch. Note the absence of an escape tower; Gemini crews, aboard their conspicuous black-and-white spacecraft, relied instead upon an ejection system. Privately, many astronauts doubted its usefulness.

required no cryogenic storage or special handling facilities. Self-igniting propellants were intrinsically safer and easier to control than the violently-reactive cryogen used by the Atlas.

In any case, Chamberlin reasoned that because the Titan II was two and a half times more powerful than the Atlas, it would be possible to relax the constraints on the spacecraft's weight. His decision to incorporate an enlarged overhead mechanical hatch in his modified Mercury, primarily as a means of emergency escape, soon expanded to fill another important requirement for lunar missions: the ability to conduct extravehicular activity (EVA), or spacewalking. Meanwhile, efforts to develop Francis Rogallo's paraglider as a recovery system were gathering pace. The Space Task Group, however, which met with Rogallo and his team in the early months of 1961, felt that too much work had still to be completed before such an experimental device could be committed to a manned spacecraft. Questions were posed over its deployment characteristics, how it was to be packaged and whether the pilot's view would be good enough to fly and land with it. They advised gathering at least six months' worth of data before making a decision on whether or not to award actual development contracts. In May 1961, three $100,000 studies were authorised to design an effective paraglider and identify its problems.

Despite the changes to the launch vehicle, the escape system, the hatch, the packaging of components and the recovery operations, Chamberlin's new spacecraft still resembled the bell-like Mercury capsule and, in its earliest form, was not expected to remain aloft for much longer than a day. Little interest was shown towards developing it further. Then, in July, the Space Task Group began looking at a so-called 'Hermes Plan', which envisaged a greatly expanded Mercury Mark II along the lines of that proposed by Chamberlin and, that same month, McDonnell's Walter Burke outlined three possible forms of an advanced spacecraft. The first simply cut hatches in the side of the capsule to improve access to components, the second – valued at $91.3 million – adhered closely to Chamberlin's proposal, whilst a third, $103.5 million suggestion envisaged a Mercury carrying not one pilot, but two.

This was not an entirely novel idea, having been brought to the table and quickly rejected in January 1959, but returned to the fore now that the capsule seemed likely to be extensively redesigned. "If we're going to go to all this trouble to redesign Mercury," said Max Faget, father of the spacecraft, "why not make it a multi-place spacecraft in the process?" In truth, Faget had already approached McDonnell several months earlier with a similar suggestion. Late in July 1961, Silverstein, Gilruth and several astronauts visited St Louis to view quarter-scale models of four basic spacecraft configurations, together with a full-size, wood-and-plastic mockup of a two-man Mercury, which Wally Schirra clambered inside. His first comment: "You finally found a place for a left-handed astronaut!"

Humour aside, the visit proved pivotal, convincing Silverstein that Mercury should be extensively upgraded into a two-man machine. This decision was accompanied by President Kennedy's commitment to a lunar landing before 1970, which prompted significant changes: the Space Task Group, based at the Langley Research Center in Virginia and originally devoted exclusively to Project Mercury, would be superseded by a Manned Spacecraft Center (MSC), to be situated near

Houston, Texas, as part of a much wider, far more complex and infinitely more expensive effort to land a man on the Moon. Rendezvous provided a means of achieving this goal far sooner than a direct-ascent method and the growing conviction throughout the summer and autumn of 1961 that rendezvous needed to be utilised in some form would provide a framework for what would become Project Gemini.

With the approval of the new project came more emphasis on the Titan II as its launch vehicle. Even though the contract with the Air Force to build the rocket had been signed scarcely a year earlier, Martin's James Decker proposed that NASA purchase nine Titans for a bargain price of $48 million, the first of which could be ready to fly by early 1963. Among the modifications needed to make it suitable for the Mercury Mark II were lengthened second-stage propellant tanks to increase its payload by 300 kg. Also, the risk of first-stage 'hardover' – a malfunction in its guidance system which could drive the gimballed engines to their extreme positions, thereby subjecting the Titan to massive dynamic overloads – could lead to the rocket breaking up before the astronauts could react. A second first-stage guidance system was added to erase this risk.

At the same time, if rendezvous was on the agenda, a rendezvous target was needed and, in August 1961, Chamberlin made his first contact with the Lockheed Missile and Space Company in Sunnyvale, California, with a view to using its highly-successful Agena-B rocket stage. Like the Titan, the Agena ran on storable hypergolics – unsymmetrical dimethyl hydrazine with an oxidiser of inhibiting red fuming nitric acid – and had a 'dual-burn' capability; in effect, it could be fired, shut down, then fired again. It also had the potential for the Mercury Mark II, after docking with it, to demonstrate advanced manoeuvres.

The growing importance of rendezvous, docking and manoeuvring was such that the new spacecraft was beginning to change into a new project in its own right and another question that it would be pressed to answer would be the effect of long-term missions on the human body. A journey across the 400,000 km gulf to the Moon and back was expected to require a flight lasting almost two weeks and Mercury Mark II might not only be able to demonstrate that such missions were survivable, but also could evaluate advanced technologies such as electricity-generating fuel cells and more stable attitude-control propellants than hydrogen peroxide. The astronauts, too, would need their own 'modifications', in the form of improved space suits to support longer missions.

Ten Mercury Mark II flights, the first in March 1963 and the last in September 1964, would be launched every two months to fly men for up to seven days and animal passengers for as long as two weeks. Investigation of the Van Allen radiation belts around Earth was a second major objective and, indeed, the first flight would be an unmanned test to ensure that the Titan and Mark II were compatible and to boost the capsule to an apogee of 1,400 km. Controlled landings would be the third goal, to be accomplished on each manned mission, most likely with the aid of a paraglider, and rendezvous and docking stood fourth. Later flights would require dual launches of the Titan II and the Agena-B, such that the Mark II could rendezvous and dock. The hope was also there that, if the spacecraft achieved all of its objectives without problems, particularly the long-duration aspect, a Mark II could be launched to dock with a liquid hydrogen-fuelled Centaur stage and boosted onto a circumlunar trajectory. Some short-lived plans even envisaged a manned around-the-Moon mission as early as May 1964. Although they did not get far beyond the drawing boards, one of Chamberlin's ideas included launching a Mark II atop a Saturn C-3 rocket and placing a manned craft on the lunar surface. Such a landing could, he suggested, be achieved late in 1966.

By the end of October 1961, however, greater emphasis had been placed on developing rendezvous techniques and flying long-duration sorties; Van Allen studies, animal flights and lunar missions were gone. There would be a dozen Mark IIs: an unmanned precursor, followed by an 18-orbit manned mission, a series of extended-duration flights of up to 14 days and, later, rendezvous and docking exercises. Two weeks after NASA formally announced its intention to proceed with Mercury Mark II, on 22 December 1961, James McDonnell signed the contract for its development, agreeing that his company would provide full-scale spacecraft mockups within six months and a mockup of the Agena-B target adaptor by October 1962. By the following March, read the provisions of the contract, McDonnell would supply the first flightworthy spacecraft, with others to follow at 60-day intervals until 12 had been delivered. In effect, this contract replaced the earlier one to procure long-lead-time items for extra Mercury capsules.

It was shortly after the awarding of the main contract that McDonnell began to subcontract out several systems which would prove instrumental in demonstrating the capabilities of the new spacecraft, by now known as 'Gemini'. One of these was the Orbit Attitude and Manoeuvring System (OAMS), which not only allowed the astronauts to 'steer' their spacecraft, but also helped them to station-keep and push away from the second stage of the Titan II. It comprised 16 small engines, fed by hypergolic mixtures of monomethyl hydrazine and nitrogen tetroxide. Each engine was mounted in a fixed position and ran at a fixed thrust level. Eight of them were rated at 11 kg of thrust and provided attitude control. These fired in pairs, permitting the Gemini to roll, pitch and yaw. The remaining eight were 'translational' thrusters, each capable of 45 kg of thrust, and were oriented in pairs to fire forward, backward, up/down and left/right. This would form the 'manoeuvring' component of the system, although the thrust level of the two forward-firing engines was reduced to 38 kg in July 1962. The re-entry controls, developed by the same subcontractor, North American Aviation's Rocketdyne Division of Canoga Park, California, consisted of two independent rings of eight 11 kg thrusters located in the nose of the Gemini, forward of the crew cabin. After the main manoeuvring system had effected retrofire, either ring could control the attitude of the spacecraft during re-entry. By the end of May, all major subcontractors had been selected to begin work.

Meanwhile, efforts to secure the Titan II and Agena required Bob Seamans and Assistant Secretary of Defense John Rubel to agree that the Air Force would act in the capacity of a contractor to NASA, whilst at the same time allowing the former "to acquire useful design, development and operational experience". The Los Angeles-based Space Systems Division of the Air Force Systems Command would handle the development of the Titan and Agena for Mark II operations and, early in January 1962, NASA issued formal instructions for work on the rockets to begin. Adapting the Titan to handle the spacecraft proved more complex than originally expected, since it required new or modified systems to ensure the astronauts' safety during the countdown and ascent. By the beginning of March, Lockheed had also started work on the development of a more advanced Agena-D, with an engine capable of being fired more than twice, which would be boosted into orbit atop an Atlas. Unlike the Agena-B, this new version boasted a radar transponder, a forward docking adaptor and an improved attitude-control system. At this time, the first – unmanned – launch of Gemini had slipped from May to August 1963, although the three inaugural flights would be spaced out at just six-week intervals.

Elsewhere, the contract for the paraglider, intended to guide Gemini to a touchdown on land, was awarded to North American on 20 November 1961, but its real future seemed less assured. Chamberlin had defended it vigorously, but Max Faget's engineering directorate within the newly-established MSC in Houston was cool to the idea, considering its reliability as lower than having main and backup parachutes. Faget instead advocated a steerable parachute, together with landing rockets to cushion the touchdown. Others, including Chris Kraft, felt that neither the paraglider nor the ejection seats were reliable and posed enormous practical obstacles to safety. The paraglider was not aided by North American's slow progress on its development, which had been unavoidable as Gemini grew from little more than a modified Mercury into an entirely new – and bigger – spacecraft. North American planned free flights of a half-size paraglider for May 1962, although this was delayed because backup parachutes were needed to avoid losing the costly test vehicle. Initial plans called for the first unmanned Gemini to land with ordinary parachutes and the second (manned) flight to utilise the paraglider, although by mid-June it became clear that it would not be available until the third mission. Still, a maiden launch in August 1963 did not seem unreasonable.

However, as 1962 wore on, it was apparent that project costs would be far higher than anticipated. Modification of the Titan II, for example, had climbed from $113 million to $164 million, owing to a multitude of changes to 'man-rate' it. These included a fully redundant malfunction-detection system, backup flight controls, an electrical network with backup circuits for guidance, engine shutdown and staging and new launch tracking hardware. Costs of developing the Agena-D similarly increased from $88 million to $106 million and the Gemini spacecraft itself ballooned from $240.5 million to $391.6 million. This cost hike came as a huge surprise, yet it encompassed McDonnell's enlarged view of what should be included in the project: from 'realistic' flight simulators and trainers in Houston and at Cape Canaveral to structural mockups for static and dynamic tests and even the development of an extra spacecraft and docking adaptor for an extended series of unmanned orbital missions (dubbed 'Project Orbit') "to investigate potential problems and to evaluate engineering changes". When the Office of Space Flight submitted its Project Gemini review to Administrator Jim Webb in May 1962, the cost of the overall programme had climbed markedly from $529 million to $744 million – and continued to grow.

The half-size emergency parachute experienced difficulties of its own. In a series

A model of the Gemini paraglider under test.

of drop tests at the Naval Parachute Facility in El Centro, California, between May and July, four failures led to an extensive redesign and placed it and the paraglider two months behind schedule. Its full-size counterpart also suffered problems and, despite some successes, all three parachutes failed during a November 1962 test and the test vehicle was destroyed when it hit the ground. Although these woes did not directly affect its potential performance, they did introduce worries. Early tests of a half-sized paraglider at Edwards Air Force Base in mid-August had failed to deploy properly after being towed to altitude by helicopter and in two subsequent attempts it was released too soon and landed too hard. A fourth try failed to deploy and a short circuit cancelled a fifth. By the end of September, even Jim Chamberlin was losing patience and ordered North American to halt all tests until it could spell out how it intended to correct the problematic electronics and pyrotechnics. After rework, the half-scale paraglider was shipped back to Edwards and, on 23 October, sailed through a perfect flight, finally demonstrating its stability.

Even the development of the Titan II rocket presented problems. In its first flight on 16 March 1962, it began to experience longitudinal vibrations, occurring 11 times per second for about half a minute. Although these did not pose a problem for the Titan, they would pose a risk for the astronauts, who would be exposed to two and a half times normal gravity and might not be able to react properly to an emergency. The vibrations, nicknamed 'pogo', partially disappeared thanks to higher pressure in the rocket's first-stage fuel tank – perhaps, engineers speculated, it was caused by oscillating pressure in propellant lines – and Martin suggested installing a surge-suppression standpipe in the oxidiser line of later Titan IIs.

Escalating costs and a spending cap limited to $660 million for 1963 eventually led to the realisation that Gemini could only go ahead with the cancellation of the paraglider, Agena and perhaps all rendezvous hardware. Surprisingly, in the subsequent rescoping of the project to take account of budget limitations, the paraglider survived almost untouched and the spacecraft itself retained most of its original features, although the Titan II testing programme was drastically reduced and the decision as to whether the Agena had a role in the project remained fluid for some months. These difficulties conspired to delay the first unmanned mission to December 1963, followed by a piloted flight three months later.

Meanwhile, development of the paraglider continued to be mired with problems. After its October 1962 success, North American refitted the half-scale test vehicle and shipped it back to Edwards for a late November flight. Minor electrical problems postponed the attempt and, when it eventually flew on 10 December, its performance was disappointing: the capsule tumbled from the helicopter, fouled the stabilising drogue parachute and the inflation of the paraglider wing only made matters worse. When the capsule spun down past 1,600 m – the minimum recovery altitude – a radio command detached the wing and allowed it to descend on its emergency parachute. Another try a few weeks later was worse still: it did not tumble this time, but the paraglider storage can was late in separating. The capsule was falling too fast when the wing started to inflate and its membrane tore. Moreover, a faulty squib switch meant the main parachute failed to deploy and the capsule crashed. Despite reporting that five distinct failures had been identified and repaired,

North American's paraglider was about to breathe its last. Chamberlin gave the project a final chance, but another attempt to deploy a half-scale wing on 11 March 1963 concluded dismally when the storage can failed to separate. The emergency parachute then failed and the paraglider, figuratively and literally, ended its days as a heap of smouldering wreckage.

For the Titan II, the key problem was overcoming its 'pogo' oscillations and Martin duly installed a surge-suppression standpipe. However, a test flight on 6 December 1962 actually worsened the pogo effect and, indeed, induced such violent shaking that the first-stage engines shut down too early. Two weeks later, another rocket with no standpipes and increased fuel-tank pressures launched successfully and exhibited lessened pogo levels. A third launch on 10 January 1963 produced similarly encouraging results and the G forces to which it would have exposed a human crew were only slightly higher than those tolerated by Mercury pilots. On the other hand, in this case, the Titan's second-stage thrust was half of what it should have been, suggesting that its engines had difficulty reaching a steady burn after the shock of ignition. This 'combustion instability' proved somewhat more complex than the pogo obstacle and led to a decision in March 1963 to increase the number of unmanned tests and reduce the total of manned flights to ten. To make matters worse, on 8 March, Gemini project cost estimates topped a billion dollars. Days later, Bob Gilruth relieved Chamberlin of his duties and replaced him with Charles Mathews.

Among Mathews' earliest moves was to insert an unmanned mission in place of one of the manned flights, largely in response to the ongoing Titan II problems. Scheduled for December 1963, the new mission would demonstrate the Titan as Gemini's launch vehicle. After the upper stage had achieved orbit, a 'boilerplate' spacecraft would remain attached and the entire assembly would be left to fall back into the atmosphere. An unmanned suborbital flight with a real spacecraft in July 1964 would then show the ability of the spacecraft to support manned missions. On this plan, the first manned mission, Gemini 3, would come in October 1964. Originally intended as a day-long, 18-orbit mission, Gemini 3 would be reduced by nervous managers to around three orbits – or five hours – and would test the spacecraft's systems. Earlier plans to fly a Rendezvous Evaluation Pod (REP) on the first manned mission were rescheduled for Gemini IV, which would run for seven days in January 1965. The new schedule implemented three-month gaps between each manned mission, in response to concerns that equipment checkouts and astronaut training required more time. Gemini V would then conduct the first rendezvous and Gemini VI would attempt a 14-day mission, to mimic the length of a full-duration lunar landing expedition. Subsequent flights would last three days apiece, each consolidating and extending rendezvous and spacewalking expertise.

Interestingly, Mathews' plan did not entirely omit the paraglider, but pushed its maiden flight back to Gemini VII, with parachutes supporting earlier missions. The plan was approved at the end of April 1963. Although paraglider tests in May and June satisfactorily proved its stabilising parachutes, a final drop on 30 July suffered a total failure and crashed. By the year's end, the paraglider was itself being challenged by the concept of the 'parasail', a flexible gliding parachute which offered a quick and relatively cheap device to achieve a touchdown on land. It could be ready, McDonnell told NASA Headquarters in September 1963, in time for Gemini VII and at a cost of just $15.7 million. It was ruled out, partly due to lack of funds, but chiefly because the paraglider's vocal supporters objected to giving up on an effort that had already consumed much time and work and was almost ready for flight testing. However, the paraglider itself was on the wane: its landing system programme was stripped of all objectives, save that it would prove the 'technical feasibility' of the concept. Parachutes would support most of the Gemini missions, with the paraglider possibly used for the last three. Although Bob Gilruth insisted that it might still fly on Gemini if its tests were successful, the first mutterings of cancellation had reared their heads.

Meanwhile, the Titan II's woes continued. A launch on 29 May 1963, carrying pogo-suppression devices for both oxidiser and fuel, burst into flames during liftoff, pitched over and broke up. Although the pogo devices themselves were absolved from blame in the incident, the flight was too brief for their effectiveness to be judged. Three weeks later, a military test of the missile from a silo at Vandenberg Air Force Base in California, although trouble-free during first-stage ascent, with pogo levels within acceptable limits, exhibited faltering second-stage thrust. Had a Gemini crew been aboard, it would have been an abortive mission. The Air Force now shifted its focus to ensuring that the Titan worked as a missile, first, before committing it as a Gemini launch vehicle, and some within NASA even came to doubt that it was the right rocket for the job.

In fact, concerns were so high that the space agency even considered adding yet another unmanned Gemini flight to test the Titan, pushing the total number of missions to 13. Designated 'Gemini 1A', it would be slotted in, sometime around April 1964, between the first and second unmanned launches and, like Gemini 1, would consist of a 'boilerplate' capsule equipped with instrumentation to examine the performance of the rocket. It would only fly, however, if the unmanned Gemini 1 failed to meet all of its objectives. Although the Gemini 1A hardware was delivered in September 1963, the extra mission had been cancelled by February of the following year, thanks to improved prospects for the Titan II. Despite the fact that two launches in August and September 1963 had gone wrong due to short circuits and guidance malfunctions and the effects of pogo were higher than expected, circumstances improved as the year drew to a close. A Titan launch on 1 November, equipped with standpipes for its oxidiser lines and mechanical accumulators on its fuel lines, reduced pogo effects well below the limit demanded by NASA. Moreover, in the next five months, the rocket would score an impressive and unbroken chain of successes, enough to 'man-rate' it in time for Gemini 1.

For all of these problems, the development of the spacecraft itself was going relatively well. Key problem areas remained the fuel cells, propulsion and the ejection seats. McDonnell had already subcontracted to General Electric to build the cells and the first serious development problem of preventing oxygen leakage through its ion-exchange membrane was soon resolved. However, resolution of this problem produced another: test units working over long periods showed degraded performance, apparently due to contamination of the membrane by metal ions from the fibreglass wicks responsible for removing water from the cell. Leakages in the tubes which fed hydrogen to the cells created another obstacle. General Electric replaced the fibreglass wicks with Dacron cloth and an alloy of titanium-palladium replaced the pure titanium tubing, although these developmental headaches pushed the project further behind schedule. NASA was sufficiently concerned to request McDonnell to conduct an evaluation of batteries to be used on Gemini 3, the first manned mission in October 1964, with fuel cells aboard, but only used for flight qualification purposes.

Ongoing problems with the fuel cells, it was realised, could restrict Gemini missions to only a few days under battery power alone. In November 1963, Charles Mathews issued instructions to adapt the electrical system to house batteries or fuel cells and, within weeks, the decision to fly Gemini 3 on batteries was official. Nonetheless, the unmanned Gemini 2 would be equipped with both systems to qualify them. A month later, Mathews decided that Gemini IV – the proposed seven-day mission – should also utilise batteries, which would have a corresponding impact on its duration, shortening it by almost half. Eventually, it would be Gemini V that would first demonstrate the use of fuel cells in space on a week-long flight, with mixed success.

The ejection seats were another concern; so much so, in fact, that the MSC even considered replacing them with an escape tower, akin to that used during Project Mercury. Simulated 'off-the-pad ejection', or 'sope', tests had been suspended late in 1962 until all components were ready. During one test, the overhead hatch failed to open and the seat shot straight through the 5 cm-thick hull, prompting John Young to remark: "That's one hell of a headache ... but a short one!"

Ultimately, the escape system involved a balloon-parachute hybrid, known as a 'ballute', which would prevent the astronaut from spinning during freefall if he had to eject from an altitude higher than the 2,000 m at which his personal parachute was supposed to open. The first tests in February 1963 were disappointing: the ballutes failed to inflate and the personal parachute did not deploy correctly, although managers felt that the problem was a dynamic one, caused by the relationship between the rocket motor's thrust vector and the shifting centre of gravity of the man-seat combination. By May, sope testing resumed and met with greater success and, the following month, dual-seat ejections were demonstrated. Nonetheless, rising costs and unending technical problems, involving the spacecraft, the Titan, the paraglider and even the Agena-D, had conspired to delay the first unmanned Gemini launch until the spring of 1964.

For the paraglider, which experienced yet more failures in December 1963 and February 1964, the end was in sight. Indeed, NASA's public stance was now that a land touchdown was riskier than an already-proven ocean splashdown, although the Air Force, which was planning its own version of Gemini in conjunction with an orbital laboratory, kept the concept alive for a time. However, the military was not keen to commit funding to the paraglider until it had first been satisfactorily demonstrated by NASA. After yet another test failure on 22 April 1964, William Schneider, Gemini's project chief at NASA Headquarters, officially announced that no more money would be spent on the paraglider. Ironically, its cancellation was actually followed by several wing-deployment successes. North American even invested its own cash in further development work and, in December, a pilot flew with the helicopter-towed vehicle and guided it to a safe landing. It was, however, too late for it to be reinstated into Gemini.

Of course, the design of the capsule itself had long since been finalised. In fact, the re-entry capsule was comparable in size to the Mercury spacecraft. A broad conical adaptor at its base, which would be shed following retrofire, held the propulsion and long-term power and life-support systems. The capsule, wrote Neal Thompson in his biography of Al Shepard, was "like a snug little sports car". The white-coloured adaptor was 2.28 m high and 3 m wide at its base and was itself split in two: an equipment section for fuel and propulsion and a retrorocket to support re-entry. Both segments were isolated from each other by means of a fibreglass honeycomb blast shield. The crew cabin, meanwhile, was a truncated, titanium-and-nickel-alloy cone, measuring 2.28 m wide at its base and 98.2 cm at its apex and topped by a short cylinder for re-entry controls and parachutes. In total, the cabin was 3.45 m high and, when mounted on the adaptor, the full Gemini stood 5.73 m tall. Inside, conditions were cramped. The Gemini's total pressurised volume was little more than 2.25 m^3 – "like sitting sideways in a phone booth," John Young said years later – and its outward-opening hatches were positioned directly above each astronaut's head. Each man was provided with a small, forward-facing window. The spacecraft was equipped with horizon sensors and an inertial reference system and the command pilot flew using a pair of hand-controllers, one for translation, the other for orientation, whilst referring to an 'artificial horizon' display.

Key to its manoeuvrability were the 16 OAMS thrusters, spaced around the capsule, and power, at least before the first demonstration of fuel cells, came from three silver-zinc batteries. During preparations for re-entry, the equipment section of the adaptor would be jettisoned, exposing the retrorocket package, whose four motors would initiate the Gemini's return to Earth. After retrofire, the package would itself be released, leaving the crew cabin, protected by an ablative heatshield and radiative shingles, to withstand the intense heat of re-entry. The centre of mass was deliberately offset to generate aerodynamic lift during re-entry and rolling the capsule using the thrusters on the nose enabled the trajectory to be controlled to aim for a specific geographic position. A parachute-aided descent would finally bring the capsule gently into the Atlantic or Pacific.

ASSASSINATION

One of the most dramatic and pervading images of the Sixties will always be the assassination of President John Kennedy on 22 November 1963, at the midpoint between the end of Project Mercury and the first unmanned flight of Gemini. Perhaps more than any other event, it marked a pivotal change in the political and social climate of the period. Perhaps, had it not occurred, co-operation between the United States and the Soviet Union in space and on the ground may have been cultivated. Maybe the escalation of involvement and conflict in Vietnam could have been averted. An entirely different Sixties could have resulted.

Kennedy had been in Texas for several days and, tanned and wearing sunglasses, had visited and been photographed at NASA's new Manned Spacecraft Center (MSC), near Houston, shortly before his murder. His decision to visit Dallas and tour the streets in an open-topped motorcade on 22 November had come about in the hope that it would generate support for his 1964 re-election campaign and help mend political fences in a state just barely won three years before.

The plan for that fateful day called for Kennedy's motorcade to travel from Love Field airport, through downtown Dallas – including Dealey Plaza, where the assassination would occur – and would terminate at the Dallas Trade Mart, where he was to deliver a speech. Shortly before 12:30 pm Central Standard Time, the motorcade entered Dealey Plaza and Kennedy acknowledged a comment from Nellie Connally, the wife of the Texan governor, that "you can't say Dallas doesn't love you". Indeed, all around him, adoring crowds thronged the streets.

As the motorcade passed the Texas School Book Depository, the first crack of a rifle sounded from one of its upper windows. There was very little reaction to the opening shot, with many witnesses believing that they had heard nothing more than a firecracker or an engine backfiring. Kennedy and Governor John Connally turned abruptly, with the latter being the first in the presidential limousine to recognise the sound for what it was. However, he had no time to respond. According to the Warren Commission, which investigated the case throughout 1964, a shot entered Kennedy's upper back and exited through his throat, causing him to clench his fists to his neck. The same bullet hit Connally's back, chest, right wrist and left thigh.

The third and final shot, captured by a number of professional and amateur photographers, caused a fist-sized hole to explode from the side of the president's head, spraying the interior of the limousine and showering a motorcycle officer with blood and brain tissue. First Lady Jackie Kennedy frantically clambered onto the back of the limousine; Secret Service agent Clint Hill, close by, thought she was reaching for something, perhaps part of the president's skull, and pushed her back into her seat. Hill kept Mrs Kennedy seated and clung to the car as it raced away in the direction of Parkland Memorial Hospital.

John Connally, though critically injured, survived, but Kennedy arrived in the Parkland trauma room in a moribund condition and was declared dead by Dr George Burkley at 1:00 pm. No chance ever existed to save the president's life, the third bullet having caused a fatal head wound. Indeed, a priest who administered the last rites told the New York Times that Kennedy was dead on arrival. An hour later, following a confrontation between Dallas police and Secret Service agents, the president's body was removed from Parkland and driven to Air Force One, from whence it was flown to Washington, DC. Vice-President Lyndon Baines Johnson, also aboard Air Force One, was sworn-in as President at 2:38 pm.

One of Johnson's earliest official acts was the establishment of the so-called 'Warren Commission' to investigate the president's death. Chaired by Chief Justice Earl Warren, the very man who had sworn Kennedy into office, the commission presented its report to Johnson in September 1964. It found no persuasive evidence of a domestic or foreign conspiracy and identified Lee Harvey Oswald, located on the sixth floor of the Texas School Book Depository, as the killer. It concluded that both Oswald and his own murderer, Jack Ruby, a nightclub owner, had operated alone and without external involvement.

Immediately after the publication of the Warren Commission's report, doubts surfaced over its findings and conclusions. Although initially greeted with widespread support by the public, a 1966 Gallup poll suggested that inconsistencies remained. An official investigation by the House Select Committee on Assassinations in 1976–79 concluded that Oswald probably shot Kennedy as part of a wider conspiracy and, over the years, countless theories have emerged, placing the blame on Fidel Castro, the anti-Castro Cuban community, the Mafia, the FBI, the CIA, the masonic order, the Soviets and others. An ABC News poll in 2003 concluded that 70 per cent of respondents felt that the assassination was the result of a broader plot, although no agreement could be reached on what outside parties may have been involved. Kennedy's death will perhaps remain one of the greatest unsolved mysteries of the modern era.

CREWLESS SUCCESSES

By late 1963, two unmanned Gemini missions, instead of one, had been timetabled before a crew would be flown. The main focus for the so-called 'Gemini-Titan 1' was to evaluate the performance of the spacecraft and rocket, and the capsule itself carried dummy equipment and ballast to match the weight and centre-of-gravity constraints of a piloted flight. Moreover, it would remain attached to the Titan throughout its mission and no plans were made for it to be recovered, so four large holes were bored into its ablative heatshield to ensure destruction during re-entry. Inside the cabin, roughly where the two pilots would sit on manned flights, a pair of pallets housed radar transponders and instruments to measure pressures, vibrations, accelerations, temperatures and structural loads. On 1 October 1963, NASA officially accepted the spacecraft from McDonnell. Three days later, Gemini 1 was shipped to Cape Canaveral for final pre-launch testing.

A flight readiness review in mid-February 1964 verified the spacecraft's systems and on 3 March it moved to Pad 19 to await the arrival of its launch vehicle. The Titan assigned to the mission, dubbed 'Gemini Launch Vehicle 1' (GLV-1), had suffered numerous problems throughout the previous year, including damaged wiring caused by faulty clamps, improperly-cleaned hydraulic tubing, a malfunctioning gyroscope and interference between several electronic components. Martin conducted a full systems acceptance test in September 1963, but GLV-1 was rejected by NASA and the Air Force's Space Systems Division when their joint inspection found severe contamination of electrical connectors throughout the rocket. A month later, after this contamination had been corrected, Gemini 1's launch vehicle was officially accepted and on 26 October its two stages were transported to the Martin airport and flown to the Cape.

Further problems pushed the launch from February into March and, eventually, into April of the following year. Lack of compatibility between the Titan and its Pad 19 support systems, together with a faulty turbopump, were the prime culprits, although contaminated oxidiser and a malfunctioning propellant valve conspired to ruin a 'wet' countdown demonstration test on 7 January. Two weeks later, the entire sequence of fuelling, countdown, ignition and shutdown commands, guidance control and telemetry were successfully demonstrated for the individual stages, with each firing for 30 seconds. By the end of January, the two stages had been mated together and on 3 March Gemini 1 arrived at the pad for installation. Liftoff, optimistically, was scheduled just four weeks later. Problems arose during electronic-electrical interference tests, which pushed the mission to no earlier than 7 April, but all other pre-flight activities proceeded smoothly.

Eventually, mission managers agreed on 8 April for launch. One second after 11:00 am Eastern Standard Time (EST), the Titan II's first stage engines ignited with a low-pitched whine and the first Gemini left the pad. Two and a half minutes into the ascent, the first stage propellants were exhausted and the second stage roared to life, explosive bolts separating the two segments of the booster. By five minutes after launch, itself expended, the second stage and Gemini 1 had achieved an altitude of 160 km and were more than 1,000 km downrange of the Cape, moving at 28,080 km/h. The combination was in orbit. So perfect was the Titan's performance that Major-General Ben Funk of the Air Force Space Systems Division described it as "a storybook sort of flight".

Within five hours, as planned, its test mission was over. The capsule and second stage, whose altitude at apogee reached 300 km, were expected to remain aloft for three and a half days; they actually managed almost four, during which time the Manned Space Flight Network, controlled by NASA's Goddard Space Flight Center in Greenbelt, Maryland, tracked them constantly. Finally, on 12 April, three years to the day since Yuri Gagarin's pioneering voyage, Gemini 1 re-entered the atmosphere to destruction, its last fragments splashing down in the South Atlantic. NASA commended the Air Force on its rocket and envisaged Gemini 2 by the end of August and an inaugural manned flight in mid-November.

Unluckily, the elements – first lightning, then hurricanes – delayed the second launch well into September. Additionally, delays in getting key components such as thrusters and fuel cells had placed the spacecraft's construction behind schedule and Gemini 2 did not even commence testing until January 1964. Also, delivery of the Titan II for the mission, which had been accepted by NASA in mid-June, was postponed due to lagging work on the spacecraft. Anticipating Gemini 2's arrival by early September, the rocket was eventually installed on Pad 19 to undergo final testing; then, in mid-August, in a severe thunderstorm, lightning struck the launch complex. No physical damage was done to the vehicle, although a number of failed parts of ground support equipment needed replacement. Even before this work had been completed, Hurricane Cleo brushed the Cape on 27 August. The Titan's second stage had been demated and placed under cover, but the still-upright first stage bore the hurricane's full force. Fortunately, Cleo's effects were lower than the rocket had been designed to withstand and little serious damage was done. By 1 September, the second stage had been reinstalled.

By now, there were suggestions of replacing GLV-2 with the GLV-3 vehicle earmarked for Gemini 3, in effect 'losing' one mission, but the Air Force eventually persuaded NASA to stick with the original Titan. Barely two weeks after Cleo, Hurricane Dora reached the Cape, forcing engineers to again demate the two stages and get them under cover. A third hurricane, Ethel, had bypassed the launch site by 14 September and preparations could resume in earnest. Gemini 2 finally arrived in Florida on the 21st, but by this time NASA's schedulers had officially given up on the already-slim chance of launching a crew before the year's end. After testing, the spacecraft was mated to the Titan II on 5 November in anticipation of a launch early the following month.

Three holds characterised Gemini 2's countdown and, as the clock hit zero at 11:41 am on 9 December, the rocket's first stage engines ignited. Barely a second later, they were automatically shut down; the Titan had apparently lost hydraulic pressure in its primary control system and had switched to its backup, which, because the rocket had not lifted off the pad, cut off the engines. It transpired that unexpectedly high pressure in one of the hydraulic lines had burst the aluminium housing of a servo valve, allowing fluid to leak out. "It was frustrating," wrote Deke Slayton, "but a good lesson. You couldn't have shut down at Atlas like that. It wouldn't have done us much good to have the Titan get off the ground, then blow up, either. So it gave us confidence in the [malfunction-detection system] which we needed a year later."

Steadily, the Titan's propellants were drained and the spacecraft made safe until a replacement valve arrived in early January 1965. The second attempt to get Gemini 2 airborne was virtually flawless: despite a problem with the sluggish fuel cell, which meant it could not be used, the Titan II roared aloft at 9:03:59 am on 19 January. Its suborbital trajectory exposed the spacecraft to the most severe re-entry heating subsequent missions could be expected to experience and it plopped safely into the South Atlantic at 9:22 am. Minor glitches had not detracted from a mission which had effectively certified Gemini to carry its first human crew.

ALL ABOARD THE 'MOLLY BROWN'

Only days after Gemini 2's splashdown, Charles Mathews revealed that late March seemed the most achievable target to launch Gus Grissom and John Young on the first manned mission. The two men and their backups, Wally Schirra and Tom Stafford, had been training since mid-April 1964; indeed, wrote Stafford, they "became virtual citizens of St Louis, Missouri ... flying in on Sunday night or Monday morning in our T-33s and T-38s, spending days at the McDonnell plant at Lambert Field then going back to Houston on Thursday or Friday". At St Louis was the Gemini mission simulator, which provided them with the sights, sounds and vibrations that they could expect on a real flight; moreover, it was adaptable for each crew, whose objectives would differ markedly. Grissom, an Air Force officer promoted from captain to major in July 1962, would become the first man to be launched into space twice. He had been one of the earliest astronauts assigned to

Grissom (in water) and Young undergo water survival training exercises in preparation for their Gemini 3 mission.

Project Gemini and the spacecraft reflected, among other things, his short stature. Whereas Grissom, at 1.6 m, could comfortably fit the cockpit, the taller Stafford "was jammed in, especially when I had to wear a pressure suit and helmet".

For Stafford and fellow 'tall' astronaut Jim McDivitt, this posed a figurative and literal pain in the neck. "Eventually," wrote Stafford, "the McDonnell engineers removed some of the insulation from the inside of the hatch to create a slight bump that gave us room for our helmets". First fitted for Gemini VI – which Stafford flew, alongside Schirra, on the first rendezvous mission – it became known, naturally, as 'The Stafford Bump'. Among the astronauts, Gemini had already earned itself the nickname 'Gusmobile', because Grissom was the only man small enough to scrunch himself into his seat and close the hatch without banging his head. The prime and backup crews spent around 35 hours apiece in the St Louis simulator in the late spring and early summer of 1964, learning its quirks and idiosyncrasies, before it was finally dismantled and shipped to MSC in Houston towards the end of July. Another, near-identical simulator was also set up at the Cape in October.

During their early months of training, the four men closely monitored the development of the 'real' spacecraft at St Louis, watching as it passed through various systems tests and inspections, then spending many hours in its cockpit. Elsewhere, in Dallas, Texas, they participated in exercises on a Gemini moving-base abort simulator, which projected their ascent profile in striking detail and enabled them to rehearse their responses to malfunctions. At Ellington Air Force Base, not far from MSC, they were dunked in a 'flotation tank' to practice getting out of a boilerplate mockup of the Gemini capsule, with and without space suits, both floating and submerged. Only weeks before launch, in February 1965, Grissom and Young rode a boat out into the Gulf of Mexico to a mockup capsule, which they had to board and run through their post-splashdown checklists, their emergency egress procedures and the opening of their one-man liferafts.

Mission planning sessions, centrifuge training at Johnsville in Pennsylvania, space suit fit checks, physical exams and preparation for the Gemini 3 experiments quickly turned their training schedule into an unending marathon. "The days just seemed to have 48 hours, the weeks 14 days and still there was never enough time," Grissom would recall later. "We saw our families just enough to reassure our youngsters they still had fathers." By the time the mission simulator had been set up at the recently-renamed 'Cape Kennedy' – the old Canaveral – in October 1964, it would become the astronauts' second home: Grissom would put in more than 77 hours of training in it, rehearsing every phase and every minute of what would be a five-hour mission, with Young slightly eclipsing him at 85 hours. As launch day neared, Grissom had sat through 225 abort scenarios, compared to 154 for his rookie pilot. By February, when queried by journalists, Grissom confirmed that, after nine months of training, he and Young were ready to go. Indeed, even Jim Webb confidently expected a launch as soon as 15 March.

Meanwhile, the development and testing of the Gemini 3 capsule was gathering pace. Its construction had been completed in December 1963 and, following six months of engineering changes and installation of equipment, it began integrated testing in the summer. McDonnell's 'in-house' testing was completed by September and on 27 December, after its own inspections and a simulated flight, NASA accepted the spacecraft for delivery. It arrived at Cape Kennedy aboard a C-124 transport aircraft on 4 January 1965 and was ready for transfer to Pad 19 early the following month. At the same time, its Titan II launch vehicle – GLV-3 – arrived in late January and was mechanically mated with Gemini 3 on 17 February.

The mission itself was too short, many felt, for any meaningful rendezvous data to be gathered. In October 1963, MSC had suggested flying the Rendezvous Evaluation Pod (REP) on Gemini 3 and releasing it into orbit to test the rendezvous radar, although this was cancelled and rescheduled for a later mission. The astronauts themselves soon got in on the act. Word leaked out in the summer of 1964 that Grissom and Young were pushing for an 'open-ended' flight, in effect giving them the option to decide how many orbits to fly. "Gus and John and the rest of us thought a 30-orbit flight, almost two days, was the next logical step after Gordo's MA-9," recalled Slayton, but on this occasion the astronauts' judgements were overruled.

The limitations of the existing tracking network and worries about erring on the side of caution on Gemini's first manned mission won the day and it remained at five hours. Even Grissom acquiesced that he felt sufficient data could be extracted from a three-orbit flight. That 'data' would come primarily in the form of demonstrations of the spacecraft's manoeuvrability, using its OAMS thrusters, and a decision was made to conduct three firings to insert Gemini 3 into a 'fail-safe' orbit, from which it could still re-enter in the event of a retrorocket failure. In reality, Grissom and Young would fly in too low an orbit to be permanent, but the fail-safe option at least ensured that the spacecraft could return promptly and, insofar as possible, that the crew would survive.

To achieve it, the aft OAMS would be fired to separate Gemini 3 from the second stage of the Titan. This would insert the spacecraft into an elliptical path of 122–182 km, whilst a second burst about 90 minutes into the mission would slightly cut the velocity and near-circularise the orbit. Then, whilst over the Indian Ocean around two hours and 20 minutes after launch, a series of 'out-of-plane' burns would be conducted to thoroughly gauge the performance of the OAMS. Finally, above Hawaii on the third pass, a pre-retrofire burst would insert the spacecraft into an elliptical re-entry orbit with a perigee of just 63 km.

Scientific and technical experiments would consume a small portion of the astronauts' time. A Panel On In-Flight Scientific Experiments (POISE) had already been established within NASA and proposed a series of investigations which would largely run themselves. Two promising candidates for Gemini 3 were those originally assigned to Al Shepard's ill-fated MA-10 mission. One explored the combined effects of radiation and microgravity on cells, the other focused on cell growth in space. The former sought to expose human blood samples to a known quantity and quality of radiation, both within the capsule and on Earth, allowing the frequency of chromosomal aberrations in the space-flown and ground-control specimens to be compared. On Gemini 3, it was mounted on the right-hand hatch, inside a half-kilogram hermetically-sealed aluminium box. To activate it, Young had to twist a handle to commence irradiation of the blood samples.

A Gemini spacecraft is prepared for launch.

The second experiment was Grissom's responsibility and was situated inside his left-side hatch. Since it was considered easier to detect the effects of microgravity in simple cell systems than more complex organisms, it consisted of the eggs of a sea urchin. These were fertilised at the beginning of the experiment and the possible changes observed at several stages of their development. Grissom was required to turn a handle 30 minutes before launch to fertilise the eggs, then four times in flight to fix the dividing cells at specific stages of growth. Each handle turn effected by Grissom would be mirrored in a control experiment on the ground.

Additionally, a third investigation, originally envisaged for MA-10, was a re-entry communications demonstration. It had been theoretically shown that by adding fluid to ionised plasma during the period of re-entry blackout, communications could be restored by lowering the plasma's frequency sufficiently to allow UHF transmissions to get through. The lengthy blackout during John Glenn's fiery plunge to Earth and Scott Carpenter's heart-stopping overshoot may have benefitted from such an experiment, despite its $500,000 price tag. It involved the fitment of a water-expulsion system, whose starter switch would be thrown by Young when the capsule descended below 90 km. Water would be automatically injected into the plasma sheath around the spacecraft in timed pulses for two and a half minutes, while ground stations monitored and recorded UHF radio reception.

Early in March 1965, a flight readiness review confirmed that, with the exception of several minor problems, Gemini 3 was ready to fly. The Titan II passed its final test on the 18th and launch was scheduled for 9:00 am on the 23rd. The countdown commenced at 2:00 am on launch morning, which dawned dull and overcast, but the decision was taken to proceed. Three hours later, Grissom and Young were awakened, sat down to the traditional low-residue breakfast of tomato juice, half a cantaloupe melon, porterhouse steaks and scrambled eggs. Young received a long good-luck telegram signed by 2,400 Orlando residents, from whose high school he had graduated years before. The two men then donned their space suits. Unlike Project Mercury, the Gemini ensembles were white, not silver, and comprised a nylon material overlaying a rubberised inner lining. The gloves were removable and were attached to rotating wrist joints, permitting full movement, and learning from the experiences of the Mercury pilots, included 'fingertip lights' to help read cockpit instruments.

The prime and backup crews had all spent the past several days in the new quarters at the Manned Spacecraft Operations Building, whose accommodations, wrote Ray Boomhower, "were a far cry from the ones the astronauts were used to at Hangar S in their Mercury days". They were comfortably furnished, quiet and even boasted a gym with exercise bicycles and punch bags. Shortly after 7:00 am on launch morning, the astronauts headed for Pad 19, where technicians inserted them into their couches in the capsule which Grissom had light-heartedly dubbed 'Molly Brown'. The name came from the unsinkable Titanic survivor popularised in the 1960 Broadway musical and 1964 movie.

Grissom hoped, it seemed, that borrowing Brown's name for Gemini 3 would ward off the demons of watery bad luck which had hit him at the end of his Liberty Bell 7 mission. "I've been accused of being more than a little sensitive about the loss of my Liberty Bell 7," he explained before launch, "and it struck me that the best way to squelch this idea would be to kid it." Kidding or not, NASA officials doubted that the name conveyed the right message, but when offered either Molly Brown or Grissom's other choice – Titanic – they quietly backed off. However, they never named it as such in official documents. In fact, at an earlier stage, Grissom had suggested the name 'Wapasha', after a Native American tribe from his home state of Indiana, although the risk of the media renaming it 'The Wabash Cannonball' meant that was also removed from the list.

Inside the spacegoing Molly Brown, a near-flawless countdown proceeded smoothly, leaving them 20 minutes ahead of schedule and giving Young reason to complain about the extra time spent lying flat on his back in his uncomfortable space suit. The overcast weather refused to lift and the clock was halted at T-30 minutes when a first-stage oxidiser line in the Titan sprang a leak. Although this problem was quickly resolved, the countdown had to be held for 24 minutes to ensure that the leak had stopped. Thankfully, by this point, the grey clouds had cleared, and at 9:24 am the Titan II took flight. "You're on your way, Molly Brown!" yelled Capcom Gordo Cooper. Liftoff, the two astronauts recalled later, was so smooth that they felt nothing; their only real cues were the startup of the mission clock on Gemini 3's instrument panel and hearing Cooper's words. In fact, the early stages of the ride were smoother than they had experienced in the moving-base simulator in Dallas.

For the first 50 seconds, Grissom kept both hands firmly on the ring which would trigger Molly Brown's ejection seats in the event of a booster malfunction. Young, wrote Ray Boomhower, also had such a ring on his side of the cabin, but, looking over, Grissom noted that the unflappable rookie had his hands calmly folded in his lap. "During this time," Young recalled later, "we didn't say a word to each other because there was so much to do so fast."

The Titan's first stage was exhaused two and a half minutes into the climb and second-stage ignition bathed the entire cabin with a flood of eerie orange-yellow light which surprised Young. The rocket, Cooper told them, had slightly exceeded its predicted thrust and the astronauts could expect a larger-than-predicted pitchdown after second-stage ignition as it began to steer a course for orbit. Five and a half minutes after launch, the second phase of the ascent was over and, with a bark like a howitzer, pyrotechnics severed Gemini 3 from the Titan. Grissom fired Molly Brown's aft OAMS thrusters to pull away from the booster. The Titan over-performed slightly, but Gemini 3 still ended up in a close-to-expected orbit of 122–175 km. For his part, Young, who was making the first flight of what would turn into a six-mission career, was simply astounded by the sense of speed and the stunning view of Earth.

THE ASTRONAUT

In some ways, Grissom and his younger pilot, John Watts Young Jr, were perfectly matched. "They were both good engineers who understood their machines," wrote fellow astronaut Mike Collins, "and liked fooling with them. They were uncomfortable with the invasion of privacy the space programme had brought into their lives and tried as hard as they could to deflect questions from themselves to their beloved machines. They were generally taciturn but both had strong opinions that could flash unexpectedly ... Neither was interested in small talk and they would endure uncomfortable silences rather than fill the void with what they considered ancillary trivia." Collins, who would fly as Young's pilot on Gemini X a year later, admitted that the "aw-shucks" demeanour and country-boy drawl cleverly concealed a sharp, talented and analytical mind that would carry him to the Moon twice, to its surface and ultimately to command of the first Space Shuttle mission.

Born in San Francisco on 24 September 1930, Young and his family moved to Cartersville, Georgia, when he was three years old and eventually settled permanently in Orlando, Florida. At around this time, he related in an interview, Young began building model aircraft. It was a hobby that would remain with him throughout high school, together, it seemed, with rockets, which he chose for a speech to his classmates in the 11th grade. Young earned his degree in aeronautical engineering, with highest honours, from Georgia Institute of Technology in 1952, receiving coveted membership of the institute's prestigious Anak Society. He joined the Navy in June of that year and, among his earliest assignments, served as fire control officer aboard the destroyer Laws. During this time, he completed a tour in Korea and a former shipmate would remember his coolness under duress.

"Though only an ensign at the time," wrote Joseph LaMantia, quoted on the website www.johnwyoung.com, "he was the most respected officer on the ship. When we sustained counter-battery fire and enemy rounds were striking the ship, it was John Young's leadership which kept us all cool and focused on returning that enemy fire ... which won the day." After Korea, Young entered flight school at Naval Basic Air Training Command in Pensacola, Florida, learning to fly props, jets and helicopters and later undertook a six-month course at the Navy's Advanced Training School in Corpus Christi, Texas. With receipt of his wings came four years' service as a pilot in Fighter Squadron 103, flying F-9 Cougars from the Coral Sea aircraft carrier and F-8 Crusaders from the Forrestal supercarrier. During these years, colleagues would describe him as "the epitome of swashbuckling aviators ... he exuded confidence coupled with uncommon ability".

This ability, indeed, would ultimately guide him into the hallowed ranks of NASA's spacefaring corps. But not yet. The selection process to pick the Mercury Seven began early in 1959, at which time Young was just starting Naval Test Pilot School at Patuxent River, Maryland; test-flying credentials were a prerequisite for astronaut training. After graduation, he worked as a project test pilot and programme manager for the F-4H weapons system at the Naval Air Test Center in Maryland, evaluating armaments, radar and bombing fire controls for both the Crusader and the F-4B Phantom fighters. During one air-to-air missile test, he and another pilot approached each other's aircraft at closing speeds of more than three times the speed of sound. "I got a telegram from the chief of naval operations," Young later quipped, "asking me not to do this anymore!" In early 1962, he would also set two time-to-climb world records.

By now a lieutenant-commander, Young's experience with the 'Phabulous'

Phantom had made him the obvious choice to set the records as part of Project High Jump. The first, on 21 February, saw him climb to 3,000 m above Naval Air Station Brunswick in Maine in 34.5 seconds; followed, six weeks later, by another attempt from Point Mugu in California, which achieved 25,000 m in 230.4 seconds. In September of that year, after leaving active naval duties as a maintenance officer in Phantom Fighter Squadron 143, he received a phone call from Deke Slayton which marked the start of an astronaut career that would span four decades. Training, though, would be arduous. "You had to learn a lot of stuff," he said later. "You probably only needed to know one per cent of all the stuff you had to learn ... but you didn't know which one per cent it was!"

As pilot of Gemini 3, Young became the first of the 1962 astronauts to fly into space. He was originally assigned to accompany Wally Schirra on backup duties for the mission, but Al Shepard's grounding turned such plans on their heads. However, Young and Grissom would work well together, providing a good basis for some famous – or infamous – banter whilst in orbit. In fact, when asked by a journalist a few days before launch if he had any qualms about flying with Gruff Gus, Young had deadpanned: "Are you kidding? I'd have gone with my mother-in-law!"

BLOODSHED IN ALABAMA

Two weeks before Grissom and Young's launch, the gradual progress of the American civil rights movement exploded into violence when 600 protestors marching from Selma to Montgomery in Alabama were attacked by club-wielding, tear-gas-spraying police. As a result, 7 March 1965 would become forever known as 'Bloody Sunday'.

At the time, Selma – seat and main town of Dallas County – had a population that was 57 per cent black, although fewer than one per cent was actually registered to vote. The vast majority of the black community lived beneath the poverty line in mundane, unskilled occupations, a situation which the Boynton family and others sought to rectify. Their efforts to achieve this had been hampered since the late Fifties by the White Citizens' Council, the Ku Klux Klan and direct violence. The situation reached a head in February 1965, when an Alabama state trooper shot Jimmie Lee Jackson as the latter tried to protect his mother and grandfather during a nocturnal demonstration.

Jackson's murder was the catalyst for the first of three Selma-to-Montgomery marches. The initial plan was for the marchers to ask Alabama Governor George Wallace if he had authorised the troopers to shoot during the demonstration, which ultimately broadened with Martin Luther King's desire to request better protection of black voting registrants from Wallace.

The reaction from the governor, disturbingly, was that the march represented a threat to public safety and he opposed it. Mounted police awaited the marchers and, in the presence of journalists, attacked them with clubs, tear gas and bull whips. Amelia Boynton, one of the organisers, was beaten and gassed and 17 other marchers were hospitalised.

Two days later, on 9 March, King organised a second march. Numbers had by now swelled to more than 2,500 in outraged reaction to the images from Bloody Sunday. However, an attempt to gain a court order to prevent the police from interfering was rejected by a federal district judge, who instead issued a restraining order to stop the march until further hearings could be held. To avoid breaking the terms of the order, King led the marchers out to the Edmund Pettus Bridge, held a short prayer session, then turned them around and disbanded. Violence, however, was not far away. That evening, three white ministers involved in the second 'march' were clubbed by white supremacists. One of the ministers, James Reeb, later died from his injuries.

After finally gaining approval for an unimpeded march, the full journey along Route 80 through rain and cold was completed from Selma to Montgomery on 24 March. Five months later, President Lyndon Johnson signed the National Voting Rights Act, which prohibited states from preventing their citizens from voting on the basis of colour or race. Previous practices of requiring voters to pass literacy tests before being cleared to cast at the ballot box were abolished. Moreover, states with a history of abuses over voting rights could not make any changes without first requesting the consent of the Department of Justice. A wind of change had taken hold in America.

CORNED BEEF SANDWICH

Shortly after their arrival in orbit, with a packed five hours ahead of them, things did not appear to be going well for the Gemini 3 crew. Twenty minutes into the mission, as Molly Brown passed out of range of the mid-Atlantic tracking station in the Canary Islands, Young noticed the oxygen pressure gauge suddenly drop. At first, he suspected a malfunction, but his attention was soon drawn to a number of peculiar readings from other instruments, suggesting that he and Grissom may have a power supply problem on their hands. Quickly, Young switched from the primary to the backup electrical convertor, which powered the dials, and the glitch vanished as abruptly as it had appeared. From the moment Young first spotted the problem to its resolution took barely 45 seconds.

The cell-growth study, to be run by Grissom, proved a dismal failure; perhaps, he said, the adrenaline was pumping a little too much and he twisted the handle too hard, broke it and ruined the whole experiment. (Ironically, the scientist on the ground, operating the control sample, also broke his handle!) For his part, Young experienced difficulties with the radiation investigation on his side of the cabin and, although he completed it correctly, the results were inconclusive. Exposed to nearly identical doses of radiation, the in-flight blood samples showed higher levels of damage than their control counterparts on the ground. After the mission, both men would blame differences between the experiment packages they flew with and trained with as the cause of the problem, but admitted that observing sea urchins did not carry the same "oh, wow" factor as manoeuvring their spacecraft and experiencing the wonders of microgravity.

At the end of the first orbit, with Molly Brown flying nose-first, Grissom fired the forward-facing OAMS thrusters for a carefully timed 74 seconds to slow down by about 15 m/sec and almost circularise the orbit. Then, passing over the Indian Ocean in darkness on the second orbit, he yawed 90 degrees to one side and fired first the forward-facing thrusters and then the aft-facing thrusters in an effort to cancel out the 3 m/sec of the previous burn, which he was almost able to do, with the residual marginally increasing the inclination of their orbit with respect to the equator. These two manoeuvres had been made 'out of plane' so as not to disturb their circular orbit.

On the third and final orbit, with Molly Brown flying base-first, a 109-second 'fail-safe' burn lowered the perigee to 72 km to ensure a successful re-entry in the event of a retrorocket malfunction. No such malfunction materialised, thankfully, and, after running through their checklists, Young fired the pyrotechnics to separate the equipment module from the adaptor and armed the automatic retrofire switch. One by one, the four braking engines ignited, another set of pyrotechnics released the spent retrorocket compartment and Molly Brown plunged, its ablative base forward, into the atmosphere. It was during this dynamic phase of the mission, at an altitude of 90 km, that another experiment – the communications task – was to begin and Young duly threw the switch on his side of the cabin as the plasma sheath broke the radio link with Mission Control. Unlike the other experiments, this one proved encouraging: at high rates of water flow, investigators later concluded, both UHF and C-band signals from the spacecraft could be received by ground stations. "We could see the whole retro pack burning up as it came in right behind us," Grissom remembered of the dramatic re-entry.

By monitoring the trajectory during re-entry, the on-board computer could predict the splashdown point and display this to Grissom, who could adjust the 'lift vector' by using the thrusters in the nose to roll left or right of the 'neutral' position in order to steer towards the target. When this indicated that they were coming in short, his efforts to 'extend' made little difference. It was later concluded that theoretical and wind tunnel predictions of Gemini's lift capability did not match its actual lift. In fact, Molly Brown would splashdown 84 km short of the intended point and 110 km from the recovery ship, Intrepid. Nevertheless, the role of an engineering test flight was to determine the vehicle's performance and this empirical data would be taken into account on future missions.

As the drogue parachute deployed, Molly Brown was oriented with its heat shield down. However, after the main canopy had inflated, Grissom threw a switch to adjust the parachute line to a two-point configuration that would angle the capsule's nose at a 45-degree angle to the horizontal. Even though both men were strapped in, this transition was so violent that it pitched them into their windows, cracking Young's helmet faceplate and punching a hole in Grissom's. Fortunately, splashdown at 2:16:31 pm was relatively smooth, although Grissom could see little through his window, as the still-attached parachute caught the wind and dragged Molly Brown's nose underwater. Fearing a similar demise as had happened to Liberty Bell 7, Grissom jettisoned the parachute and Gemini 3 bobbed upright. This time, he had not lost his spacecraft ... but, alas, with the swelling waves, quickly lost his breakfast.

That breakfast had, of course, been augmented somewhat by Young's crafty corned beef sandwich, one of the few events of the mission still remembered decades later. "I was concentrating on our spacecraft's performance," Grissom recalled after the flight, "when suddenly John asked me: 'You care for a corned beef sandwich, skipper?' If I could have fallen out of my couch, I would have! Sure enough, he was holding an honest-to-john corned beef sandwich!" As Grissom sampled the treat, bits of rye bread began to float around the pristine cabin, forcing him to put it away. His only complaint was that there was no mustard on it. Still, it proved somewhat tastier than Gemini 3's staple of reconstituted apple sauce, grapefruit juice and chicken bits.

In his autobiography, Deke Slayton admitted that he had given permission for Young to carry the sandwich, but in view of the complaints NASA later received over its 'frivolous' astronauts' antics, he was obliged to render a formal, though mild, reprimand. For Grissom, though, it would be a highlight of the mission. It did not affect Young's career and on 6 April, barely two weeks later, he and Grissom were assigned to the backup crew for Gemini VI, the rendezvous mission, scheduled to take place in the autumn.

With Intrepid still some distance from them, it is hardly surprising that Grissom refused to open Molly Brown's hatches until Navy swimmers from a rescue helicopter had affixed a flotation collar to the spacecraft. The splashdown point was in the vicinity of Grand Turk Island in the Atlantic. Although the spacecraft proved lousy as a boat, its performance in orbit had been nothing short of outstanding. "I do know that if NASA had asked John and me to take Molly Brown back into space the day after splashdown, we would have done it with pleasure," said Grissom. "She flew like a queen, did our unsinkable Molly, and we were absolutely sure that her sister craft would perform as well." Still, the seasick Grissom was first to leave the capsule and Young kidded him about his failure to adhere to the old saying about captains being last to leave. Without missing a beat, Grissom replied "I just made you captain as I got out!" Indeed, in a little more than a year's time, Young would captain his own Gemini into orbit.

ONWARD, UPWARD ... AND OUTSIDE

As remarkable a success as Gemini 3 had been, it had been overshadowed, five days earlier, by Alexei Leonov's spacewalk. Admittedly, NASA had its own plans for astronaut Pete Conrad to perform a short, 'stand-up' EVA on Gemini V that summer, but the Voskhod achievement encouraged the agency to move its own spacewalk forward to the next Gemini, by now redesignated with Roman numerals as 'Gemini IV'. This proved good news for the crew of that flight, Jim McDivitt and Ed White, but continuing problems with the certification of General Electric's fuel cells had already halved their mission from seven to four days. The long-duration flight would now be rescheduled for Gemini V, as would the Rendezvous Evaluation Pod (REP), although McDivitt and White would still come close to the five-day record set by Valeri Bykovsky at the end of Vostok 5 and would mark an enormous leap for the United States.

It would come as something of a disappointment for Wally Schirra and Tom Stafford, who, on 15 April 1965, finished their duties backing up Grissom and Young and were named as the prime crew of Gemini VI, planned for later that year. At one stage, this mission was expected to conduct not only the first rendezvous with an Agena-D target, but Stafford would make the United States' first EVA. The spacewalk was subsequently moved to Gemini V and Pete Conrad, before the Voskhod 2 surprise prompted NASA to provide extravehicular suits for White instead. Then, to Stafford's chagrin, the planners began talking of accelerating the rendezvous, with McDivitt and White performing station-keeping manoeuvres with the second stage of their Titan II. "It looked as though we weren't going to be left with any new challenges," Stafford wrote, "though I was puzzled by the claim that station-keeping with an upper stage was a 'rendezvous'."

Schirra viewed things differently. Conservative in his approach, he was reluctant to complicate his already-complex rendezvous mission with an EVA. He and Stafford both knew how crucial rendezvous was to a lunar landing; if they could not make it work, the chances of meeting John Kennedy's end-of-the-decade goal would be severely jeopardised. Little did either man know at the time, but Gemini VI would indeed achieve the world's first space rendezvous ... albeit with a somewhat different target.

On the ground, meanwhile, the new Mission Operations Control Room (MOCR) in Houston was slated to assume day-to-day command of future missions, taking over from the updated Mercury Control at Cape Kennedy, which had run Gemini 3. From the MOCR, Lead Flight Director Chris Kraft would oversee three shifts – his own, focusing on the flight schedule, that of Gene Kranz, to track systems performance, and finally that of John Hodge, to manage real-time mission planning – and the force of controllers tiered beneath them included experts on virtually every component within the Gemini capsule and its Titan II booster. They, in turn, had direct links to their own support teams, scattered in plants and factories throughout the United States.

McDivitt and White were assigned to Gemini IV on 27 July 1964, with Frank Borman and Jim Lovell as their backups. The prime crew had known each other since their days at the University of Michigan, from which McDivitt graduated in 1959 with a bachelor's degree in aeronautical engineering, the same year that White received his master's in the same field. Borman and Lovell first met during the 1962 selection process and would fly two missions together; in fact, both born in March 1928, both blue-eyed, both of equivalent rank in their respective services, they were about as close to twins as unrelated astronauts could get.

Except, perhaps, for the close parallels between McDivitt and White. Barely a year separated the two men in age and both were married to women named Pat. Both had earned their aeronautical engineering degrees from the same institution and in the same year, after which both completed test pilot training at Edwards Air Force Base in California and secured approximately the same amount (around 2,000 hours) of experience in jets. Both also wore the gold leaf of a major in the Air Force at the time of Gemini IV. Above all, both had applied for, and been selected to join, the second group of astronaut candidates in September 1962. "Jim and I have

The Mission Operations Control Room at the Manned Spacecraft Center (MSC) in Houston.

been following right along together," White once said. "It seems that every time we got together we were taking examinations of some kind." Their biggest test still lay ahead of them.

Among their earliest jobs was a thorough review of the Gemini IV spacecraft, whose construction in St Louis had met with delay following a shortage of parts. Not until November 1964, when the Gemini mission simulator became available in Houston, could they begin direct training ... and lobbying. In fact, it has often been remarked that, without McDivitt and White's determination in getting the EVA added to their mission, the G4C extravehicular suit might not have been available to be assigned to Gemini IV. This determination, wrote Barton Hacker and James Grimwood in their 1977 history of Project Gemini, showed that the astronauts' role in the decision-making process "went far beyond that of the normal test pilot in determining what was to be done and when". Still, when NASA announced in May 1965 that White would indeed venture beyond the pressurised confines of his spacecraft's cabin, some observers felt it was little more than a ploy to keep up with the Soviets.

In reality, spacewalking had been a major goal of the project virtually from its conception and the public had drawn a link between EVA and Gemini IV since McDivitt and White's assignment to the mission. At the July 1964 press conference to announce their selection, Gemini's deputy manager Kenny Kleinknecht had announced that one of the astronauts might open the hatch and stick his head outside. Even earlier, in January, a plan for EVA operations had flagged Gemini IV as the possible first flight to incorporate some kind of extravehicular activity, although, at the time, the availability and development of the required equipment presented a real question mark. As the year progressed, the situation improved: the AiResearch Manufacturing Company was contracted to build a shoebox-sized chest pack for White's space suit, the David Clark Company received specifications to build the bulky all-white ensemble itself and McDonnell set to work modifying the Gemini IV capsule to support an extravehicular option.

Altitude chamber tests of the spacecraft in November provided an opportunity to quieten the naysayers in MSC's Crew Systems Division, who felt a spacewalk should not be attempted until the astronauts had endured 'realistic' simulations on the ground. However, McDonnell had their own reservations. They did not want to risk injuring astronauts in the altitude chamber and, said John Young, many within NASA were none too happy about "putting guys in vacuum with nothing between them but that little old lady from Worcester, Massachusetts [the David Clark seamstress] and her glue pot and that suit". Nevertheless, the test went ahead, at 12,000 m altitude conditions, although the first EVA attempt left something to be desired when the astronauts could not close the hatch properly.

The suit itself was basically the same as that worn by Grissom and Young on Gemini 3, with the exception that it had redundant zippers, a pair of over-visors for visual and physical protection, automatic-locking ventilation settings and a heavier outer covering. Shortly after Voskhod 2, efforts gathered pace when MSC Director Bob Gilruth and his deputy, George Low, reviewed a hand-held manoeuvring device, which finally convinced the higher echelons of management that an EVA on Gemini IV was a realistic option.

By the end of April 1965, a model spacecraft had been installed in MSC's vacuum chamber for advanced testing, and in mid-May Gilruth received the staunch support of Bob Seamans, who in turn described the plan to Jim Webb and his deputy, Hugh Dryden. One note of contention came from NASA's manned spaceflight chief, George Mueller, who doubted that the EVA hardware could be ready in time for June, but was appeased on 19 May when Charles Mathews announced that all equipment for White's excursion was ready to go. Webb supported the plan for Gemini IV, but Dryden felt it gave the impression of a knee-jerk reaction to Voskhod 2. However, after Webb asked Seamans to prepare a report on the need for an early spacewalk, Dryden relented and gave his approval. That approval, scribbled in the corner of Seamans' report, was given on 25 May. Nine days later, Gemini IV was ready to fly.

Questions remained, however, over when to announce the EVA: before, during or after the event. Despite early plans in April to announce it at a news conference, if it was approved, NASA's policy of openness obliged the agency to include it in their Gemini IV press kit. When the latter was published on 21 May, it included reference to a "possible extravehicular activity", which the press learned had become a certainty after its final approval by Webb and Dryden. Nor would it be a relatively puny case of White pushing open the hatch and poking his head into space: he would actually leave Gemini IV and manoeuvre around with the aid of the hand-held

Interior view from Gemini IV, showing McDivitt in the foreground and White behind. Note the cover over White's tinted visor.

propulsion device. This device, it was further added, could be used to move himself over to inspect the just-jettisoned second stage of the Titan II. Gilruth and Low had first latched onto this idea shortly after Gemini 3 arrived in orbit, when Capcom Gordo Cooper suggested that Grissom try to 'rendezvous' with the Titan, and lent their support for such a station-keeping exercise on McDivitt and White's mission.

The plan was for McDivitt to match his spacecraft's velocity with that of the second stage – a relatively short distance away and in the same orbital plane – and evaluate his ability to station-keep. Lack of a rendezvous radar on Gemini IV made the task more complex and, although Martin installed flashing lights on the GLV-4 rocket's second stage, the two astronauts would have to rely upon their own eyes as navigational aids. Further, they had no way of rehearsing such a station-keeping exercise in the ground-based simulators; at least, that is, until McDonnell engineered a mockup view of the target against a starry backdrop.

In addition to station-keeping and the spacewalk, Gemini IV would attempt the longest American manned spaceflight to date. As far back as August 1964, Charles Mathews had announced that the unavailability of fuel cells would render it a four-day, battery-reliant flight. Another reason, McDivitt explained, could have been that only enough food was packed aboard the spacecraft "for two normal people" for four days. Unable to resist, backup pilot Jim Lovell quipped: "And these two ain't normal!"

However, Chuck Berry, in charge of all Project Gemini medical matters, was reluctant to give his blessing for such a long mission: cardiovascular problems, he noted, had cropped up in the final Mercury-Atlas flights and it was feared that McDivitt and White's bodies would be subjected to the same kind of physiological strain as that imposed by prolonged bedrest, followed by immediate and vigorous physical exercise. Four days in weightless conditions, Berry argued, could decondition the men to such an extent that they might not be able to withstand the stresses of re-entry, perhaps even losing consciousness. As a result, bungee cords requiring a force of 32 kg to fully extend would be carried for McDivitt and White to exercise their upper-body muscles.

For all of its goodies – rendezvous, long-duration flight, spacewalking – one of the things that Gemini IV lost was the chance for its astronauts to give it a name. Gus Grissom's choice of 'Molly Brown' had not gone down well with NASA management and, despite McDivitt's proposal of a patriotic 'American Eagle', a firm stance was maintained: the mission would be known simply as 'Gemini IV'. (Ironically, in the wake of Gemini 3, even President Johnson congratulated Grissom on the success of 'Molly Brown', remarking that she "was as unsinkable as her namesake.") Instead, McDivitt and White insisted that they wear American flags on the sleeves of their space suits, making them the first United States astronauts to do so. Their crew patch, though, bore only the name 'Gemini IV'. At a press conference, when asked if he intended to name his spacecraft, McDivitt responded: "Don't know. What's playing on Broadway these days?"

DISCIPLINE, PERSISTENCE, DEDICATION

These three words alone could sum up Edward Higgins White II. The son of a West Point graduate and Air Force major-general, he was born on 14 November 1930 in San Antonio, Texas, and his parents instilled in him from a very early age the values of self-discipline, persistence and an absolute single-minded determination to achieve his goals. "Flying was his birthright," wrote Mary C. White in her biography of him, published by NASA, and, indeed, he was aboard an old T-6 training aircraft with his father – and taking its controls – at the tender age of 12. In fact, White's father remained active as a pilot during 35 years of Air Force service.

Throughout his childhood, White travelled to bases scattered across the United States, from the East Coast to Hawaii, and, despite the semi-nomadic lifestyle, excelled both academically and as an athlete. In fact, it was only whilst enrolled in Western High School in Washington, DC, when he came to investigate college admission policies, that his lack of continuous residency posed an obstacle. With an extensive history of family service in the military, "there never seemed to be any question," White said later, "that I would go there too". He was admitted to the Military Academy at West Point to study for a bachelor of science degree and excelled in academics and athletics: serving as a half-back on the football team, making the track team and setting a new record in the 400 m hurdles. His athletic credentials were so impressive that he narrowly missed selection (by just 0.4 seconds) for the United States' track team in the 1952 Olympics.

Whilst at West Point, White met his future wife, Pat Finnegan, and upon graduation in 1952 followed in his father's footsteps by enlisting in the Air Force. Initial flight instruction in Florida and receipt of his wings were followed by assignments in Germany, where he piloted F-86 Sabre and F-100 Super Sabre fighters and completed the Air Force Survival School in Bad Tolz. His aviation career took a new path in 1957, when he read about plans to hire astronauts and "something told me: this is it – this is the type of thing you're cut out for. From then on, everything I did seemed to be preparing me for spaceflight". By now married and the father of two children, White returned to the United States and enrolled in a master's programme at the University of Michigan at Ann Arbor, specialising in aeronautical engineering. He was convinced that such a qualification would give him the academic edge over other astronaut candidates. It was whilst in Michigan that he met an undergraduate named Jim McDivitt.

White completed his degree in 1959, the same year that the Mercury Seven were introduced to the world. His next step on the road towards the hallowed membership of NASA's corps was to achieve test-piloting credentials, after which he was assigned to Wright-Patterson Air Force Base in Dayton, Ohio, evaluating research and weapons systems and making recommendations for improvements to aircraft design and construction. Whilst in Ohio, he also flew cargo aircraft on parabolic flights to prepare the Mercury astronauts for their weightless missions. "Two of my passengers were John Glenn and Deke Slayton," he said later. "Two other passengers of mine were Ham and Enos, the chimpanzees." In flying such missions, White would estimate that he "went weightless" at least 1,200 times before his selection to join NASA's astronaut corps.

When the call went out in April 1962 for volunteers for the second astronaut team, then-Captain White's was one of the first applications and five months later his perseverence proved successful. However, neither he, nor McDivitt, could have anticipated the sheer outpouring of adoration they received when they moved into the El Lago neighbourhood in Harris County, Texas: groups of children asking for their autographs and screaming "Astronauts in the house!" before either man had even begun training! Despite his credentials and drive, White did not see himself as a hero, but he certainly stood out for the old heads of Project Mercury. They regarded him as "a man who, when asked an intelligent question, will answer thoughtfully and to the point ... but will rarely volunteer information".

Basic training included helicopter airdrops in pairs into the Panama rainforest, where they spent days with iguana, boa constrictor and palm hearts as their foodstuffs. White had always pursued physical exercise with a passion akin to religious faith: volleyball, handball, squash and golf were the staples of his sporting diet, together with daily long-distance jogs, bicycle rides to work and even squeezing a rubber ball whilst running to build strength in his hands and arms. He set up a climbing rope in the backyard of his El Lago home and was said to perform 50 sit-ups and 50 press-ups, back-to-back, without so much as a sharp intake of breath. Without doubt, he was the most physically fit of all of the astronauts – the Mercury

Seven included – and this conditioning would prove essential in undertaking America's first spacewalk.

Physicians, in fact, remarked that they could not find the slightest hint of fat on White's 77 kg frame. His appetite, though, was voracious, and it was said that he "could put away two full-course dinners at one sitting and then ask for dessert with a straight face!" His almost superhuman agility was remembered clearly by fellow astronaut Neil Armstrong's wife Janet; their backyards were separated by a tall fence. One night in early 1964, a fire broke out in the Armstrongs' home and White, heroically, was first on the scene with a water hose. "Still to this day," James Hansen wrote in his biography of Armstrong, "Janet vividly recalls the image of Ed White clearing her six-foot fence. 'He took one leap and he was over.'"

After basic astronaut training, White was assigned to monitor the design and development of the Gemini flight controls, a task he appreciated "because it involves the pilot's own touch – the human connection with the spacecraft and the way he manoeuvres it". As part of his work, White campaigned and succeeded in securing a standard hand controller to be used in all of NASA's manned spacecraft. "It seemed inconceivable to me," he said, "that ... an astronaut would fly toward the Moon in an Apollo using one kind of stick, them climb into the LEM [Lunar Excursion Module, later renamed the Lunar Module] and use a different kind of controller to land him on the Moon." Landing on the Moon and becoming the first to set foot on its surface was immensely important to White and, sadly, his death in January 1967 means that no one will ever know if he could have achieved his most exalted goal.

"His goal," said his father, "is to make that first flight". He would have a lot of competition.

FALSE START

Physical conditioning would prove crucial for Ed White's survival on the United States' first spacewalk, as fellow astronaut Gene Cernan would discover on his own excursion a year later. During training, White spent a total of 60 hours in vacuum chambers, rehearsing opening the hatch, pushing himself outside and moving around in a mockup of the space suit that he would wear, all at simulated altitudes of up to 55 km. The suit itself weighed 14 kg, cost over $30,000 to construct and comprised no fewer than 22 layers to provide protection from the intense heat of direct sunlight to the frigid cold of orbital darkness and, Time magazine told its readers, "as a pressure force to keep White's body from exploding in the near-vacuum of space". Estimates that the suit could be punctured by a high-velocity micrometeorite were placed at about 10,000 to one, although tests blasted it with splinters of plastic fired at speeds of 7.6 km per second – and it held its own. The hand-held manoeuvring unit, which White would position just below his midriff, consisted of two cylinders of compressed oxygen, belted to a handle which also acted as a trigger to send jets of air through a pair of hollow tubes. White would position it as necessary to aim its impulse through his centre of mass in order to make a specific movement.

As the launch drew nearer and information about the forthcoming spacewalk trickled out, some sections of the media expressed scepticism that it had been a long-planned exercise and felt it was a hastily-concocted stunt to catch up with the Russians. Chris Kraft, the lead flight director for Gemini IV, angrily suppressed such talk. "We're not playing Mickey Mouse with this thing," he snapped. "We're trying to carry out flight operations. I don't think it's very fair to suggest we're carrying out a propaganda stunt."

In spite of the doubts, the media response to the mission was enormous. All 800 seats in the MSC's main auditorium proved woefully inadequate for the 1,100 journalists who requested accreditation before the launch and NASA was forced to lease a nearby building at a total cost of $96,165 per year, plus $181,000 for modifications, television monitors and chairs, to handle the overspill. This 'Gemini News Center' would track each mission over the next 17 months and provide a base for over a thousand newspaper, magazine, radio and television representatives, as well as five dozen public relations groups from industry.

Twelve hours before liftoff, a Martin team began the lengthy effort to fuel the Titan II, while the backup crew of Frank Borman and Jim Lovell oversaw checks inside the capsule itself – flipping switches to their launch positions, testing communication circuits and handling routine chores for McDivitt and White. The prime crew was duly awakened at 4:10 am on 3 June 1965 and marched crisply through their medical checks, steak-and-eggs breakfast, suited-up and were at the foot of Pad 19 by 7:07 am. Among the procedures put into place in support of White's EVA – which would necessitate the depressurisation of the entire cabin, thus exposing McDivitt to vacuum, too – was prebreathing pure oxygen to flush nitrogen from their bloodstreams and avoid attacks of the bends.

Once aboard the spacecraft, White's faceplate fogged, but he quickly cleared the problem by switching on his suit fan. Then, barely half an hour before the scheduled launch, as the erector was being lowered, it stuck at a 12-degree angle from the Titan. A second attempt was made to raise then lower it, but it stuck again. Eventually, engineers discovered an improperly-fitted connector in a junction box, replaced it and the erector lowered. After a 76-minute delay, Gemini IV speared for the heavens at 10:16 am, to the synchronised yells of "Beautiful!" from both McDivitt and White. Despite initial pogo effects, which caused the astronauts to stutter their words over the communications link, the Titan quickly calmed down and gave them a perfect ride to orbit. They would later describe it as exhibiting little noise, but stressed that the near-perfect silence was shattered abruptly when pyrotechnics jettisoned the first stage and the second stage ignited.

Monitoring the ascent and, indeed, the entire mission, were 300 flight controllers, technicians, engineers, physicians, and scientists supervised by Chris Kraft in the new MOCR. In the Real Time Computer Complex, five IBM 7094-11 computers each processed 50,000 bits of telemetry per second from the vehicle. Also watching the launch, thanks to live television coverage from the Intelsat 1 ('Early Bird') satellite in geostationary orbit, were citizens of a dozen European nations.

Within minutes, Gemini IV had entered an initial elliptical orbit of 163–282 km and, almost immediately, McDivitt set to work on the first task of the mission: to station-keep with the Titan II's second stage. It was whilst attempting this manoeuvre that they encountered problems. Firstly, despite having been outfitted with flashing, 2.5 million-candlepower lights, the stage had not been designed as a rendezvous target and when the capsule entered orbital nighttime, it was rendered almost invisible to them. Moreover, it was tumbling and the astronauts were cautious about getting too close. Difficulty in judging distances by eyesight alone complicated matters yet further.

When they first spotted the booster, propellant streaming from its nozzle, McDivitt estimated it to be 120 m from them, whereas White felt it was 70 m away. McDivitt cancelled the motion imparted by the separation manoeuvre and thrusted towards the target, but after two OAMS burns he was surprised to observe that the Titan seemed to move 'away' and 'downward'. A few minutes later, he pitched the spacecraft nose-down and pulsed the thrusters again, with no success. Approaching orbital nighttime by this point, he reported that he could see the flashing lights, but that the gap seemed to have increased, he guessed, to around 600 m. For a while, circumstances improved and McDivitt felt he was gaining on the stage, but with the early streaks of dawn its lights dimmed and vanished from view. At length, realising that he was wasting precious propellant, McDivitt asked Chris Kraft which objective was more important – rendezvous or EVA – and was assured that the latter was the mission's main task.

Their biggest obstacle of all, as rendezvous expert (and fellow astronaut) Buzz Aldrin would explain after the flight, was a lack of understanding of basic orbital mechanics. "When they emerged into daylight, the booster was below and ahead of Gemini IV," wrote Tom Stafford, who was following the mission closely in anticipation of his own rendezvous on Gemini VI. "Jim's instinctive move was to thrust toward it, as though he were flying formation in a jet airplane. By doing so, of course, he increased the speed – and moved into a higher orbit even further behind the booster. The only way to get even close to the Titan, in these circumstances, would have been to fire thrusters retrograde – against the direction of travel – slowing the Gemini and dropping its orbit." It was an early lesson: adding speed raises altitude, moving a spacecraft into a higher orbit than its target. However, paradoxically, the faster-moving spacecraft actually slows in comparison to the target, since its orbital period – a direct function of its distance from the centre of gravity – also increases. To catch up with a target ahead, future crews would drop into a lower orbit, then rise back up to meet it.

It was an early lesson, admittedly, but not an easy one. "It's a hard thing to learn," wrote Deke Slayton, "since it's kind of backward from anything you know as a pilot." Added engineer André Meyer: "We just didn't understand or reason out the orbital mechanics involved. As a result, we all got a whole lot smarter and really perfected rendezvous manoeuvres."

INSIDE MAN

To be fair, both McDivitt and White only received a minimum amount of rendezvous training and the primary focus of their mission was the EVA itself. For this task, James Alton McDivitt would be the Inside Man, although, fully-suited, he would be exposed to vacuum throughout White's spacewalk.

The first Roman Catholic to be launched into space, son of an electrical engineer and the first American astronaut to command a crew on his first flight, McDivitt was born in Chicago, Illinois, on 10 June 1929. Described as "whippet-lean" by Time magazine, his 1.8 m frame was one that was forced to squeeze its way uncomfortably into the Gemini capsule before the Stafford Bump eased the tall astronauts' suffering. Yet his background, unlike White's, did not immediately mark him out as an obvious spacefarer. After graduating from high school in Kalamazoo, Michigan, he worked for a year as a furnace repairman, then drifted into college in 1948, vaguely describing his ambitions for the future as either a novelist or an explorer. Two years later, he completed his education and opted to enter the Air Force, discovering a love of aviation whilst flying 145 combat missions over Korea. His achievements were rewarded with three Distinguished Flying Crosses and five Air Medals.

After Korea, in 1957, McDivitt was sent by his parent service to read for an aeronautical engineering degree at the University of Michigan, where he proved himself to be a straight-A student, graduating first in his 607-strong class. Whilst at Michigan, he met Ed White for the first time, though he could hardly have guessed how far their friendship would endure. He was selected for the Experimental Test Pilot School at Edwards Air Force Base and seemed a likely candidate to fly the X-15 rocket-propelled aircraft, but applied instead for the 1962 astronaut class. Despite Time having labelled him as "a superb pilot and a first-class engineer", McDivitt approached his NASA career from a purely practical and technical standpoint. "There's no magnet drawing me to the stars," he was quoted as saying. "I look on this whole project as a real difficult technical problem – one that will require a lot of answers that must be acquired logically and in a step-by-step manner."

OUTSIDE MAN

The fruitless station-keeping exercise had led to a 42 per cent depletion of their fuel supply, which would correspondingly reduce the extent to which McDivitt could manoeuvre the spacecraft while White was outside. It forced the astronauts to continue with their primary mission, the EVA, and leave rendezvous for Gemini VI. McDivitt, aware that his partner was tired and hot after the rendezvous attempt, told the Kano tracking station that he wanted the spacewalk postponed from the second to the third orbit. Chris Kraft agreed and the astronauts spent some time relaxing, admiring the view of the Gulf of Mexico and Florida and chatting to Gus Grissom, the Houston capcom. Then they dived headlong into the 54-item checklist to prepare the EVA equipment. At length, White snapped a gold-tinted faceplate onto his helmet, hooked up the 7.4 m umbilical to provide him with oxygen and a communications link to McDivitt and, with the aid of a small mirror, strapped the 3.7 kg chest pack into place. He checked his camera gear three times, wanting to make sure he did not leave the lens cap stuck on. "I knew I might as well not come back if I did," White said later.

Zip-gun in hand, Ed White tumbles through space during his EVA.

Since the EVA would place the entire cabin into vacuum, the astronauts had to steadily reduce its pressure from the normal 3.51 bars to 2.55 bars. Depressurisation commenced over Carnarvon in Australia, but quickly hit a snag when the overhead hatch refused to unlatch. A spring had failed to compress properly. Four hours and 18 minutes after launch, at 2:34 pm, White cranked a ratchet handle to loosen a set of prongs lining the opening of the hatch, raised it to 50-degrees-open and poked his head into the void.

After receiving assurance from Kraft that he was good to go, White pushed himself from his seat and caught his first glimpse of Earth from the ethereal vantage point known only to the spacewalker – with barely a helmet faceplate and 166 km of emptiness separating him from his home planet. 'Below', he beheld the intense blue of the Pacific and, coming up to the east, Hawaii. Losing no time, he tested the hand-held manoeuvring device and found that it responded crisply to his commands, as he 'squirted' it to propel himself firstly underneath the capsule, then to its top. Within a short time, the manoeuvring device's gas supply was gone and for the remainder of his 21 minutes outside, White twisted and hand-pulled himself backwards and forwards along his tether. The umbilical imparted the force to the spacecraft, which reacted in response. "One thing about it," noted McDivitt as he fired the thrusters to hold the craft stable, "when Ed starts whipping around that thing, it sure makes the spacecraft hard to control." Also tricky was the fact that the umbilical kept drawing White towards the adaptor section and he had no desire to contaminate his suit with the toxic residue from burning monomethyl hydrazine and nitrogen tetroxide.

The spacecraft was nearing the California coastline when Capcom Gus Grissom asked for photographs. "Get out in front where I can see you again," McDivitt called and White duly complied. It is hard to comprehend that in little more than a quarter of an hour, White had 'walked' from the central Pacific, crossed California and, very soon, he and McDivitt were gliding high above Houston, talking to Grissom, somewhere, directly beneath them. "There's Galveston Bay right there!" McDivitt yelled with excitement. "Hey, Ed, can you see it on your side of the spacecraft?" White concurred and snapped a picture with a 35 mm camera affixed to his hand-held pack. McDivitt, using a 70 mm Hasselblad, also took pictures, venturing "they're not very good". On the contrary, these actually turned into some of the most iconic images from the annals of the early manned space effort. A 16 mm movie camera also recorded scenes of White in motion, bouncing backwards and forwards over a cloud-studded, blue-and-white Earth.

From his seat in the MOCR, Grissom was having a hard time trying to contact the two men. Every time McDivitt or White spoke, the spacecraft's voice-activated system cut off messages from Mission Control – and they talked a lot during those exhilarating minutes. When McDivitt called Houston to ask if there was anything they wished to say, Kraft pushed his communications switch, something he rarely did, and ordered, "The Flight Director says 'Get back in'!" The spacecraft was heading toward Earth's shadow and White had to be inside by then with the hatch closed. As White returned to the cabin, he described it as "the saddest moment of my life". His last view was of the entire southern portion of Florida and the islands chain of Cuba and Puerto Rico. McDivitt turned up the interior lights to guide his partner to safety in case they hit orbital darkness before he was in. White pushed his feet back through the hatch, onto his seat and finally under the instrument panel. He had walked across America – and then some – in barely 21 minutes.

White's return to the capsule was not entirely smooth, however, and his pulse rate soared from 50 to 178 beats per minute at the end of the spacewalk. He closed the hatch over his head and reached for the handle to lock it, quickly realising that it would be as hard to seal as it had been to open. As he pushed on the handle, McDivitt pulled onto him to give him some leverage and, eventually, the hatch was secured. The official ending time of the first American spacewalk was 3:10 pm, some four hours and 54 minutes into the mission and 36 minutes between the opening and closure of the hatch. Repressurisation started two minutes later. White had long since exceeded the cooling capacity of his space suit, resulting in severe condensation inside his helmet and sweat streaming into his eyes. The hatch problems led to a decision from the MOCR not to re-open it again to discard unneeded equipment.

After the mission, White would recount that the hand-held manoeuvring unit worked in the pitch and yaw axes, more or less as it had done during ground simulations. In the roll axis, however, he considered it more difficult to control without using excessive fuel. He experienced no sensations of vertigo or disorientation; nor, indeed, did he feel any inkling of the tremendous 28,100 km/h at which he was moving. White's excursion also demonstrated that astronauts would be able to cross from the Apollo lunar module back to the command module, if necessary, in the event that the two spacecraft could not dock properly after ascent from the Moon's surface.

Meanwhile, aboard Gemini IV, White relaxed and McDivitt began powering down some of the spacecraft's systems to conserve electrical power and OAMS fuel, intending to drift for the next two and a half days. Plans called for the men to sleep alternate periods of four hours each, although this would prove difficult with the constant crackle of radio chatter from MOCR, frequently bumping into each other and an inability to turn down the volumes on their headsets. In spite of the drama of the past few hours, they had barely begun their 98-hour mission. It would be an uncomfortable and tedious slog.

RECORD BREAKERS

The men's physical and psychological wellbeing was of paramount concern. Fear of dehydration led physicians to remind them regularly to drink water – at least 1.2 litres a day – because their space suits' cooling systems evaporated perspiration as it formed, thus increasing the loss of body fluids. Their food sounded appealing, but in reality its freeze-dried or dehydrated nature and the need to mix it with water and knead it until mushy, lessened its attractiveness. Still, beef pot roast, banana pudding, fruitcake and even a Roman Catholic treat of fish on Friday for McDivitt formed the basis of their four-day diet. They would also recall space sandwiches, "covered with waxy-tasting stuff to keep the crumbs from getting in your eyes, ears and nose", undoubtedly less desirable than Gus Grissom's corned beef option. Spaghetti dishes, too, required rehydration by water pistol. "You cut the other end with a pair of scissors," McDivitt recalled later, "put the tube in your mouth and squeezed the stuff." Indeed, it provided much-needed sustenance, rather than desirable food.

Sanitation on such a long mission presented its own obstacles. Both men would return to Earth with four-day beards, neither having been able to shave, and 'washing' was effected with little more than small, damp cloths to mop their faces. Urine was dumped overboard, while faeces were stored in self-sealing bags with disinfectant pills. Living amidst all of this, they had 11 experiments to perform. Photography of selected land and near-shore regions for geological, geographical and oceanographical studies undoubtedly proved the most enjoyable and 207 images were acquired with a hand-held 70 mm Hasselblad 500-C camera. Among the most

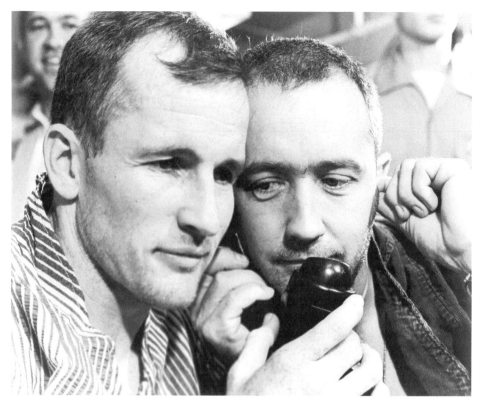

White (left) and McDivitt speak to President Johnson after the flight.

visually stunning were terrain images of north-western Mexico, the south-western United States, North Africa, the Bahamas and the Arabian peninsula, although weather photographs captured a broad range of meteorological phenomena, including cellular cloud patterns, layers of clouds in tropical disturbances, lines of cumulus covering the oceans and vast thunderheads. The Hasselblad also proved essential for a series of two-colour images of Earth's limb, part of efforts to better define the daylit horizon with red and blue filters.

Elsewhere, a proton-electron spectrometer monitored the radiation environment encountered through the South Atlantic Anomaly region (an intense 'pocket' of Earth's ionosphere) and a tri-axis magnetometer measured the magnitude and direction of the local geomagnetic field with respect to the spacecraft. Five dosimeters, scattered throughout Gemini IV, kept watch on radiation levels, particularly as McDivitt and White passed through the South Atlantic Anomaly. In other areas, a bone demineralisation experiment revealed the first signs of mass loss in astronauts exposed to long periods of weightlessness and both men agreed that systematic exercise programmes were a necessity on future flights. A bungee cord was provided, but even the super-fit White found that his desire to do strenuous work dwindled as the mission dragged on, perhaps due to lack of sleep.

Rest, it seemed, was a precious commodity and one which both McDivitt and White found hard to capture. During their 33rd orbit, two days into the mission, Gus Grissom told them that they had a relatively free 18 hours and advised them to get as much sleep as they could. He recommended that one of them unplug their headset entirely to ensure uninterrupted rest. At other times, the chatter was incessant. Grissom radioed to McDivitt on one occasion that his son's Pee Wee League team, the Hawks, had defeated the Pelicans 3–2, and to White that his son had scored a hit in a Little League game. The astronauts talked to their wives, with McDivitt asking Pat if she was behaving herself and assuring her that "about all I can do is look out the window". White's wife, the second Pat, commented that her husband seemed to be "having a wonderful time" on his EVA and advised him to drink plenty.

On their third day in orbit, the spacecraft's IBM computer failed. It was supposed to have been updated during a pass over the United States and McDivitt was asked to switch it off and then back on again. However, he quickly discovered that he could not bring it back to life. Attempts to try different switch positions came to nothing. Ironically, only days earlier, IBM had published an advert in the Wall Street Journal, praising its computers as being so reliable that even NASA used them. The failure caused no great alarm, but it did mean that a computer-controlled re-entry would now be impossible and, in Gemini IV's final orbits, Chris Kraft advised McDivitt that ground computers would help steer the spacecraft for him. As the 7 June return to Earth neared, the astronauts were told to brace themselves for an 8 G re-entry, which McDivitt, only days short of his 36th birthday, joked was "too much for an old man like me!"

Although in good spirits, neither astronaut felt particularly comfortable. McDivitt told Chuck Berry that he felt "pretty darn woolly", needed a bath, and, when asked if there was anything else he needed, replied "Yeah, my computer!" After the pre-retrofire checklist, the Hawaii capcom counted them down to the OAMS 'fail-safe' burn at 11:56 am, which reduced Gemini IV's perigee to just 80 km. The burn lasted two minutes and 41 seconds and used most of the remaining propellant. They jettisoned the equipment section shortly before making contact with the station in Mexico. McDivitt initiated the retrofire one second late. The capsule hit the ocean at 12:12:11 pm, some four days and two hours since liftoff. Their splashdown point was about 725 km east of Cape Kennedy and, despite being slightly long of its target, McDivitt and White were soon joined by frogmen and landed by helicopter on the deck of the aircraft carrier Wasp at 1:09 pm. Their sturdy spacecraft was also safely aboard the carrier by 2:28 pm.

Re-entry, McDivitt recounted later, was the prettiest part of the flight. "We saw pink light coming up around our spacecraft," he said. "It got oranger, then redder, then green. It was the most beautiful sight I have ever seen." The two men were described by Time as being heavily bearded and sweaty, their faces lined with fatigue, although that did not prevent McDivitt from letting out a whoop of joy. Medical examinations revealed that White, whose normal heart rate was 50 beats per minute, registered 96 whilst lying supine aboard the Wasp; this climbed to 150 when the table was tilted slightly. McDivitt, on the other hand, was found to have flecks of caked blood in his nostrils, probably attributable to the dryness of his mucous membranes after inhaling pure oxygen for so long.

Both men had lost weight – McDivitt shed 1.8 kg, White some 3.6 kg – although, summing up, Chuck Berry was more than satisfied that they were in good physical shape. Gemini IV and the condition of its astronauts promised, he said, "to knock down an awful lot of straw men. We had been told that we would have an unconscious astronaut after four days of weightlessness". Clearly, that was not the case. As if further demonstration were needed, a day after splashdown, still aboard the Wasp, White noticed a group of marines and midshipmen having a tug-of-war and joined them for 15 minutes. Although 'his' team lost, the astronaut certainly appeared to be the epitome of health and fitness.

On the day of Gemini IV's splashdown, the two men received congratulations from President Johnson, together with joint promotions from majors to lieutenant-colonels and NASA's Distinguished Service Medal. Elsewhere, the University of Michigan awarded them both with newly-created honorary doctorates in astronautical science. Also promoted to the same rank, Johnson announced, were Gordo Cooper and Gus Grissom. "I can hardly get used to people calling me 'Colonel'," wryly observed Ed White. "I know in a million years, I'll never get used to people calling me 'Doctor'!" (The spot promotions may have been at least partly inspired by a remark made by Grissom. When asked if there were any differences between American astronauts and Soviet cosmonauts, Gruff Gus had replied: "Yeah. They get promoted and we don't!")

Before McDivitt and White could take their new titles, however, they both needed to take a shower. After four days without washing, White wondered what all the fuss was about. "I thought we smelled fine," he said of his and McDivitt's 'distinct aroma'. "It was all those people on the carrier that smelled strange!"

"HOW MANY BOYS, LBJ?"

Days after Gemini IV hit the waves of the Atlantic, the United States Military Command in South Vietnam announced that its troops would shortly begin fighting alongside South Vietnamese forces against the pro-communist North. It was the beginning of a long, bloody and infamous phase of American history that would see a markedly different United States by the end of the decade compared to that which John Kennedy had inherited in 1961.

Yet Kennedy himself was at least partly to blame for the steady escalation in the conflict in south-east Asia, as was his presidential successor, Lyndon Baines Johnson (LBJ). In fact, one of the issues Kennedy had faced during the 1960 election was a perceived 'missile gap' between America and the Soviets and in his inaugural address he had promised to "pay any price, bear any burden, meet any hardship, support any friend, oppose any foe, in order to assure the survival and success of liberty". Following the crisis of the Bay of Pigs and the erection of the Berlin Wall, however, Kennedy feared that if his administration did nothing to halt the incessant advance of communism from North into South Vietnam, the United States would lose all credibility with its allies.

His initial move was to increase American troop numbers and, although he felt that President Ngo Dinh Diem must ultimately defeat the communist insurgents, the poor organisation, corruption and incompetent leadership of the South Vietnamese military made this an unlikely prospect. A key facet in the effort to isolate the populace from the communists was the Strategic Hamlet Programme, established in 1961, yet even its provisions of improved education and healthcare did little to prevent its infiltration by guerrilla factions. Moreover, the South Vietnamese peasantry resented being uprooted from their ancestral villages. As a result, by 1963, the project had all but collapsed.

That summer, some Washington policymakers were predicting that Diem's inability to subdue the communists might force him to make a deal with North Vietnamese premier Ho Chi Minh. His main concern seemed to be the need to fend off military coups against himself and, according to Attorney-General Bobby Kennedy, "Diem ... was difficult to reason with". Early suggestions to encourage a coup and forcibly remove Diem from power were ultimately discarded, for fear of the potential destabilising effects of such an action, and it was decided instead that his younger brother, the hated chief of the secret police, Ngo Dinh Nhu, be removed instead. Only weeks before Kennedy's assassination, the CIA advised anti-Diem generals that the United States would support his removal from office. On 2 November, during a failed attempt to escape, Diem and Nhu were captured and summarily executed in the back of an armoured personnel carrier.

Kennedy, who had not approved such an act, was visibly shocked and in the wake of his own assassination three weeks later the South Vietnamese situation bounced from one military junta to another, all of which were regarded as nothing more than puppets of the United States. Their existence served only to encourage North Vietnam to consider the leaders of the South as slaves of colonialism. Indeed, in the wake of the Diem and Nhu murders, the number of American 'military advisors' multiplied to 16,300 to cope with rising guerrilla violence. Fears that the pro-communist Vietcong retained de facto control of the South Vietnamese countryside prompted the United States to adopt a different policy of pacification and "winning over hearts and minds".

In the aftermath of Kennedy's assassination, Johnson asserted his own support for continued military operations in South Vietnam, before the end of 1963 pledging $500 million in aid to Saigon. In August of the following year, the destroyer Maddox, on an intelligence mission along the North Vietnamese coastline, fired upon several torpedo boats in the Gulf of Tonkin. Two days later, an alleged attack on the Maddox and another destroyer, the Turner Joy, prompted the Americans to initiate air strikes, marking their first large-scale military involvement in the conflict. Although the second incident was later discovered to be an error, it led Congress to pass the Gulf of Tonkin Resolution, giving Johnson the authority to assist any south-east Asian nation under threat of communist aggression. Over the years, the Gulf of Tonkin 'error' has prompted some observers to argue that Johnson deliberately misled the American populace to gain support for greater involvement in Vietnam. Others refute this, but suggest that Robert McNamara and the Pentagon were in the mood to retaliate and presented their 'evidence' of the attacks to support their case.

In February 1965, an attack on a Marine barracks in Pleiku led to the ignition of Operations Flaming Dart and Rolling Thunder, a pair of vigorous aerial bombing campaigns to force North Vietnam to terminate its support for the Vietcong. The campaigns would last for three years, depositing, by November 1968, over a billion kilograms of missiles, rockets and bombs into North Vietnam, along the Ho Chi Minh Trail in Laos and Cambodia and onto Vietcong installations in South Vietnam. Ultimately, it failed.

A month after the Pleiku attack, 3,500 Marines were deployed to South Vietnam to help defend the United States' air bases, effectively beginning the ground war, with public opinion overwhelmingly supporting the move as part of a global conflict against communism. By the end of the year, troop numbers had swelled to 200,000. It was a commitment which many would regret as the Sixties wore on and the death toll rose, with the phrase "Hey, hey, LBJ! How many boys did you kill today?" chanted by protesters on the steps of the Pentagon in late 1967.

By the second half of the Sixties, with almost half a million American troops in Vietnam, many of the astronauts were developing itchy feet to return to active military service. "We were uncomfortable wearing the hero image," wrote Gene Cernan, "while our buddies were bleeding, being captured and dying in a real shooting war for which we had been trained." Some of them, indeed, approached Deke Slayton with a view to taking leave of absence from NASA, polishing up their carrier landing qualifications and returning to the front line ... only to be awakened to the harsh reality. They were free to leave, if they wished, Slayton told them, but he offered no guarantees of a job when they returned.

"The Pentagon hammered in the final nail," continued Cernan. "We could return to active duty if we wanted to, and even fly, but never – ever – would we be allowed into combat." The negative propaganda impact of the Vietcong capturing American astronauts in a combat zone was too unpalatable for the Johnson administration to bear. "Vietnam," wrote Cernan, "would not be our war."

WISEGUY

"That's a vagina," quipped Charles 'Pete' Conrad Jr. "Definitely a vagina." The psychiatrist noted his response without a word, perhaps realising, perhaps not, that he was the victim of yet another wisecrack from the gap-toothed, balding Navy lieutenant. Yet, despite his comments about each of the Rorschach cards shown to him, Conrad was not entirely obsessive about the female genitalia. He had actually been tipped-off the night before by another astronaut candidate, Al Shepard: what the NASA psychiatrists were really looking for was male virility. "I got the dope on the psych test," Shepard had assured him. "No matter what it looks like, make sure you see something sexual." So Conrad did.

His key concern, though, that spring in 1959, had been the impact that this crazy 'Project Mercury' idea might have on his career. Instead of logging hours in the Navy's new F-4 Phantom fighter, he spent a week at the Lovelace Clinic in Albuquerque, New Mexico, much of his time focused on the provision of stool, semen and blood samples and the collection of 24-hour bagfuls of urine. On the evening before a major stomach X-ray, told not to drink alcohol after midnight, Conrad had sat up until 11:57 pm draining a bottle with Shepard and another naval aviator called Wally Schirra to loosen themselves up for the next day. Conrad doubted that Lovelace's invasive tests had anything remotely to do with spaceflight: the physicians, he told Shepard, seemed far more interested in "what's up our ass" than in their flying abilities. Shepard had warned him to be careful – to give the right answers to questions and to remember that Lovelace's staff were watching their every move.

In spite of his frustration, Conrad persevered. He followed Shepard's advice, saw the female anatomy in every Rorschach card, deadpanned to a psychologist that one blank card was upside down, pedalled a stationary bicycle for hours, sat in a hot room for an age, then dunked his feet into ice-cold water and argued with one of the physicians that he considered it pointless to have electricity zapped into his hand through a needle. However, all this torture, Conrad felt, would at least give him the opportunity to lay his entire naval career on the line for just one chance to fly something even faster: to ride a rocket, outside Earth's atmosphere, "at a hell of a lot more Machs than anything he was flying right now". Flying higher and faster, and pushing his own boundaries, had been the story of Conrad's life.

Born in Philadelphia, Pennsylvania, on 2 June 1930, the offspring of a wealthy family which made its fortune in real estate and investment banking, Conrad's father insisted that he be named 'Charles Jr' – "no middle name" – although his strong-willed mother, Frances, felt that this tradition of Charleses should be broken. Frances liked the name 'Peter', wrote Nancy Conrad in her 2005 biography of her late husband, and although it never became his official middle name, Charles Conrad Jr would become known as 'Peter' or 'Pete' for the rest of his life. His fascination with anything mechanical reared its head at the age of four, when he found the ignition key to his father's Chrysler and reversed it off the drive. Later, in his teens, he worked summers at Paoli Airfield, mowing lawns, sweeping and doing odd jobs for free flights. Aged 16, he even repaired a small aircraft single-handedly. Conrad was an engineer and tinkerer at heart.

Education-wise, he would partly follow in his father's footsteps: the private Haverford School, from which he was expelled, then the Darrow School in New York, where Conrad's dyslexia was identified and where he shone. Although his father intended him to attend Yale University, he actually enrolled in 1949 at Princeton, with a Reserve Officers Training Corps (ROTC) scholarship from the Navy to pay for his studies in aeronautical engineering. Graduation in 1953 brought him not only his bachelor's degree, but also a pilot's licence with an instrument rating, marriage (to Jane DuBose) and entrance into naval service.

He breezed through flight training, earning the callsign 'Squarewave' as a carrier pilot. In 'Rocketman', his widow wrote that Al Teddeo, executive officer of Fighter Squadron VF-43 at Naval Air Station Jacksonville, Florida, had his doubts when he first met the young, seemingly-wet-behind-the-ears ensign one day in 1955. Those doubts were soon laid to rest when Teddeo discovered that Conrad could handle with ease any manoeuvre asked of him. Tactical runs, strafing runs, spin-recovery tests; Ensign Conrad did it all. "Hell, we refuelled three times till I just had to get back to my desk," Teddeo recalled years later. "It was like telling a kid at the fair that it was time to go home."

Next came gunnery training at El Centro, California, and transition from jet trainers to the F-9 Cougar fighter, before reporting to Pax River in 1958 to qualify as a test pilot. Later that same year, he received, along with over a hundred others, classified instructions to attend a briefing in Washington, DC. Conrad was told to check into the Rice Hotel under the cover name of 'Max Peck'. Only when he got there did he find that another 35 'Max Pecks' were also there – including an old naval buddy, Jim Lovell. Neither Conrad nor Lovell would make the final cut for the Mercury selection, but their day would come three years later.

Whereas Lovell was cast aside for a minor liver ailment, however, Conrad's cause for failure proved a little ironic. "Unsuitable for long-duration flight," read the explanatory note. He had, it seemed, shown a little too much cockiness and independence during testing; characteristics which were at loggerheads with the panel's notion of a good, all-rounded, level-headed astronaut. Six years after reading those words, Conrad and his Gemini V command pilot, Gordo Cooper, would rocket into space and set a new record ... for long-duration spaceflight!

"EIGHT DAYS OR BUST"

Although Gemini V, the first to carry and utilise fuel cells for electrical power, had long been planned to fly for seven or even eight days, the success of its predecessor and the performance of Jim McDivitt and Ed White had emboldened NASA to move up their estimates for the first lunar landing from 1970 to 1969 and, perhaps, said Joe Shea, as early as mid-1968. Both Gemini IV astronauts would remain very much part of the unfolding action: White was named within weeks to the backup command slot for Gemini VII, an assignment rapidly followed by the coveted senior pilot's seat on the maiden Apollo voyage. McDivitt, too, would go on to great things: commanding Apollo 9, a complex engineering and rendezvous flight to pave the way for the first Moon landing. He would even be offered, but would refuse, the chance to walk on the lunar surface himself.

First, though, came the adulation. After an initial Houston reception, they headed for Chicago, where a million people greeted them and showered them in tickertape along State Street and Michigan Avenue. This was followed, in Washington, DC, by another parade down Pennsylvania Avenue to the Capitol, receptions in the Senate, meetings with foreign diplomats and even a free trip to Paris to upstage the appearance of Yuri Gagarin and a mockup of Vostok 1 at 1965's Air Show. It is unknown to see such scenes as tickertape parades for astronauts today and, perhaps, the only ones in the foreseeable future may be for the men and women who return to the Moon or become the first to tread the blood-red plains of Mars.

In the Sixties, however, every mission was heroic. Moreover, despite the appalling workload and the inevitable strain the astronaut business placed on marriages and families, every man who left Earth's atmosphere was a fully-fledged hero. Not for nothing did Gerry and Sylvia Anderson name their five Thunderbird heroes after five of the heroes of the Mercury Seven: Alan, Virgil, John, Scott and Gordon. For one of those heroes, Gordo Cooper, and his rookie pilot, Pete Conrad, the reality in the build-up to their mission was one of exhausting 16-hour workdays, plus weekends, and a tight schedule to launch on 1 August 1965, eight weeks after McDivitt and White splashed down. Cooper and Conrad and their backups, Neil Armstrong and Elliot See, had only been training since 8 February, giving them less than six months to prepare for the longest mission yet tried. "We realised they needed more time," wrote Deke Slayton. "I went to see George Mueller to ask him for help and he delayed the launch by two weeks."

Despite the pressure, Cooper and Conrad found time to give some thought to names for their spacecraft, even though NASA had officially barred them from doing so. Due to its pioneering nature, the two men wanted to call Gemini V 'The Conestoga', after one of the broad-wheeled covered wagons used during the United States' push westwards in the 18th and 19th centuries. Their crew patch, in turn, would depict one such wagon, emblazoned with the legend 'Eight Days or Bust'. This was quickly vetoed by senior managers, who felt it suggested a flight of less than eight days would constitute a failure, and Conrad's alternative idea – 'Lady Bird' – was similarly nixed because it happened to be the nickname of the then-First Lady, wife of President Johnson. Its possible misinterpretation as an insult could provoke unwelcome controversy which NASA could ill-afford. The astronauts, however, would not be put off and Cooper pleaded successfully with Jim Webb to approve the Conestoga-wagon patch, although the administrator greatly disliked the idea. The duality of the word 'bust' as denoting both a lack of success and the female breasts did not help matters, either ...

Preparations for Gemini V had already seen Conrad gain, then lose, the chance to make a spacewalk. According to a January 1964 plan, the Gemini IV pilot would depressurise the cabin, open the hatch and stand on his seat, after which an actual 'egress' would be performed on Gemini V (Conrad's mission), a transfer to the back of the spacecraft and retrieval of data packages on Gemini VI and work with the Agena-D target vehicle on subsequent flights. Following the Voskhod 2 success, however, plans for a full egress were accelerated and granted to Ed White. The result: instead of 'Eight Days or Bust', Gemini V would come to be described by Cooper and Conrad as 'Eight Days in a Garbage Can'; they would simply 'exist' for much of their time aloft, to demonstrate that human beings could survive for at least the minimum amount of time needed to get to the Moon and back. (The maximum timespan for a lunar mission, some 14 days, would be an unwelcome endurance slog earmarked for the Gemini VII crew.)

Yet the Conestoga mission did have its share of interesting gadgets: it would be the first Gemini to run on fuel cells, would carry the first production rendezvous radar and was scheduled to include exercises with a long-awaited Rendezvous Evaluation Pod (REP). Originally, it was also intended to fly the newer, longer-life OAMS thrusters, although these were ready ahead of schedule and incorporated into Gemini IV. Only weeks after Cooper, Conrad, Armstrong and See began training, on 1 April 1965 fabrication of the Gemini V capsule was completed by McDonnell,

A tired and heavily-bearded Conrad (left) and Cooper aboard the recovery ship after the flight.

tested throughout May in the altitude chamber and finally delivered to Cape Kennedy on 19 June. Elsewhere, GLV-5 – the Titan booster assigned to launch the mission – was finished in Baltimore, accepted by the Air Force and its two stages were in Florida before the end of May. Installation on Pad 19 followed on 7 June, the day of McDivitt and White's splashdown, and Gemini V was mounted atop the Titan II on 7 July. Five days later, the last chance for an EVA on the mission and, indeed, on Geminis VI and VII, was rejected by NASA Headquarters. There seemed little point in repeating what White had already done and, further, Cooper and Conrad, not wishing to be encumbered by their space suits for eight days, had campaigned vigorously for greater comfort in orbit by asking to wear helmets, goggles and oxygen masks. The launch of Gemini V was scheduled for 19 August.

It would be a false start. Thunderstorms ominously approached the Cape, rainfall was copious and a lightning strike caused the spacecraft's computer to quiver. The latter, provided by IBM, had caused concern on Gemini IV and, this time around, had been fitted with a manual bypass switch to ensure that the pilots would not be left helpless again. The attempt was scrubbed with barely ten minutes remaining on the countdown clock and efforts to recycle for another try on 21 August got underway. On this second attempt, no problems were encountered. Aboard Gemini V, Cooper turned to Conrad. "You ready, rookie?" Conrad, white as a sheet, replied that he was nervous. Surely the decorated test pilot who had flown every supersonic jet the Navy owned wasn't scared? Conrad milked the silence in the cabin for a few seconds, then burst out laughing. "Gotcha!" he said with his trademark toothy grin. "Light this son-of-a-bitch and let's go for a ride!" And ride they did. At 8:59:59 am, Cooper and Conrad were on their way.

Ascent was problematic when noticeable pogo effects in the booster jarred the men for 13 seconds, but smoothed out when the second stage ignited and were minimal for the remainder of the climb. Six minutes after launch, as office workers across America snoozed away their Saturday morning, Gemini V perfectly entered a 163–349 km orbit. Nancy Conrad wrote that her late husband compared the instant of liftoff to "a bomb going off under him, then a shake, rattle and roll like a '55 Buick blasting down a bumpy gravel road – louder than hell".

Hitting orbit made Cooper the first man to chalk up two Earth-circling missions. (Gus Grissom, of course, had piloted a suborbital flight on Liberty Bell 7, before commanding the orbital Gemini 3.) However, Gemini V would shortly encounter problems. The flight plan called for the deployment of the 34.5 kg REP, nicknamed 'The Little Rascal', from the spacecraft's adaptor section, after which Cooper would execute a rendezvous test, homing in on its radar beacon and flashing lights. Before the REP could even be released, as Gemini V neared the end of its first orbit, Conrad reported, matter-of-factly, that the pressure in the fuel cells was dropping rapidly from its normal 58.6-bar level. An oxygen supply heater element, it seemed, had failed. Nonetheless, as they passed over Africa on their second orbit, Cooper yawed the spacecraft 90 degrees to the right and, at 11:07 am, explosive charges ejected the REP at a velocity of some 1.5 m/sec. Next, the flight plan called for Gemini V to manoeuvre to a point 10 km below and 22.5 km behind the REP, although much of this work was subsequently abandoned. However, Chris Kraft's ground team was becoming increasingly concerned as the fuel cell pressures continued to decline and when a pressure of 12.4 bars was reached this was insufficient to operate the radar, radio and computer. Kraft had little option but to tell the astronauts to cancel their activities with the pod.

It seemed likely that a return to Earth would be effected and Kraft ordered four Air Force aircraft to move into recovery positions in the Pacific for a possible splashdown some 800 km north-east of Hawaii. A naval destroyer and an oiler in the region were also ordered to stand by. Keenly aware of the situation, Cooper radioed that a decision needed to be made over whether to abort the mission or power down Gemini V's systems and continue, to which Kraft told him to shut off as much as he could. All corrective instructions proved fruitless: neither the automatic or manual controls for the fuel cell's oxygen tank heater would function. Nor could the heater itself, located in the adaptor section, be accessed by the crew. Cooper and Conrad even manoeuvred their spacecraft such that the Sun's rays illuminated the adaptor, in the hope that it might stir the system back to life. It was all in vain.

By now, most of their on-board equipment – radar, radio, computer and even some of the environmental controls – had been shut down and, as Gemini V swept over the Atlantic on its third orbital pass, there was much speculation that a re-entry would have to be attempted before the end of the sixth circuit, since its flight track thereafter would take it away from the Pacific recovery area. Then, as the astronauts passed within range of the Tananarive tracking station in the Malagasy Republic, off the east coast of Africa, Cooper reported that pressures were holding at around 8.6 bars, suggesting, Kraft observed, that "the rate of decrease is decreasing". As he spoke, the oxygen pressures dropped still lower, to just 6.5 bars, and fears were high that if they declined much further, Gemini V would need its backup batteries to support another one and a half orbits and provide power for re-entry and splashdown. The astronauts were asked to switch off one of the fuel cells to help the system and as they entered their sixth orbit the pressures levelled-out at 4.9 bars.

Capcom Jim McDivitt asked Cooper for his opinion on going through another day under the circumstances. "We might as well try it," replied Cooper, but Kraft remained undecided. After weighing all available options, including the otherwise satisfactory performance of the cabin pressure, oxygen flow and suit temperatures, together with the prestige to be lost if the mission had to be aborted, he and his control team emerged satisfied that oxygen pressures had stabilised at 4.9 bars. If there were no more drops, Gemini V would be fine to remain in orbit for a 'drifting flight', staying aloft just long enough to reach the primary recovery zone in the Atlantic, sometime after its 18th orbit. Admittedly, with barely 11 amps of power, only a few of the mission's 17 experiments could be performed, but Kraft felt "we were in reasonably good shape ... we had the minimum we needed and there was a chance the problem might straighten itself out". As Cooper and Conrad hurtled over Hawaii on their fifth orbit, he issued a 'go' for the mission to proceed.

With the reduced power levels, the REP, which kept the spacecraft company up until its eighth orbit, was useless for any rendezvous activities. "That thing's right with us," Cooper told Mission Control during their sixth circuit of Earth. "It has been all along – right out in back of us." Two orbits later, Conrad turned Gemini V a full 360 degrees, to find that the pod had re-entered the atmosphere to destruction. Nonetheless, Gemini V's radar did successfully receive ranging data from the REP for some 43 minutes.

As the mission entered its second day, circumstances improved and oxygen pressures climbed. "The morning headline," Kraft radioed the astronauts on 22 August, referring to a newspaper, "says your flight may splash down in the Pacific on the sixth orbit." Having by now more than tripled that number of orbits, Conrad replied that he was "sorry" to disappoint the media. Despite the loss of the REP, on their third day aloft Cooper conducted four manoeuvres to close an imaginary 'gap' between his spacecraft and the orbit of a phantom Agena-D target. This 'alternate' rendezvous had been devised by the astronaut office's incumbent expert, Buzz Aldrin. Cooper fired off a short burst from the aft-mounted OAMS thrusters to lower Gemini V's apogee by about 22 km, then triggered a forward burn to raise its perigee by some 18 km and finally yawed the spacecraft to move it onto the same orbital plane as the imaginary target. One final manoeuvre to raise his apogee placed Gemini V in a co-elliptical orbit with the phantom Agena. Were it a 'real' target, he would then have been able to guide his spacecraft through a precise rendezvous. Such exercises would prove vital for Gemini VI, which was scheduled to hunt down a 'real' Agena-D in October 1965, and one of the greatest learning experiences, said Chris Kraft, "is being able to pick a point in space, seek it out and find it".

Notwithstanding the successes, the glitches continued. On 25 August, two of the eight small OAMS thrusters jammed, requiring Cooper to rely more heavily on their larger siblings and expend considerably more propellant than anticipated. It was at around this time that Gemini V broke Valeri Bykovsky's five-day endurance record and Mission Control asked Cooper if he wanted to execute "a couple of rolls and a loop" to celebrate; the laconic command pilot, however, declined, saying he could not spare the fuel and, besides, "all we have been doing all day is rolling and rolling!" When the record of 119 hours and six minutes was hit, Kraft blurted out a single word: "Zap!" Gordo Cooper, with an additional 34 hours from Faith 7 under his belt, was now by far the world's most flown spaceman. His response when told of the milestone, though, was hardly historic: "At last, huh?"

The dramatic reduction of available propellant made the last few days little more than an endurance run. Kraft told the astronauts to limit their OAMS usage as much as possible and many of their remaining photographic targets – which required them to manoeuvre the spacecraft into optimum orientations – had to be curtailed. Still, a range of high-quality imagery was acquired. The hand-held 70 mm Hasselblad flew again to obtain photographs of selected land and near-shore areas and, of its 253 images, some two-thirds proved useful in post-mission terrain studies. These included panoramas of the south-western United States, the Bahamas, parts of south-western Africa, Tibet, India, China and Australia. Images of the Zagros Mountains revealed greater detail than was present in the official Geological Map of Iran. Cooper and Conrad also returned pictures of meteorological structures – including the eye of Hurricane Doreen, brewing to the east of Hawaii – together with atmospheric 'airglow'. In addition, they took pictures of the Milky Way, the zodiacal light and selected star fields. Other targets included two precisely-timed Minuteman missile launches and infrared imagery of volcanoes, land masses and rocket blasts.

The scientific nature of many of these experiments did not detract – particularly in the eyes of the Soviet media – from the presence of a number of military-sponsored investigations. Cooper and Conrad's flight path carried them over North Vietnam 16 times, as well as 40 times over China and 11 times over Cuba, prompting the Soviet Defence Ministry's Red Star newspaper to claim that they were undertaking a reconnaissance mission. The situation was not helped by President Johnson's decision, whilst the crew was in orbit, to fund a major $1.5 billion Air Force space station effort, known as the Manned Orbiting Laboratory (MOL). Among the actual military experiments undertaken by Gemini V were observations of the Minuteman plume and irradiance studies of celestial and terrestrial backgrounds, together with tests of the astronauts' visual acuity in space to follow up on reports that Cooper had made after his Faith 7 mission. Large rectangular gypsum marks had been laid in fields near Laredo, Texas, and Carnarvon, Australia, although weather conditions made only the former site visible.

Cardiovascular experiments performed during the mission would reveal that both men lost more calcium than the Gemini IV crew, although principal investigator Pauline Beery Mack expressed reluctance to predict a 'trend', since "a form of physiological adaptation may occur in longer spaceflight". Medically, Chuck Berry's main concerns were fatigue and his advice was that they get as much sleep as possible. "I try to," yawned Conrad at one stage, "but you guys keep giving us something to do!" All in all, they managed between five and seven hours' sleep at a time and expressed little dissatisfaction with Gemini V's on-board fare: bite-sized, freeze-dried chunks of spaghetti and meatballs, chicken sandwiches and peanut cubes, rehydratable with a water pistol. An accident with a packet of shrimp, though, caused a minor problem when it filled the cabin with little pink blobs. Conrad even tried singing, out of key, to Jim McDivitt at one point.

Years later, Conrad would recall that the eight-day marathon was "the longest thing I ever had to do in my life". He and Cooper had spent the better part of six months training together, so "didn't have any new sea stories to swap with one another … there wasn't a whole lot of conversation going on up there". Nancy Conrad would recall her late husband describing how the confined cabin caused his knees to bother him – their sockets felt as if they had gone dry – and that he would have gone "bananas" if asked to stay aloft any longer. (Ironically, on two future missions, Conrad would stay aloft for much, much longer … but on those occasions, his tasks would include a couple of meandering trots around the lunar surface and floating inside a voluminous space station.) He found it hard to sleep, hard to get comfortable and the failures meant he and Cooper spent long periods simply floating with nothing to do. After the flight, he told Tom Stafford that he wished he had taken a book, and this gem of experience would be noted and taken by the crew assigned to fly the 14-day mission.

Nancy Conrad described Cooper's irritation at losing so much of his mission. He was far from thrilled that the two main tasks for Gemini V, rendezvous and long-duration flight, were becoming little more than "learning-curve opportunities" and suggested throwing an on-board telescope in the Cape Kennedy dumpster when it twice refused to work. Later, when the spacecraft was on minimum power and the astronauts were still expected to keep up with a full schedule, Cooper snapped "You guys oughtta take a second look at that!" As for physical activity, he grimaced that his only exercise was chewing gum and wiping his face with a cleansing towel.

On the ground, Deke Slayton was concerned that such an attitude would not help Cooper's reputation with NASA brass. Indeed, Gemini V would be his final spaceflight and, although he would later complain bitterly about 'losing' the chance to command an Apollo mission, some within the astronaut corps would feel that Cooper's performance and strap-it-on-and-go outlook had harmed his career. Tom Stafford was one of them. "Gordo ... had a fairly casual attitude towards training," he wrote, "operating on the assumption that he could show up, kick the tyres and go, the way he did with aircraft and fast cars."

To spice matters up still further, worries about the fuel cells continued to plague Gemini V's final days. Their process of generating electricity by mixing hydrogen and oxygen was producing 20 per cent too much water, Kraft told Conrad, and there were fears that the spacecraft was running out of storage space. This water excess might back up into the cells and knock them out entirely. In order to create as little additional water as possible, the astronauts powered down the capsule from 44 to just 15 amps and on 26 August Kraft even considered bringing them home 24 hours early, on their 107th orbit. However, by the following day, the water problem abated, largely due to the crew drinking more than their usual quota and urinating it into space, and a full-length mission seemed assured.

Eitherway, they had long since surpassed Bykovsky's Vostok 5 record. In fact, by the time Cooper and Conrad splashed down, they would have exceeded the Soviets on several fronts: nine manned missions to the Reds' eight, a total of 642 man-hours in space to their 507 and some 120 orbits on a single mission to their 81. At last, after eight years in the shadows – first Sputnik, then Gagarin, Tereshkova, Voskhod 1 and Leonov – the United States was pulling ahead into the fast lane of the space race. When it seemed that Gemini V might come home a day early and miss the scheduled Sunday 29 August return date, mission controllers in Houston even played the song 'Never on Sunday', together with some Dixieland jazz.

The astronauts also had the opportunity on the last day of the mission to talk to an 'aquanaut', Aurora 7 veteran Scott Carpenter, who was on detached duty to the Navy. Carpenter, who had broken his arm in a motorcycle accident a year before and been medically grounded by NASA, was partway through a 45-day expedition in command of Sealab II, an underwater laboratory on the ocean floor, just off the coast of La Jolla, California. The Sealab effort, conceived jointly by the Navy and the University of California's Scripps Institution of Oceanography, sought to discover the capacity of men to live and work effectively at depth. In doing so, Carpenter became the first person to place 'astronaut' and 'aquanaut' on his career resumé. Yet, unlike Cooper and Conrad, his chances of returning to space were non-existent. He had not impressed senior NASA managers with Aurora 7 and, indeed, the partial success of an operation to repair the injury to his arm meant he would remain grounded anyway. He resigned from NASA in early 1967.

The music, the chat with Carpenter and even Conrad's dubious singing did little to detract from the uncomfortable conditions aboard the capsule. As they drifted, even with coolant pipes in their suits turned off, the two men grew cold and began shivering. Stars drifting past the windows proved so disorientating that they put covers up. Sleep was difficult. Chuck Berry had wired Conrad with a pneumatic belt, a blood-pressure-like cuff, around each thigh, which automatically inflated for two minutes of every six throughout the entire mission. The idea was that, by impeding blood flow, it forced the heart to pump harder and gain its much-needed exercise. Berry felt that if Conrad came through Gemini V in better physical shape than Cooper, who did not wear the belt, a solution may have been found for 'orthostatic hypotension', the feelings of lightheadedness and fainting felt by some astronauts after splashdown.

For the two astronauts, that splashdown could not come soon enough. By landing day, 29 August, their capsule had become cluttered with rubbish, including the litter of freeze-dried shrimp, which had escaped earlier in the mission. The appearance of Hurricane Betsy over the prime recovery zone prompted the Weather Bureau to recommend bringing Gemini V down early and Flight Director Gene Kranz agreed to direct the Lake Champlain to a new recovery spot. At 7:27:43 am, Cooper fired the first, second, third, then fourth OAMS retrorockets, then gazed out of his window. It felt, he said later, as if he and Conrad were sitting "in the middle of a fire". Since it was orbital nighttime, they had no horizon and were entirely reliant upon the cabin instruments to control re-entry. In fact, Gemini V remained under instrument control until they passed into morning over Mississippi.

Cooper held the spacecraft at full lift until it reached an altitude of 120 km, then tilted it into a bank of 53 degrees; whereupon, realising that they were too high and might overshoot the splashdown point, he slewed 90 degrees to the left to create more drag and trim the error. Although experiencing a dynamic load of 7.5 G after eight days of weightlessness, the astronauts did not, as some had feared, black out. The parachute descent was smooth. No oscillations were evident and the 7:55:13 am splashdown, though 170 km short of the planning spot, was soft. As would later be determined, the computer had been incorrect in indicating that they would overshoot. A missing decimal point in a piece of uplinked data had omitted to allow for Earth's rotation in the time between retrofire and splashdown. In fact, Cooper's efforts to correct the false overshoot had progressively drawn them short of the recovery zone. "It's only our second try at controlling re-entry," admitted planning and analysis officer Howard Tindall. "We'll prove yet that it can be done."

Gemini V had lasted seven days, 22 hours, 55 minutes and 14 seconds from its Pad 19 launch to hitting the waves of the western Atlantic and the crew was safely aboard the Lake Champlain by 9:30 am. With the exception of the failed REP rendezvous, and one experiment meant to photograph the target, all of Cooper and Conrad's objectives had been successfully met. Yet more success came when Chuck Berry realised that, despite the days of inactivity with little exercise aboard the capsule, the astronauts were physiologically 'back to normal' by the end of August, clearing the way for Frank Borman and Jim Lovell to attempt a 14-day endurance run on Gemini VII in early 1966. First, though, Wally Schirra and Tom Stafford would fly Gemini VI for one or two days in October and complete the first rendezvous with an

Agena-D target. The mission – or, rather, missions – that would follow would snatch victory from the jaws of defeat and set aside another obstacle on the path to the Moon. But not before suffering a major setback of its own.

CHANGED PLANS

Gordo Cooper and Pete Conrad had flown the equivalent of a minimum-duration lunar flight and, indeed, one of them would tread its dusty surface in a little over four years' time. Apart from stiff joints, heavy beards, a tendency to itch and an aroma, like that of McDivitt and White, which seemed somehow 'different' from everyone else on the recovery ship, they were fine. Cooper, whose heart averaged 70 beats per minute, had come through Gemini V in better shape than Faith 7. After eight days in a half-sitting, half-lying position, both men managed to do some deep-knee bends aboard the recovery helicopter, hopped onto the deck of the Lake Champlain without assistance and walked without wobbling.

They had, NASA flight surgeons determined, come through the mission less fatigued than the Gemini IV crew. This was at least partly because Cooper and Conrad got around six hours' sleep per night during the early portion of their flight. However, their weight loss was perplexing: Cooper had lost 3.4 kg and Conrad 3.8 kg whilst aloft. They had, admittedly, only eaten 2,000 calories per day, rather than the scheduled 2,700, and drank their quotas of water, but both regained the lost weight within days. Still, neither exhibited signs of orthostatic hypotension and Chuck Berry asserted that "we've qualified man to go to the Moon".

Those plans received an abrupt setback on 25 October 1965.

When Wally Schirra and Tom Stafford were assigned to Gemini VI, they were told by Deke Slayton that their two-day mission would feature the world's first rendezvous with a Lockheed-built Agena-D target. Their backups, Gemini 3 fliers Gus Grissom and John Young, had been picked because Slayton "wanted a veteran backup crew to help with training". In his autobiography, Stafford would recall that Young sat actively through simulations with them, whereas Grissom was often absent, racing cars or boats. After three years working on Gemini, Grissom now had his sights set on commanding the first Apollo mission. Following Gemini VI would come Gemini VII, sometime early in 1966, flown by Frank Borman and Jim Lovell and backed-up by Ed White and Mike Collins. The decision to fly the first rendezvous mission ahead of the long-duration 14-day flight had come about because of ongoing Agena problems; Charles Mathews wanted assurances that, if anything went wrong on Gemini VI, there would be enough time to resolve it before resuming rendezvous practice from Gemini VIII onwards.

The first Gemini-Agena Target Vehicle (GATV), numbered '5001', was shipped to Cape Kennedy in May 1965, purely as a non-flying test article, and three months later its successor – '5002' – was officially earmarked for Schirra and Stafford's mission. However, doubts over its reliability lingered. Its main engine, some felt, could not be trusted to execute manoeuvres with a docked Gemini and, although Schirra lobbied for it to go ahead, opposition within NASA to firing it was strong.

Jim Lovell (seated left) and Frank Borman study plans before the mission. Backup pilot Mike Collins stands at left.

Next, Schirra pushed for a firing of the Agena's less powerful secondary propulsion system, although this was not initially incorporated into the Gemini VI flight plan. To be fair, rendezvous techniques were very much in their infancy in 1965, as demonstrated by the unsuccessful attempt of Jim McDivitt to station-keep with the second stage of his Titan II. Buzz Aldrin, however, was a rendezvous specialist, having completed a doctorate in the field before becoming an astronaut, and he joined forces with Dean Grimm of NASA's Flight Crew Support Division to plan a so-called 'concentric rendezvous' technique for Gemini-Agena missions.

Their plan was for the target to be launched, atop its Convair-built Atlas rocket, into a 298 km circular orbit, after which Schirra and Stafford would be despatched into a lower, 'faster', elliptical orbit. "Two hundred and seventy degrees behind the Agena," wrote Stafford, "you'd make a series of manoeuvres that would eventually raise the orbit of the Gemini to a circular one below the Agena. Then you'd glide up below the Agena on the fourth revolution. At that time the crew would make a series of manoeuvres to an intercept trajectory, then break to station-keeping and docking." This docking would occur over the Indian Ocean, some six hours into the mission, after which Schirra and Stafford would remain linked for seven hours and return to Earth following their battery-restricted two-day flight. The astronauts wanted to relight the Agena's engine whilst docked, but NASA managers vetoed it as too ambitious.

During their training at McDonnell's St Louis plant during the last half of 1965, the astronauts practiced manoeuvres again and again, plotting them on boards. In total, they did more than 50 practice runs and spent many hours rehearsing the actual docking exercise with the Agena-D in a Houston trainer. "Housed in a six-story building," wrote Schirra, "it consisted of a full-scale Gemini cockpit and the docking adaptor of the Agena. They were two separate vehicles in an air-drive system that moved back and forth free of friction. We exerted control in the cockpit with small thrusters, identical to those on the spacecraft. We could go up and down, left and right, back and forth. The target could be manoeuvred in those planes as well, though it was inert. It would move if we pushed against it, just as we assumed the Agena would do in space."

On one such training session, Schirra hosted Vice-President Hubert Humphrey in the pilot's seat. The vice-president asked Schirra if their voices could be heard from outside the trainer. When Schirra replied that, no, it was sound-proofed, Humphrey asked if Schirra minded him having a ten-minute nap. When Humphrey awoke, he asked Schirra to tell him what had happened so that he could tell the people outside. "I was a fan of Hubert Humphrey from that day on," wrote Schirra.

Although barred from naming Gemini VI, Schirra sketched a design for a patch which he and Stafford could wear. It featured the constellation of Orion, which, navigationally, was to play an important part in the rendezvous. "The patch would be six-sided," Schirra wrote, "since six was the number of our mission. Orion also appears in the first six hours of right ascension in astronomical terms, a quarter of the way around the celestial sphere."

In anticipation of this dramatic mission, processing of Gemini VI's flight hardware ran smoothly. In April 1965, its launch vehicle, GLV-VI, became the first Titan to be erected in the new west cell at Martin's vertical testing facility in Baltimore, Maryland. The rocket's two stages arrived in Florida at the beginning of August and were placed into storage. Schirra and Stafford's spacecraft arrived at about the same time and was hoisted atop a timber tower for electronic compatibility testing with GATV-5002. Such exercises would later become standard practice in readying Gemini-Agena missions. It was the last Gemini to run on batteries, thus limiting Schirra and Stafford to no more than 48 hours in space, although, by September, NASA was pushing for just one day if all objectives were completed. Even Gemini VI's experiments – two rendezvous tests (in orbital daytime and nighttime), one medical, three photographic and one passive – were considered secondary to the proximity operations with the Agena. Said Schirra: "On my mission, we couldn't afford to play with experiments. Rendezvous [was] significant enough!"

However, so much reliance was being placed on the radar, the inertial guidance platform and the computer that Grimm and Aldrin found the pilot's role was seriously impaired; if these gadgets failed, wrote Barton Hacker and James Grimwood, so too would the whole mission. Grimm persuaded McDonnell managers to rig up a device which could allow the astronauts to simulate trajectories, orbital insertion and spacecraft-to-Agena rendezvous paths. As a result, Schirra and Stafford were able to participate in no fewer than 50 simulations, conferring with Aldrin on techniques and procedures. Stafford would also recall the admirable efforts of 'Mr Mac' himself – James McDonnell, founder of the aerospace giant which bore his name – who, upon learning that the astronauts needed more time on the training computers, complied. "Mr Mac was always behind the programme," Stafford wrote. In fact, he and Schirra invited McDonnell to dinner on the night before the Agena's, and their own, planned liftoff.

Early on 25 October, out at the Cape's Pad 14, a team from General Dynamics oversaw the final hours of the Atlas-Agena countdown. The Atlas booster, tipped with the slender, pencil-like Agena, was scheduled to fly at precisely 10:00 am. Meanwhile, Al Shepard, by now the chief astronaut, woke the Gemini VI crew and joined them for breakfast and suiting-up. Schirra, struggling to give up smoking, lit up a Marlborough during the ride to Pad 19. He felt, wrote Stafford, "he could survive a twenty-four-hour flight without getting the shakes". One and a half kilometres to the south, General Dynamics launch manager Thomas O'Malley pressed the firing button for the Atlas-Agena at 10:00 am and the first half of the GTA-VI mission was, it seemed, underway. The countdown had gone without a hitch and, with 140 flights behind it since 1959, the performance of the Agena was unquestioned. The plan was for it to separate from the Atlas high above the Atlantic, then fire its own 7,200 kg-thrust engine over Ascension Island to boost itself into orbit. The complex orchestra of synchronised countdowns would culminate at 11:41 am with Schirra and Stafford's own liftoff to initiate the rendezvous. Then, things began to go wrong.

The Agena apparently separated from the Atlas, but seemed to wobble, despite the efforts of its attitude controls to stabilise it. Right on time, downlinked telemetry confirmed that its engine had indeed ignited . . . and then nothing more was heard. It

Tom Stafford (standing) and Wally Schirra suit-up.

had reached an altitude of some 230 km and was 872 km downrange of Cape Kennedy. Fourteen minutes after launch, it should have appeared to tracking radars in Bermuda, but was nowhere to be seen; except, that is, for what appeared to be five large fragments. Aboard Gemini VI, the astronauts, fully-suited, aboard a fully-fuelled rocket and ready to go, were puzzled. "Maybe it's the tracking station," Schirra surmised. "Let's wait for Ascension Island." As time dragged on, their countdown was held at T-42 minutes, but no sign of the Agena was forthcoming. Ascension Island saw nothing. "No joy, no joy," came an equally dismal report from the Carnarvon station in Australia and NASA's public affairs officer Paul Haney was forced to tell listeners at 10:54 am that the target vehicle was almost certainly lost. The Gemini VI launch was scrubbed.

In fact, problems had become apparent very soon after the Atlas-Agena left its pad. At 10:06 am, just six minutes into its ascent, Jerome Hammack at the Pad 14 blockhouse was convinced that something was wrong. So too was the Air Force officer in charge of the launch. Although early analysis of the partial telemetry data gave little inkling of what had happened, an explosion seemed the most plausible explanation. "Later investigation," wrote Tom Stafford, "concluded that the Agena had exploded, thanks to an oxidiser feed sequence that had been changed."

In Houston, Flight Director Chris Kraft, together with Bob Gilruth and George Low, surveyed the damage. It was clear that if a rendezvous mission was to take place at all, a delay of several weeks simply to identify the cause of the accident would be unavoidable. "And if it turns out to be a major design failure in the Agena," Time magazine drearily told its readers, "the Gemini programme is in deep trouble." Critics argued that the Agena, with a satisfactory track record as a missile, had been extensively modified for its Gemini role and many of these modifications had never been tested in space. A disappointed Schirra and Stafford were quietly extracted from their capsule, to be told by Al Shepard: "Boys, what we need is a good party." It was the perfect answer, to cheer everyone up, so the three men, together with Grissom and Young, headed off into town.

"For a day of so," wrote Deke Slayton, "we thought about recycling Agena 5001, the ground test bird that hadn't ever come up to specs." However, a lengthy investigation into what went wrong with the 5002 target would still need to be carried out, the results of which would not be clear for weeks. Moreover, a new Agena would not be ready until early 1966. A perfect alternative, however, was on the horizon. Immediately after the Agena's loss, Frank Borman overheard a conversation between McDonnell officials Walter Burke and John Yardley: the former suggested launching Gemini VII as Schirra and Stafford's 'new' rendezvous target. A study of sending Geminis up in quick succession had been done months earlier and seemed an ideal option, but for one detail. Burke sketched his idea onto the back of an envelope, but Borman doubted the practicality of installing an inflatable cone onto the end of Gemini VII to permit a physical docking. Moreover, George Mueller and Charles Mathews dismissed the entire idea, since it would require the launch of both Geminis within an impossibly tight two-week period.

Other managers thought it could be done. Joseph Verlander and Jack Albert proposed stacking a Titan II and placing it into storage until another had been assembled. The Titan's engine contractor, Aerojet-General, had stipulated that the vehicle must remain upright, but this could be achieved with a Sikorsky S-64 Skycrane, after which the entire rocket could be kept on the Cape's disused Pad 20. Immediately after the first Gemini's launch from Pad 19, the second Titan stack could be moved into position and sent aloft, conceivably, within five to seven days. The plan, however, held little appeal and received little enthusiastic response, with most attention focused on swapping the 3,553 kg Gemini VI for the 3,670 kg Gemini VII, thereby making good of a bad situation by at least using the Titan II combination already on the pad to fly Borman and Lovell's 14-day mission.

In the next few days, as this was discussed in the higher echelons of NASA management, it became evident that if the two spacecraft were swapped, the earliest that Borman and Lovell could be launched would be 3 December. However, if the Gemini VII spacecraft were to prove too heavy for the GLV-VI Titan, a delay until around 8 December would become necessary to erect the more powerful GLV-VII. It was then envisaged to launch Schirra and Stafford to perform their rendezvous mission with another Agena sometime in February or early March 1966.

As these plans crystallised, Burke and Yardley posed their joint-flight idea to Bob Gilruth and George Low, who could find few technical obstacles, with the exception, perhaps, that the Gemini tracking network might struggle to handle two missions simultaneously. Even Mathews, when presented with the option, could find few problems, although Chris Kraft's initial response was that they were out of their minds and it could not be done. Then, having second thoughts, he asked his flight controllers for their opinions, and the most that they could object to was that the Gemini tracking network might struggle to handle two missions. Kraft called his deputy, Sigurd Sjoberg, to discuss the possibility further with the Flight Crew Operations Directorate, headed by Deke Slayton. News filtered down, eventually, to Schirra and Stafford, who heartily endorsed it.

The prospects for Burke and Yardley's plan steadily brightened when it became clear that the heavy Gemini VII – which, after all, was intended to support a mission seven times longer than Gemini VI – could not be lofted into orbit by Schirra and Stafford's Titan: it simply was not powerful enough to do the job. Yet the question of tracking two vehicles at the same time remained. Then, another possibility was aired. Could the tracking network handle the joint mission if Gemini VII were regarded as a passive target for Gemini VI? Borman and Lovell would launch first, aboard Gemini VII, and control of their flight would proceed normally as the Gemini VI vehicle was prepared to fly.

As soon as controllers were sure that Gemini VII was operating satisfactorily, they would turn their attention to sending up Gemini VI; in the meantime, Borman and Lovell's flight would be treated like a Mercury mission, wrote Deke Slayton, "where the telemetry came to Mission Control by teletype, letting the active rendezvous craft have the real-time channels that were available". This mode would continue until 'Gemini VI-A' – so named to distinguish it from the original, Agena-rendezvousing Gemini VI mission – had completed its tasks and returned to Earth. After Schirra and Stafford's splashdown, Borman and Lovell would again become the focus of the tracking network.

Before NASA Headquarters had even come to a decision, the rumour mill had already informed the press, some of whom reported the possibility of a dual-Gemini spectacular. On 27 October, barely two days after the Agena failure, Jim Webb, Hugh Dryden, Bob Seamans and other senior managers discussed the idea and George Mueller asked Bob Gilruth to confirm that it would work. The answer was unanimously in the affirmative and Webb issued a proposal for the joint flight to the White House. He informed President Johnson that, barring serious damage to Pad 19 after the Gemini VII launch, Schirra and Stafford's Titan could be installed, checked out and flown within days to rendezvous with Borman and Lovell. Johnson, residing at his ranch in Austin, Texas, approved the plan on 28 October and his press secretary announced it would fly in January 1966. At NASA, however, December 1965 was considered more desirable.

As October turned to November, preparations gathered pace. Aerojet-General set to work implementing steps by which, contrary to its stipulation, the Titans could be handled in a horizontal position, whilst the Air Force destacked GLV-VI from the pad and placed it in bonded storage under plastic covers at the Satellite Checkout Facility. On 29 October, Gemini VII's heavy-lift Titan was erected on Pad 19. Guenter Wendt's first reaction when he saw the short, nine-day Gemini VI-A pad-preparation schedule, was "Oh, man, you are crazy!" The Gemini VI-A spacecraft, meanwhile, was secured in a building on Merritt Island. Although Schirra and Stafford's mission would essentially not change, that of Borman and Lovell was slightly adjusted to circularise its orbit and mimic the Agena's flight path as closely as possible. Elsewhere, the Goddard Space Flight Center was busy setting up and altering tracking station layouts to enable simultaneous voice communications with both capsules.

At one point, ideas were even banded around for an EVA, in which Lovell and Stafford would spacewalk to each other's Geminis and land in a different craft. However, Borman had little interest in such capers. His target was a 14-day mission and he had no desire to do anything that would compromise it. Also, he wished to use the new 'soft' suits that could be doffed in flight. If a spacewalk were to be added to the flight plan, he and Lovell would have to wear conventional space suits, which would make a 14-day mission an even greater chore. In any case, for Lovell and Stafford to exchange places they would have to detach and reconnect their life-support hoses in a vacuum, leaving them with nothing but their backup oxygen supplies for a while. The bottom line for Borman was that whilst an external transfer might have made great headlines, "one little slip could have lost the farm". Coupled with the fact that Stafford, as one of the tallest of the New Nine, sometimes had difficulty egressing from the capsule during ground tests, the decision was taken to eliminate EVA from the joint mission.

TWINS?

"I had pointed Frank Borman at one of the Gemini long-duration missions from the very begininng," Deke Slayton wrote in his autobiography, "because of his tenacity." That tenacity, some argued, had also led to his removal from the right-hand pilot's seat, alongside Gus Grissom, on the original Gemini V. It has been speculated that the two men's strong personalities might have made them incompatible as a commander-pilot duo and the no-nonsense, decisive Borman was instead directed to lead Gemini VII. Some astronauts regarded him as obnoxious and Gene Cernan labelled him a "tight-assed son-of-a-bitch", but none questioned his abilities or impeccable leadership skills. Indeed, he remains one of only five American astronauts to have commanded a crew on his very first mission. One day, in the not too distant future after Gemini VII, his talent and credentials would also lead him to command the first human expedition to the Moon.

Frank Frederick Borman II was born in Gary, Indiana, on 14 March 1928. As a child, he suffered from numerous sinus problems, caused by the cold and damp weather, so his father moved the family to the better climate of Tucson, Arizona, which became Borman's hometown. Like many future astronauts, he could trace his fascination with aviation from an early age and began flying at the age of 15. From Tucson High School, Borman studied for a bachelor's degree at the Military Academy in West Point, followed by a master's in aeronautical engineering from the California Institute of Technology.

After graduation from West Point in 1950, ranking eighth in his class, he entered the Air Force, serving as a fighter pilot in the Philippines and later as an instructor attached to various squadrons across the United States. During one practice dive-bombing run, he ruptured an eardrum, leading him to fear that he may never fly again. However, Borman recovered. Before coming to NASA in September 1962, he also graduated from the Air Force's Aerospace Pilot School as an experimental test flier and served for a time as an assistant professor of thermodynamics and fluid dynamics at West Point. At the time of his selection, he had more experience in jet aircraft – some 3,600 hours – than any of the others in the New Nine.

Clearly, even among the Nine, Borman stood out. During the selection process, his absolute devotion to the West Point military code of Duty-Honour-Country and his unwavering commitment to whatever mission he was assigned led some psychologists to shake their heads in disbelief. Surely nobody could be that uncomplicated, they thought. Yet that was Borman. Like Gus Grissom, he did not dabble in small talk and, in true military fashion, made whatever decisions needed to be made, stuck by them and told his crew afterwards.

In many ways, James Arthur Lovell Jr – whom Pete Conrad had nicknamed 'Shaky' for his bubbling stores of nervous energy – was virtually Borman's twin. Both were born within two weeks of each other, both held equivalent ranks within different services, both were fair-haired and blue-eyed and both were selected as astronaut candidates together. Lovell was born on 25 March 1928 in Cleveland, Ohio and his fascination with rockets, like Borman's with aviation, manifested itself at a young age. In his book on the Apollo 13 mission, co-authored with Jeffrey Kluger, Lovell recounted sheepishly visiting a chemical supplier in Chicago one day in the spring of 1945 to buy chemicals with which he and two school friends could build a rocket for their science project. Despite the boys' admonitions over wanting to create a liquid-fuelled device, like those of Robert Goddard and Hermann Oberth,

Frank Borman performs a visual acuity test during Gemini VII.

their teacher guided them instead towards a solid-propelled one, loaded with potassium nitrate, sulphur and charcoal. Days later, after packing the gunpowder ingredients into a shell of cardboard tubes, a wooden nosecone and a set of fins, the boys took their rocket into a field, lit the fuse and ran like hell.

"Crouching with his friends," Lovell wrote in third-person narrative about his exploits, "he watched agape as the rocket he had just ignited smouldered for an instant, hissed promisingly and, to the astonishment of the three boys, leapt from the ground. Trailing smoke, it zigzagged into the air, climbing about 80 feet before it wobbled ominously, took a sharp and surprising turn, and with a loud crack exploded in a splendid suicide."

Lovell's interest in the workings and possibilities of such projectiles eclipsed that of his two friends, who regarded this as little more than a lark, but his family situation made it unrealistic to hope that a career in rocketry was within his grasp. The Lovells had moved to Milwaukee, Wisconsin, when he was a young boy and his father's death in a 1940 car accident placed enormous pressure on his mother to make ends meet. The military and, in particular, the Navy, seemed an attractive alternative. (Lovell's uncle, in fact, had been one of world's earliest naval aviators during the First World War.) He was accepted, eventually, into the Navy, which offered to pay for two years of an undergraduate degree, provide initial flight training and six months of active sea duty. Lovell jumped at the chance and, within months, was registered as an engineering student at the University of Wisconsin at Madison. He would complete his studies in 1952, receiving a bachelor's degree from the Naval Academy at Annapolis.

Whilst at the academy, he met Marilyn Gerlach, whom he married barely three hours after his graduation ceremony ... and who had typed up his carefully-prepared thesis on liquid-fuelled rocketry. Flight training consumed much of the next two years, after which Lovell was attached to Composite Squadron Three, based in San Francisco, whose speciality included nighttime takeoffs and landings on aircraft carriers at the height of the Korean conflict. Several months later, he was flying F-2H Banshee jets from the Shangri-La aircraft carrier over the Sea of Japan, routinely swooping in to land on its deck. On one occasion, however, a routine flight went seriously wrong. Moreover, the flight was his first mission in darkness.

The only means of determining where the carrier was at night, Lovell wrote, was a beamed, 518-kilocycle signal from the Shangri-La, which allowed the Banshee's automatic direction finders to guide him home. However, poor weather forced the ship to cancel the mission of Lovell, his teammates Bill Knutson and Daren Hillery and their group leader Dan Klinger; in fact, Klinger had not even left the deck of the Shangri-La when the flight was terminated. Unfortunately for Lovell, his direction finder had picked up the signal of a tracking station on the Japanese coast – which also happened to be transmitting at 518 kilocycles – and, far from guiding him back to the Shangri-La, was actually taking him further away. Around him, he saw nothing but a "bowl of blackness".

Perhaps the homing frequencies had changed, Lovell thought. At once, he turned to the list of frequencies on his kneeboard, but upon switching on his small, jury-rigged reading light, "there was a brilliant flicker – the unmistakable sign of an overloaded circuit shorting itself out – and instantly, every bulb on the instrument panel and in the cockpit went dead". His options seemed dire: ask the Shangri-La to switch its lights on, which was hardly advisable and would prove hugely embarrassing, or ditch in the icy sea. Then, in a story repeated by Tom Hanks, who played Lovell in the 1995 movie 'Apollo 13', he saw a faint greenish glow, like a vast 'carpet', stretching out below and ahead of him. It was the phosphorescent algae churned up in the Shangri-La's wake and it guided him back to the company of his two wingmen, Knutson and Hillery, and a safe, though hard, landing which he later described as "a spine-compressing thud".

For his efforts, the sweat-drenched Lovell was given a small bottle of brandy, downed in a single gulp, and the opportunity to fly his next nocturnal mission ... the very next night. This time, thankfully, his automatic direction finder behaved flawlessly. Eventually, he accumulated no fewer than 107 carrier landings and became an instructor in the FJ-4 Fury, F-8U Crusader and F-3H Demon jets, before moving to the Navy's Test Pilot School at Patuxent River. He graduated first in his class, ahead of Wally Schirra and Pete Conrad. Less than two years later, in early 1959, he was one of 110 military test pilots ordered to attend a classified briefing in Washington, DC. Like Conrad, he would be turned down for Project Mercury, but secured admission into the exalted ranks of NASA's astronaut corps, together with Frank Borman, in September 1962.

THREE HUNDRED HOURS IN SPACE

Since their assignment as Gemini VII's prime crew on 1 July 1965, Borman and Lovell had been intensely focused on their primary objective: to spend 14 days – a total of 330 hours – in space, thereby demonstrating that astronauts could physically and psychologically withstand a maximum-length trip to the Moon. The results from the two previous long-duration flights, Gemini IV and V, had been mixed. Jim McDivitt and Ed White had returned fatigued after four days, while Cooper and Conrad had hardly enjoyed their eight days sitting in an area the size of the front seat of a Volkswagen Beetle. Sleeping in shifts of four or five hours apiece had proven impractical, Borman and Lovell learned, so they resolved to sleep and work together. Moreover, they felt that their 'work' time would not benefit from a rigid plan, opting instead for a broader outline which they could adapt in orbit.

Their 'days' would consist of two work sessions, roughly coinciding with Houston's 'morning' and 'afternoon' time zone and fitting in well with the three flight control shifts which would monitor Gemini VII. Storage space, not just for experiments and equipment, but also for foodstuffs, was at a premium. To make the best use of this space, Kenny Kleinknecht accompanied the astronauts to McDonnell's St Louis plant and decided that waste paper from meals could be kept behind Borman's seat for the first week and behind Lovell's for the second.

Suits proved another concern. Several months before, McDonnell had begun an effort to determine if ordinary Air Force flight garments – wired with medical monitoring equipment, communications headsets and oxygen bottles – could be worn as a lighter, more comfortable alternative to the bulky pressure ensembles. In fact, astronauts Gordo Cooper and Elliot See had tested such suits in June 1965 at a simulated 36,000 m in the altitude chamber, with positive results. Then, in July, McDonnell engineer James Correale suggested a lightweight suit akin to Gemini 3's G3C garment. It would not allow astronauts to continue a mission if the cabin lost pressure, but would provide them with enough margin of safety to get to a recovery area. Of course, from an environmental-control point of view, Gemini operated more efficiently with suits off, but neither NASA nor McDonnell was keen to leave them so vulnerable.

Work on Correale's suit was begun by the David Clark Company in August, with engineers removing as much 'corsetry' as possible from the 10.7 kg ensemble. Replacing its fibreglass helmet was a soft cloth hood, which utilised zips rather than a neck ring to attach it to the torso, and the entire suit could be removed easily and laid on the sides of the Gemini seats, without having to be stowed away. When complete, it weighed some 7.3 kg. It would be removed no sooner than the second day of the mission, to allow time for Gemini VII's life-support systems to be monitored and verified as satisfactory. However, it would be worn during critical phases such as rendezvous, re-entry and splashdown. The suits were delivered in

November, only a few days before Borman and Lovell were due to launch.

Also ready was one of the largest complements of experiments – primarily medical ones – ever carried aloft. Of the 20 investigations, eight would focus on the physical and physiological responses of the two men. They ranged from calcium-balance studies to in-flight sleep analysis with a portable electroencephalogram to examining the effects of spaceflight on the chemistry of body fluids. (For the EEG, Borman would have two spots shaved on his head and dipilatory rubbed on to accommodate its sensors. Lovell was not involved in this experiment.) They had to closely monitor and keep records of their food and liquid intakes, and 'outputs', not only throughout their time in orbit, but also for nine days before launch and four days after splashdown. Their meals were prepared and weighed, gram by gram, by a nutritionist from the National Institutes of Health. Nine experiments were reflights from McDivitt's and Cooper's missions, plus three new ones: an in-flight transmitter to be aimed at a laser beacon at the White Sands Test Facility in New Mexico to evaluate optical communications, together with landmark-contrast measurements of shorelines and a study of the usefulness of stellar occultations for navigation.

Although Gemini VII would primarily serve as a passive rendezvous target, the spacecraft itself needed some last-minute modifications to support its 'extra' mission. In early November, acquisition and orientation lights, a radar transponder, a spiral antenna and a voltage booster were installed. Further, the decision to fly a joint mission with Gemini VI-A reduced the amount of fuel that Borman and Lovell could use for experiments and station-keeping.

LONG HAUL

At 2:30 pm on 4 December, precisely on time, Gemini VII roared into orbit. "We're on our way, Frank!" yelled Lovell as the Titan rolled and pitched in its ascent trajectory, achieving orbit five and a half minutes later and establishing itself in a 160 km path around the globe. Unfortunately, in spite of its historic nature, it proved to be the least-watched launch to date; many American viewers being outdoors on the bright Saturday afternoon, out Christmas shopping or watching football games.

A minor pressure loss in a fuel cell was soon rectified by applying pressure from the cabin oxygen tank to the fuel cell oxygen tank and Borman and Lovell succeeded, in their first few minutes of orbital flight, to manoeuvre their capsule and fly in formation with the Titan's discarded second stage. Borman yawed Gemini VII some 180 degrees, at a rate of three or four degrees per second, to face the stage, which he reported was venting its last vestiges of propellant in the form of snowflake-like particles. He manoeuvred to a point 60 m ahead of the Titan, then performed a series of OAMS pulses to approach it, before taking up position around 15–18 m directly ahead of it in terms of their orbital motion. After the flight, Borman, who described the spent stage as "bigger than the devil", would recall that his quick 'out-and-back' station-keeping procedure seemed to solve the problem experienced by McDivitt in June, since it took "a lot of the orbital mechanics out of the situation".

By now heading eastwards over the North Atlantic, observation of the Titan became more difficult as it passed right in line with the Sun. Borman fired the OAMS thrusters again to move north-of-track, to get the glare out of his line of sight, but actually created a pattern of criss-crossing paths with the stage and its debris cloud of frozen propellant particles. Using their eyes, a set of four tracking lights on the Titan and a docking light on Gemini VII itself, the astronauts managed to station-keep for around 15 minutes as the rocket tumbled violently and vented frequently. As Borman flew, Lovell performed one of the military-sponsored experiments, taking infrared readings of the Titan with a small photometric instrument on his side of the cabin. The resultant movie images showed white plumes pouring from the stage, whose erratic movements, they recalled later, were both translational and rotational.

"A couple of times," Borman said, "we got in a little too close and I backed out, because you just do not dare get as close as you do the way this thing is spewing." In the aftermath of the Gemini VI-A rendezvous a few days later, he would consider the Titan station-keeping a much more difficult and unpredictable exercise. For his part, Lovell, who would command his own Gemini rendezvous with an Agena-D less than a year later, felt that the Titan's tracking lights were of limited use in judging range and range rates. "We had four lights on," explained Borman, "and I'll be darned if I will try to judge distance by four lights – or by 50 lights! You have got to have illumination or you have got to have a stable vehicle." Gemini VII had neither. By the time Borman finally executed a 'breakout' manoeuvre at 2:51 pm to permanently pull away from the Titan he found that he had expended seven per cent more fuel than anticipated. A little over 20 minutes later, the astronaut saw the stage pass within a couple of degrees of the Moon, then saw it again on their second orbit and again about two and a half hours after launch. By this time, they reported that it was "surrounded by a billion particles" of frozen propellant from its engine bell.

With the station-keeping behind them, Borman and Lovell settled down to eight days of experiments before the Gemini VI-A launch on 12 December. Three hours and 48 minutes into the mission, Lovell fired the OAMS in a major perigee-lifting burn lasting 76 seconds, which boosted the low point of Gemini VII's orbit from 140 to 193 km and also brought them back into close proximity with the Titan. "We had come back into the vicinity of the booster," said Borman. "Just about midway through the burn, the booster venting that was still occurring suddenly lit up – became lit up. It looked like we were flying through a lot of foreign objects or debris. I was afraid we were going to hit something." In response, Lovell halted the perigee burn a few seconds early and a trailing strap attached to the rear of Gemini VII whipped forward and slapped against his window; at first, Borman thought that they had hit some debris. They pulsed the OAMS for a few more seconds, getting quite close to the planned 'delta-V' for the burn, and settled down to the first of their medical experiments. First were the cardiovascular conditioning cuffs, snapped onto Lovell's legs. From then on, virtually every bodily action – from thinking to breathing to urinating to defecating – would be monitored.

By 7:00 pm, less than five hours into the mission, they turned to routine housekeeping and at 9:30 pm ate their first meal in orbit. The only real concern,

Spectacular view of the Andes from Gemini VII.

judging from the space-to-ground chatter, was a problematic fuel-cell warning light, which intermittently blinked on and off. When the two men came to sleep, Borman found his G5C suit to be much warmer than anticipated, forcing him to turn the control knob to its coolest setting. Next morning, Capcom Elliot See ran through systems checks, their experiment load for the day, football scores, the news that two airliners had collided over New York ... and the theme song of Gemini VII's prime recovery ship, the Wasp: 'I'll Be Home For Christmas'. To Borman, it seemed that he and Lovell were safer in space than people on Earth.

As the flight wore on, conditions became somewhat less comfortable, with both men complaining of stuffy noses and burning eyes. The cabin, Borman reported, was too warm. Removing their suits helped, yet even that had been a matter of some debate on the ground. Days earlier, on 29 November, Bob Gilruth had requested approval from NASA Headquarters for the astronauts to remove their suits after the second sleep period and only don them at critical junctures, such as rendezvous and re-entry. By the time Gemini VII launched on 4 December, the plan had been amended slightly: one of them had to be suited at all times, insisted George Mueller and Bob Seamans, but the other could remove his garment for up to 24 hours. Both men, however, had to be fully-suited for rendezvous and re-entry operations. Still, the intense discomfort was there and, as the mission wore on with no major environmental-control issues, the rationale behind the one-suit-on/one-suit-off decision became unsupportable.

Even with his suit unzipped and gloves off, Borman sweated heavily, while the unsuited Lovell remained dry. After 24 hours, Lovell asked to sleep unsuited, to which Borman agreed, despite his own discomfort. Lovell, the larger of the two, had more difficulty getting out of his suit in the confined cabin and, although he donned some lightweight flight coveralls for a few minutes, he removed them just as quickly, due to the intense warmth. After four days of this torment, Borman asked the flight controller on the Coastal Sentry Quebec tracking ship to ask Chris Kraft about the chances of both men taking off their suits. Capcom Gene Cernan discussed the request, firstly, with Deke Slayton, before approaching Kraft, but there was little option but to ask Lovell to put his suit back on so that Borman could remove his. Concern was mounting, however, about how alert the astronauts would be for the Gemini VI-A rendezvous if they were so hot and uncomfortable. Bob Gilruth certainly favoured both men having their suits off at the same time and Chuck Berry, looking at the biomedical data, saw clear signs that blood pressures and pulse rates were closer to normal when Borman and Lovell were unsuited. Eventually, on 12 December – the very day that Schirra and Stafford were due to fly – NASA Headquarters finally agreed to allow the Gemini VII crew to remove their uncomfortable suits.

In spite of their discomfort, the two men got along well, even singing Top 40 hits to each other to pass the time. More musical accompaniment came from Houston controllers, who sent up tunes on a radio band which would not interfere with voice communications, and by the end of the mission Gemini VII's cabin echoed to Bach, Handel, Glinka and Dvorak. The astronauts' patience was, however, tried on a number of occasions, most notably when a urine bag broke in Borman's hands. "Before or after?" asked Chuck Berry. When Borman affirmed it was the latter, Berry replied "Sorry about that, chief". After the flight, Lovell would describe their living and working conditions in a similar manner to Cooper and Conrad: like sitting in a men's toilet for a fortnight without access to a shower. This did not bode well for the physicians: after splashdown, one of their tasks was to examine calcium loss in space and they would be obliged to not only sift through Borman and Lovell's liquid and solid waste, but also microscopically analyse the contents of their underwear ...

The cramped nature of the cabin was further exacerbated by the equipment for their 20 scientific and medical experiments. One of these was a hand-held sextant, which enabled them to sight stars setting on Earth's horizon and determine that they could navigate their position in space without relying on a computer. (This would prove particularly important when Borman and Lovell next flew together in December 1968, on the first manned lunar mission.) As part of one of their military investigations, they tracked a Minuteman missile launch and acquired infrared imagery of the plasma sheath of ionised air that was created when its warhead plunged back into the denser atmosphere. Other tasks were somewhat less successful. A blue-green laser beam, fired from a transmitting station in Hawaii, could not be kept in sight for long enough to effect experimental voice communications. As useful as these tasks were for future technologies, the monotony of the mission was even affecting flight controllers. "What a helluva bore," one of them yawned as Borman and Lovell drifted into their second week aloft.

That second week, though, would be one of the most dramatic yet seen. It would begin by scraping its knuckles on a near-disaster and end triumphantly ... to the sound of 'Jingle Bells'.

"SITTIN' HERE BREATHING"

The immediate aftermath of a launch, Jack Albert said later, was normally something of an anticlimax. Except, that is, on 4 December 1965, when spirits remained high. Another Gemini would be despatched in just a few days' time, and, judging from the minimal damage sustained by Pad 19, one major obstacle standing in the way of the joint mission had dissolved. The next day, the two stages of GLV-VI had been erected and by sundown the Gemini VI-A spacecraft was added. A computer problem quashed hopes to launch on 11 December, but the installation of a replacement part brightened prospects for Schirra and Stafford to fly a day later.

On the morning of the 12th, the astronauts awoke, showered, breakfasted and suited-up just as they had seven weeks earlier, albeit on this occasion Schirra dispensed with smoking a cigarette. Launch was scheduled for six seconds past 9:54 am and the countdown clock ticked perfectly toward an on-time liftoff. Precisely on cue, the Titan's first-stage engines ignited with a familiar whine. Then, after less than 1.2 seconds, they shut down. Instantly, Schirra, his hand clasping the D-handle which would have fired his and Stafford's ejection seats and boosted them to safety, faced a life-or-death decision. The mission clock on the instrument panel had started running, as it would in response to the vehicle lifting off, but Schirra could feel no movement in the rocket. If the Titan had climbed just a few centimetres from the pad at the instant of shutdown, there was a very real risk that its 150,000 kg of volatile propellants could explode in a holocaust, known, darkly, among the astronauts, as a Big Fucking Red Cloud (BFRC).

In his autobiography, Stafford remembered vividly the moment that the behemoth came alive and, just as vividly, the instant at which its roar ceased. "The sound of the engines died even though the clock started and the computer light came on, both indications that we had lifted off," he wrote. "But I could feel that we hadn't moved. More important, there was no word from [Capcom] Al Bean, confirming liftoff, which was critical." In fact, it was the feeling of stillness in the

Titan that convinced Schirra not to risk ejecting. Kenneth Hecht, head of the Gemini escape and recovery office, was surprised that he did not eject, but in reality, neither Schirra nor Stafford had much confidence in the seats and, as test pilots, instinctively desired to remain with their 'bird' for as long as possible. Stafford felt that the 20 G acceleration of an ejection would have left him with, at best, a cricked neck for months. Moreover, there was a very real risk of death. "Given that we'd been soaking in pure oxygen for two hours," Stafford wrote, "any spark, especially the ignition of an ejection seat rocket, would have set us on fire. We'd have been two Roman candles shooting off into the sand and palmetto trees." Yet Schirra would not have put them in undue danger. "If that booster was about to blow," he said, "if we really had a liftoff and settled back on the pad, there was no choice. It's death or the ejection seat."

In emotionless tones, the unflappable Schirra reported that propellant pressures in the Titan were lowering and Martin's test conductor, Frank Carey, responded in a similarly calm manner with "Hold kill", a missile-testing term denoting a shutdown. Although Schirra knew that the rocket had not left the pad and that the mission clock – which should have started at the instant the Titan began to climb – was wrong, his 'gutsiness' that morning would win him deserved praise from his fellow astronauts. Had he and Stafford ejected, the entire rendezvous would have been over. There would have been no way that Gemini VI-A could have been readied for another launch attempt in less than the six remaining days of Borman and Lovell's mission. Moreover, with the increasing likelihood that another Agena-D would not be ready until the spring of 1966, the crucial step of proving rendezvous as a means of getting to the Moon would have been seriously jeopardised.

When the smoke had cleared, and after receiving assurances that the ejection seat pyrotechnics had been safed, Guenter Wendt and his team returned to the capsule to begin extracting the two disappointed astronauts. "It took 90 minutes to raise the erector and get us out, a lot longer than it should have," wrote Stafford. "Although he had kind words for Guenter and the pad crew, Wally was furious." The families of the two men were also understandably anxious and, from then onwards, it would become standard practice to have another astronaut present with them during a launch attempt. The day itself was already a bad one: a Cape Kennedy rescue helicopter had crashed in the nearby Banana River and Randy Lovelace – bane of the astronauts' lives during their selection – and his wife had been killed in a private aircraft crash in Aspen, Colorado.

Later that same afternoon, President Johnson told Jim Webb that he was "greatly disturbed" by the abort, although he was assured that enough time remained to identify the Titan glitch, fix it and get Gemini VI-A into orbit before the end of Borman and Lovell's mission. That glitch did not take long to find: an electrical tail plug had dropped prematurely from the base of the rocket and activated an airborne programmer – a clock in Gemini VI-A's cockpit which should not have started until liftoff. The plug was supposed to require 18 kg of 'pull' in order to separate, but had rattled loose from its housing. Although it had been installed properly, tests revealed that some plugs did not fit as snugly as others and pulled out more easily.

Then, as engineers pored over engine trace data, it became clear to Ben Hohmann of the Aerospace Corporation that the Titan's oxidiser pressure and overall thrust had begun to decline before the plug fell out. Subsequent analysis of oscilloscope wiggles identified a blockage in the gas generator and, eventually, an Aerojet technician found the answer: a thimble-sized dust cover had been accidentally left on its fuel inlet port during processing. Months earlier, when the engine was still at Martin's plant in Baltimore, the gas generator had been removed for routine cleaning and when the check valve at its oxidiser inlet was detached, a plastic cover was installed to keep dirt out. As checkout of the engine proceeded, the dust cap was overlooked and forgotten. To be fair, its location would have been almost impossible to find. However, had the initial tail plug dropout not stopped the launch, the gas generator blockage certainly would have done. "It was serendipitous that we shut down," said Joe Wambolt, then a Gemini propulsion engineer, in an interview published years later in Quest magazine, "because the other engine was not going to thrust." In his autobiography, Wally Schirra wrote that, had he known of this 'second' brewing problem at the time, he probably would have chosen to fire the ejection seats.

By 13 December, the gas generator had been cleaned and replaced and the launch was provisionally targeted for the 16th, just two days before Borman and Lovell were due to return to Earth. However, Elliot See radioed the Gemini VII crew with the news that, barring any further problems, the 15th seemed a more likely launch date.

In addition to demonstrating the steely nerves of the Gemini VI-A crew – one of Schirra's first messages had been "We're just sittin' here breathing" – the abort also verified, in the most dramatic manner possible, that the Titan's malfunction detection system worked. Sensing no upward movement, it had correctly and automatically closed the valves to prevent more fuel entering the combustion chambers and had duly shut down the engines. Catastrophe had been averted. In the Soviet Union, Nikolai Kamanin, fuming over his nation's failure to catch up with the Americans, admitted in his diary that, despite the abort, a successful Gemini rendezvous was only a matter of days away. Indeed it was.

FIRECRACKERS, FOOTBALLING AND THE F-86

It was third time lucky when Wally Schirra – newly raised from commander to captain in June 1965, part of President Johnson's spot-promotion of active-duty military astronauts – and Air Force Major Thomas Patten Stafford Jr boarded their spacecraft under clear blue skies on the warm morning of 15 December. Launch at 8:37:26 am was perfect, the Titan behaving flawlessly and inserting them into an elliptical orbit of 160–260 km. The bald-headed Stafford was another of NASA's 1962 astronaut intake whom Deke Slayton described as "too green" to have been a realistic candidate for the Project Mercury selection. Indeed, as Slayton and his six colleagues were announced in April 1959, Stafford was barely graduating from test pilot school and his height would have rendered him ineligible anyway. Further, had it not been for the decision to increase the height limit for the larger Gemini spacecraft, Stafford would not have been picked at all.

In time, he would become one of NASA's most accomplished astronauts, flying four times into space, including a lunar voyage and command of the joint Apollo-Soyuz mission with the Soviets. His habit of trying to speak faster than he could think led fellow astronauts to nickname him 'Mumbles'. During Al Shepard's training to command Apollo 14, Stafford would also take charge of the astronaut office as its chief. He was born in Weatherford, Oklahoma, on 17 September 1930, the son of a dentist father and a teacher mother, becoming an avid reader and an enthusiastic watcher of each silvery DC-3 airliner which frequently soared above his childhood home. After the Pearl Harbour bombings, Stafford took a paper round to buy parts and build his own balsa wood model aeroplanes and in 1944 he took his first flight in a two-seater Piper Cub. That flight alone, he wrote, "made me eager to become a fighter pilot and help win the war".

The young Stafford would not engage in combat in the Second World War, but his dream would one day come true. In high school, he excelled in football, eventually becoming captain, although he recounted in his autobiography that he was far from perfect: shooting out streetlights with a BB gun, throwing a firecracker into the police station and attempting, with his friends, to disrupt their English lessons with a cleverly orchestrated symphony of coughing. "The neighbours could always tell when I had been caught," he wrote. "I would be out front painting the fence as a punishment, like Tom Sawyer."

His footballing abilities, though, drew the attention of the University of Oklahoma's coach, although Stafford had also applied for, and would receive, a full ROTC scholarship from the Navy to study there. He had already undertaken some military training in 1947 as part of the Oklahoman National Guard and was even called to temporary duty when the small town of Leedey was hit by a tornado. Stafford also worked on manoeuvres to plot howitzer targets and his calculations contributed to his battery receiving an award for the most outstanding artillery unit. The following year, 1948, brought both success and tragedy: acceptance into the Naval Academy, tempered by the death of his father from cancer.

During four years at Annapolis, he was assigned to the battleship Missouri, where he met another midshipman named John Young. "We would have laughed," Stafford wrote, "at the suggestion that someday we would become astronauts flying in space and circling the Moon together." After graduation in 1952 with a bachelor's degree, his decision to opt for an Air Force, rather than naval, career was inspired by his eagerness to fly the F-86 Sabre jet. He achieved his coveted silver wings from Connally Air Force Base in Waco, Texas, late the following year. By now married to Faye Shoemaker, Stafford underwent advanced training in the F-86 – "the hottest thing in the sky" – and the T-33 Shooting Star. He was then assigned to an interceptor squadron, based in South Dakota, and later moved to Hahn Air Base in Germany as a flight leader and flight test maintenance officer for the Sabre.

Few opportunities for promotion almost led Stafford to resign from the Air Force in 1957 and he even drafted application letters to numerous airlines … before deciding to stay in the service when he first saw the F-100 Super Sabre jet and the forthcoming F-104 Starfighter. "If I went to an airline," he wrote, "I'd be flying the equivalent of cargo planes and could say goodbye to high-performance fighters." He tore up the letters and was promoted to captain the following year, together with selection for the Air Force Experimental Test Pilot School at Edwards. His time there, he remembered, saw him working harder than ever before. "Each morning's flight generated a pile of data from handwritten notes, recording cameras, oscilloscopes and other instruments. We had to reduce this data to a terse report that we submitted to the instructors and we had a test every Friday." From such schools, pilot astronauts were, are and will continue to be drawn.

Stafford graduated first in his class in May 1959, stayed on at Edwards as an instructor and, over the next couple of years, oversaw a number of newer test pilot candidates, including Jim McDivitt, Ed White, Frank Borman and Mike Collins. He also met a visitor from the Navy's test pilot school, an aviator named Pete Conrad. Additionally, he co-authored two flight test manuals: the Pilot's Handbook for Performance Flight Testing and the Aerodynamics Handbook for Performance Flight Testing. By the spring of 1962, he was due for a permanent change of station and confidently expected to study for an advanced master's degree in a technical field, but was picked to attend Harvard Business School; a business administration credential, he realised, would benefit both his military career and any subsequent plans he had. In April, Stafford also learned that NASA was recruiting its second class of astronauts and submitted his application. Five months later, and three days after starting his master's degree at Harvard, he received the call from Deke Slayton that would truly change his life.

RENDEZVOUS!

Gemini VI-A's launch on 15 December was precisely timed so that its Titan would insert it into an orbital plane which closely coincided with that of Borman and Lovell. Trajectory planners had calculated that a liftoff 6.471 seconds past 8:37 am provided ideal conditions for a rendezvous during their fourth orbit. The Gemini VII crew saw only cloud when they tried to spot the launch, but once the Titan climbed out of the weather Borman and Lovell had an oblique view of its contrail from their vantage point out over the Atlantic.

From within the capsule, however, Schirra and Stafford's rise from Earth was dramatic. In his autobiography, Stafford would recount feeling little discomfort as G forces climbed beyond five, then seven, peaking at nearly eight, and his first view of the planet's horizon as the Titan's second stage inserted them into a preliminary orbit. The G loads caused him to feel some pain in his gut and pressure on his lungs, forcing him to take short, sharp breaths – then, all at once, five minutes and 35 seconds after launch, the forces went from eight down to zero. The two men raised their visors, took off their gloves and finally removed their helmets, stowing them below their knees.

Six hours of work awaited them. At orbital insertion, they were trailing Gemini VII by almost 2,000 km. An hour and a half after launch, Schirra pulsed the OAMS thrusters to increase the apogee to 272 km and close the distance to some 1,175 km; this was followed at 10:55 am by a 'phase-adjustment' burn, whose purpose was twofold. Firstly, it reduced the distance between them and the target and secondly, it raised Gemini VI-A's perigee to 224 km. Most importantly, it established the timing for the subsequent chase. Half an hour later, Schirra turned the spacecraft 90 degrees to the 'right' – in a southward direction – and again fired the OAMS to push Gemini VI-A into the same plane as Borman and Lovell. By this point, three hours after launch and entering their third orbit, they had narrowed their distance still further to 483 km. At 11:52 am, Capcom Elliot See told them that they should soon be able to establish radar contact with Gemini VII. Indeed, a flickering signal was replaced by a solid lock at a range of 434 km.

A little under four hours into the flight, in the so-called Normal Slow Rate (NSR) manoeuvre of the rendezvous sequence, Schirra pulsed the aft-mounted thrusters for 54 seconds to slightly increase Gemini VI-A's speed and enter an orbital path of 270 x 274 km, co-elliptic with and fixed 27 km below the target, which was now 319 km ahead. Stafford, meanwhile, busied himself with a circular slide rule and heavily-crosshatched plotting chart on his lap, checking the computer's analysis of the radar data and relaying information to Mission Control. Shortly thereafter, they placed their spacecraft in the 'computer' – or automatic – rendezvous mode and Schirra dimmed the interior lights to aid his visibility. At 1:41 pm, he announced his first visual sighting of what he thought was a star: "My gosh, there is a real bright star out there. That must be Sirius." It wasn't. It was Gemini VII, glinting in the sunlight, just 100 km away. They lost sight of it briefly when it entered Earth's shadow, but when their eyes adjusted they identified its blue tracking lights. Twelve minutes later, when the target was 60 km ahead and the geometry was correct, Schirra initiated the terminal phase manoeuvre designed to close the range to 3 km. He then executed a pair of mid-course correction burns and, at 2:27 pm, just 900 m from their target, started pulsing Gemini VI-A's forward thrusters to steadily reduce the closure rate.

Closer and closer they drifted, until Schirra and Stafford were just 40 m from Borman and Lovell, with no relative motion between them. Back on Earth, in the MOCR, flight controllers erupted in applause and waved small American flags, while Chris Kraft, Bob Gilruth and other senior managers fired up celebratory cigars. Unlike Vostoks 3 and 4, which had merely drifted past each other at a distance of several kilometres as a result of being in slightly different orbits in August 1962, Wally Schirra had achieved a 'real' rendezvous. He defined it thus: "I don't think rendezvous is over until you are stopped – completely stopped – with no relative motion between the two vehicles, at a range of approximately 120 feet [40 m]. That's rendezvous!"

At one point during the rendezvous Stafford had been confused, however. After being rivetted to his plotting board, he suddenly looked up, glanced out his window, and saw randomly moving stars. Thinking Schirra had lost control, he barked out that they had blown it. Quickly, however, Schirra reassured him that the 'stars' were not stars, but merely John Glenn's fireflies: frozen particles reflecting sunlight. The two men laughed. (Later in the mission, using fast ASA 4000 film in a Hasselblad,

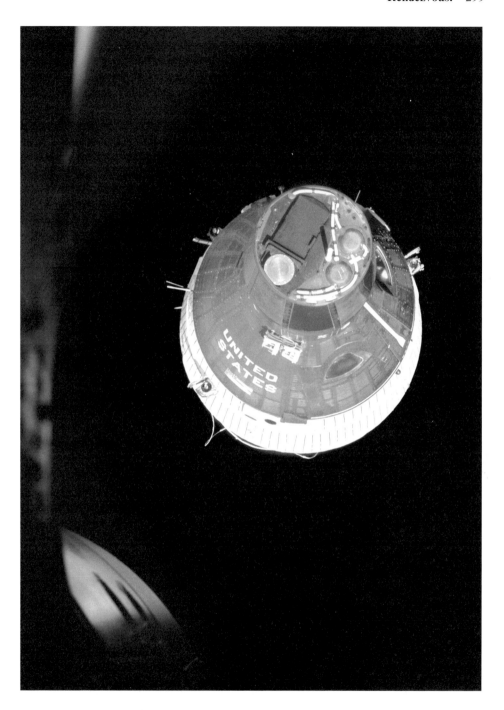

Gemini VII in space, as seen by Schirra and Stafford.

they engaged in astronomical photography for one of the principal investigators, Jocelyn Gill. However, not all of the images were astronomical; some were urine dumps and when they returned to Earth, Gill looked at one beautiful constellation of 'stars' and asked them what it was. Without missing a beat, Schirra looked at the shimmering cloud of just-dumped urine droplets and deadpanned: "Jocelyn, that's the constellation Urion!")

From their vantage point in the 'passive' spacecraft, Borman and Lovell had expressed fascination at the thruster bursts and spurts emerging from Gemini VI-A. At one stage, they were startled to see a tongue-like jet some 12 m in length! Both crews would report cords and stringers 3–5 m long streaming and flapping behind their respective spacecraft; these turned out to be the remains of covers from the shaped explosives which severed the Geminis from the final stage of their respective Titan IIs. The rendezvous had cost Schirra barely 51 kg of fuel and only 38 per cent had been expended in total from Gemini VI-A's tanks, leaving him plenty in reserve to fly a tour of inspection of, and station-keep with, Gemini VII for the next five hours and three orbits. At one stage, Schirra manoeuvred as close as 30 cm, allowing he and Stafford to hold up a 'Beat Army' card in their window to torment Borman, a West Point graduate. In response, Borman held up a 'Beat Navy' card. Gemini VI-A moved to the rear of its sister craft to examine the stringers, then came nose-to-nose, and so stable were both Geminis that, for a time, neither command pilot had to even touch his controls.

During manoeuvres, Schirra found the spacecraft responded crisply, allowing him to make velocity inputs as low as 3 cm per second, which he and Stafford concluded were fine enough to execute a docking with an Agena-D or any other target. "I took my turns flying," recounted Stafford, "having convinced the Gemini programme managers to add a second manoeuvre controller to the pilot's side." As the crews' workday drew to a close, Schirra flipped Gemini VI-A into a blunt-end-forward orientation and pulsed the OAMS thrusters to separate. After a meal and sleep, Schirra awoke to a stuffy head and runny nose, which made him glad that the mission was flexible and, assuming all of the tasks were completed, had the option to return to Earth after 24 hours. Moreover, Gemini VII's fuel cells needed the attention of mission controllers and would benefit from having the additional tracking burden of Gemini VI-A out of the way.

But not before a final 'gotcha' from Schirra and Stafford. It was nine days before Christmas, after all …

As the two spacecraft went their separate ways, the MOCR controllers and Borman and Lovell were initially alarmed to hear Stafford report that he saw "an object, looks like a satellite, going from north to south, probably in polar orbit … Looks like he might be going to re-enter soon. Stand by one. You just might let me try to pick up that thing." Then, over the communications circuit, came the sound of the Gemini VI-A astronauts playing 'Jingle Bells'. The 'object', it seemed, was the familiar, jolly, red-suited, white-bearded old man himself, making his annual 're-entry' to deliver his payload of presents to terrestrial children. "You're too much!" radioed Capcom Elliot See.

In fact, Mickey Kapp, producer of Bill Dana's 'José Jiménez in Orbit' album, had provided Schirra with a small, four-hole Hohner harmonica just days before launch. Schirra had secured it in one of the pockets of his space suit with dental floss. "I could play eight notes," he wrote, "enough for 'Jingle Bells'. It may not have been a virtuoso performance, but it earned me a card in the musicians' union of Orlando." (Schirra would also receive a tiny gold harmonica from the Italian National Union of Mouth Organists and Harmonica Musicians.) Not to be outdone, Francis Slaughter of the Cape's Flight Crew Operations Office, had fitted small bells to the boots of Stafford's suit ... supposedly as a joke, but little realising they would provide backing rhythm for Schirra's Christmas soiree. Today, the tiny harmonica and Stafford's bells are enshrined in the Smithsonian.

A little more than a day after launch, Schirra placed his spacecraft into an inverted, 'heads-down', attitude, to provide better observation of Earth's horizon. At an altitude of 100 km, to ensure that Gemini VI-A did not overshoot its splashdown point, he set its banking angle at 55 degrees left and held it steady until the computer took control at 85 km above the ground. "We were going backwards, heads-down," wrote Stafford, "so I had a great view of the horizon and the cloud-covered Gulf of Mexico, and a clear sense that we were really moving fast." The astronauts duly switched off the computer at 24 km, deployed the drogue parachute at 14 km and the main canopy blossomed out at 3.2 km. Impact with the Atlantic, in the first successful demonstration of a controlled re-entry, took place at 10:28:50 am on 16 December, at 23 degrees 35 minutes North latitude and 67 degrees 50 minutes West longitude, merely 13 km from its pre-planned splashdown point. It was fortunate that it was so successful, for Gemini VI-A's descent was in full view of live television beamed from the recovery ship Wasp, transmitted via the Early Bird communications satellite. An hour later, displaying a thumbs-up of a job well done, Schirra, then Stafford, strode down the Wasp's red carpet to the strains of a band playing 'Anchors Aweigh'.

For Borman and Lovell, almost three days awaited them before their own splashdown. They started by removing their suits. The novelty of being in space had now worn very thin and, years later, Borman would describe the time after Gemini VI-A's departure as "a tough three days" in which the two bearded, exhausted and uncomfortable men "simply existed ... in a very, very cramped space". At the suggestion of Cooper and Conrad, they had taken novels. Borman spent some time reading Mark Twain's 'Roughing It', which proved apt, and Lovell dived into part of 'Drums Along The Mohawk' by Walter D. Edmonds, a text about the American pioneers.

Even Mike Collins, who served as Lovell's backup on Gemini VII, wondered how they managed to endure it. In his autobiography, Collins admitted the Gemini was so small that, on the ground, he could not sit in the simulator for more than three hours at a time and even with the relative freedom of weightlessness – which allowed Borman and Lovell to float, restore circulation and avoid bedsores – it was an uncomfortable existence. "The cockpit was tiny, the two windows were tiny, the pressure suits were big and bulky and there were a million items of loose equipment which constantly had to be stowed and restowed," he wrote, adding that "no one who has never seen a Gemini can fully appreciate what it's like being locked inside one for two weeks."

As the mission wore on, the Post Office passed up a request for them to mail their Christmas cards and parcels early. Lovell complained that he had "a stack of stuff up here", to which the capcom replied that he should have sent his presents home with Gemini VI-A. With their homecoming looming, they were reminded by Chuck Berry to elevate their feet and pump their legs, to which Borman announced that they were eager to get out of Gemini VII as soon as possible.

Retrofire, described graphically by Lovell, commenced as they flew over the Canton Islands on their 206th orbit. "Retrofire has a unique apprehension in the fact that both of us are aviators and we understand the apprehension in flying," he said. "If you have an accident in an airplane, something's going to happen: you hit something or it blows up. Now, in liftoff and re-entry, a space vehicle is like an airplane. Something's happening. But if the rockets fail to retro, if they fail to go off, nothing's going to happen. You just sit up there and that's it. Nothing happens at all. That's the unique type of apprehension, because you know that you've gotten rid of the adaptor, you know that you're going to have 24 hours of oxygen, ten hours of batteries and very little water. So you play all sorts of tricks to get those retros to fire."

Fortunately, the four retrorockets fired in tandem and to perfection. As their descent commenced, Capcom Elliot See told them to fly a 35-degree left bank until Gemini VII's computer guidance assumed control. "You have no control over how close you're going to get to the target," Lovell recalled later. "Your only control is how good that computer is doing or how good your [centre of gravity] was when you set up the computer and the retrofire time." Borman rolled Gemini VII into a heads-down orientation, allowing him to use the horizon as an attitude guide, but could see nothing and was forced to rely upon his instruments and Lovell's called-out adjustments. After the flight, Borman would endorse the need for two pilots to fly a Gemini, since there was no practical way to follow the instruments and monitor the horizon at the same time.

The dynamic loads after 330 hours in weightlessness came, they recounted, "like a ton", even though G forces reached a peak of 3.9, barely half as much as a typical Mercury re-entry. Drogue parachute deployment jolted the two men, rocking the spacecraft by 20 degrees to either side, after which the main canopy opened. When Gemini VII hit the ocean at 9:05 am, Borman, unable to see any recovery helicopters, felt that he had lost a bet with Wally Schirra – that he could land the closest to his planned impact point. In reality, he had landed just 11.8 km off-target. In view of their lengthy stay in space, the two men were surprisingly fit, although Borman felt a little dizzy and both walked with a slight stoop on the deck of the recovery ship, Wasp. "The most miraculous thing," reported a jubilant Chuck Berry, "was when they could get out of the spacecraft and not flop on their faces; and they could go up into the helicopter and get out on the carrier deck and walk pretty well." They were, added Berry, in better physical condition than Cooper and Conrad had been. Lovell's cardiovascular cuff revealed that less blood pooled in his legs than Borman and both maintained their total blood volumes.

As Chris Kraft and his flight control team fired up more cigars on the afternoon of 18 December 1965, the prospects for a lunar landing before the end of the decade had grown steadily brighter. Alexei Leonov's triumphant spacewalk in March had been followed by the United States' decisive response: no fewer than five Gemini missions – ten men blasted aloft, in total – whose endurance records had shown that astronauts could survive two-week flights to the Moon and back with few physical or psychological problems. They could rendezvous and survive the rigours of working outside their spacecraft in pressurised suits ... or so it seemed. The next year, 1966, would see five more flights, closing out the programme in advance of the first Apollo mission, and all were destined to push the envelope still further by physically docking Geminis onto Agena-D targets and having astronauts spacewalk from craft to craft to install and remove experiments.

It would be an ultimately successful, though risky, year. Indeed, Deke Slayton would describe it as "NASA's best". However, it would begin inauspiciously. In its third month, aboard Gemini VIII, the man who would someday be first to set foot on the Moon almost became one of the first to die in space. In its sixth month, the man who would one day be the last Apollo astronaut to set foot on the Moon would come close to losing his own life as the dangers of EVA became terrifyingly clear. Before that, on the gloomy, overcast morning of 28 February 1966, fire and death would rain down over St Louis, Missouri.

5

Onward and Upward

TRAGEDY

The failure of the first Agena-D target vehicle on 25 October 1965 left the Gemini effort in a quandary. Its role in the following year's rendezvous and docking missions was crucial, yet its reliability had been brought into serious doubt. Efforts to resolve its woes spanned four months and the first few weeks after the failure were spent identifying the cause. Investigators quickly focused on the Agena's engine. By November, a 'hard-start' hypothesis – in which fuel was injected into the combustion chamber ahead of the oxidiser, effectively causing it to 'backfire' – had been generally accepted by the engineers. However, this problem was itself deeply rooted in NASA's original specification for the Agena-D to be able to restart itself up to five times during a single mission.

In order to achieve multiple restarts, oxidiser began flowing first, then a pressure switch restricted fuel flow until a given amount of oxidiser had reached the combustion chamber. This had the advantage of enhancing the engine's startup characteristics, but was also extremely wasteful, with numerous instances of oxidiser leaks. As a result, oxidiser was often expended before fuel. In an effort to rectify this wastage problem, engine subcontractor Bell Aerosystems removed the pressure switch, allowing fuel to enter the combustion chamber ahead of the oxidiser. However, the investigative panel for the Agena-D speculated that fuel in the chamber – perhaps in considerable quantities – might have caused the engine to backfire when the oxidiser arrived, resulting in an explosion.

Meanwhile, in mid-November, a two-day symposium on hypergolic rocket ignition at altitude convened to discuss the failure and identify corrective actions. The data from the accident implied that oscillations and mechanical damage had been induced after engine ignition and temperature drops pointed towards a fuel spillage of some sort. When the Agena's electrical circuitry failed, the engine stopped, but a valve responsible for managing fuel tank pressures remained open. As the fuel stopped flowing, pressures built up inside the tanks, which ruptured and destroyed the vehicle. Although the symposium was unsatisfied that this represented the definitive cause of the accident, it had little other data upon which to base its judgements. One of its recommendations was that future Agena-D engines should be modified and tested at simulated altitudes closer to those at which it would operate: 76 km. (Previously, it had only been tested at simulated altitudes of 34 km.)

In response, Lockheed formed a Project Surefire Engine Development Task Force to carry out the modification programme, which continued to arouse debate until the week before Gemini VIII's scheduled March 1966 launch. By that time, the crew assigned to a subsequent mission, Gemini IX, had lost their lives in a blaze which nearly claimed their spacecraft, too. Civilian Elliot See, a deeply religious former General Electric pilot who had performed engine testing before coming to NASA, was paired with Air Force Major Charlie Bassett – "a terrific stick-and-rudder man," according to Gene Cernan – to fly a three-day mission and practice rendezvous, docking and spacewalking.

At 7:35 am on 28 February 1966, the men and their backups, Cernan and Tom Stafford, took off from Ellington Field near Houston in a pair of T-38s and flew in tandem to McDonnell's St Louis plant. "Prime and backup were not allowed to fly in the same airplane," wrote Mike Collins, "lest a crash wipe out the entire capability in one specialty." In other words, in Gemini IX's case, See could fly with anyone but Stafford and Bassett with anyone but Cernan. Their schedule called for them to spend ten days in St Louis, practicing rendezvous procedures in the simulator, as well as viewing their just-completed Gemini IX spacecraft. Weather conditions in St Louis that morning were bad, with low cloud, poor visibility, rain and snow flurries, and at 8:48 am Lambert Field airport – located 150 m from the McDonnell plant – prepared to support two instrument-guided landings. It was standard practice to rely upon instruments under such appalling weather conditions.

The two sleek T-38s – tailnumbered 'NASA 901' (See and Bassett) and 'NASA 907' (Stafford and Cernan) – descended through the murky clouds at 8:55 am, directly above the centreline of the south-west runway, far too low and flying too fast to land. Stafford, who had been concentrating on remaining in position on See's right wing, decided to ascend and attempt a flyaround. However, See inexplicably announced his intention to enter a tight turn and make another approach. Normally careful, considered and judicious, it has been speculated over the years that he wanted to beat the backup crew to the runway. If this was indeed the case, it surely demonstrated that See had the competitive nature typical of an astronaut. Sadly, good luck was not on his side that day.

Stafford was surprised as See's T-38 disappeared from view, exclaiming to Cernan: "Goddamn! Where the hell's he going?" Breaking through the clouds, heading directly for the corrugated-iron Building 101, which contained Gemini IX, See realised he could not land successfully. He lit his afterburners, broke hard right and pulled back on the stick, but at 8:58 am the T-38 grazed the top of the building, gashed open the roof – losing a wing as it did so – and cartwheeled into a nearby parking lot, whereupon it exploded.

Inside Building 101, McDonnell foreman Domien Meert watched aghast from his desk in the subassembly room as a sheet of flame flared across the now-exposed ceiling. Workers dived for cover under benches as honeycomb shards from the T-38's

The original Gemini IX crew, Elliot See (left) and Charlie Bassett. Scheduled for a three-day mission in May 1966, they were both killed in St Louis just 11 weeks before launch.

shattered wing hit the Gemini X spacecraft, still under construction. Elliot See, who had been thrown from the fuselage, was found dead in the parking lot, his parachute half-opened, while Charlie Bassett – one of the most promising of the third group of astronauts, selected in October 1963 – had been decapitated. His severed head was later retrieved from the rafters of the very building in which his spacecraft was being readied for flight.

Stafford and Cernan were oblivious to the tragedy. They were simply ignored by air traffic controllers, left to their own devices and, wrote Cernan, "annoyed that the tower was being so vague in its communications". Eventually, with a near-empty fuel gauge pushing him close to declaring an emergency, Stafford set NASA 907 down on the runway without incident and turned onto the taxiway. He was puzzled by an odd question from the tower: "Who was in NASA 901?" When Stafford told them that the Gemini IX prime crew was aboard the other T-38, he was advised that McDonnell Aircraft had "a message" for him. Minutes later, after opening his canopy, Stafford was told by James McDonnell himself that See and Bassett had both been killed.

The men's remains, wrote Cernan, were unrecognisable and their identification was not helped by the fact that all four astronauts had placed their NASA badges and personal papers in a baggage pod aboard See and Bassett's T-38 before leaving Houston. Only by checking with the men who were still alive was it possible to determine which ones had died.

It seemed impossible to imagine 28 February 1966 as a day for miracles, but it was to say the very least fortuitous that no deaths were sustained on the ground. If See had been a little lower at the moment of impact, he would have hit Building 101 in a head-on collision, destroying Gemini IX and potentially killing hundreds of McDonnell staff who were skilled in the art of manufacturing spacecraft The United States' plan to reach the lunar surface before the end of the decade would have evaporated.

Later that afternoon, James McDonnell climbed onto the roof to survey the damage. Next day, his company's 37,000 employees returned to work and, as planned, on 2 March, Gemini IX was loaded aboard a C-124 transport aircraft and flown to Cape Kennedy. At around the same time, the entire astronaut corps gathered at Arlington National Cemetery to watch as the remains of 38-year-old Elliot McKay See Jr and 34-year-old Charles Arthur Bassett II were laid to rest.

NASA immediately created a seven-man investigative board, chaired by Al Shepard, which examined every parameter and detail relating to See and Bassett's tragic final flight. The T-38, it was found, was in perfect operational order and the men's physical and psychological state was fine. Their flying abilities, on paper at least, were exemplary and both had renewed their instrument flying certificates within the last six months. The appalling weather was certainly a contributory factor in the disaster, but the board's final conclusion that pilot error was to blame did not surprise Deke Slayton.

"Of all the guys in the second group of astronauts," he wrote, "Elliot was the only one I had any doubts about. I had flown with him and the conclusion was just that he wasn't aggressive enough. Too old-womanish ... he flew too slow – a fatal problem in a plane like the T-38, which will stall easily if you get below about 270 knots." Slayton had named See as Neil Armstrong's pilot on the Gemini V backup crew, but had not felt confident to keep the pairing together for Gemini VIII, particularly since the latter would feature a lengthy EVA. "He wasn't in the best physical shape," Slayton wrote, adding that "I didn't think he was up to handling an EVA. I made him commander of Gemini IX and teamed him up with Charlie Bassett – who was strong enough to carry the two of them."

Neil Armstrong, who worked with See on Gemini V, has said little about his qualifications as an astronaut, but certainly found it difficult to blame his comrade for his own death. "It's easy to say ... what he should have done was gone back up through the clouds and made another approach," Armstrong told biographer James Hansen. "There might have been other considerations that we're not even aware of. I would not begin to say that his death proves the first thing about his qualification as an astronaut."

Regardless, years later, Slayton would admit that he had allowed himself to "get sentimental" about giving See a mission and that, ultimately, it was a bad call. Within hours of the tragedy, he had telephoned Tom Stafford to tell him that he and Cernan were now the Gemini IX prime crew. Three weeks later, Jim Lovell and Buzz Aldrin were named as their backups. It is interesting that the deaths of See and Bassett proved pivotal in deciding the identities and shaping the futures of the men who would someday be the first to set foot upon the Moon. By backing up Gemini IX, for example, Aldrin eventually wound up as pilot on the very last Gemini mission. Slayton admitted in his autobiography that without this twist of fate, it would have been "very unlikely" that Aldrin would have gone on to join the first Apollo lunar landing crew. Indeed, Neil Armstrong and Dave Scott, set to fly Gemini VIII on 16 March 1966, would become the only Gemini crew who would both someday walk on the lunar surface.

For Armstrong, his astronaut career would be the culmination of a lifetime of aviation which had carried him to the edge of the atmosphere in rocket-propelled aircraft, into the world's first spaceflying corps and almost into suborbital space aboard a revolutionary machine called 'Dyna-Soar'.

DEMISE OF THE DYNA-SOAR

It is July 1966. At Cape Canaveral Air Force Station in Florida, a 41-year-old test pilot named Jim Wood is moments away from becoming the United States' 17th man in space. Alone, pressure-suited and tightly strapped into the tiny cockpit of a stubby winged ship called 'Dyna-Soar', he will shortly be boosted by a Titan III-C rocket onto a suborbital trajectory to evaluate the world's first reusable manned spacecraft. It was a precocious forerunner of the Space Shuttle and, indeed, had it flown, many observers believe that it could have revolutionised – perhaps even 'routinised' – today's space travel.

Wood's mission never happened. Nor did Dyna-Soar itself reach fruition, although at the time of its cancellation in December 1963 it was supposedly eight months away from performing a series of airborne drop-tests from a modified B-52 bomber. Much criticism has been levelled at then-Secretary of Defense Robert McNamara, accusing him of poor judgement in killing a project with so much promise, so close to completion and whose contractors had already spent nearly half of its $530 million development budget. To be fair, Dyna-Soar faced immense problems of its own, the most important of which was its ill-defined purpose.

The spacecraft was viewed from two different perspectives during its genesis in the late Fifties: as a research vehicle to explore hypersonic flight regimes or, as the Air Force preferred, as a fully functional military glider capable of delivering live warheads with precise, pilot-guided accuracy onto targets anywhere on Earth. Ambitious plans were even afoot for the inspection and maybe destruction of enemy satellites in orbit, as well as the carriage of reconnaissance cameras, side-looking radar and electronic intelligence sensors. Assuming a manned flight sometime in the summer of 1966, it was hoped that Dyna-Soar would evolve into this advanced weapons system by the mid-Seventies. This would offer military strategists a route around the problem that 'conventional' ballistic missiles might no longer be able to strike hardened targets with sufficient accuracy. Moreover, the 'boost-glide' flight profile of Dyna-Soar – able to cover velocities between Mach 5 and 25 – was perceived as a better alternative to using complex, air-breathing turbojet or ramjet engines, which were difficult to develop and could only operate a lower speeds. Indeed, according to studies conducted by the Rand Corporation, vehicles flying below Mach 9 might be rendered vulnerable to Soviet air defences by 1965.

In the most paranoid days of the Cold War, Dyna-Soar thus provided the United States with a seemingly invincible means of attacking and snooping on enemy targets from any direction and, when flying at low altitudes, gave barely a three-minute warning of its arrival. Additionally, it could sweep across Soviet territory at altitudes in the range of 40 to 80 km, providing a much better imaging resolution than was possible with the best spy satellites and its data could be in the hands of Pentagon officials within hours.

Size-wise, this astonishing machine was a third as large as today's Shuttle: 10.6 m long, with a wingspan of 6 m. Powered by a Martin-built 'trans-stage' engine, capable of 32,660 kg of thrust, it would have ridden into space atop the Titan III-C. This choice of launch vehicle changed significantly as Dyna-Soar's own purpose fluctuated. "It was originally scheduled to be launched on the Titan I," said Neil Armstrong, one of its original pilots, "[then] when the Titan III was introduced, with additional [solid] rocket engines strapped on the side, it [became] an orbital vehicle." Armstrong left the project in mid-1962 to join NASA's astronaut corps and later revealed that Dyna-Soar's main aim was for hypersonic research, hence its acronym: the 'dynamic soarer'. Its 72.48-degree delta wings were flat-bottomed and its aft fuselage was 'ramped' to give directional stability at transonic speeds. This would have provided a sufficient hypersonic lift-to-drag ratio to permit a cross-range capability of around 3,200 km. In other words, a diverted landing from Edwards Air Force Base in California meant that it could conceivably touch down anywhere in the continental United States, Japan or even Ecuador.

In fact, thanks to a unique set of wire-brush 'skids', it could even land on compacted-earth runways little more than 1.6 km long. Typically, it would have been launched from Cape Canaveral by the Titan III-C, then remained affixed to the rocket's third and final stage. This would have acted as a restartable 'trans-stage', capable not only of inserting Dyna-Soar into a 145 km orbit, but also adjusting its altitude and inclination. The trans-stage could boost its velocity, thus frustrating ground-based efforts to predict its overflight path during bombing or spying missions.

Emergency aborts during a Dyna-Soar launch, though, did not fill Armstrong with confidence. "There was a question [of] what kind of abort technique would be practical to try to use in case there was a problem with the Titan," he recalled years later. "It was determined, rather than a 'puller' rocket, [we had a] 'pusher' rocket to push the spacecraft up to flying speed from which it could make a landing, but it wasn't known at that time what might be practical, how much thrust would be needed and how much performance would be needed. We had the F-5D aircraft, which I determined could be configured to have a similar glide angle for the Dyna-Soar for similar flight conditions and devise a way of flying the aircraft to the point at which the pusher escape rocket would burn out, so you would start with the identical flight conditions that the Dyna-Soar would find itself after a rocket abort from the launch pad. Then, you only had to work out a way to find your way to the runway and make a successful landing. I worked on that project for a time and found a technique that would allow us to launch from the pad at Cape Canaveral and make a landing on the skid strip. We practiced that and I believe that [NASA test pilots] Bill Dana and Milt Thompson both continued after I transferred from Edwards to Houston. There was a NASA report written about the technique. It was a practical method. I wouldn't like to have to really do it in a real Dyna-Soar!"

During its development, the Air Force also hoped to conduct synergistic exercises in which the spacecraft could dip into the upper atmosphere, employ its aerodynamic manoeuverability to change its inclination relative to the equator and refire the trans-stage to boost itself back into orbit. This tricky task was provisionally pencilled-in for the fifth manned test, sometime in the spring of 1967, after which pilots would have begun evaluating its precision-landing capabilities. In the event of trans-stage problems, a solid-fuelled abort motor, derived from the Minuteman missile, was attached to Dyna-Soar and would have separated the pair, performed an emergency retrofire and initiated re-entry. On the other hand, assuming a successful mission had been completed, the trans-stage would have been jettisoned over the Indian Ocean and the spacecraft would commence its long glide through the atmosphere to touch down at Edwards. Later missions, intended to complete two or even three Earth orbits, were expected to fly at higher altitudes of around 180 km. During re-entry, the pilots would test Dyna-Soar's controllability at various pitch angles during a range of hypersonic and thermal flight regimes.

It was not, however, completely controllable by its pilot throughout the entire speed range and a fly-by-wire augmentation system was provided to run in four automatic modes. A side-arm controller offered him the ability to perform pitch and roll inputs and, through conventional pedals, to execute yaw manoeuvres; the Titan III-C would even have been able to be flown manually during part of its initial boost in orbit! Throughout re-entry, the guidance computer – capable of storing up to ten airfield locations – would have provided continuous updates on Dyna-Soar's display to advise on issues such as angle-of-attack, banking angles and structural limitations.

Physically, the spacecraft was based on a Rene 41 steel truss, which compensated for thermal expansion within the heated airframe during re-entry. Dyna-Soar was roughly divided into four parts: a pilot's cockpit, pressurised central section and two unpressurised equipment bays. Each internal compartment was encased in a 'water wall' to offer passive cooling during re-entry, allowing their pressure shells to be constructed from conventional aluminium. Additional cooling within each compartment brought temperatures down still further to around 46°C.

In order to withstand the fiery plunge into the atmosphere, Dyna-Soar's belly and wing leading edges were coated with molybdenum and its nosecap tipped with zirconium. Theoretical predictions expected the wing edges to reach temperatures of 1,550°C, the nosecap around 2,000°C and the belly some 1,340°C during the most extreme re-entry profiles. During all flight phases, the pilot had clear views through two side windows, but the three-piece forward windshield was covered during ascent, orbital operations and most of re-entry. Interestingly, the cover – which guarded against thermal extremes – was not scheduled to be blown off until Dyna-Soar reached a speed of Mach 6, just in time for landing. Tests conducted by Neil Armstrong in the modified Stingray fighter, however, showed that, in the dire eventuality that the cover failed to jettison properly, landings could be safely performed using only the two side windows for visibility. Clearly, this was far from ideal and stands testament to the skills of the men chosen to fly Dyna-Soar: Wood, Armstrong, William 'Pete' Knight, Milt Thompson, Al Crews, Hank Gordon and Russ Rogers.

The cockpit in which the pilot would have sat provided all the instrumentation and life-support gear to fly the vehicle, together with an ejection seat which could only be used at subsonic speeds. Behind the cockpit, the central section – pressurised with 100 per cent nitrogen – would have carried a 450 kg instrumentation package, fitted with data recorders for more than 750 temperature, pressure, loads, systems performance, pilot biometrics and heat flux sensors. Finally, at the rear of Dyna-Soar were the equipment bays, containing liquid hydrogen and nitrogen supplies, hydrogen peroxide tanks and power system controls.

The leading contenders for building the spacecraft in the late Fifties were Bell and Boeing, both of which offered the capability to cover the entire requirement range from low-Mach speeds to orbital velocity using a single vehicle. Ultimately, despite Bell's expertise in designing winged spacecraft of this type, Boeing was chosen as Dyna-Soar's prime contractor in June 1959. Original plans called for three 'waves' of operations: a series of suborbital flight tests, followed by orbital and eventually operational weapons-delivery missions. By the autumn of 1962, critical design reviews of Dyna-Soar's subsystems had been completed and significant break-throughs achieved in high-temperature materials and fabrication of components for the 'real' airframe. Publicly, it seemed worth the wait. When a mockup of the spacecraft was displayed at Las Vegas in September of that year, it quickly grabbed the imagination of America. Writing in Reader's Digest after the event, John Hubbel described it as looking "like a cross between a porpoise and a manta ray" and enthused that Dyna-Soar was one of the most important aviation triumphs since the Wright Brothers' first flight.

Many others agreed with him. The Mercury capsules were blunt and uninspiring in comparison to this sleek, futuristic spaceplane. Despite the public adoration of Dyna-Soar, however, the project was in serious trouble and barely months away from cancellation. Robert McNamara had already expressed serious concerns that it lacked direction or specific purpose – the Department of Defense regarded it as a hypersonic research vehicle, the Air Force as a strategic bomber – and very little attention had been paid to precisely what missions it would undertake. McNamara made his opinions clear during a series of reviews of the project during the course of 1963, before ultimately cancelling it at year's end in favour of a military space station called the Manned Orbiting Laboratory. Ultimately that, too, wasted millions of dollars and never bore fruit.

More than four decades later, Dyna-Soar is recognised as an ambitious, far-sighted endeavour, literally at the cutting-edge of the technology of its day. Its legacy remains visible in many elements of the Shuttle's design and capabilities – from the concept of the 'rocket-glider' to the payload bay, from landing on pre-determined runways to its reusability – and its extensive wind-tunnel testing provided valuable engineering data for later projects. In the words of a recent NASA study of the X-planes, of which Dyna-Soar was one, "very few vehicles have contributed more to the science of high-speed flight – especially vehicles that were never built!"

QUIET CIVILIAN

A year before Robert McNamara finally axed the Dyna-Soar, Neil Alden Armstrong's test-piloting career took a different turn ... one that would someday guide him to the lunar surface. Armstrong's life was one of movement. Born in Wapakoneta, Ohio, of Scots-Irish and German descent, on 5 August 1930, his father was a government worker and the family moved around the state for many years: from Warren to Jefferson to Moulton to St Mary's, finally settling permanently in Wapakoneta in 1944. By this time, Armstrong was an active member of his local Boy Scout group and his mind was filled with dreams of flying. "I began to focus on aviation probably at age eight or nine," he told NASA's oral history project in 2001, "and [was] inspired by what I'd read and seen. My intention was to be an aircraft designer. I later went into piloting because I thought a good designer ought to know the operational aspects of an airplane."

When Armstrong enrolled at Purdue University in 1947 to begin an aeronautical engineering degree, he became only the second member of the family to undergo higher study and famously had learned to fly before he could drive. (Years later, he recalled first flying solo aged just 16. Alas, his early logbook entries were lost in a fire at his Houston home in 1964.) Under the provisions of the Holloway Plan, Armstrong committed himself to four years of paid education in return for three years of naval service and a final two years at university. He was summoned to active military duty in January 1949, reporting to Naval Air Station Pensacola in Florida for flight training. Over the next 18 months, Armstrong qualified to land aboard the aircraft carriers Cabot and Wright. A few days after his 20th birthday, he was officially classified as a fully-fledged naval aviator.

His initial assignments were to Naval Air Station San Diego, then to Fighter Squadron 51, during which time he made his first flight in an F-9F Panther – "a very solid airplane" – and later landed his first jet on an aircraft carrier. By late summer in 1951, Armstrong had been detailed to the Korean theatre and would fly 78 missions in total. His first taste of action came only days after arrival, whilst serving as an escort for a photographic reconnaissance aircraft over Songjin. Shortly thereafter, whilst making a low-altitude bombing run, his Panther encountered heavy gunfire and snagged an anti-aircraft cable. "If you're going fast," he said later, "a cable will make a very good knife." Armstrong somehow managed to nurse his crippled jet back over friendly territory, but the damage was of such severity (a sheared-off wing and a lost aileron) that he could not make a safe landing and had to eject. Instead of a water rescue, high winds forced his ejection seat over land, close to Pohang Airport, and he was picked up by a jeep driven by an old flight school roommate, Goodell Warren.

By the time Armstrong left naval service in August 1952, he had been awarded the Air Medal, a Gold Star and the Korean Service Medal and Engagement Star. For the next eight years, however, he remained a junior lieutenant in the Naval Reserve. After Korea and his departure from the regular Navy, Armstrong completed his degree at Purdue in 1955, gaining coveted admission to the Phi Delta Theta and Kappa Kappa Psi fraternities and meeting his future wife, home economics student Janet Shearon. They were married in January 1956 and their union would endure for almost four decades, producing three children, one of whom – a daughter, Karen – tragically died in her infancy.

Armstrong's aviation career, meanwhile, expanded into experimental piloting when he joined NACA, the forerunner of NASA, and was initially based at the Lewis Flight Propulsion Laboratory in Cleveland, Ohio. Whilst there, he participated in the evaluation of new anti-icing aircraft systems and high-Mach-number heat-transfer measurements, before moving to the High Speed Flight Station at Edwards Air Force Base in California to fly chase on drops of experimental aircraft. There, aboard the F-100 Super Sabre, he flew supersonically for the first time. On one occasion, flying with Stan Butchart in a B-29 Stratofortress, Armstrong was directed to airdrop a Douglas-built Skyrocket supersonic research vehicle. Upon reaching altitude, however, one of the B-29's four engines shut down and its propellor began windmilling in the airstream. Immediately after airdropping the Skyrocket, the propellor disintegrated, its debris effectively disabling two more engines. Butchart and Armstrong were forced to land the behemoth B-29 using the sole remaining engine.

His first flight in a rocket-propelled aircraft came in August 1957 aboard the Bell X-1B, reaching 18.3 km, and three years later he completed the first of seven missions aboard North American's famous X-15, to the very edge of space. On one of these flights, in April 1962, just a few months before joining NASA's astronaut corps, he reached an altitude of 63 km. However, during descent, he held up the aircraft's nose for too long and the X-15 literally 'bounced' off the atmosphere and overshot the landing site by some 70 km, but he returned and achieved a safe touchdown. Although he was not one of the handful of X-15 pilots who actually reached space, exceeding 80 km altitude, Armstrong's abilities in the rocket aircraft have been widely praised. The late NASA research flier Milt Thompson called him "the most technically capable" X-15 pilot.

In November 1960, by now flying for NASA as a civilian research pilot, Armstrong was chosen for the Dyna-Soar effort, ultimately leaving the project in the summer of 1962 as the selection process for the second group of astronauts got underway. At around the same time, he last flew the X-15, achieving a peak velocity of Mach 5.74. When his name was announced by NASA in September, he became one of only two civilian astronauts. Although Deke Slayton later wrote that nobody pressured him to hire civilians, fellow selectee Jim Lovell felt that Armstrong's extensive flying history within NACA and NASA rendered him a likely choice to make the final cut. In fact, Armstrong's application arrived a week after the 1 June 1962 deadline, but, according to Flight Crew Operations assistant director Dick Day, "he was so far and away the best qualified ... [that] we wanted him in".

After admission into the New Nine, Armstrong came to be regarded as by far the quietest and most thoughtful. "When he said something," recalled Frank Borman, "it was worth listening to." His Apollo 11 crewmates Mike Collins and Buzz Aldrin would both characterise his nature as "reserved" and Dave Scott, the man who would fly with him aboard Gemini VIII, described him as "cool, calm and energised", who never operated in a frantic manner, but who could identify and resolve problems quickly, efficiently and smartly. All of these qualities would prove vital on his first spaceflight, when he would come close to losing his life.

FIRST AMONG EQUALS

In a strange kind of way, Neil Armstrong's work during his first couple of years as an astronaut had helped pave the way for the selection of Dave Scott, his 33-year-old colleague aboard Gemini VIII. Deke Slayton had given Armstrong responsibility for mission operations and training and entrusted him to devise a system which would identify how many astronauts would be needed at any given time. It was this schematic, wrote James Hansen in his biography of Armstrong, that "allowed Slayton to determine when additional astronauts needed to be brought in ... culminating in Houston's announcement in June 1963 that NASA was looking for a new class of astronauts".

Slayton's decision, he recounted, "was based on planning documents that were starting to arrive from [NASA] Headquarters". The 1962 astronaut intake, together with the remaining members of the Mercury Seven who were still on flight status, barely provided enough seats for Project Gemini. Early predictions for Apollo envisaged at least 12 missions in Earth and lunar orbit before a landing on the Moon could be attempted. That required Slayton to accept "a minimum of ten new astronauts in 1963 and ... as many as twenty". Fourteen newbies duly arrived in Houston in October of that year.

Although the selection criteria changed slightly – a more rigid upper age limit of 34 and the elimination of the need for the new astronauts to be test pilots being the main differences – David Randolph Scott still fitted perfectly into the classic mould of a spacefarer. Tall, athletic and with a middle name honouring the Air Force base on which he was born, it has often been said that it was no accident that Scott was the first of the 1963 group of astronauts to fly into space ... and eventually would become the first of his class to command a mission. "In some circles," noted Andy Chaikin in his book about Project Apollo, "there was a joke that if NASA ever came out with an astronaut recruiting poster, Dave Scott should be on it." Further, Chaikin added, even astronauts who did not get on with Scott placed him at the top.

Despite the near-disaster which befell it, Gemini VIII marked the only mission in which the entire crew would one day set foot on the Moon; Armstrong as the first man to tread its dusty surface at the Sea of Tranquility and Scott in command of Apollo 15 in July 1971.

Scott was born on Randolph Air Force Base in San Antonio, Texas, on 6 June 1932, the son of an Air Force officer and progeny of a strict, frugal military family who instilled in him the virtues of personal discipline and devotion to setting and achieving ambitious life goals. In his autobiography, jointly co-authored with cosmonaut Alexei Leonov, Scott recalled watching Jenny biplanes soaring over Randolph as a three-year-old boy and was fascinated that aboard one of them was his father, Army Air Corps flier Tom Scott. From that tender age, the young Scott set his sights on someday becoming a pilot.

With a military father, the family moved many times during his childhood, from Texas to Indiana, abroad to the Philippines and, in 1939, back to the United States. Scott's father was posted overseas after Pearl Harbour to support the war effort and the young boy developed an interest in model aircraft. Then, when his father returned at the end of the conflict, Scott received his first flying lesson.

Despite a desire to study at West Point, Scott won a swimming scholarship to read mechanical engineering at the University of Michigan in 1949. After a year in Michigan, he was summoned for a physical at West Point, passed and headed to upstate New York to begin preparations for a military career. In his autobiography, Scott would credit his four years at West Point – plus his own upbringing – as "the most valuable and formative ... of my life".

Ultimately, in 1954, he graduated fifth in his class and opted to join the Air Force, receiving initial flight training at Marana Air Force Base in Tucson, Arizona. He subsequently moved to Texas to begin working on high-performance jets, followed by gunnery preparation and assignment to a fighter squadron in Utrecht, Holland. Whilst in Europe, Scott flew F-86 Sabre and F-100 Super Sabre jets under a variety of weather conditions and, in October 1956, when Soviet tanks rolled into Budapest, his squadron was placed on high alert for the first time.

Three years later, as the Mercury Seven were introduced to the world's media, Scott watched from afar with scepticism; why, he wondered, were they abandoning such promising military careers? His own focus was upon gaining an advanced degree in aeronautics from Massachusetts Institute of Technology (MIT) and achieving admission to test pilot school. His work at MIT, he later wrote, "was like trying to drink water from a high-pressure fire hydrant ... Compared to the hard grind of MIT, the five or six years I had spent flying fighter jets felt like playing." He and his wife, Lurton, also had a newborn daughter, Tracy.

As part of his master's degree, Scott was introduced to the new field of 'astronautics' – "my first exposure to space" – and his dissertation focused upon the mathematical application of guidance techniques and celestial navigation. This undoubtedly proved of benefit during Projects Gemini and Apollo, both of which depended heavily upon rendezvous and docking. Scott passed his final exams shortly after John Glenn's Friendship 7 mission, hoping for reassignment to test pilot school ... only to be detailed instead as a professor of aeronautics and astronautics at the Air Force Academy. Fortunately, a conversation with a sympathetic superior officer led to a change of orders and Scott reported to the Experimental Test Pilot School at Edwards Air Force Base in California.

Graduation was followed, in mid-1963, by the lengthy application process to join NASA's astronaut corps. In his autobiography, Scott recalled undergoing cardiograms, running on treadmills and enduring hypoxia evaluations, in which he was starved of oxygen to assess his physical response. The psychologists were especially difficult to please. When asked about his MIT days and how he had liked Boston, Scott replied that he found New Englanders cold, aloof and "a little hard to get to know" ... only to discover that the stone-faced psychologist was a born-and-raised Bostoner!

It obviously had little adverse impact on his application, however, and in October 1963 Scott and 13 other candidates – Edwin 'Buzz' Aldrin, Bill Anders, Charlie Bassett, Al Bean, Gene Cernan, Roger Chaffee, Mike Collins, Walt Cunningham, Donn Eisele, Ted Freeman, Dick Gordon, Rusty Schweickart and Clifton 'C.C.' Williams – were notified of their assignment to NASA. Of these, eight were test pilots, whom Slayton intended to use "for the more immediately difficult work" as solo command module pilots for the Apollo missions, while the others represented a mixture of operational military fliers, engineers or researchers. The latter, added Slayton, "would get their chance, too, but on the development end of things". The excitement of the accelerating lunar effort, Scott wrote, was palpable, even in the wake of President John Kennedy's assassination a few weeks later.

The new astronauts worked together surprisingly well, with the Air Force pilots kidding their Navy counterparts that they couldn't bring their jets down without a thump. In return, the naval aviators retorted that the Air Force fliers needed far too much runway on which to land. Despite the steadily burgeoning number of astronauts, they were still offered deals on cars, small Life magazine contracts and "the banks all wanted to have our accounts". Unlike the Mercury Seven, however, their privacy was better respected and their wives and children were pestered less by journalists from other publications.

After initial training, Scott was assigned guidance and navigation as his area of responsibility and in June 1965 served as one of the backup capcoms for Gemini IV and America's first spacewalk. Less than three months later, at the end of August, Deke Slayton caught up with him for "a word". Now that Gemini V was over, Slayton said, Neil Armstrong was free from his backup duties and would be teamed with Scott for Gemini VIII, scheduled for March 1966. The mission, Slayton added, would involve a lengthy, two-hour EVA for Scott. The boy from Randolph Air Force Base could not have been more delighted.

TRAINING

Of course, wrote Scott, he had to be sure that Armstrong, as Gemini VIII's command pilot, was happy with his selection. He need not have worried. "When I caught up with him in a corridor outside the VIP room," Scott related, "he gave me a big grin and held out his hand. That was all I needed."

The official announcement came on 20 September, with Pete Conrad and Dick Gordon assigned as their backups. Gemini VIII would be the most complex mission ever attempted, involving all of the tasks practiced by previous flights ... and more: three days in orbit, rendezvous – and, for the first time, physical docking – with an Agena-D target, combined manoeuvres, scientific experiments, a precision re-entry and a tricky spacewalk. The sheer complexity of the flight, wrote Scott, "was reflected in our mission badge ... a rainbow of colours refracted from twin white stars through a prism to form the Gemini symbol together with a Roman 'VIII' ... to reflect its many objectives".

Scott's EVA was part of a goal that simply had to be perfected before the early Apollo missions and the Department of Defense, through the Air Force, had already invested heavily in an Astronaut Manoeuvring Unit (AMU) to be tested by Charlie Bassett on Gemini IX. Before that, Dave Scott would wear a suit with a chest pack (known as the Extravehicular Life Support System) to feed oxygen from the spacecraft's supply and from a backpack (dubbed the Extravehicular Support Package) located on the spacecraft's rear adaptor section. This backpack also provided a radio and 8 kg of propellant for a zip-gun manoeuvring tool – similar to the hand-held device used by Ed White – and was connected to Gemini VIII's systems by an 8 m oxygen-hose tether.

When Scott reached the rear of the spacecraft and backed himself into the backpack, he would add a lightweight 23 m tether to the hose, allowing him to move up to 30 m into space. The exercise was extremely risky and if Scott encountered problems, there would be no way for Armstrong to aid him: the command pilot therefore requested a realistic model of the Gemini VIII adaptor for training, to rehearse donning the backpack, together with practice runs in an altitude chamber. Scott, meanwhile, worked closely with Ed White and by the eve of launch had completed more than 300 zero-gravity aircraft parabolas and over 20 hours on an air-bearing table, donning and doffing the bulky suit again and again. It was hot, hard, strenuous work. "I remember Dick [Gordon] turning to me once," wrote Scott, "drenched in sweat and joking 'Isn't this glamorous?'"

Conversations with White had certainly identified the need for Scott to maintain

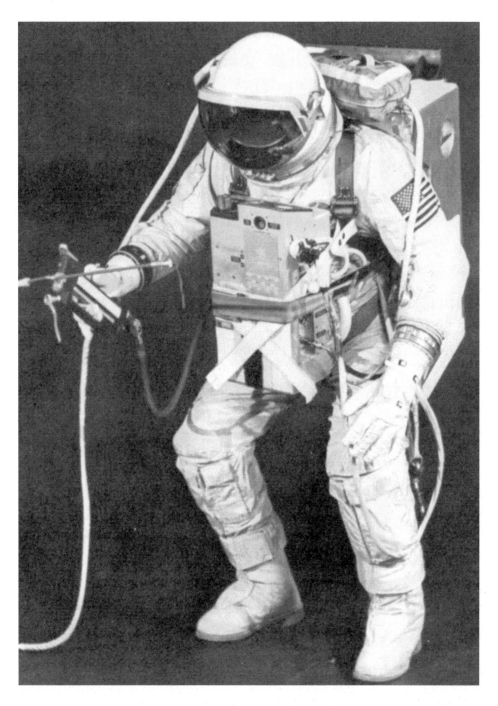

The space suit and equipment scheduled to be used by Dave Scott on Gemini VIII.

his physical fitness and strength, particularly in light of the difficulties in closing the hatch on Gemini IV. A consequence of this was that a lever was added to Gemini VIII's hatch, allowing it to be closed with less physical force. Nonetheless, White cautioned, the space suit itself was both stiff and heavy, requiring strength and stamina to move around for two hours. Scott began jogging, playing handball and lifting weights in the gym ... and, in his autobiography, would recall how this brought him face to face with Neil Armstrong's sly humour.

Armstrong's theory on exercise, John Glenn once said, was that a human being had only a finite number of heartbeats and should not waste them frivolously. One day during Gemini VIII training, as Scott pumped iron in the gym, Armstrong turned up, set the exercise bicycle onto its lowest possible tension setting and began pedalling, telling him: "Attaboy, Dave! Way to go!"

Work on the 6 × 7 m air-bearing table – which pumped highly-compressed air through holes in its floor to remove friction – allowed Scott to literally 'fly' across its surface using the zip-gun, which he described as one of the most valuable parts of his EVA training. The gun had 15 times more propellant than White's device and, instead of oxygen, used freon, a refrigerant with a correspondingly higher density. This multiplied the gun's impulse, although Scott worried how the gas might behave in space. During one test at low temperatures, for example, the freon caused the gun's poppet valve to stick open when triggered. Had this occurred in space, the escaping gas might have caused Scott to tumble uncontrollably. New seals were fitted to solve the problem and two new shut-off valves provided additional safety.

The space suit's other equipment presented its own problems. One key obstacle was the risk that an ejector in the chest pack could freeze and block the flow of oxygen from both Gemini VIII's supply and from the backpack. In response, engineers incorporated 20-watt heaters near the ejector. Other obstacles included overcoming the jumble of umbilicals, tethers and jumper cables whilst donning the chest pack inside the cramped cabin, although by December 1965 Scott was satisfied that he could do the task satisfactorily. The explosion of Wally Schirra and Tom Stafford's Agena-D, on the other hand, did not fill Armstrong or Scott with confidence. Mission Control shared their concerns, telling them to "get out fast" if the target showed the slightest hint of a malfunction whilst docked.

Despite the EVA experience gained by Ed White, the Gemini VIII excursion would eclipse it in complexity. Not only would Scott be outside for two hours and 40 minutes – nearly ten times longer than White – but the flight plan also required him to move to the rear of the spacecraft to don his backpack. He would spacewalk through orbital daytime and nighttime, retrieve an emulsion package, activate a micrometeoroid collector on the Agena and test a reactionless power wrench. As a result, in addition to aircraft parabolas and trial runs on the air-bearing table, he was submerged into a large water tank with a mockup zip-gun to practice moving around in a neutrally-buoyant environment. Today, such underwater exercises are common as spacewalkers rehearse procedures outside the International Space Station and are regarded by many as the closest analogue to the real thing.

Training for the spacewalk was, however, only part of the Gemini VIII mission; Armstrong and Scott's 70 hours aloft also featured a full plate of experiments. One

Artist's concept of Dave Scott during his EVA.

of these was the Zodiacal Light Photography investigation, which sought to capture a series of electronically-controlled exposures using a modified 35 mm Widelux camera with a rotating lens. "This was to observe ... a very faint glow on the horizon on Earth seen just after sunset and before sunrise," Scott wrote. To accustom their eyes to the subtle zodiacal light, caused by dust circling towards the inner Solar System, Armstrong and Scott would turn down the lights in their T-38 cockpit during cross-country night flights.

Also aboard Gemini VIII was a hand-held spectrometer to deduce cloud-top heights from the absorption of light by the oxygen band, together with a nuclear emulsion experiment to determine cosmic radiation effects on bromide crystals, a micrometeoroid investigation affixed to the Agena and a series of synoptic terrain photography tasks for meteorological research.

"SMOOTH AS GLASS"

Meanwhile, preparation of the Gemini VIII flight hardware got underway in mid-August 1965, when the Agena-D target, tailnumbered '5003', arrived at Cape Kennedy and, after final assembly, commenced pre-launch testing in October. Then, in early January of the following year, the two stages of the Titan II were delivered to Florida, mated and installed on Pad 19. Leak checks of the second stage engine on 7 February turned up small cracks in the thrust chamber manifold, a problem solved by rewelding, but by the 10th the rocket and Gemini VIII were fully mated and had undergone electrical compatibility tests.

Final preparation of the Agena-D occurred in tandem, the Atlas having been erected on Pad 14 at the Cape in early January. By this time, procedural and design changes – a result of Project Surefire – had been fully implemented. Two weeks later, the pencil-like Agena-D was mated to its docking adaptor and on 1 March was mounted atop the Atlas. The only problem in the last few days was an overfilling of its propellant tanks, which required the replacement of its regulator and relief valve and pushed the launch date from 15 March to the 16th.

In spite of the renewed vigour injected into the lunar effort by the spectacular Gemini VII/VI-A rendezvous late the previous year, morale in the astronaut corps suffered a devastating blow with the deaths of Elliot See and Charlie Bassett in the crash of their T-38 jet on 28 February 1966. Two days later, although their minds were almost wholly focused on their upcoming mission, Armstrong and Scott and their wives joined a huge crowd of mourners at Seabrook Methodist Church for See's memorial service and at Webster Presbyterian Church for that of Bassett. A sense of foreboding pervaded the Gemini project. "People started ... fearing we were in a run of bad luck," wrote Dave Scott. "The memorial services in Houston ... were sad, depressing affairs." The following year, 1967, would be worse still.

Armstrong and Scott and their backups, Pete Conrad and Dick Gordon, had little time to waste on contemplation; a few days after the See and Bassett memorials, they were in quarantine at Cape Kennedy as launch drew closer. At 7:00 am on the morning of 16 March, Armstrong and Scott awoke and proceeded through the time-honoured ritual of a pre-flight physical and breakfast of filet mignon, eggs and toast with butter and jelly. Clear blue, cloud-speckled skies over the marshy Florida landscape promised a perfect opportunity for a launch. As they suited up, a watch belonging to aviation pioneer Jimmy Mattern – who had unsuccessfully attempted a round-the-world solo feat in 1933 – was strapped around the wrist of Armstrong's suit. Not to be outdone, Scott, in his own nod towards aviation heritage, carried pieces of wood and cloth from an old Douglas World Cruiser, the New Orleans, which first flew around the world in 1924. Both artefacts had been borrowed from the museum at Wright-Patterson Air Force Base.

Shortly after insertion into his seat, a glitch with Scott's parachute harness threatened a delay. One of the technicians discovered some epoxy in the catcher mechanism of the harness, but the efforts of Pete Conrad and pad fuerher Guenter Wendt finally got it unglued. "Pete ... rushed around until he found a dentist's toothpick with which to try and clear the connection out," recalled Scott. "I remember looking back and seeing Pete sweating like mad digging this stuff out." Conrad's toothpick did the trick. The hatch was finally closed and Gemini VIII's countdown proceeded.

One hundred minutes before launch, at 10:00 am, as their cabin atmosphere was steadily being purged with pure oxygen, the first stage of the mission got underway as the Atlas-Agena lifted-off successfully from Pad 14. Unlike Schirra and Stafford's experience the previous October, the rocket performed near-flawlessly. Despite following a trajectory that was slightly low and to the south of its intended flight path, the Atlas' sustainer engine compensated by pushing itself back on track, ultimately carrying the 6 m-long, pencil-like Agena-D into a circular orbit, 298 km high. Advised of the successful insertion, Armstrong told launch controllers that he and Scott were ready to go.

"Cradled in my contoured seat it felt almost as if I was being held in someone's arms," wrote Scott, comparing the sensation to taking a brand-new Ferrari out on the open road for the first time. The pure oxygen atmosphere introduced a cool cleanness and freshness to the cabin.

Liftoff occurred precisely on time at 11:41 am, conducted under the auspices, for the first time, of a flight director other than Chris Kraft. The distinctly English-accented, tweed-suited and pipe-smoking John Hodge, who had led one of three teams during the Gemini IV mission, was in charge. "The Titan was smooth," continued Scott, describing a few shudders and evidence of pogo oscillations, but overall "a solid feeling, a sharp kick in the tail". His heart rate in the moments preceding liftoff peaked at 128 beats per minute; Armstrong's touched 146. Post-flight aeromedical studies would see the difference simply as a 'keying-up' of Armstrong's physical awareness, rather than an indicator of undue stress.

Launch and ascent aboard the Titan, Armstrong told James Hansen, was "very definite; you knew you were on your way when the rocket lit off. You could hear the thrust from the engines, at least at low altitudes, but the noise did not interfere with communications ... The G levels got to be pretty high in the first stage of the Titan – something like 7 G". As the rocket headed higher into the rarefied atmosphere and

Launch of an Atlas-Agena target vehicle.

its second stage took over, the two men saw bright red and yellow debris from a severed joining strap through their windows.

Riding the second stage, Scott wrote, was "smooth as glass". Later, his first experience of weightlessness came when a small metal washer hovered in front of his eyes. As he released his checklist, he was amused to see it drift across the cabin. Nothing, however, could have prepared him for his first glimpse of Earth. Armstrong rolled Gemini VIII and the men beheld the deep blue of the Mediterranean, together with Italy and, in the distance, the outline of the Middle East and the Red Sea, laid out, map-like, before them. Although he had reached 63 km in his final X-15 flight and had seen the curvature of Earth, Armstrong, too, was quite unprepared for the view. In his autobiography, Scott remembered the difficulty other astronauts had when describing the sheer beauty and grandeur of the home planet; now he knew how they felt. Perhaps, someday, he thought, a poet or an artist would be blasted into the heavens. Maybe he or she could describe it better than a test pilot. "But I wanted to capture what I could," he wrote. "I pulled out my camera and took my first photographs from space."

Incredible as the view was, there was little time to dwell upon it. Within minutes, they established themselves in a 160 × 172 km orbit, trailing the Agena by 1,963 km. Scott's spacewalk was looming in barely 24 hours' time and, before that, they had the Agena-D rendezvous and docking to perform. A minor radiator problem did not prevent the astronauts from undertaking some sightseeing over the Pacific, a cloud-covered Hawaii, Baja California, the naval base in San Diego and the area around Edwards Air Force Base in the Mojave Desert, where Armstrong and Scott had both worked and studied years earlier.

Their first thruster firing came at 1:15 pm, a little more than 90 minutes after launch, which slightly lowered their apogee. A break for lunch took longer than anticipated, requiring them to Velcro-patch the food packages to Gemini VIII's 'ceiling' whilst they executed a second burn on their second orbit to raise their perigee. When they came to eat lunch, it was hardly home cooking: Armstrong's chicken and gravy casserole, despite having been rehydrated, was dry in places and the astronauts' brownie cookies were stuck together and crumbly. "We had been running on adrenalin," wrote Scott. "There had been no time for food – no thought of it either." Nor was there time to spend worrying. A third manoeuvre, executed high above the Pacific Ocean at 2:27 pm, placed Gemini VIII into the same orbital plane as the Agena, albeit imprecisely.

"A fundamental requirement of rendezvous," Armstrong told James Hansen, "is to get your orbit into the same plane as the target's orbit, because if you're misaligned by even a few degrees, your spacecraft won't have enough fuel to get to its rendezvous target. So the plan is to start off within just a few tenths of a degree of your target's orbit. That is established by making your launch precisely on time, to put you in the same plane under the revolving Earth as is your target vehicle."

Armstrong then fired the aft OAMS thrusters, producing a horizontal velocity change of 8 m/sec, after which Capcom Jim Lovell requested that he add an extra 0.6 m/sec to his velocity. The adjustment, Armstrong said later, "was a pretty loose burn ... without much preparation", but the two astronauts quickly moved onto their next task: activating and testing the spacecraft's rendezvous radar. Westinghouse, the company responsible for developing the device, had promised that it would be able to acquire its target at a distance of around 343 km.

They detected the Agena with their radar long before achieving a visual sighting, whilst still some 332 km away. A little under four hours into the mission, high above Madagascar, another burn adjusted their orbit. This prepared them with near-perfection to begin the 'terminal' phase of the rendezvous. An hour later, at 4:21 pm, during their third orbit, Scott reported his first visual sighting of the Agena as a speck in the distance, 140 km away, its rendezvous beacon flickering against the black sky. "Gradually, she became a sleek, silver tube," he wrote, "a spectacular sight."

As both spacecraft drifted into orbital darkness, the Agena disappeared from view, although Armstrong and Scott were still able to discern its blinking acquisition lights. When the target was at the proper angle, ten degrees 'above' them, Armstrong recalibrated the platform of Gemini VIII's inertial reference system for a translational manoeuvre. Next, he pitched the nose upwards 31.3 degrees and canted the spacecraft 16.8 degrees to the left, finally braking Gemini VIII by eyesight alone as Scott called out radar range and range rates. Several smaller thruster spurts brought them to a position just 46 m from the Agena, with no relative velocity and, after 30 minutes of inspections to ensure that the target had not been damaged during launch, they were given the go-ahead to execute the world's first-ever docking manoeuvre.

Moving his spacecraft at barely 8 cm/sec, Armstrong gingerly pulsed towards the Agena, announcing the onset of the 'station-keeping' phase of the rendezvous at 5:40 pm. Years later, he told James Hansen that station-keeping at such close proximity to the target posed no major problems; conversations with Wally Schirra and Frank Borman had assured him that flying two vehicles close together was very easy to do once the correct position had been achieved.

For the next 25 minutes, he and Scott electronically checked the target's systems, antennas and lights using radio command, before nudging their spacecraft closer; so close, in fact, that they could read a small, illuminated instrument panel above its docking cone. After receiving a go-ahead to dock from Keith Kundel, the capcom on the Rose Knot Victor communications ship, Armstrong pulsed Gemini VIII's thrusters and achieved physical contact at 6:15 pm. An electric motor aboard the Agena retracted the docking cone, pulling the spacecraft's nose about 60 cm into the target and connecting their electrical systems. By now close enough to read, the Agena's display confirmed a green 'rigid' sign, indicating that both vehicles were mechanically and electronically mated.

"Flight, we are docked," Armstrong exulted, "and it's really a smoothie. No noticeable oscillations at all." Seconds later, as the realisation of what had been done finally set in, sheer pandemonium broke out in Houston. Armstrong and Scott's 'smoothie' had cleared another hurdle on the road to the Moon.

That smoothie would rapidly give way to a far rockier road ... one that would come close to claiming the two astronauts' lives.

"TUMBLING END-OVER-END"

Gemini VIII was not aided by a distinct lack of available tracking stations across its flight path, which resulted in some very spotty communications with Mission Control. "We could communicate with Houston," wrote Scott, "for only three five-minute periods every 90 minutes as we passed over secondary tracking stations." Two ship-based stations were aboard the Rose Knot Victor and the Coastal Sentry Quebec, plus a land-based site in Hawaii. Shortly before one such loss of contact, at 6:35 pm, Armstrong and Scott were advised by Jim Lovell that, if problems arose whilst docked, they should deactivate the Agena and take control with the Gemini. "Just send in the command 400 to turn it off," Lovell told them, "and take control with the spacecraft." For now, however, the only problems seemed to be difficulties verifying that the Agena was receiving uplinked commands and a glitch with its velocity meter, used to specify the magnitude of a burn by its main engine.

Twenty-seven minutes after docking, Scott commanded the Agena to turn them 90 degrees to the right and Armstrong reported that the manoeuvre had "gone quite well". His call came seconds before Gemini VIII passed out of radio range of the land-based Tananarive station in the Malagasy Republic. Working alone, the astronauts transmitted an electronic signal to start the Agena's tape recorder. Shortly thereafter, their attitude indicators showed them to be in an unexpected, almost imperceptible, 30-degree roll. "Neil," called Scott, "we're in a bank".

Perhaps, they wondered, the Agena's attitude controls were playing up or its software load was wrong. Since Gemini VIII's OAMS was now switched off and both men could see the Agena's thrusters firing, they reasoned that the target's controls must be at fault. They could not have known at the time that one of their own OAMS thrusters – the No. 8 unit – had short-circuited and stuck in its 'on' position. Unaware, Scott promptly cut off the Agena's thrusters, whilst Armstrong pulsed the OAMS in a bid to stop the roll and bring the combined spacecraft under control. For a while, his efforts succeeded.

After four tense minutes, the docked spacecraft slowed and steadied itself. Then, as Armstrong worked to reorient them into their correct horizontal position, the unwanted motions began again ... much faster than before and, wrote Scott, "on all three axes". Perplexed, the men jiggled the control switches of the Agena, then of the Gemini, off and on, in a fruitless effort to isolate the problem. At around this time, Scott noticed that Gemini VIII's own attitude propellant had dropped to just 30 per cent, clearly indicative of a problem with their own spacecraft. "It was clear," Scott related, "we had to disengage from the Agena, and quickly."

Undocking presented its own problems, not least of which was the very real risk that the two rapidly rotating vehicles could impact one another. However, Scott, demonstrating an intuitive test pilot's awareness of the importance of recording all pertinant data, pre-set the Agena's recording devices such that ground controllers would still be able to remotely command it. "I knew that, once we undocked, the rocket would be dead," he wrote. "No one would ever know what the problem had been or how to fix it." Scott's prompt action saved the Agena and preserved it for subsequent investigations and tasks.

Still out of radio contact with the ground, Armstrong moved onto the next step of the flight rules and undocked from the Agena. He then fired a long burst of Gemini VIII's translational thrusters to pull away, only to discover that the spacecraft, now free, began to spin more wildly, in roll, pitch and yaw axes. It was now much worse than before, because the stuck-on No. 8 thruster was no longer turning the entire combination, only the Gemini. High above south-east Asia, they came within range of the Coastal Sentry Quebec, which received Scott's urgent radio transmission at 6:58 pm: "We have serious problems here ... we're tumbling end-over-end. We're disengaged from the Agena."

Aboard the ship, Capcom Jim Fucci acknowledged the call and enquired as to the nature of the problem. Both men were relieved to hear Fucci's voice. "He was an old NASA hand," wrote Scott, "very experienced." Quickly, yet with characteristic calmness, Armstrong reported that he and Scott were "in a roll and we can't seem to turn anything off ... continuously increasing in a left roll". Fucci duly passed the

Gemini VIII during rendezvous activities with the Agena.

report over to Houston: Gemini VIII was suffering "pretty violent oscillations". The three-way conversation with Mission Control meant that it was some seconds before Flight Director John Hodge picked up all of the details; Fucci having to repeat that "he's in a roll and he can't stop it".

Armstrong quickly threw circuit breakers to cut electrical power and hence the flow of propellant to the attitude thrusters, including the No. 8 unit. However, with no friction or counterfiring thruster to stop it, the spinning continued. At its worst, it reached 60 revolutions per minute. Everything in the cabin – checklists, flight plans, procedures charts – were hurled around by the centrifugal force. The unfiltered Sun flashed in the astronauts' windows with alarming regularity, said Scott, "like a strobe light hitting us in the face".

Such a rotation rate placed the men at serious risk of physically blacking out and, indeed, both had difficulties reading their instruments properly. "It was rather like the feeling you get as a kid when you twist a jungle rope round and round and then hang on it as it spins and unfurls," wrote Scott. "In space, it was not a good feeling." Physician Chuck Berry would later note that the astronauts experienced two conditions brought on by the rapid rotation: a complete loss of orientation caused by the effects on their inner ears (the 'coriolis effect'), coupled with 'nystagmus', an involuntary rhythmic motion of the eyes.

The incessant rotation and the depletion of Gemini VIII's attitude propellant had already alerted Armstrong and Scott to a problem with their own spacecraft, but at the time they did not know that the short circuit in the OAMS had caused the No. 8 thruster to become stuck 'on' and caused the rapid drop in fuel. Even had they known, there would have been no time to ponder it. Armstrong's responsibility as the command pilot was to ensure the safety and success of the mission. Scott's spacewalk, docked activities with the Agena and most of the experiments were now off the agenda; the safe return of the crew was paramount.

Armstrong decided that his only available course of action was to use Gemini VIII's 16 re-entry controls. It was easier said than done. The re-entry control switch was in the most awkward position imaginable, directly above Armstrong's head, and, worse, was on a panel with around a dozen toggles. "With our vision beginning to blur," wrote Scott, "locating the right switch was not simple." Fortunately, both men had carried their years of test-piloting experience into the astronaut business and intuitively knew every switch, literally, with their eyes closed. "Neil knew exactly where that switch was without having to see it," Scott continued, but admitted "reaching above his head ... while at the same time grappling with the hand controller ... was an extraordinary feat."

The effort to reduce the spacecraft's rates to zero with the re-entry controls, though ultimately successful, consumed 75 per cent of the fuel. "We are regaining control of the spacecraft slowly," Armstrong reported, as the spinning stopped within 30 seconds. The flight, however, was over. Mission rules decreed that the re-entry controls, once activated, would require an immediate return to Earth at the next available opportunity. Just ten hours into its three-day mission, Gemini VIII was on its way home.

PACIFIC RETURN

On the ground, the television networks – which had cancelled their showings of 'Batman' and, ironically, 'Lost in Space' – were deluged with complaints from viewers as attention turned to a dramatic recovery effort. Original plans called for Gemini VIII to land in the Atlantic and be picked up by the aircraft carrier Boxer; however, the earlier-than-expected return called instead for a splashdown in the western Pacific during their seventh orbit.

The timing was strict. Gemini VIII's flight path had precessed so far westwards that it would be another full day before Armstrong and Scott could reach a location from which they could be easily recovered. Consequently, a naval destroyer named the Leonard F. Mason, based off the coast of Vietnam, was directed to intercept the new splashdown point, 800 km east of Okinawa. It would be the only Gemini splashdown in the Pacific, in a landing zone designated 'Dash 3'. "I looked it up in our manuals," wrote Scott. "Dash 3 was a secondary landing zone in the South China Sea. It was over 6,000 miles away from our primary landing site."

It was far from ideal. By this time, John Hodge's 'blue' flight control team had been at their consoles for 11 hours and a second ('white') team, headed by Gene Kranz, reported for duty to supervise the end of the mission. Kranz's team had more experience in recovery procedures than that of Hodge and, had Gemini VIII run to its intended three-day length, he would have overseen re-entry and splashdown anyway. It made sense, therefore, for Kranz to take the helm.

The news of an impending return was met with grim resignation by Armstrong and Scott, who ran through their pre-retrofire checklists with the capcoms at the Coastal Sentry Quebec, Rose Knot Victor and Hawaiian tracking stations. Unlike Gemini V, which had been nursed through a lengthy mission, despite problems, the situation in which Armstrong and Scott found themselves was compounded by a dangerously-low propellant load. By this time, having tested each of the OAMS thrusters in a now-stable Gemini VIII, Armstrong had identified the glitch with the No. 8 unit, which Scott later described as not exhibiting "a consistent, linear problem ... it was really screwed up". In fact, the thruster had been off when it should have been on, and vice-versa, on several occasions. The cause, however, would have to wait for the post-flight investigation.

Loading the re-entry flight program into Gemini VIII's 4,000-word-memory computer was difficult, particularly as it was already overloaded from the rendezvous with the Agena. This required Scott to erase the rendezvous and docking programs, then feed the re-entry data into the computer by means of a keypad and an on-board device known as an auxiliary tape memory unit. As he worked to punch in a series of nine lines of seven-digit numbers, Scott was relieved that the unflappable Jim Fucci, aboard the Coastal Sentry Quebec, was there to watch his every move. "He read off those numbers as if he was talking about taking a stroll in the park," Scott wrote. "I entered them quickly so that I could transmit them back to verify with him before we lost contact again."

Gemini VIII's retrorockets duly ignited at 9:45 pm, whilst out of radio contact, high above a remote part of south-central Africa. Worse, retrofire was conducted during orbital darkness, giving Armstrong and Scott no horizon by which to judge alignment. Minutes later, over the Himalayas, the spacecraft entered the tenuous upper atmosphere and as it continued to descend through the steadily thickening air, Scott reported that he could see nothing but a pinkish-orange glow through his window ... then haze and, finally, minutes before splashdown, the glint of water! Ten hours and 41 minutes after leaving Cape Kennedy, at 10:22 pm, the spacecraft hit the Pacific with a harsh thump and Scott yelled "Landing Safe!"

Throughout all this – during the launch, rendezvous, docking, crisis with and without the Agena, re-entry and splashdown – Jimmy Mattern's watch, tightly strapped around Armstrong's wrist, continued to tick faithfully ...

Despite having suffered severe space sickness and, now, seasickness as the spacecraft's windows rhythmically rolled and pitched with each wave, the astronauts swiftly proceeded through their post-splashdown checklist, shutting down electrical systems, placing switches and valves into their correct positions and activating their high-frequency communications antenna. Only now did Armstrong and Scott regret not taking Mission Control's advice to swallow meclizine motion sickness tablets before re-entry. "When Mission Control told us about three-foot waves," Scott wrote, "they had forgotten to mention the 20-foot swells!"

Scott called the search-and-rescue team from Naha Air Base in Okinawa by their callsign 'Naha Rescue One', but was met with silence on the radio. Both men were hot in their suits, particularly Scott, whose ensemble had extra layers to provide radiation protection on his spacewalk. Fumes from the ablated heat shield, too, left them nauseous. Within half an hour, a C-54 aircraft, flown by Air Force pilot Les Schneider, which had spotted Gemini VIII's descent and splashdown, arrived on the scene. Its crew visually checked the spacecraft, marked its landing co-ordinates and dropped three pararescue swimmers and an emergency liferaft. For the Naha Rescue One team, which was more accustomed to missions in Vietnam, Laos and Cambodia in those war-charged times, 16 March 1966 was a distinctly different and highly memorable day.

Notwithstanding the rough swells, the pararescue swimmers, themselves queasy, affixed a flotation collar to the spacecraft, then signalled the C-54 with a 'thumbs-up' that Armstrong and Scott were alive and well. This was duly radioed to other aircraft in the area, to the Leonard F. Mason, then to Hawaii, to NASA's Goddard Space Flight Center and finally to Mission Control in Houston, from where public affairs officer Paul Haney announced the news to an anxious world. Meanwhile, the encounter between the antiquated C-54 and the state-of-the-art Gemini was, said Neil Armstrong, "the most unusual rendezvous in aviation history".

Three hours after splashdown, in the small hours of 17 March, the two astronauts and their spacecraft were safely aboard the Mason. The rough seas, though, had made the hoisting of Gemini VIII difficult, to such an extent that it kept crashing against the side of the destroyer, denting its nose at one point. The Mason's crew, wrote Scott, had initially been less than happy about being given the task of recovering Gemini VIII. They had just completed a seven-week tour in Vietnam and been given a brief spell of liberty in Okinawa. However, their spirits rose as the realisation set in that the astronauts were safe. In spite of their tiredness and the

Scott and Armstrong, surrounded by recovery swimmers after performing the Gemini project's first – and only – splashdown in the Pacific.

effects of nausea, Armstrong and Scott managed smiles and greetings for the crew and were found to be healthy, suffering from minimal dehydration.

They were, however, shaken by what had actually come close to disaster ... as, indeed, had many within NASA. Deputy Administrator Bob Seamans had been advised of the crisis over the telephone whilst at the reception to the prestigious Robert H. Goddard Memorial Dinner and swore that he would never again be caught in such a position during the critical phase of a future mission.

At the same time, publicly, NASA was reluctant to over-emphasise the near-disaster, particularly if it wanted continued funding for a Moon landing by 1970. When Life magazine proposed titling its Gemini VIII article as 'Our Wild Ride in Space by Neil and Dave', its editor-in-chief received a firm request from Armstrong to change it to something less melodramatic. Ultimately, bound by an ongoing contract, the magazine agreed and would publish watered-down headlines for Gemini VIII and subsequent missions.

In spite of the troubles, President Lyndon Johnson reassured the American public that his administration remained firmly committed to John Kennedy's goal of bootprints on the Moon before the end of the decade. Some have argued over the years that Armstrong's coolness was pivotal in his selection to command Apollo 11, although some isolated individuals within the astronaut office speculated that his status as a civilian test pilot had contributed to the failure.

Indeed, Walt Cunningham, later to fly Apollo 7, would criticise what he saw as flaws in both astronauts' performance, while Tom Stafford felt that the decision to undock from the Agena was a flawed one. Gene Kranz, on the other hand, perceived the crisis as the result of a broader training failure – malfunction procedures did not cover the problems encountered whilst the Gemini and Agena were docked – and both Frank Borman and Wally Schirra praised Armstrong and Scott's actions as having prevented disaster. Indeed, without their safe return and the knowledge of what had happened, an erroneous assumption that the Agena was to blame could have diseased the final days of Gemini and made it very difficult for Apollo, with its emphasis on rendezvous and docking, to proceed. "It could have been a showstopper," admitted Dave Scott.

Gene Cernan, though, rationalised the critics' thinking. "Screwing up was not acceptable in our hypercompetitive fraternity," he told James Hansen. "Nobody got a free ride when criticism was remotely possible. Nobody." Still, Gemini VIII did little damage to either man's career. Definitive testament came two weeks after the flight, when the Gemini VIII Mission Evaluation Team "positively ruled out" any errors on the astronauts' part and, indeed, Bob Gilruth himself praised them for their "remarkable piloting skill". Scott was promoted to lieutenant-colonel and assigned a seat on an Apollo crew within days, while Armstrong received the backup command slot for Gemini XI. Still, the quiet civilian was demoralised by what he saw as only a partial success.

Had he been "smarter", Armstrong said later, he might have figured out the problems earlier, perhaps saving Scott's EVA and some of the mission's other objectives. Many of Gemini VIII's experiments – the zodiacal light photography task, the growth of frogs' eggs, the synoptic terrain studies, the nuclear emulsions

Armstrong (left) and Scott with crewmen aboard the recovery ship Mason.

and the cloud spectrophotography – were left incomplete and some have speculated over the years that, had Scott's EVA been underway when the spinning started, he may have seen the burst from the stuck-on No. 8 thruster and warned Armstrong to shut off its propellant.

However, others considered it fortuitous that Scott's EVA had never come to pass. It "had seemed terribly complex and dangerous," wrote Mike Collins. The need for Scott to get outside, manoeuvre himself to Gemini VIII's adaptor section and worry about swapping connectors and keeping track of tethers was, in Collins' mind, too risky at such an early stage. "My own EVA scheme on Gemini X was far from ideal," he wrote, "in that I had to stuff everything into an already crowded cockpit, but at least I could make nearly all my preparations inside the pressurised cocoon ... Not so with Dave's complicated gear."

Other naysayers have added that, during the uncontrollable spinning, Scott may have been whirled around so violently on his tether as to have hit the side of Gemini VIII, almost certainly producing fatal injuries ...

ROCKET ARMCHAIRS AND FIREPROOF PANTS

One saving grace of the crisis was that Scott had the presence of mind, before undocking, to switch over command of the Agena to Mission Control. The result: the Gemini VIII-Agena Target Vehicle (GATV-VIII) could – and would – be reused during a subsequent mission. Four months later, Gemini X's John Young and Mike Collins would fly part of their own rendezvous, docking and spacewalking extravaganza with the Agena. In the days after Armstrong and Scott splashed down, the rocket's main engine was fired ten times, its various systems were vigorously tested and it successfully received and executed more than 5,400 commands. By 26 March, its electrical power had been exhausted and it could no longer be effectively controlled, but by this stage it had been raised into a higher orbit to permit inspection by the Gemini X crew.

Before Young and Collins could complete their mission, however, came Gemini IX; stricken, it seemed, by bad luck since the dull, chill February day when its prime crew lost their lives in St Louis. Days after the deaths of Elliot See and Charlie Bassett, their backups, Tom Stafford and Gene Cernan, were appointed to replace them. With a launch scheduled for mid-May, Stafford would record the shortest turnaround between flights of any space traveller thus far, blasting off just five months after his Gemini VI-A splashdown. Newly-promoted to become the 'new' Gemini IX backups were Jim Lovell and Buzz Aldrin, who, by following Deke Slayton's three-flight crew rotation system, were now in prime position to fly the Gemini XII mission in November 1966.

Gemini XII, the last flight in the series, was originally to be the preserve of Stafford and Cernan in their capacity as See and Bassett's backups. In fact, in his autobiography, Cernan recalled trips to McDonnell's plant in St Louis to inspect and train on the Gemini IX capsule ... yet finding himself, in rare moments of spare time, drifting down the line of almost-complete spacecraft to take a wistful look at the skeletal form of Gemini XII, his and Stafford's ship. Years later, Cernan would still recall his desire to know every switch, every circuit breaker, every instrument, every bolt and rivet, inside the Gemini before he and Stafford took this engineering marvel into the heavens.

The prime and backup crews for Gemini IX were announced in early November 1965 and, indeed, with Stafford still busy preparing for his mission with Wally Schirra, Cernan was forced to train alone with See and Bassett until early the following year. His role not only shadowed Bassett, but prepared himself for the possibility, however remote, of actually flying the mission and conducting a lengthy EVA wearing an Air Force contraption known as the Astronaut Manoeuvring Unit (AMU). It looked, Cernan wrote, "like a massive suitcase" that was "so big that it would be carried aloft folded up like a lawn chair and attached within the rear of the Gemini". (In fact, the Air Force's project officer for the AMU, Major Ed Givens, was selected by NASA as an astronaut candidate in April 1966.)

Having manoeuvred himself over to the device, Bassett would "slip onto a small bicycle-type seat, strap on the silver-white box and glide off into space, manoeuvring with controls mounted on the armrests". Sounding very much like something from a Buck Rogers episode, the AMU had evolved through seven years of developmental work, with its focus on military tasks associated with a Pentagon-sponsored space station called the Manned Orbiting Laboratory. "The possibility of using it to send someone scooting off to disable an enemy satellite," wrote Cernan, "wasn't mentioned in public because we weren't supposed to be thinking about the militarisation of space."

For NASA's purposes, however, the 75 kg AMU provided an essential tool in understanding how effectively astronauts could work and manoeuvre outside the confines of their spacecraft. When he was named to Gemini IX, Bassett was tasked with an EVA that would span at least one 90-minute circuit of the globe and would be able to control his movements and direction by means of 12 hydrogen peroxide thrusters. The AMU was also equipped with fuel tanks, lights, oxygen supplies, storage batteries and radio and telemetry systems. The device would be controlled by knobs on the end of the AMU's twin arms – a left-hand one providing direction of motion, a right-hand one for attitude – although, for safety, Bassett would remain attached to Gemini IX by a 45 m tether throughout the spacewalk.

Undoubtedly raising Cernan's hopes for his own mission was the possibility that, if Bassett's excursion went without a hitch, plans were afoot for a more autonomous AMU spacewalk on Gemini XII, perhaps untethered. In the days before enormous water tanks became the norm for EVA training, Bassett and Cernan spent much of their time physically conditioning themselves. Both men recognised that vast reserves of strength and stamina would be required to handle the demands of a spacewalk encased inside a bulky pressurised suit and resorted to lengthy spells in the gym, games of handball and hundreds of press-ups. "Before long," Cernan wrote, "we grew Popeye-sized forearms."

Their suits needed to be somewhat different from that worn by Ed White on Gemini IV, partly in recognition of the demands of the AMU, as well as to provide additional comfort and protection. The new ensembles included a white cotton long-john-type undergarment for biosensors, a nylon 'comfort' layer, a Dacron-Teflon link net to maintain the suit's shape and several layers of aluminised Mylar and nylon for thermal and micrometeoroid protection. Guarding them from the searing hydrogen peroxide plumes from the AMU (one of which would jet directly between

Bassett's legs!) were the heat-resistant 'trousers' of the suit. These were composed of 11 layers of aluminised H-film and fibreglass, topped by a metallic fabric woven from fibres of the alloy Chromel R. One day during training, Bassett and Cernan watched as a technician charred the material with a blowtorch for five minutes, telling them that despite the intense temperature of the AMU's exhausts, they would remain comfortable within their suits.

As Cernan continued his training as Bassett's understudy, the pair – indeed, the foursome, if one also counted See and Stafford – spent so much time working together than a relationship akin to family developed. Despite their intense focus on Gemini IX, Stafford and Cernan undoubtedly looked forward to their own rendezvous, docking and spacewalking adventure with their own Gemini, their own Agena and their own AMU, towards the end of 1966. All that changed on the morning of 28 February, when it became clear that Cernan's first journey into space would come much sooner, more unexpectedly and more horrifyingly, than he could have ever imagined or wished.

THE SUBSTITUTE

Cernan's grandparents emigrated to America shortly before the outbreak of the First World War; on his mother's side, they were Czechs from a Bohemian town south of Prague, while on his father's side were Slovak peasantry from a place close to the Polish border. Their children, Rose Cihlar and Andrew Cernan, would produce the child who would someday gaze down on Earth through the faceplate of a space suit, would see the sheer grandeur of the lunar landscape and would become one of only a handful of men to go prospecting in the mountains of the Moon.

Eugene Andrew Cernan, a self-described "second-generation American of Czech and Slovak descent", was born in Chicago, Illinois, on 14 March 1934. As a young boy, he learned from his father how machinery worked, how to plant tomatoes, how to hammer a nail straight into a board and how to repair a toilet; all of which instilled in him an ethos "to always do my best at whatever I put my hand to". In high school, that ethos led him to play basketball, baseball and football, for which he was even offered scholarships, but eventually he headed to Purdue University in 1952 to read electrical engineering.

Four years later, Cernan graduated and was commissioned a naval reservist, reporting for duty aboard the aircraft carrier Saipan. After initial flight training, he received his wings of gold as a naval aviator in November 1957 and gained his first experience of flying jets aboard the F-9F Panther. He was subsequently assigned to Miramar Naval Air Station in San Diego and attached to Attack Squadron VA-126, during which time he performed his first carrier landing aboard the aircraft carrier Ranger, flying the A-4 Skyhawk. Then, in November 1958, Cernan participated in his first cruise of the western Pacific, flying Skyhawks from the Shangri-La aircraft carrier, when, "armed to the teeth and ready for a fight", he frequently encountered Chinese MiG fighters in the Straits of Formosa.

Shortly thereafter, the Mercury Seven were introduced to the world and Cernan heard about, and for the first time wondered about, the role of these new 'astronauts'. In his autobiography, he noted that he met just two of NASA's requirements – age and degree relevance – and had little of their experience and no test-piloting credentials. "By the time I earned those kind of credentials," he wrote, "the pioneering in space would be over." Still, the germ of a new interest, to become a test pilot and fly rockets, implanted itself in the young aviator's brain.

In the early summer of 1961, now married to Barbara Atchley, Cernan was approaching the end of his five-year commitment to the Navy when he was offered the opportunity to attend the service's postgraduate school for a master's degree in aeronautical engineering. It offered him a route into test pilot school. When NASA selected its second group of astronauts in September 1962 Cernan knew that, although he held the right educational credentials, becoming a test pilot was still years away. Ultimately, however, the decision was made for him when one of his superiors recommended him to NASA for its third astronaut class.

As 1963 drew to a close, and by now the father of a baby daughter, Tracy, whose initials he would one day etch into the lunar dust at the valley of Taurus-Littrow, Cernan was repeatedly summoned to an unending cycle of physical and psychological evaluations and interviews by the space agency. Like so many others before him, he checked into Houston's Rice Hotel under the assumed name of 'Max Peck' and sat, "like a prisoner before the parole board", at an interview with such famous men as Al Shepard, Wally Schirra and Deke Slayton. The questions were awkward. "Someone asked how many times I had flown over 50,000 feet," Cernan wrote. "Hell, for an attack pilot like me, who had spent his life below 500 feet, that was halfway to space!" How to turn the question to his advantage? He flipped it around, telling them that he had flown very low and "if you're going to land on the Moon, you gotta get close sometime".

He was also getting close to actual selection, as friends began calling to enquire as to why FBI agents had visited them with questions about Cernan's character, his background, his military record, his educational record, his parking tickets and his disciplinary records. At the same time, he was close to completing his master's thesis, focusing on the use of hydrogen as a propulsion system for high-energy rockets. Then, just a few weeks before John Kennedy's assassination, he received the telephone call from Deke Slayton that would truly change his life. Little did he know that one of his Navy buddies, Ron Evans, rejected by NASA on this occasion, would himself be hired in 1966 and the two of them would someday travel to the Moon together.

Cernan's first two years as an astronaut were spent mired in technical assignments … and, despite being just one of a much larger gaggle of prospective spacegoing pilots, he and his colleagues still benefitted from the Life magazine deal, which nicely supplemented their military salaries. During the early Gemini flights, he occupied the 'Tanks' console in Mission Control, overseeing pressurisation and other data for the Titan II's fuel tanks. Then, one day towards the end of 1965, a technician tapped on his office door and told Cernan that Slayton wanted him to get fitted out for a space suit. The reason was inescapable: a flight assignment, surely, was just around the corner.

On 8 November, it was official: Cernan and Stafford would support Elliot See and

Charlie Bassett, with an expectation that they could then rotate into the prime crew slot for the Gemini XII mission. Four months later, just promoted to lieutenant-commander by the Navy, Cernan had a new assignment. He and Stafford were now the Gemini IX prime crew and it would be Cernan, not Bassett, who would evaluate the AMU rocket armchair during one of the trickiest and most hazardous spacewalks ever attempted.

THIRD TIME LUCKY

Five days before See and Bassett were killed in St Louis, the AMU was delivered to Cape Kennedy for testing. Initial inspections were worrisome: with nitrogen pressurant leaks from its propulsion system and oxygen leaks from its integral life-support unit. However, by mid-March, engineers had rectified these glitches and the rocket armchair was once more on track for Gemini IX's launch, planned for 17 May 1966. Right from the start, in terms of complexity, its three days aloft would mark a quantum leap even over the ambitious Gemini VIII.

Newly bumped from backup to prime crew, Tom Stafford and Gene Cernan would tackle a flight that even an internal NASA memo had dubbed "really exciting" and which, if successful, would generate "experience one would not ordinarily expect to get in less than three missions". Key tasks, aside from the lengthy EVA, would be a simulation, using the Agena, of how an Apollo command and service module would rendezvous and dock with the lunar module. Stafford and Cernan would then fire the Agena's main engine to boost themselves into a higher orbit. After the completion of the Agena rendezvous activities, Cernan would perform his spacewalk.

On 2 March, the Gemini IX spacecraft – which had so narrowly avoided destruction on the factory floor of McDonnell's Building 101 – was shipped to Cape Kennedy and its Titan II rocket was erected at Pad 19 three weeks later. By the end of the month, the spacecraft had been attached to the tip of the Titan and electrical and mechanical compatibility tests got underway in anticipation of the mid-May launch. Elsewhere at the Cape, the Atlas booster which would be used to loft Stafford and Cernan's Agena into orbit was installed on Pad 14. By early May, the Agena itself, tailnumbered '5004', had arrived at the launch complex and was mated to the Atlas.

In the small hours of 17 May, Flight Director Gene Kranz arrived at his console to oversee the launches of the Atlas-Agena and, 99 minutes later, of Gemini IX. Meanwhile, in the crew quarters at Cape Kennedy, Stafford and Cernan were awakened, underwent standard medical checks and sat down to breakfast with Deke Slayton and Al Shepard. In his autobiography, Cernan would recount keeping "a stone face, all business, but butterflies stirred in my stomach". He strung a religious medal around his neck, bearing a silver disk with the image of Our Lady of Loreto and the legend 'Patroness of Aviation, Pray for Me', then settled into a couch to have his biosensors and space suit fitted.

The heightened sense of anxiety was not helped when Slayton took Stafford aside for a private 'word'; Cernan would not learn until later what their conversation had been about. It was a conversation that Slayton would have with many a Gemini command pilot whose mission featured an EVA. Cernan's spacewalk would be an exceptionally dangerous one, Slayton told Stafford, and if something went wrong and he was unable to get back inside Gemini IX, NASA could ill-afford to have a dead astronaut floating in orbit. In such a dire situation, somehow, Stafford would have to bring Cernan's corpse back to Earth.

In his autobiography, Stafford recalled staring at Slayton in astonishment. "To bring him back," he wrote, "the hatch is going to be left partially open because the attachment point for the umbilical is inside the spacecraft near the attitude hand controller." Such an awkward re-entry would not be survivable. In reality, he told Slayton, when the explosive bolts blew at the base of the Titan, signalling liftoff, it was Stafford, as Gemini IX's command pilot, who would call the shots and make the difficult decisions if something should go wrong.

Cernan also knew that the only realistic option for Stafford would be to cut him loose, close the hatch and return to Earth alone. He understood the risks equally as well as Stafford and Slayton. "I knew Tom would be unable to pull me back inside if I couldn't get myself out of trouble," he wrote. "He would work like the devil to rescue me, but eventually would have to abandon me. We both knew it."

Slayton would have a similar conversation a few weeks later with Gemini X's command pilot, John Young, and would receive a similar reception. "There was no way," Young recounted in a 1996 interview, "if anything happened to somebody going outside a Gemini that you could get them back in." The seat was too narrow and it was impossible for the command pilot to reach over and pull an inflated, rigidised space suit with an immobile person inside back into the right-hand seat with enough overhead clearance to close the hatch. It is more than fortunate, therefore, that such an eventuality never came to pass.

By the time Stafford and Cernan arrived at Pad 19 and were strapped inside their spacecraft, all eyes were on the impending Atlas-Agena launch and a fervent hope pervaded the Cape that there would be no repeat of the Gemini VI debacle. All seemed to be going well and, at precisely 10:12 am, the rocket thundered aloft. Aboard Gemini IX, Stafford and Cernan were exuberant as the final hurdle before their own launch at 11:51 am was cleared ... or so it seemed.

One hundred and twenty seconds after liftoff, wrote Cernan, "one of the two main engines on the Atlas went weird". The No. 2 engine wobbled, then inexplicably gimballed into a full-pitchdown position, spinning the entire rocket into an uncontrollable tumble. All attempts by the rocket's stabilisation system to correct the problem were useless. Ten seconds later, as intended, the engines shut down and the needle-like Agena separated on time, but, Cernan continued, "it was too late, too low, too fast and all wrong". So wrong, in fact, that the 216-degree pitchdown had effectively pointed the Agena back towards Cape Kennedy, with a climbing angle just 13 degrees above horizontal. Worse yet, guidance was lost and the Agena plopped into the Atlantic, 198 km off the Cape, at 10:19 am.

Thirteen million dollars' worth of hardware was gone, all the result, it later became clear, of a short in a servo control circuit. Atop the Titan on Pad 19,

Stafford's first reaction, understandably, was "aw, shit", as the second Atlas-Agena of his astronaut career vanished. He and Cernan quickly inserted the safety pins back into their ejection seats' safe-and-arm devices and Guenter Wendt's team began the laborious process of extracting them from the spacecraft. Despite the disappointment, good fortune glimmered on the horizon. Gemini IX would still fly its mission, thanks to a decision made late the previous year.

A rendezvous with Gemini VIII's Agena was out of the question, since its orbit had not decayed sufficiently to be reachable by Stafford and Cernan. However, late in 1965, following the loss of Gemini VI's Agena, NASA had ordered General Dynamics to furnish a backup Atlas. In response, McDonnell prepared an alternate rendezvous vehicle, known as the Augmented Target Docking Adaptor, or ATDA. It had to be ready, the agency stipulated, within two weeks of an accident and ongoing Agena engine problems brought it close to being used on Gemini VIII. Early in February 1966, the ATDA arrived at Cape Kennedy and was placed into storage for the very eventuality that NASA now faced with Gemini IX. Within hours of the failure, NASA formally approved the use of the ATDA and its Atlas, tailnumbered '5304', for launch on the first day of June.

The tube-shaped ATDA, nicknamed 'The Blob' by the astronauts, looked very much like the Agena from the front and possessed a docking collar covered by a fibreglass cone; the latter was to be jettisoned shortly after arrival in orbit. Unfortunately, the ATDA did not have the Agena's rear fuel tanks and powerful rocket engine, just two rings of thrusters to help with rendezvous and proximity operations. To ensure that the ATDA's Atlas did not succumb to a similar failure, the cause of the 17 May mishap had to be pinpointed. Within a week, it was clear that a pinched wire in the autopilot had been responsible for the short circuit, necessitating additional work on the rocket's electrical connectors.

Following a brief return to Houston for additional simulator training, Stafford and Cernan were back in Florida in good time for the 1 June launch attempt. Nothing would stop them this time: even if the ATDA and its Atlas were lost, they intended to use the final stage of their Titan as a rendezvous target. Shortly after five that morning, they were awakened to black clouds and the knowledge that Hurricane Alma brewed somewhere in the distance. The weather had little impact on the proceedings. At 10:00 am, the Atlas lumbered off Pad 14 and within six minutes had inserted The Blob almost perfectly into a 298 km orbit. 'Almost' perfectly, that is, because telemetry data quickly indicated that the cone covering The Blob's docking collar had only partially opened and had failed to separate.

A brief conference confirmed that this problem was not insurmountable and the newly-renamed Gemini IX-A remained on schedule. Stafford and Cernan had a six-minute 'window', between 11:38 and 11:44 am, to launch, after which they would rendezvous with The Blob on their third orbit and dock high above the United States. (Conducting rendezvous progressively 'earlier' in a mission was deemed to offer the closest analogue for lunar orbital rendezvous operations.) A glitch in the Gemini's inertial guidance system halted the proceedings, setting them three minutes behind schedule in an already-tight countdown. Finally, when it could not be rectified in time, the launch was scrubbed.

Another attempt could not be made until at least 3 June, giving technicians sufficient time to refuel the Titan, check the computers and identify and resolve the glitch. Launch on the 3rd would be scheduled for 8:39:50 am, precisely timed as The Blob hurtled directly above the Cape. That morning, the two astronauts again headed for their spacecraft, Stafford in no mood for humour, having already been nicknamed 'The Mayor' of Pad 19 because he had spent so much time there over the past eight months. Cernan wondered, indeed, if Stafford was jinxed. "I straightened him out," Stafford recounted in his autobiography. "Schirra and Cernan were the jinxes. I was fine!"

Some of the pad personnel still could not resist, however, hanging a large sign on the door to the gantry elevator which read 'Tom and Gene: Notice the 'down' capability for this elevator has been removed. Let's have a good flight.' Stafford and Cernan's backups, Jim Lovell and Buzz Aldrin, had even composed and hung their own poetic verse over the Gemini's hatches. It read: 'We were kidding before / But not anymore / Get your ... uh ... selves into space / Or we'll take your place'. Humour aside, Cernan later wrote, "it would be a cold day in hell before Buzz Aldrin flew as the pilot of Gemini IX instead of me".

The potential for another glitch reared its head in the closing minutes when mission controllers transmitted a final update to the inertial guidance system and it again refused to respond. This time, however, it was decided to override it with another successfully-received trajectory update from 15 minutes earlier. Cernan described the liftoff as "just ... different" and nothing at all like he expected it to be. "I sensed movement," he wrote in his autobiography, "a feeling of slow pulsation and then heard a low, grinding rumble as that big rocket started to lift away from Earth in agonisingly slow motion."

That slow-motion start quickly gave way to the increasing sensation of tremendous speed as the Titan headed away from the Cape and thrust Stafford and Cernan, both gritting their teeth, towards orbit. As he saw and felt things never experienced before, Cernan wished he were a poet and could adequately describe what was happening. Eight minutes after launch, hurtling through the high atmosphere under the push of the rocket's second stage, the astronauts found talking was restricted to grunting as 7.5 G imposed huge pressures on their lungs.

That sensation was soon replaced, when the second stage shut down, by one that Cernan had never known before: the onset of zero gravity. "A few nuts and bolts left behind by workers oozed out of their hiding places," he wrote. "Dust particles and a piece of string did a slow dance before my nose. My hands drifted up in the weightlessness and my legs, wrapped in those metal pants, became featherlight." Glancing through his tiny window, Cernan beheld the unmistakable shape of Africa, speckled with white clouds, and a distant glint of ocean. There was little time to gawp. He and Stafford had a date with an alligator.

THE ANGRY ALLIGATOR

Following insertion into a 158 × 267 km orbit, Gemini IX-A's computers set to work determining the rendezvous flight path. Forty-nine minutes into the mission, an inaugural manoeuvre raised their perigee to 232 km, prompting Cernan to remark that he "felt that one, Tom!" A second firing corrected phase, height and out-of-plane errors and established them in an orbit of 274 × 276 km, after which they checked their spacecraft's systems and stowage lists, removed their helmets and gloves and readied cameras for the rendezvous ahead.

The astronauts acquired their first spotty radar readings at a distance of 240 km from The Blob and had a solid 'lock' at 222 km. This led to visible relief on the part of the radar's Westinghouse builders, who had worried that the unstabilised ATDA and its changing radar reflectivity would cause its acquisition to wobble. Three hours and 20 minutes after launch, the astronauts were rewarded with their first glimpse of the target – now just 93 km away – and as they drew closer saw its flashing acquisition lights. Thinking that the shroud must have jettisoned successfully (the lights could not be seen otherwise), Stafford began slowing Gemini IX-A's approach profile ... and the reality became clear: the shroud was actually gaping half-open, like an enormous pair of jaws. "It looks," he told Mission Control, "like an angry alligator."

Initial hopes that he might be able to nudge it with his spacecraft's nose to fully open the jaws were rejected as too risky by Flight Director Gene Kranz and Stafford was forced instead to station-keep less than 12 m away. It was clear, he reported, that the ATDA's explosive bolts had fired, but two neatly-taped lanyards stubbornly held the shroud in place. The high tensile strength of these lanyards made it unadvisable to nudge the jaws. Moreover, Gemini IX-A's parachutes were housed in its nose and damaging them was unthinkable.

On the ground, at a strategy meeting that night with Bob Gilruth and Chris Kraft, backup pilot Buzz Aldrin suggested sending Cernan outside to manually clip the lanyards with a pair of surgical scissors. Astronauts Jim McDivitt and Dave Scott, in Los Angeles at the time, were despatched to the Douglas plant to examine a duplicate ATDA and determine if this could be done. Their consensus: it was possible, but would leave many sharp edges which could tear Cernan's suit. Also, the tumbling of the ATDA, the almost-complete lack of spacewalking experience and the dangers of the explosive bolts holding the lanyards together posed their own risks. "Gilruth and Kraft were aghast," wrote Deke Slayton and the suggestion, though not entirely outrageous, would lead to the first of many discussions about Aldrin's suitability to fly Gemini XII.

In the meantime, efforts by controllers to tighten and relax The Blob's docking cone, in the hope that the action might free the shroud, were unsuccessful. "That only pushed out the bottom part of the shroud," wrote Cernan, "and forced the other end, which was open, to partially close. Contracting the collar had the reverse effect, and to us, it seemed that those moving jaws were opening and closing." The alligator, quite literally, was laughing at their misfortune.

After the mission, it would become clear that the problem centred on the fact that the Agena, the ATDA and the shroud were built by three different organisations, namely Lockheed, McDonnell and Douglas. Before McDonnell technicians had made a final inspection on the ATDA at Cape Kennedy, a Douglas engineer had supervised a practice run, with the exception of the lanyards which controlled the electrical disconnect to the explosive bolts. In the interests of safety, the lanyards were not hooked up for the test.

Crucially, the Douglas engineer was then forced to leave to return home and tend his pregnant wife, telling his McDonnell counterpart to "secure the lanyards". Consequently, on launch day, the McDonnell crew followed procedures published by Lockheed, which had themselves been copied from Douglas documentation. The instructions referred to a blueprint which was not present and the absence of the engineer meant that those technicians responsible for fixing the ATDA's shroud simply wondered what to do with the dangling lanyards and decided that their best and safest bet was to tape them down. It was those taped-down lanyards which had now ruined Stafford and Cernan's target in orbit.

Over the years, some historians have commented that having several companies managing different parts of the same vehicle was simply a classic extension of the metaphor 'too many cooks spoil the broth' and, indeed, an investigation into the ATDA fiasco would later conclude that future simulations should be practiced completely, experienced people should remain 'on the job' and written instructions should be followed exactly.

In the meantime, five hours into the Gemini IX-A mission, Stafford nosed his spacecraft 'down' by 90 degrees and fired his forward thrusters for 35 seconds to enter an elliptical equi-period orbit with the ATDA. Simulating a failed radar, they then plotted their position with an on-board sextant, notepad and pencil, checked their results against a pre-planned chart solution and commenced a series of four manoevures to bring themselves back into a station-keeping stance with the target. It was far from easy and, wrote Cernan, represented "a bitch of an exercise that demanded unimagined mental and physical effort". Nonetheless, six and a half hours after launch, they were finally in the vicinity of The Blob, only to depart again shortly thereafter for a third exercise. To prepare for this, at 3:55 pm, a little over seven hours into the mission, Stafford again pulsed the OAMS thrusters to reduce speed and widen the gap between Gemini IX-A and the ATDA.

By now exhausted, the two astronauts checked their systems, took an opportunity to gobble some toothpaste-like mush of chicken and dumplings – "No crumbs that way," wrote Cernan, but "not much taste, either" – and tried with little success to sleep. Awakened in the small hours of 4 June to begin their second day in orbit, they were almost immediately immersed in the third rendezvous: reducing the size of their orbit to again intercept the still-laughing Blob. By rendezvousing with an object 'beneath' them, Stafford and Cernan would mimic the procedures to be followed by an Apollo command module pilot tasked with rescuing a lunar module stuck in a low orbit around the Moon.

Phase and height adjustments, followed by an OAMS burst, placed Gemini IX-A in an orbit of 307 × 309 km and within three hours the astronauts had reduced the gap between themselves and the target to just 28 km. At this stage, by now still 'above' and 'ahead' of the ATDA, Stafford nosed 19 degrees down and yawed 180 degrees to the left. "The mental perception was that we were falling straight down to Earth," Cernan recalled years later, "and we did not even see the gator until we were within three miles of it." Stafford, too, later admitted to sensations of mild vertigo.

Stafford and Cernan's first close-up glimpse of The Blob, which Tom Stafford rather appropriately nicknamed "an angry alligator".

At this point, Stafford spotted what appeared to be "a pencil dot on a sheet of paper" and would point out that, had it not been for the radar, the rendezvous would have failed. The rendezvous was completed at 6:21 am and Stafford and Cernan withdrew from the ATDA at 7:38 am, this time for good.

The two men felt justifiably proud: they had conducted no fewer than three rendezvous in less than a day. However, their work had taken its toll. Both were exhausted, as, indeed, was their spacecraft, whose fuel supply had dwindled from 311 kg at launch to less than 25 kg after the marathon rendezvous effort. Ahead, later on 4 June, lay Cernan's spacewalk, but Stafford donned his command pilot's cap and told Mission Control that the excursion should be postponed. "We've been busier'n left-handed paper-hangers up here," he drawled. "I'm afraid it would be against my better judgement to go ahead and do the EVA at this time ... Perhaps we should wait until tomorrow morning." For the first time, Cernan wrote, a pair of astronauts had seemingly 'questioned' their duties and, although not a military organisation, some within NASA felt that they were quitting. Yet there was little doubt that Stafford and Cernan were best placed to know the situation inside their spacecraft and Capcom Neil Armstrong duly responded that their recommendation had been accepted. Armstrong would later describe Stafford's actions as reflecting "exceptionally good judgement".

As a result, the EVA was moved to 5 June and the remainder of the day was spent focusing upon Gemini IX-A's experiments and ensuring that both men were fully rested. Stafford and Cernan's experiment load consisted of seven tasks, one of which was a medical study to measure their reactions to stress by recording their intake and 'output' of bodily fluids before, during and after the mission. Codenamed 'M-5', it required their wastes to be collected and labelled; a complicated, tricky and messy process whose physical requirements, Stafford growled, amounted to those required for doing a rendezvous and a half.

Elsewhere, Edward Ney's zodiacal light photography experiment was originally planned to be used during the EVA, but problems forced it to be used instead from inside Gemini IX-A. It consisted of a hand-held camera, equipped with automatic triggering, to obtain images of atmospheric airglow, the zodiacal light, the Milky Way and the celestial field. Overall, Stafford and Cernan would return home with 44 useful images, together with 160 Hasselblad terrain photographs which would prove useful in applications from geology to oceanography. The remaining experiments – including retrieving a micrometeorite collector and controlling the AMU rocket armchair – were assigned to Cernan's spacewalk. That spacewalk, which began early on 5 June, would force managers and astronauts to rethink everything they thought they knew about extravehicular activity.

HELLISH SPACEWALK

Original plans, dating back to before the deaths of Elliot See and Charlie Bassett, called for the Gemini IX spacewalker to spend at least two hours outside, remove the AMU from its housing at the back of the spacecraft's adaptor and test it. He was also supposed to retrieve a micrometeorite package from the Agena, although this was scratched from the flight plan when the target ended up at the bottom of in the Atlantic on 17 May. A subsequent plan to remove a micrometeorite detector from the ATDA was also called off when it proved impossible to dock.

Still, preparations for the excursion were intense. Early on 5 June, Stafford lowered Gemini IX-A's orbit while Cernan pulled his chest pack down from a shelf above his left shoulder, strapped it on and plugged in a 7.5 m umbilical tether which would provide him with oxygen, communications and electrical power. Years later, he would describe removing the umbilical from its container and attaching it to his suit as akin to unleashing a garden hose in a space no larger than the front seat of a car. Obviously, since the whole cabin would be reduced to vacuum, Stafford also had to be protected and both men laboriously clicked their helmet visors shut, pulled on heavy gloves and pressurised their suits until they went, in Cernan's words, "from soft to rock-hard around our bodies".

Yet Cernan's suit had much more insulation and protection than that of Stafford. "Out where I was going," he wrote, "the temperature in unfiltered sunlight would be many times hotter than any desert at high noon on Earth, while the nighttime cold could freeze steel until it was as brittle as glass." Approaching dawn on their 31st orbit, the men received permission to go ahead and at 10:02 am Cernan twisted the handle above his head and the huge hatch swung outwards.

Words clearly defied even the normally-chatterbox Cernan at this point as he pushed himself 'upwards', stood on his seat and rode "like a sightseeing bum on a boxcar" towards the California coastline. Hollering "hallelujah" at the top of his voice, he would later describe the glorious, ever-changing sight as like "sitting on God's front porch", as orbital darkness gave way, almost instantaneously, to the first stirring of a shimmering dawn.

As was typical in space, there was little time to sightsee. With Stafford holding onto his foot to steady him, Cernan set to work positioning a 16 mm Maurer movie camera on its mounting and retrieving a nuclear emulsion package which recorded radiation levels and measured the impact of space dust. Next, he affixed a small mirror onto the docking bar on Gemini IX-A's nose, such that Stafford could watch as he made his way towards the AMU at the rear of the spacecraft. Unlike Ed White, Cernan was not equipped with a hand-held zip-gun and he quickly set to work on his next task: to evaluate his ability to manoeuvre himself around by tugging at his snake-like tether.

It would, he wrote in his autobiography, teach him new lessons about Newton's laws of motion. "My slightest move would affect my entire body, ripple through the umbilical and jostle the spacecraft," Cernan explained. "Since I had nothing to stabilise my movements, I went out of control, tumbling every which way, and when I reached the end of the umbilical, I rebounded like a bungee jumper, and the snake reeled me in as it tried to resume its original shape." As he looped around Gemini IX-A, the experience was comparable to wrestling an octopus and Cernan's only chance at controlling his motions came when he managed to grab the tether tightly at the point at which it emerged from the hatch.

After half an hour of helplessness – and by now having broken the spacewalk

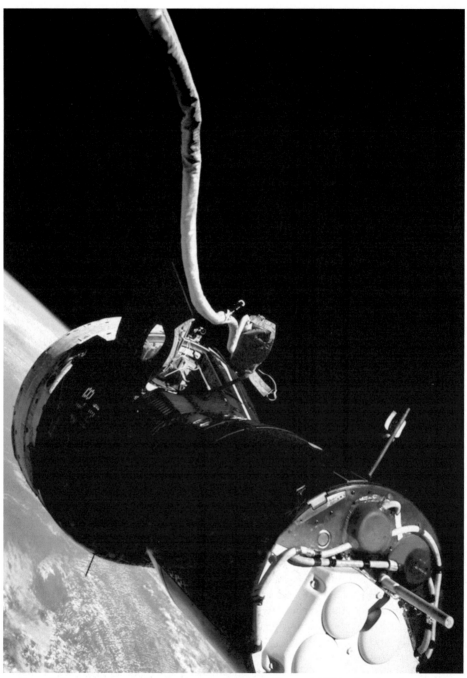

Two of the few photographs acquired during Gene Cernan's EVA, showing the nose of Gemini IX-A, the open pilot's hatch and the snake-like tether at left and the astronaut himself at right.

endurance times of both Alexei Leonov and Ed White – he somehow seized a handrail and pulled himself towards Gemini IX-A to rest. Clearly, he told Stafford and Mission Control, future spacewalkers would need propulsion and more handholds; otherwise they would be unable to prevent themselves from flopping around like rag dolls on the end of their tethers. Cernan's rest break was brief: he had to reach the back of the spacecraft before the arrival of orbital dusk to checkout and strap on the AMU, exchange his oxygen umbilicals for those attached to the rocket armchair and commence the next phase of his spacewalk.

His move to the rear of Gemini IX-A was much harder than he could have anticipated. The stiff, bulky suit fought his every move and lacked the two crucial ingredients – flexibility and mobility – that he now desperately needed. Nonetheless, Cernan laboured, hand-over-hand, along a small rail, halting at times to loop his tether through tiny eyelets and thus keep it from damage. Finally, he reached the adaptor at the back of the spacecraft and, swinging himself around it, disappeared from view in Stafford's mirror. The Sun, too, vanished as Gemini IX-A entered orbital darkness over South Africa.

Working in near-pitch blackness, Cernan flicked on a pair of lights – only one of which worked, yielding a glow little more effective than a candle – and prepared to activate the AMU. Thirty-five meticulous steps lay between him and achieving the goal of becoming the first human satellite; steps ranging from pushing buttons to opening valves and disconnecting, then reconnecting, his oxygen supply. His heart rate, which had reached 155 beats per minute when he arrived at the adaptor section, showed no signs of slowing as Cernan puzzled over why he had been able to accomplish the task with ease in a parabolic aircraft and yet the real thing was leaving him exhausted, drenched with sweat and almost blind. At last, he flipped the last switch and prepared to take the AMU on its maiden outing.

All was far from being well. A hundred minutes into the spacewalk, Cernan was scarcely able to see through his fogged-up visor – the suit's environmental control system was struggling and failing to absorb the humidity and exhaled carbon dioxide – and his heart rate soared to 195 beats per minute. Unable to wipe the stinging sweat from his eyes, he had no choice but to rub his nose on the inside of his visor just to make a 'hole' through which he could see. He also tried increasing the oxygen flow to his suit in a bid to clear the visor, without success.

Cernan's lack of visibility could hardly have come at a more inappropriate time, precisely as he was completing the intricate procedure of readying the AMU to fly. At one stage, he even had to rely on the reflection in a polished metal mirror on his wrist and on his sense of touch through his thickened gloves for guidance. Merely turning knobs, without adequate leverage, was virtually impossible. So too was telescoping and folding out the AMU's armrests – getting them extended into place was, he wrote years later, "akin to straightening wet spaghetti".

It was at this point that one of Gemini IX-A's experiments – known as 'D-14', a UHF/VHF polarisation study – met its untimely end at Cernan's hands. The instrument comprised an extendable antenna in the adaptor section, which had been used successfully during the first portion of the mission to measure inconsistencies of the electron field along Gemini IX-A's flight path. The astronauts had operated it five times whilst above Hawaii and once over Antigua, but Cernan's struggles with his suit and the AMU caused him to accidentally break it.

Eventually, after much tugging and twisting, he found success, slid onto the saddle and strapped himself into place. His next step was to disconnect himself from the tether and reconnect himself to the backpack's life-support and communications supplies. From his position, inside the concave steel adaptor at the rear of Gemini IX-A, he temporarily lost communications with Stafford, who could barely hear Cernan's crackled garble that he was unable to see in front of his own eyeballs. From the command pilot's seat, Stafford was now worried for his colleague's safety, advising Mission Control that communications had degraded and Cernan's visibility through his visor was so poor that the AMU test was risky.

On the ground, the physicians were coming to similar conclusions: data from Cernan's biomedical sensors clearly indicated that he was exhausted, expending energy at a rate equivalent to running up a hundred stairs per minute and his heart was pumping three times faster than normal. Cernan knew that their judgement could spell the end of his spacewalk ... an eventuality that, as a pilot who had been training for more than six months, he had no wish to contemplate. At length, the decision was snatched out of his hands.

The onset of orbital dawn over the Pacific brought the garbled news from Stafford: "It's a no-go ... because you can't see it now. Switch back to the spacecraft electrical umbilical." The Hawaii capcom concurred with his judgement. Obviously disappointed that he had not only lost his chance to fly the AMU, but that the Air Force's $10 million rocket armchair was destined to burn up in the atmosphere, Cernan unstrapped and clawed his way back to his hatch. To protect the interior of Gemini IX-A from solar radiation, he had left it partially closed and was now blinded by the Sun as he struggled to find it.

Finally gripping and pulling open the hatch, Cernan twisted himself and pushed his feet through the opening. Stafford manually reeled in the umbilical, then grabbed one of his suited ankles to anchor him back inside the cabin. As he tried to get back inside, Cernan inadvertently kicked the Hasselblad camera that Stafford had been using to photograph the EVA and it drifted off into space. "There went my still pictures," he wrote later, "but I did retrieve the movie camera."

Scrunching himself painfully into his seat, still fighting against the stiffness of the suit, he quickly found that he could not close the hatch. Eventually, with Stafford's help, the pair managed to yank it down and Cernan pumped the handle until the hatch was secure. In his autobiography, he would admit that the pain was so intense that he cried aloud – "but only Tom really knows" – and was close to losing consciousness. Then, as Stafford began repressurising Gemini IX-A's cabin, Cernan felt the rigidity of the suit begin to soften and he was finally able to breathe properly and remove his helmet. The United States' second spacewalk was over in two hours and eight agonising minutes.

Exhausted, the now-beetroot-faced Cernan was doused with weightless droplets fired by Stafford from a water pistol and strips of skin of his swollen hands tore away as he removed his gloves. He looked, wrote Stafford, "like he'd been baked in a sauna too long". However, with the exception of the reaction he might get from the other astronauts – had he screwed up? and would he ever fly again? – Cernan really did not care. He had endured the most traumatic spacewalk to date … and, astonishingly, had lived!

Less than a day later, at 9:00 am on 6 June, Gemini IX-A was bobbing in the Atlantic. Cernan described his first fiery re-entry through the atmosphere as "like a meteoric bat out of hell" and compared the spacecraft to having the aerodynamic characteristics of a bathtub as it plummeted Earthward. They splashed down safely just 700 m from the intended point. So close were they to their prime recovery vessel, the aircraft carrier Wasp, that they were able to offer and acknowledge thumbs-up signals. An hour after hitting the Atlantic, they and their spacecraft were safely aboard.

In his autobiography, however, Cernan would relate that their first moments after splashdown were not entirely idyllic, when rough waves and strong winds gave the impression that Gemini IX-A's hull had been ruptured. In fact, a harder-than-anticipated landing had ruptured a drinking water line, spilling its contents into the cabin. Still, the discomfort and disappointment was sweetened by the splashdown. It was, wrote Stafford with justifiable pride, "the closest-to-target landing of any manned spacecraft in history" prior to the Shuttle.

Gene Cernan's harrowing EVA would teach a harsh, yet valuable lesson to those engineers, managers and even astronauts who perceived extravehicular activity as a proverbial walk in the park. Why, some journalists asked him in the weeks that followed, was his spacewalk so difficult in comparison to Ed White's graceful stroll? The key differences, of course, were that White had been equipped with a hand-held propulsion device and that, other than floating around, he was not actually given any specific tasks.

Yet Cernan's problems – the shortcomings of his suit's environmental controls, the fogging of his visor, the difficulties encountered when getting back into the spacecraft, the need for handholds, the impossibility of moving without a propulsion device – highlighted an urgent need for such issues to be rectified before the closure of the Gemini chapter in November 1966. Apollo managers, then hard at work preparing for the first flight of their spacecraft in the fourth quarter of 1966, also took heed: future Moonwalkers could not operate on the lunar surface for many hours under such life-threatening conditions. It is quite remarkable, therefore, that by the time Cernan's backup, Buzz Aldrin, completed his own EVAs on Gemini XII, the problems would have been virtually resolved.

"NOTHING SPECIAL"

Deke Slayton and others have freely admitted that they were forced to rethink the practicalities of EVA in the seven-week interval between Geminis IX-A and X. Fortunately, the latter mission would star as its spacewalker a man who, perhaps more so than any other astronaut, knew the G4C pressure suit literally inside out. Michael Collins, who described himself as "nothing special", "lazy" and "frequently ineffectual", would later gain eternal fame as 'the other one' on the Apollo 11 crew.

Before that, as John Young's pilot on Gemini X, he would become the first man to make two extravehicular activities, the first man to physically touch another vehicle in space ... and, alas, the first spacewalker to bring home absolutely no photographic record of his achievement.

Slayton saw Young and Collins as a perfect team. Both were obsessive hard-workers, but in contrast to Young's reserved and publicity-shy nature, the gregarious Collins was "smooth and articulate". Prior to his selection in October 1963, to improve this smoothness, the Air Force sent its astronaut applicants to 'charm school', in which Collins learned more social skills essential for spacefarers: wearing knee-length socks "that go on forever", abhorring hairy legs and needing to hold hands on hips in a particular way "because people you don't want to talk about hold 'em the other way!"

With a father, uncle and elder brother who would all rise through the ranks to become generals, it was obvious that Collins would follow in their military footsteps. He entered the world in Rome on 31 October 1930, becoming the first American astronaut born outside the United States, and throughout his childhood was often on the move: from Italy to Oklahoma, to Governor's Island in Upper New York Bay, to Maryland, to Ohio, to Puerto Rico and to Virginia. Whilst in Puerto Rico, Collins took his first ride in a twin-engined Grumman Widgeon, although he would admit that as graduation from West Point neared in 1952, his "love affair with the airplane had been neither all-consuming nor constant".

Nonetheless, he graduated from the Military Academy in the same class as fellow astronaut-in-waiting Ed White and his eventual choice of the Air Force as his parent service was based on two factors. The first was sheer wonder over where aeronautical research would lead in years to come ... whilst the second was simply to avoid accusations of nepotism, "real or imagined", since his uncle happened to be the Army's chief of staff at the time! As a cadet, Collins completed initial flight training in Mississippi aboard T-6 Texans, before moving on to jets at Nellis Air Force Base in Nevada, flying the F-86 Sabre.

Nuclear-weapons-delivery training followed at George Air Force Base in California, as part of the 21st Fighter-Bomber Wing, and Collins transferred with the detachment to Chaumont-Semoutiers Air Base in France in 1954. Two years later, whilst participating in a NATO exercise, he was forced to eject from his F-86 when a fire erupted behind his cockpit. He met Pat Finnegan in the officers' mess and, despite their differing religious beliefs – she being a staunch Roman Catholic, he a nominal Episcopalian – the couple married in 1957.

Subsequent work as an aircraft maintenance officer, during which "dismal" time he trained mechanics, was followed by a successful application to join the Experimental Flight Test Pilot School at Edwards in August 1960. It involved flying on a totally new level. "Fighter pilots can be impetuous; test pilots can't," Collins recounted years later. "They have to be more mature, a little bit smarter ... more deliberate, better trained – and they're not as much fun as fighter pilots." By this time, he had accumulated over 1,500 hours in his logbook, the minimum requirement for a prospective student at the exalted school. (In fact, his class included future astronauts Frank Borman and Jim Irwin.)

An exhausted Cernan puts on a brave face for Tom Stafford's camera after finally removing his helmet. The world's longest EVA to date had uncovered a chilling reality: that spacewalking was hazardous and by no means routine.

Two years later, when John Glenn completed America's first orbital spaceflight, Collins took notice and submitted his application for the 1962 astronaut intake. He underwent the full physical and psychological screening process, narrowly missing out on selection and, despite his disappointment, moved on to study the basics of spaceflight, flying the F-104 Starfighter to altitudes of 27 km and receiving his first taste of weightlessness. He had barely returned to fighter operations when, in June 1963, NASA announced its intent to choose more astronauts. Years later, Deke Slayton would write that the 1962 selection panel considered Collins a good candidate who had been "held back to get another year of experience".

Initial instruction as part of the third class of 14 spacefarers, whom the press widely dubbed 'The Apollo Astronauts', included lunar geology, a subject for which Collins had no great enthusiasm or interest; ironic, perhaps, in view of where his career would eventually take him. Although he felt, like Slayton, that the New Nine was probably the best all-round astronaut group yet chosen, Collins admitted that the Fourteen were the best-educated: with average IQs of 132, an average 5.6 years in college and even an ScD among them.

Completion of initial training led to assignment to oversee the G4C extravehicular suit and he would express annoyance at being left out of the loop in May 1965 when a closed-door decision was made to give Ed White a spacewalk on Gemini IV. In his autobiography, Collins described the suit and the astronaut's relationship with it, as "kind of love-hate ... love because it is an intimate garment protecting him 24 hours a day, hate because it can be extremely uncomfortable and cumbersome". The suit, and the timeline for which astronauts were to get fitted for it, provided a never-ending source of rumour as to who would be assigned next to a mission slot.

Recognition for this work came in June 1965 with a backup assignment, teamed with Ed White, to Gemini VII. Despite falling ill with viral pneumonia shortly thereafter, Collins recovered promptly and performed admirably, even taping a 'Home Sweet Home' card inside Jim Lovell's window on launch morning. His eventual assignment, with John Young, to Gemini X came in January 1966, by which time White had been named to the first Apollo mission. "I was overjoyed," wrote Collins. "I would miss Ed, but I liked John, and besides I would have flown by myself or with a kangaroo – I just wanted to fly."

SUCCESS AT LAST

Gemini X overcame many of the obstacles encountered by its predecessors; for during the course of Young and Collins' three days aloft, the techniques of rendezvous, docking, spacewalking and boosting themselves into a higher orbit would all be successfully accomplished. Nor would Gemini X perform just 'any' rendezvous. Its crew would conduct joint activities with not one, but two targets: with their own Agena and, after raising their orbit to around 400 km, with the now-dead Agena left in space by Armstrong and Scott back in March 1966.

"We were going to be navigational guinea pigs," wrote Collins, "and were going to compute on-board our spacecraft all the manoeuvres necessary to find and catch our first Agena, instead of using ground-computed instructions to get us within range of our own radar." To help them, Gemini X would have an expanded computer memory, known as 'Module VI', which required Collins to use a portable sextant to measure angles between target stars and Earth's horizon. "Combining Module VI data with a variety of charts and graphs carried on-board," he continued, "we would be able to determine our orbit and predict where we would be at a given future time relative to our Agena target."

The downside was that Module VI would operate in a totally different manner to Apollo's navigation system, rendering it, in Collins' words, a "technological dead-end". Still, he and Young were being granted the test pilot's dream job: to compute their own manoeuvres, autonomously of Mission Control, and the two men embraced it warmly. Yet there would be no avoiding the fact that completing all of their objectives in just three days would be a tough call. Young even approached Charles Mathews with a view to extending Gemini X to its four-day consumables limit, but his request was flatly rejected.

Young's concerns, wrote Barton Hacker and James Grimwood, were twofold. Firstly there was the issue of whether he could slow the docked vehicles – his own Gemini and Agena-X – sufficiently to avoid hitting Agena-VIII and secondly was the very real possibility that he might be unable to find the second target using only on-board optical equipment. "The problem with an optical rendezvous," Young said later, "is that you can't tell how far away you are from the target. With the kind of velocities we were talking about, you couldn't really tell at certain ranges whether you were opening or closing."

Moreover, since Agena-VIII was by now out of action and totally passive, Young and Collins and their Agena would need to be despatched from Cape Kennedy at very precise times, which presented its own problems should there be a launch failure or delay. By the opening months of 1966, a plan had crystallised: Gemini X would rendezvous and dock with its Agena during its fourth orbit, after which the second day of the mission would be dedicated to experiments and Agena-VIII operations would be deferred to the third day. Since many of its planned experiments could not be conducted whilst Gemini X and the Agena-X were docked, a happy balance of compromise would have to be struck between the engineering and scientific communities. In the eventuality that Young and Collins' Agena, like that of Stafford and Cernan, should meet an unhappy end, it was decided to press on with the Agena-VIII rendezvous instead.

Preparations for Gemini X had gone relatively smoothly both before and after the assignment of the crew and their backups (Jim Lovell and Buzz Aldrin) in January 1966. In fact, when Young and Collins were named to the mission, their launch date did not change: they were scheduled for 18 July ... and on 18 July, without a single delay, they would fly. The only change came in March, when Lovell and Aldrin were shuffled to the backup slot on Gemini IX and their places were taken by rookies Al Bean and Clifton 'C.C.' Williams.

Bean, who would one day become the fourth man to walk on the Moon, has been seen by some historians as an astronaut whose talent was overlooked by Deke Slayton; indeed, he would be the last of the surviving members of the 1963 class to fly in space. However, in his autobiography, Slayton vehemently disagreed. "Al was just a victim of the numbers game," he wrote. "I would only point to the fact that he was the first guy from his group assigned as a crew commander. I was confident he could do the job if anything happened to John Young." Ultimately, Bean would draw two of NASA's best missions: a lunar landing and command of the second Skylab crew in 1973.

Meanwhile, the Titan II's second-stage fuel tank had to be replaced with one previously earmarked for Gemini XI when a leaking battery caused some corrosion of its dome, but the remainder of the processing flow ran smoothly. After formal acceptance by Martin in mid-April 1966, the Titan's two stages were shipped to Cape Kennedy late the following month, mated together and installed on Pad 19 on 7 June. Two days later, the Gemini X spacecraft itself arrived at the launch complex

The crew of Gemini X: the gregarious Mike Collins (left) and reserved John Young.

and was attached to the top of the booster. Processing of the mission's Agena target vehicle – designated GATV-5005 – proved similarly faultless and it was accepted by the Air Force and moved to the Eastern Test Range on 16 May. Nine days later, it was mated to its target docking adaptor and, following several weeks of further testing, was taken to Pad 14 and mounted atop its Atlas carrier rocket on 1 July.

There was no avoiding the reality that the Gemini X double rendezvous would be by far the most complex ever attempted. A late afternoon liftoff was necessary to intercept the Agena-VIII and, in the final few weeks, Young and Collins found themselves shifting their sleep patterns accordingly. "For the last couple of days," Collins wrote, "we were staying up until 3 or 4 am and sleeping until noon. Granted, we were staying up studying, but somehow the late hours carried with them a connotation of leisure and relaxation."

They were awakened in their Cape Kennedy crew quarters around noon on 18 July 1966 and their Agena lifted-off without a hitch at 3:39 pm, barely two seconds late. It was followed promptly into orbit by the two astronauts at 5:20 pm. Collins would describe 'feeling' the rumble of the Titan, with minute sideways jerks, virtually no sensation of speed and, at first, a noticeable increase of little more than 1 G. The vibrations intensified as the rocket continued to climb, with mild pogo effects causing "a high-frequency quivering of body and instrument panel".

The violent shock of staging – as the Titan's first stage, exhausted, fell away and its second stage picked up the thrust – was soon replaced by a sense of serenity as Gemini X headed towards its initial orbit. At one point, spectators at the Cape thought the rocket had exploded. In fact, it had. "An instant after the two stages separated," Collins related, "the first-stage oxidiser tank ruptured explosively, spraying debris in all directions with dramatic, if harmless, visual effect." The G forces gradually began to climb, peaking at close to 7 G, before ending abruptly when the Titan's second stage shut down, precisely on time.

On the ground, Capcom Gordo Cooper flipped the wrong switch, opening the communications loop and allowing Young and Collins to overhear him summoning all launch personnel to a debriefing in the ready room. Collins radioed that he and Young were otherwise engaged and would be unable to attend.

There was much to do. Immediately after unstrapping, Collins unstowed the Kollsman sextant to begin the lengthy optical navigation procedure and Young manoeuvred the spacecraft into position to begin its six-hour pursuit of the Agena-X. Minor difficulties were encountered when Collins mistook the thin line of atmospheric airglow for Earth's horizon, then could not get the lens of the sextant to work properly. He tried an Ilon instrument instead, with limited success.

Eventually, Gordo Cooper told the two men that they would have to rely on ground computations instead. More trouble, however, was afoot. When Young pulsed the OAMS thrusters to adjust Gemini X's orbit to 265×272 km, he did not realise that the spacecraft was slightly turned, introducing an out-of-plane error, and needed to perform two large midcourse corrections. It was not Young's fault, Deke Slayton wrote, because "there had been a mistake in loading the initial guidance program into the Gemini computer, so the spacecraft was never quite where the instruments said it was".

The computer had actually yielded a figure some 2 m/sec greater than that radioed up by Mission Control. Collins' own slide-rule measurements agreed with Gemini X's computer and, satisfied, Young opted to follow this double-checked on-board solution. Ten minutes into the OAMS manoeuvre, he quickly discovered that he had 'overthrust' the spacecraft and was heading into an orbital path several kilometres 'behind' and 'above' the Agena. He tried to 'dive' Gemini X back onto its correct trajectory through sheer brute force and succeeded in saving the manoeuve, but at the expense of losing a large quantity of fuel. Radar acquisition of the Agena-X was achieved during their second orbit. Young's all-out effort to reach the target consumed 180 kg of propellant – three times more than any previous mission, leaving just 36 per cent remaining – but a successful docking was 'in the bag' at 11:13 pm, almost six hours after Gemini X's launch.

The excessive fuel usage prompted Flight Director Glynn Lunney to abandon plans for 'docking practice', in which Young would have pulled away and then redocked with the Agena, and the astronauts were instead advised to press on with the Agena-VIII pursuit. In support of this goal, the Agena-X's main engine roared to life, precisely on time, increasing the combined spacecraft's velocity by 420 km/h and setting them on course. "At first," Young explained later of the first 'space switch', "the sensation I got was that there was a pop, then there was a big explosion and a clang. We were thrown forward in the seats. Fire and sparks started coming out of the back end of that rascal. The light was something fierce and the acceleration was pretty good." With abrupt suddenness, and precisely on time, the Agena-X's engine shut down with what Young described as "a quick jolt". Since the Gemini and Agena were docked nose-to-nose, the firing propelled the astronauts 'backwards', producing so-called 'eyeballs-out' acceleration forces, as opposed to the 'eyeballs-in' experienced in a launch from Earth.

Comparing notes, Young and Collins both felt that the Agena's acceleration and their change of direction – now travelling 'backwards' – were much greater than they had anticipated. The closest terrestrial analogy to the swift onset of thrust, they thought, was that it was similar to riding the afterburner of a J-57 jet engine, which Young had done as an F-8 Crusader aviator and Collins had experienced during his days flying the F-100 Super Sabre.

The first burn had effectively increased Gemini X's apogee to 763 km, allowing the two astronauts to gaze down upon Earth from a greater vantage point than had ever been witnessed by human eyes; far higher than the 473 km achieved by Pavel Belyayev and Alexei Leonov aboard Voskhod 2. It also marked the first time that a manned spacecraft had employed the propulsion system of another vehicle to power its own flight, a technique which had important implications for future orbital refuelling. As they busied themselves with calculations for the remainder of the pursuit and rendezvous, they snapped photographs through the windows, including some of the Red Sea, and both would later recount that their home planet from an altitude of 763 km possessed a definite curvature.

Sleep posed its own problems for Collins. "My hands dangle in front of me at eye level," he wrote, "attached as they are to relaxed arms, which seem to need gravity to hold them down." Fearful that they may accidentally trip switches whilst asleep, he debated whether to tuck them behind his head (uncomfortable) or in his mouth (ill-advised), before finally telling himself not to worry. Such worries were further allayed by the two window shades, onto which had been pasted "photos of two voluptuous, wildly beautiful girls". Nursing a painful knee, Collins popped a couple of aspirin and tried with limited success to nap.

After their first restless night, the astronauts were awakened early on 19 July by updated computations for their next Agena burn, which Young duly executed for 78 seconds to reduce Gemini X's velocity by 300 km/h and lower its apogee to 382 km. This second firing of the rocket's engine impressed them both: after almost a day in microgravity, the 1 G acceleration, said Young, was "the biggest 1 G we ever saw!" Since the thrust was against the direction of travel, it produced a braking effect, after which a third firing circularised the orbit by raising the perigee to 377 km, barely 17 km 'below' Agena-VIII. At this point, they were trailing their target by less than 1,800 km.

In addition to the rendezvous commitment, the two astronauts also busied themselves with their load of experiments. Twenty minutes after reaching orbit, they activated the tri-axis magnetometer to measure radiation levels in the South Atlantic Anomaly, an area in which the lowermost portion of Earth's Van Allen belts dip to within a few hundred kilometres of the surface. Elsewhere, two other devices – a beta

"The light was something fierce," said John Young of the first Agena-X firing, "and the acceleration was pretty good." Unfortunately, the astronauts' view 'outside' was somewhat impaired by the bulk of the target vehicle.

spectrometer and a bremsstrahlung spectrometer – also monitored radiation dosages. "I only know that if I ever develop eye cararacts," wrote Collins, "I will try to blame it on Gemini X."

The radiation measurements, however, left investigators on the ground surprised and relieved. In fact, they were so disbelieving of the small numbers reported by the astronauts – 0.04 rad here, 0.18 rad there – that they mildly accused them of having their detectors switched off. In truth, radiation levels, even passing through the South Atlantic Anomaly, were much lower than anticipated. Several other experiments could only be done outside Gemini X and their deployment was one of Collins' main tasks during the first of two EVAs on the mission.

Midway through preparations for his first excursion, Deke Slayton came on the radio, asking the crew to "do a little more talking from here on". Neither Young nor Collins was in any mood for idle conversation, having their hands full with systems checks, Agena commands, navigation and a 131-step checklist to get everything ready for the EVA. Young's ho-hum response to Slayton's request was typical: "Okay, boss. What do you want us to talk about?"

Gemini X's cabin atmosphere was completely vented to vacuum at 4:44 pm, approaching orbital dusk, and Collins poked his head out of the hatch three minutes later. His first period of extravehicular activity actually took the form of standing up on his seat, during which time he set up a 70 mm general-purpose camera to study stellar ultraviolet radiation. "It was not enough simply to lift the camera above Earth's atmosphere," Collins explained later. "It had also to be moved outside the spacecraft, because the protective glass of the spacecraft windows screened out most of the ultraviolet rays which the astronomers wished to measure." As a result, Collins aimed the camera towards the southern Milky Way and scanned it from Beta Crucis to Gamma Velorum. In total, he acquired 32 images.

The view, even from his vantage point, hanging out of the hatch up to his waist, was astonishing. "The stars are everywhere," Collins wrote, "above me on all sides, even below me somewhat, down there next to the obscure horizon. The stars are bright and they are steady." Since it was now orbital nighttime, he could see little of Earth, save for the eerie bluish-grey tint of water, clouds and land. Minutes later, approaching dawn and with a rollful of 20-second ultraviolet exposures completed, he prepared to move onto his next task.

This involved photographing swatches of red, yellow, blue and grey on a titanium panel to determine the ability of film to adequately record colours in space. Unfortunately, Collins was unable to complete this objective in its entirety, when his eyes suddenly filled with tears; as, indeed, did Young's. "That's all we need," Collins related, "two blind men whistling along with the door open, unable to read checklists or see hatch handles or floating obstructions." They wondered aloud if a new anti-fog compound on the inside of their visors, added in the wake of Gene Cernan's experience, had caused the irritation. Mission Control's suggestion that perhaps urine had contaminated their oxygen supply was rejected by the astronauts, who described the odour as nothing of the sort.

Eitherway, the first period of EVA on Gemini X ended six minutes early, at 5:33 pm on 19 July. In total, Collins had been 'outside' for 49 minutes. It would later become clear that lithium hydroxide, used to scrub exhaled carbon dioxide from the cabin atmosphere, was mistakenly being pumped into the astronauts' suits. Switching off a troublesome compressor device, wrote Deke Slayton, solved the problem. "I was crying a little all night," Young quipped later, "but I didn't say anything about it ... I figured I'd be called a sissy!"

Alongside the Agena-VIII rendezvous commitment, the astronauts' third day in space was consumed with further experiments, though, thankfully, M-5, the much-loathed bioassays of body fluids task, which required them to collect, store and laboriously label their urine, was deleted from Gemini X's roster. In fact, Charles Mathews had removed M-5 on 12 July, barely a week before launch, together with an experiment to measure the ultraviolet spectral reflectance of the lunar surface as a means of devising techniques to protect astronauts' eyes from ultraviolet reflectivity. Other work included measurements of the ion and electron structure of Gemini X's wake after separating from Agena-X and terrain and weather photography targets.

Undocking from their Agena came at 2:00 pm on 20 July, after almost 39 hours linked together, which both men found tiresome. Young would describe the sight through his window as a little like backing down a railway line in a diesel engine with "a big boxcar in front of you ... the big drawback of having the Agena up there is that you can't see the outside world". Their view with Agena-X attached, Young added, was "just practically zilch". After undocking, the view returned and, as Gemini X drew closer to Agena-VIII, Collins connected his 15 m umbilical and prepared to go outside for his second EVA.

Forty-five hours and 38 minutes into the mission, Young reported what he thought was his first visual sighting of Armstrong and Scott's Agena, which actually turned out to be their own Agena-X! The other Agena, in fact, was still 176 km away, prompting Mission Control to tell the astronauts that it was "a pretty long range". Young, a little embarrassed, perhaps, at the mix-up, replied: "You have to have real good eyesight for that." In fact, he and Collins did not get their first glimpse of Agena-VIII until they were 37 km away; the target resembled "a dim, star-like dot until the Sun rose above the spacecraft nose".

Two hours later, at 4:26 pm, Young initiated his final closure on the Agena, with Collins – whom the command pilot had nicknamed 'Magellan' during training for his navigational prowess – busily computing figures for two midcourse correction burns. The target, which had by now been aloft for four months, appeared very stable and Young was able to station-keep at a distance of just a few metres 'above' it, before moving in to inspect a micrometeoroid package that Collins was to retrieve. The rendezvous was accomplished entirely manually; the long-dead Agena was giving off no radar signals and its lights had long since stopped working. "They really just had to eyeball their way in," recounted Deke Slayton.

Shortly thereafter, approaching orbital dawn at 6:01 pm, Collins cranked open his hatch and floated outside. In doing so, he became the first person to complete two EVAs. His first real movements 'outside' the Gemini were, he found, equally as difficult as those of Cernan: all tasks took considerably longer than anticipated. However, on this occasion, he was equipped with a zip-gun which aided his move over to the Agena-VIII and he crisply removed its micrometeoroid detector. He then moved to Gemini X's adaptor section and attached his zip-gun to the nitrogen fuel supply, before returning to the cabin as Young closed to within a couple of metres of Agena-VIII. Collins pushed himself away from the spacecraft, floated over to the Agena and grasped its docking collar with his gloved hands.

Discussions with Gene Cernan had already taught him that moving around and holding himself steady would be difficult and, indeed, Collins quickly lost his grip and drifted away from the Agena. Perhaps recalling Cernan's problems, he decided not to rely on the tether and instead used squirts from the zip-gun – successfully – to propel himself 5 m or so back towards Gemini X, then back to the Agena. He was then able, with relative ease, to grab the micrometeoroid detector, known as experiment 'S-10', although he decided against installing a replacement device for fear that he might lose the one he had just retrieved.

His efforts accidentally sent Agena-VIII into a slight gyrating motion, which caused some consternation for Young, who was tasked with keeping the two spacecraft close together. Not only did Young have to keep a close eye on three independent 'bodies' – the Gemini, the Agena and Collins – but he was also trying to keep sunlight from falling onto the right-hand ejection seat in the cabin. If it heated up too much, its pyrotechnics might have fired, taking both of Gemini X's hatches and Young's own seat with it. Meanwhile, Collins moved himself, hand-over-hand, along the umbilical back to the Gemini and handed S-10 to Young.

Not so successful was Young's effort to photograph Collins' work. In fact, not a single image or frame of footage exists from what was a quite remarkable spacewalk. "I had four cameras," he told an interviewer two decades later, "and none of 'em were working!" Nor was Collins able to photograph a human being's first contact in orbit with another spacecraft; his hand-held Hasselblad camera inadvertently drifted off as he was struggling to control his tether.

Next came a demonstration of the tether itself, extended to its full 9 m length, but this ended almost as soon as it had begun, when Mission Control told Young that he could not afford to use any more fuel for station-keeping. Young instructed Collins to return inside, which proved difficult because the spacewalker had gotten tangled in the umbilical and the rigidity of his suit prevented him from 'seeing' or even 'feeling' where it had wrapped itself. Finally returning to his seat after 39 minutes outside, Collins had to be helped out of the snake-like tangle by Young. Closing the hatch, thankfully, posed no difficulties. The EVA was, however, disappointingly short in duration; a consequence of Gemini X's dwindling fuel supply.

Also disappointing was the loss of the micrometeoroid collector, which apparently drifted outside and away into space during Collins' ingress back into Gemini X. It had contained 24 sample slides of materials including nitrocellulose film on copper mesh, together with copper foil, stainless steel, silver-coated plastic, lucite and titanium-covered glass, with the intention of exposing them to the harsh environment of low-Earth orbit.

Fuel woes were by now causing concern on the ground and the capcom asked them to confirm that they were not using any of their thrusters. An hour after Collins' return inside, at 7:53 pm, the two men briefly reopened his hatch for three minutes to throw the tether – whose presence in the cramped cabin had made "the snake house at the zoo look like a Sunday school picnic", according to Young – and the chest pack into space. Time was still tight and at 8:58 pm, they performed their next task: pulsing their OAMS thrusters to reshape their orbit, reducing their perigee to 106 km and rendering the point of re-entry more precise. After a few end-of-workday tasks, including more synoptic terrain and weather photographs, Young and Collins retired for their final night's sleep in space.

Seven hours after awakening, at 3:10 pm on 21 July, Gemini X began its fiery plunge back through the atmosphere, high above the Canton Island tracking station on its 43rd orbit. During re-entry, Young was able to 'steer' his spacecraft's banking angles by computer solutions and they splashed down in the western Atlantic at 4:07 pm, barely an hour shy of three full days since launch and just 5.4 km from the prime recovery vessel, the amphibious assault ship Guadalcanal. So close, in fact, were they to the intended splashdown point that the ship's crew were able to watch Gemini X descending through the clouds beneath its parachutes.

After the installation of a flotation collar by pararescue swimmers, Young and Collins were helicoptered away to the ship for a red-carpet welcome, a band playing 'It's a Small, Small World' and the inevitable medical checks. Their mission, together with that of Gemini IX-A, had addressed many questions which demanded answers before Apollo could go to the Moon: the demonstration of rendezvous methods, orbit-shaping manoeuvres to avoid trapped-radiation hazards in deep space and, of course, docking. Using Agena-X as a switch engine – in effect, a kind of 'space tug' – also had untold ramifications for future spacecraft and the establishment of orbiting laboratories. In the next few days, the Agena participated in a number of remotely-commanded manoeuvres, boosting its apogee at one stage up to 1,390 km and eventually returning to a near-circular 352 km orbit.

Still posing an obstacle, though, was EVA, to which Collins now added his opinion: more and better restraints and handholds were definitely needed for spacewalkers to work effectively outside and operating in a rigidised suit was far more difficult than previously thought. "A large percentage of the astronaut's time," Collins related after the flight, "is ... devoted to torquing his body around until it is in the proper position to do some useful work." With only two more Gemini missions remaining, time was tight to get such problems resolved.

'ANIMAL'

Upon the safe return of Young and Collins, barely six months remained to fly the final two Geminis. In order to focus resources entirely upon the Apollo effort, 31 January 1967 had been mandated as the deadline for the end of Project Gemini. Judging from the rate of launches thus far, NASA management felt confident that flying Gemini XI on 7 September and Gemini XII on 31 October was achievable. Those two missions, both lasting three to four days, would perfect each of the techniques demonstrated thus far: rendezvous, docking, EVA and using the Agena-D target to adjust their orbits.

First up would be Gemini XI's Pete Conrad and Dick Gordon, who had been named to the mission on 21 March 1966, only days after finishing their previous stint as Neil Armstrong and Dave Scott's backups on Gemini VIII. The two men were an almost-perfect match, sharing a friendship that long pre-dated their NASA days, back to a time in the late Fifties when they were roommates aboard the aircraft carrier Ranger. A decade later, as astronauts, they earned a reputation for being cocky and fun-loving – Gordon, indeed, was such a ladies' man that Conrad nicknamed him 'Animal' – yet both were intensely focused.

The respect in which Gordon was held as a test pilot and naval aviator preceded his time at NASA; in fact, when he applied unsuccessfully to join the 1962 astronaut class, he was already on first-name terms with Al Shepard, Wally Schirra and Deke Slayton. In 1971, Slayton would consider it one of his most difficult tasks trying to choose between Gordon and Gene Cernan to command the final Apollo lunar mission; Gordon would lose, but by barely a whisker.

Richard Francis Gordon Jr was born in Seattle, Washington, on 5 October 1929, attending high school in Washington State with dreams of the priesthood, rather than any aspiration to fly. Upon receiving his bachelor's degree in chemistry from the University of Washington in 1951, his focus had shifted somewhat to professional baseball or a career in dentistry. Gordon had settled firmly on the latter when the Korean War broke out and, in 1953, he joined the Navy and discovered his life's true calling: aviation.

He would win top honours for his precise aerial manoeuvres, which guided him through All-Weather Flight School to jet transitional training to the all-weather fighter squadron at the Naval Air Station in Jacksonville, Florida. It was shortly after being selected to join the Navy's test pilot school at Patuxent River, Maryland, in 1957 that he and Conrad met and became lifelong friends. The pair would frequently while away raucous nights in bars and nightspots, knocking back beer and shots, then show up at the flight line six hours later, models of sobriety. "They were not only good pilots," wrote Deke Slayton, "but a good time."

After graduation, Gordon test-flew F-8U Crusaders, F-11F Tigers, FJ-4 Furies and A-4 Skyhawks and also served as the first project pilot for the F-4H Phantom II. Later, he moved on to become a Phantom flight instructor and helped introduce the aircraft to both the Atlantic and Pacific Fleets. His expertise and reputation as one of the hottest F-4H fighter jocks in the world reached its zenith when Gordon used it to win the Bendix transcontinental race from Los Angeles to New York in May 1961. By doing so, he established a new speed record of almost 1,400 km/h and completed the epic coast-to-coast journey in barely two hours and 47 minutes.

In light of such astounding professional accomplishments, it came as a surprise to many – not least Pete Conrad – when Gordon did not make the final cut for NASA's 1962 astronaut intake. The intensely competitive Gordon would describe his reaction as "pretty pissed off", but he plunged directly into applying for aviation jobs, intending to retire from naval service. One night in the bar he was met by the just-selected Conrad, whose widow Nancy later described the encounter that would change Gordon's career.

"Still crying in your beer, Dickie-Dickie?"

"Just crying for you, Pete, ya poor dumb sumbitch. Stuck in a garbage can in space with some Air Force puke while I'm out smoking the field in my Phantom."

"So, Dick. They're gonna fill out this Gemini program now that Apollo's approved. At least ten more slots. I think you oughtta apply again."

"And why would I do that?"

"Because you miss me."

A few months later, in October 1963, Gordon was picked as an astronaut. Three years after that, to his surprise and great joy, he would fly right-seat alongside his long-time Navy buddy. And three years after that, they would also fly to the Moon together. It would bring back memories of a picture of a flight-suited Conrad that he had sent to Gordon in 1962, just after his own selection. On the back, he had written: 'To Dick: Until we serve together again'.

'M EQUALS 1"

In spite of Young and Collins' success, one particular aspect of Gemini XI caused the most worry: the desire to complete rendezvous and docking with the Agena-D target on their very first orbit! To be fair, 'desire' is probably the wrong word. The necessity for rendezvous and docking at such an early stage would be essential when an Apollo command and service module was approached by the ascent stage of a lunar module after the Moon landing.

"There was a lot of concern that it wasn't going to be successful," Gordon told Neil Armstrong's biographer James Hansen. "For the Apollo application, the desire was to rendezvous as rapidly as possible because the lifetime of the LM's ascent stage was quite limited in terms of its fuel supply." To accomplish this first-lap rendezvous and docking, Gemini XI would have to meet a launch window barely two seconds long and Conrad and Gordon's three days aloft would also feature a lengthy spacewalk, experiments ... and the physical tethering of their spacecraft to the Agena target.

The purpose of the latter, said backup command pilot Neil Armstrong, was to "find out if you could keep two vehicles in formation without any fuel input or control action". Armstrong had been teamed with rookie astronaut Bill Anders to shadow Conrad and Gordon's training regime and the foursome frequently found themselves in the Cape Kennedy beach house over the summer months of 1966, running through rendezvous and docking procedures, trajectories and flight plans. Before too many more years had passed, all four of them, on three separate missions, would apply this knowledge on flights to the Moon.

Conrad's assignment to Gemini XI was inextricably linked to the mission's demonstration of achieving high altitudes. In mid-1965, he learned of plans to fly the spacecraft around the Moon and, even when such a mission was rejected by Jim Webb and Bob Seamans, had pushed vigorously to use some of his Agena fuel to carry Gemini XI into a high orbit. Among the earliest champions of Conrad's scheme were the scientists – flying to high altitude, he argued, would allow the astronauts to acquire high-resolution imagery of weather patterns and possibly benefit other experiments. Fears that the Van Allen radiation belts might scupper such a mission were allayed when Conrad despatched Anders – a qualified nuclear

Dick Gordon (left) and Pete Conrad demonstrate the tether experiment at their post-flight press conference.

engineer – to Washington to devise and argue ways of avoiding them. Indeed, radiation data gathered by Young and Collins on Gemini X had turned out to be barely a tenth of pre-flight estimates.

The plan to fly Gemini XI to high altitude thus approved, another of its key objectives – tethering the spacecraft to its Agena as a means of demonstrating better station-keeping – bore fruit. Under direction from NASA, McDonnell engineers recommended that a nylon or Dacron line no longer than about 50 m could produce a reasonable amount of cable tension. The station-keeping requirement for such an experiment quickly expanded, however, to encompass a means of inducing a form of artificial gravity. Ultimately, a 30 m Dacron tether was chosen for the task and the only serious concern was how Gemini XI could be freed from the Agena. It was decided to fire a pyrotechnic charge and, failing that, to snap a break line in the tether with a small separation velocity.

Particularly worrisome, though, was achieving rendezvous with the Agena on Conrad and Gordon's very first orbital pass. (Within NASA, it was known as an 'M = 1' rendezvous.) This had been suggested several years earlier by Richard Carley of the Gemini Project Office, in response to Apollo managers' concerns that such practice was necessary to provide a close approximation of lunar orbit rendezvous. During Gemini X, John Young had expended so much fuel that it seemed overly ambitious, but Flight Director Glynn Lunney eventually appeased naysayers that Mission Control could provide enough backup data on orbital insertion and manoeuvring accuracy that Conrad and Gordon would have all they needed to perform it. Still sceptical, William Schneider, deputy director for Mission Operations, bet the review board's chairman James Elms a dollar that a first-orbit rendezvous could not be done economically.

Despite the greater success of Mike Collins in performing his EVAs on Gemini X, concerns remained. Practicing spacewalking techniques in a realistic environment had led, in mid-1966, to the adoption of 'neutral buoyancy' – submerging fully-suited astronauts in a water tank – as the closest terrestrial analogue to the real thing. In his autobiography, Collins related its introduction in the closing weeks before his own mission, but had little time to spare practicing under such conditions. Gene Cernan, perhaps best-placed to observe the difficulties of EVA, was one of the first to undergo neutral buoyancy training, and found that it did indeed approximate his efforts in space.

Yet refining techniques also encompassed an astronaut's body positioning and the inclusion of more practical aids, including handholds, shorter umbilicals (Young and Collins recommended a 9 m tether, rather than 15 m, to avoid entanglement, which was accepted) and better foot restraints in the Gemini adaptor section. This last proposal led to two suggestions from McDonnell – a spring clamp, like a pair of skis, and a 'bucket' type restraint – with NASA finally opting for the latter, which came to be nicknamed the 'golden slippers'.

Rendezvous, docking, tethered station-keeping and spacewalking … and a full plate of 12 scientific experiments would thus fill Conrad and Gordon's three days in space. Two of the experiments – Earth-Moon libration region imaging (S-29) and dim-light imaging (S-30) – were newcomers to Project Gemini, together with seven already-flown investigations into weather, terrain and airglow photography, radiation and microgravity effects, ion-wake measurements, nuclear emulsions and ultraviolet astronomy and a trio of technical objectives focusing on mass determination, night-image intensification and the evaluation of EVA power tools. By the end of August 1966, the Manned Spacecraft Center had announced that all of the experiments were ready to fly.

Elsewhere, preparations for Gemini XI had become closely entwined with those of the previous mission, Gemini X. The Titan II propellant tank for the latter had become corroded by leaking battery acid during transfer from Martin's Denver to Baltimore plants in September 1965. It had been replaced with the tank originally allocated to Conrad and Gordon's flight; the Gemini XI crew were correspondingly assigned the propellant tank previously earmarked for Gemini XII. This finally reached Baltimore for checkout and integration in January 1966.

Six months later, Conrad and Gordon's Titan was in place on Pad 19 at Cape Kennedy. At around the same time, the Gemini spacecraft itself was hoisted into position and the hot weeks of August were spent connecting cabling and replacing some leaking fuel cell sections.

By this stage, Gemini XI's launch date had slipped by two days to 9 September, but as the fuelling of its Titan II booster progressed, a minute leak was found in the first-stage oxidiser tank. Technicians quickly set to work rectifying the problem: a sodium silicate solution and aluminium patching plugged the leak and launch was rescheduled for the following day. That attempt, too, came to nothing, as Conrad and Gordon were en-route to Pad 19, when it was learned that their Agena-XI's Atlas booster was experiencing problems with its autopilot. Hoping to resolve the glitch in time, the General Dynamics test conductor called a hold in the countdown to check it out.

The Atlas engineers out at Pad 14 reported that they were receiving faulty readings and were in the process of running checks to determine if the autopilot component needed replacement. Ultimately, the checks took too long to meet the launch window on 10 September and, after an hour of troubleshooting, the attempt to launch the two vehicles that day was called off. Another attempt, William Schneider announced, would be made on the 12th. It later became clear that the fault was caused by a fluttering valve, coupled with unusually high winds and an over-sensitive telemetry recorder. Fortunately, none of the Atlas' components needed to be changed and managers cleared the mission to fly.

Early on 12 September, Conrad and Gordon were sealed inside their spacecraft by Guenter Wendt's closeout team at Pad 19. Despite a minor oxygen leak which required the reopening and resealing of Conrad's hatch, the launch of their Agena-XI target went ahead on time at 8:05 am. Still, there was little time to waste, for Gemini XI had been granted barely two seconds in which to launch; its almost-impossibly-short 'window' dictated by the requirement to rendezvous with the Agena on the astronauts' first orbit. It demonstrated, if nothing else, the growing maturity of America's space effort. "Rocketeers of the Forties, Fifties and early Sixties," wrote Barton Hacker and James Grimwood, "would have been aghast at the idea of having to launch within two ticks of the clock."

Aghast or not, Conrad and Gordon's liftoff was perfect, coming at 9:42:26.5 am, just half a second into its mandated two-second launch period. Six minutes later, on the fringes of space, the two astronauts received the welcome news from Mission Control: their ascent and the performance of their Titan II had been 'right on the money' and they were cleared for their $M = 1$ rendezvous. To kick off its first sequence, just 23 minutes after launch, Conrad pulsed Gemini XI's thrusters in a so-called 'insertion-velocity-adjust-routine' – or 'Ivar' – manoeuvre, correcting their orbital path and placing them on track to 'catch' the Agena, then 430 km away.

Conrad's next manoeuvre was more tricky, occurring as it did outside of telemetry and communications range. At the appointed moment, he performed an out-of-plane manoeuvre of about one metre per second, then pitched Gemini XI's nose 32 degrees 'up' from his horizontal flight plane. This completed, the two men activated their rendezvous radar and ... just as predicted, they received an immediate electronic 'lock-on' with the Agena. By the time they re-established radio contact with the ground, they were just 93 km from their quarry.

Capcom John Young, seated at his console in Houston, sent final numbers through the Tananarive tracking station to the astronauts and, as Gemini XI neared the apogee of its first orbit, Conrad ignited the OAMS to produce 'multi-directional' changes – forward, 'down' and to the right – in support of the rendezvous' terminal stage. All at once, less than 40 km away, the Agena flashed into view with orbital sunrise over the Pacific Ocean, almost blinding them as it did so. The two men scrambled for sunglasses and Conrad pulled his spacecraft to within 15 m of the target. Making landfall over California, and a little more than 85 minutes since launch, they had achieved the world's quickest orbital rendezvous. Moreover, they still retained some 56 per cent of their fuel reserves. William Schneider lost his bet with James Elms, writing that "I never lost a better dollar".

"Mr Kraft," the jubilant astronauts called to Flight Director Chris Kraft, "would [you] believe M equals 1?" On 12 September 1966, he certainly did.

HIGH RIDE

Amidst such euphoria, Conrad's actual docking with Agena-XI at 11:16 am seemed anticlimatic, although both astronauts were able to practice the docking-and-undocking exercise which had eluded John Young two months earlier. Pulling loose from the target, then redocking, said Conrad, was much easier in space than on the ground and the astronauts managed it in daylight and darkness. It also gave Dick Gordon the distinction of becoming the first astronaut not in a command position to perform a docking manoeuvre.

This was followed by an ignition of the Agena's main engine to boost their orbital altitude to the highest yet achieved by humans. Facing 90 degrees away from the flight path, Conrad fired the engine to add 33 m/sec to their velocity, as a test-run, and confirmed to Young that this was "the biggest thrill we've had all day". It was but the prelude for one of their main tasks on 14 September: increasing the high point of their orbit to no less than 1,370 km. For now, however, the astronauts rested, powering down Gemini XI, tucking into their first meal and bedding down for their first night's sleep in space.

Despite its audacious nature, their only problem was a pair of dirty windows. This had, in fact, plagued each Gemini to date and even covers which could be jettisoned after launch had been little help. Capcom Al Bean told Gordon to rub half of Conrad's window with a dry cloth during his spacewalk on 13 September. Four hours before they were scheduled to open Gemini XI's hatch, they began preparing their suits and equipment; only to realise that, so thorough was their training, they actually needed barely 50 minutes to have all of their gear up and running.

By this point, Gordon could conceivably have gone outside, so Conrad called a halt, which left them sitting idly, fully kitted-out in their equipment. An hour later, they hooked up Gordon's environmental support system and conducted several oxygen-flow checks, which proved a mistake because it dumped oxygen into the cabin and its excess was then vented into space. They could ill-afford this rate of oxygen loss and Conrad told Gordon to switch back to Gemini XI's systems. Gordon, by this point uncomfortably warm, was relieved to get back onto the interior environmental control system. After all, the extravehicular system's heat exchanger had been designed to operate in a vacuum, rather than inside a pressurised cabin.

The main problem, wrote astronaut Buzz Aldrin, actually lay in Conrad and Gordon's impatience in skipping through the formal, six-page, hundred-plus-step sequence of donning and preparing their equipment. "As a result," Aldrin related in 1989, "they upset cabin pressurisation when they checked out Gordon's oxygen umbilical and he became overheated long before his EVA began."

The men considered asking Flight Director Cliff Charlesworth to let Gordon begin his EVA one orbit early, but due to issues with tracking and lighting, decided to stick with the pre-planned schedule, a decision that they would come to regret. As they prepared to fit Gordon's sun visor onto his helmet, it proved a stubborn chore; Conrad eventually fastened one side, but could not reach over to snap the other side in place, leaving Gordon hot, bothered and in need of rest. After struggling for a few more minutes, Gordon eventually snapped the right side in place – cracking the sun visor in the process – and was thoroughly winded by the time he cranked open the hatch and stood on his seat at 9:44 am on 13 September, just two minutes past one full day into the mission ... and precisely on time.

Instantly, Gordon's exit into space was accompanied by everything else inside the cabin that was not tied down. Standing on his seat, his first activity was to deploy a handrail – a fairly easy task – after which he removed the S-9 nuclear emulsion package from outside the spacecraft and passed it in to Conrad, who stuffed it into his footwell. Next, Gordon tried to install a camera in a bracket to photograph his own movements, but this proved difficult. To resolve it, Conrad let enough of the umbilical slide through his gloved hand to let Gordon float above the camera, thump it with his fist and secure it in place.

The spacewalker's next task was to attach the 30 m Dacron tether, housed in Agena-XI's docking collar, onto Gemini XI's nose. As Gordon pushed himself forward, he missed his goal and drifted in an arcing path above the adaptor and around in a semi-circle, until he reached the back of the spacecraft. However, Conrad had released only a couple of metres of the 9 m umbilical, so he pulled Gordon back to the hatch to start his trek again.

At length, Gordon reached the target and grabbed some handrails to pull himself astride Gemini's nose – prompting Conrad to yell "Ride 'em, cowboy!" – but the exercise proved more difficult in space than it had in ground tests. Aboard the zero-G aircraft, Gordon had been able to push himself forward, straddle the Gemini's re-entry and recovery section and wedge his feet and legs between the docking adaptor and the spacecraft to hold himself in place, thus leaving his hands free to attach the tether and clamp it to the docking bar. However, in the vacuum of space, he found himself constantly fighting against his suit to keep himself from floating away; a situation rendered all the more difficult by the lack of any kind of 'saddle' or 'stirrup' to help him. All Gordon could do was hold on with one hand and try to operate the tether clamp with the other.

He struggled vigorously for six minutes, finally securing the line and setting the stage, at last, for the tethered flight experiment which would come later in the mission. Yet it was clear to Conrad that Gordon was encountering severe difficulties and the differences between EVA practice in terrestrial conditions and the real thing were profound: the spacewalker, soaked with sweat and eyes stinging, was reduced to groping his way blindly around. He tried to remove a mirror on Gemini XI's docking adaptor, to help Conrad see him at the back of the spacecraft, but it would not move, and had no chance to wipe the windows either.

As Gordon neared the hatch, Conrad helped as much as he could, discussing procedures for getting to the spacecraft adaptor to store his zip-gun. It was obvious, though, that Gordon was exhausted: when they passed next over the Tananarive tracking station, Conrad told John Young that he had "brought Dick back in ... he got so hot and sweaty, he couldn't see". Unlike Gene Cernan, however, Gordon had no trouble whatsoever getting back inside Gemini XI or closing the hatch. Disappointingly, the spacewalk had lasted just 33 of its intended 107 minutes and one of its key tasks – experiment D-16, a power tool evaluation which had also evaded Dave Scott on Gemini VIII – was lost. Clearly, many of the complexities of EVA still remained to be resolved.

Having said that, Gordon's exhaustion did not disrupt the remainder of the mission. Flight planners had learned to schedule periods of less-vigorous activity immediately after heavy workloads and Conrad and Gordon's next task involved leisurely repacking equipment and restoring some semblance of order to the cabin. Additionally, communications with Mission Control dwindled to little more than short transmissions about spacecraft systems and medical checks, which gave them much-needed respite. Conrad test-fired a sluggish thruster, the men ate a meal and photographed the atmospheric airglow.

These moments of relatively quiet time would not last. Ahead of them lay the so-called 'high ride' – their Agena-assisted climb to a record-breaking altitude of 1,370 km. To prepare themselves, they donned their suits, closed their visors and secured as much cabin equipment as possible, as if in readiness for re-entry, and focused their attention on the Agena-XI. A problem quickly appeared. As they made a pre-firing check of the target, it became clear that it was not properly accepting commands; instructions, in fact, had to be transmitted twice before they were acknowledged. Conrad expressed his concerns to Capcom Al Bean, who told them that, in fact, the Agena was responding correctly, but Gemini XI's displays were at fault. "Heck of a time to have a ... glitch like that show up," Conrad grumbled, but was assured that everything remained 'go' for the burn.

Forty hours and 30 minutes after Gemini XI's launch, at 2:12 am on 14 September, Conrad fired the Agena's engine for 26 seconds, adding a blistering 279.6 m/sec to their speed. Both men were electrified by the burn. "Whoop-de-doo!" Conrad yelled. "[That's] the biggest thrill of my life!" Like the experience of John Young and Mike Collins two months earlier, Conrad and Gordon were pushed 'forward' against their seat harnesses and, gradually, saw the Earth change from a vast expanse of blue and white beneath them ... into something more planet-like, with a very distinct curvature. "I'll tell you," Conrad told the capcom in Carnarvon, Australia, "it's go up here and the world's round ... you can't believe it ... I can see all the way from the end, around the top ... about 150 degrees."

Beneath them, Conrad told Al Bean, the men beheld the intense and striking blues of the oceans, the sprinkling of clouds and the astonishing clarity of Africa, India and Australia. "Looking straight down," he radioed, "you can see just as clearly. There's no loss of colour and details are extremely good." To cope with the adverse radiation effects of the Van Allen belts, Gemini XI's high-apogee orbits were timed to take place over Australia, where levels were calculated to be relatively low. Over Carnarvon, indeed, Conrad reported that on-board dosimeters read barely 0.2 rads per hour. "Sounds like it's safer up there than a chest X-ray," replied Bean. In fact, Conrad added, Gemini XI experienced less radiation at 1,370 km than Young and Collins had endured in a longer period of time at 830 km.

It was at this altitude – 1,370 km high, the highest yet attained by humanity – that the flight's imaging experiments, notably the synoptic terrain and weather photography objectives, produced some of their most stunning results. In total, Conrad and Gordon clicked more than 300 exposures and their descriptions of the sheer clarity of the eastern hemisphere filled principal investigators with excitement and anticipation.

Not until Apollo 8's journey to the Moon in December 1968 would humans travel to higher altitudes and four decades later, Conrad and Gordon retain the record for the highest-ever Earth orbit attained by humans. "As the coupled craft soar toward their record apogee," Time magazine told its readers a week after Gemini XI's splashdown, following NASA's release of the pictures, "the curvature of the Earth's horizon becomes more pronounced and the Earth assumes an unmistakably globelike shape. Though the pictures are sharp and show geological features plainly, the Earth seems devoid of life; it offers no visible evidence of its teeming population, its great cities, its bridges or its dams."

Two orbits later, on their 26th revolution, as Gemini XI passed over the United States, Conrad again fired the Agena's engine for 23 seconds to lower their apogee to 304 km and reduce their speed by 280 m/sec. After a bite to eat, at 8:49 am, Gordon opened his hatch, high above Madagascar, for his second EVA. This time, he stood

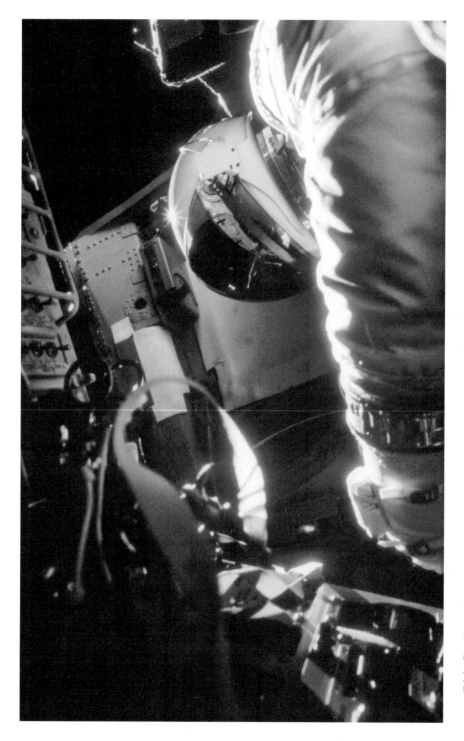

Dick Gordon, photographed by Pete Conrad shortly before opening Gemini XI's hatch to throw out unwanted equipment.

on the 'floor' of Gemini XI, poked his helmeted head outside and watched the sunset. Secured by a short tether, he could at least use both hands to mount cameras easily in their brackets and remained 'outside' for no less than two full hours. "Most enjoyable" was his summary of the stand-up EVA and, indeed, flight surgeons commented that from the biosensor data, it was uneventful.

Whilst outside, Gordon experienced two nighttime passes, photographing several star fields with the S-13 ultraviolet camera and his view was so unimpaired that he was able to coach Conrad about which way to direct Gemini XI. Although the Agena's stabilisation proved somewhat erratic, the linked vehicles remained sufficiently stable to yield excellent results in about a third of Gordon's photographs. Indeed, as the spacecraft passed over the United States, the skies were so clear that both men were able to marvel at the view of Houston. Drifting across Florida, then out over the Atlantic, Gordon suddenly broke the silence to tell Conrad that he had just fallen asleep outside. He was not the only one: Conrad, too, had dozed off inside the spacecraft.

"That's a first," radioed John Young. "First time sleeping in a vacuum."

Returning inside after two hours and eight minutes, Gordon's exhaustion was not, like that of Cernan, caused by over-exertion and battling against his rigid suit, but instead by sheer concentration on his tasks.

Their next step was the tethered vehicle exercise, which could be attempted by two different means. One of these assumed the position of a 'pole', always pointing towards Earth's centre – a so-called 'gravity gradient' attitude – in which Conrad and Gordon would have backed their spacecraft out of the Agena's docking cone slowly until the 30 m tether became taut. If properly positioned, a slight thrust of just 3 cm/sec would have kept the tether taut and the joined 'pole' would have drifted serenely around the globe, each spacecraft maintaining the same relative position and attitude.

Should this have been unsuccessful, the men were tasked with trying a 'spin-up', or 'rotating', mode, a technique studied by McDonnell engineers. In this case, after physical undocking, Conrad would fire Gemini XI's thrusters to induce a rotation of one degree per second and as the tethered pair circled Earth, their mutual centre-of-gravity would lie at a specific point on the tether, around which they would do a slow, continuous cartwheel. Centrifugal force would, it was theorised, keep the tether taut and the two spacecraft apart, with the tether itself providing centripetal force to keep them both in equilibrium.

Over Hawaii, Conrad and Gordon separated from the Agena and began the cautious attempt to start the gravity gradient demonstration. There was enough initial tension in the tether to upset the target and cause the Gemini to move to the 'right' – towards the Agena's docking adaptor – and Conrad quickly adjusted his motion. Then, as he backed away, the tether stuck, probably in the stowage container, when just 15 m had been released. Conrad pulsed the OAMS thrusters to free the tether, but it quickly became hung up on a patch of Velcro and he was forced to shift Gemini XI out of vertical alignment to peel the tether off the Velcro pad. This disturbed the Agena again and there still remained about 3 m of tether to be pulled out.

On the ground, engineers began to worry. The Agena should have taken around seven minutes to stabilise itself; when it took longer, they began to suspect that something was wrong with its attitude-control system and opted to abandon the gravity gradient attempt and adopt the 'spin-up' mode instead. However, when Conrad tried to initiate the rotation, another problem arose when he could not get the tether taut; it seemed to rotate counter-clockwise. "This tether's doing something I never thought it would do," he told John Young. "It's like the Agena and I have a skip rope between us and it's rotating and making a big loop ... Man, have we got a weird phenomenon going on here!"

Although the spinning line was curved, it also had tension, and for several minutes Conrad and Gordon jockeyed Gemini XI's thrusters to straighten the arc. Eventually, the tether straightened and became taut and Conrad rolled the spacecraft and fired the thrusters to begin the slow cartwheeling motion. At first, it seemed that he had 'stretched' the tether, which had a big loop in it, but steadily, as both astronauts gritted their teeth, centrifugal force took over and it smoothed out. A 38-degree-per-minute rotation rate was obtained and remained steady throughout a nightside orbital pass.

Moving into dawn, they were asked by the Hawaii capcom to accelerate their spin-up rate and, with some reluctance, Conrad agreed. Suddenly, Gordon shouted "Oh, look at the slack! It's going to jerk this thing to heck!" When the added acceleration started, the tether tightened, then relaxed, causing a 'slingshot' effect which seesawed both astronauts up to 60 degrees in pitch. In response, Conrad steadied the spacecraft and, to his surprise, the Agena stabilised itself. Their rotation rate checked out at 55 degrees per minute and they were able to test for a tiny amount of artificial gravity: when they placed a camera against Gemini XI's instrument panel and let go, it moved in a straight line to the rear of the cockpit and parallel to the direction of the tether. Neither of the astronauts, however, felt any physiological effect of gravity. After three hours of docked operations, they jettisoned Gemini XI's docking bar and pulled away.

Despite understandable disappointment that the gravity gradient technique could not be fully demonstrated, the spin-up mode at least proved that station-keeping could be done economically. After undocking, Conrad originally intended to decrease his spacecraft's speed, allowing him to pull ahead of the Agena, but was advised instead to prepare for a 'coincident-orbit' rendezvous, whereby he would follow the target by 28 km in its exact orbital path. This would demonstrate their ability to station-keep at very long range with little fuel usage.

As a result of the plan change, Gemini XI's separation manoeuvre was adjusted; instead of a retrograde firing, Conrad and Gordon 'added' speed and height to their orbit, such that Agena-XI passed 'beneath' and in front of them. Next, they fired their OAMS to place themselves in the same (coincident) orbit as the Agena, trailing it. Three-quarters of a revolution around Earth, Conrad decreased his 'forward' speed and, as expected, Gemini XI dropped into the Agena's orbital lane, 30 km behind it, with no relative velocity between the pair.

At the same time, the men set to work on another of their scientific tasks: the night-image intensification experiment (D-15), which sought to evaluate the usefulness of equipment which scanned ground-based objects onto the instrument panel. Conrad aimed Gemini XI at specific targets, including towns and cities, cloud formations, lightning flashes, horizons and stars, airglow, coastlines and peninsulas, while Gordon described his view on the monitor into a tape recorder. Unfortunately, Conrad's dirty window prevented him from seeing much and, indeed, the glow from the monitor meant that he never became adequately dark-adapted. Nonetheless, the men returned with astonishing recollections, including the lights of Calcutta, whose shape almost exactly paralleled official maps of the city.

Turning their attention back to the Agena, they asked ground controllers on the Rose Knot Victor tracking ship how far they were from the target and were advised that their distance remained 30 km, closing very slowly. A second rendezvous, beginning at 3:09 am on 15 September, was near-perfect: Conrad tilted Gemini XI's nose 53 degrees above level flight and fired the forward thrusters. This placed them in a lower orbit than the Agena, ready to catch up with it, and they took some time to tend to their experiments. An hour later, Conrad fired the aft thrusters to raise his spacecraft's orbit, then began to brake Gemini XI until he finally reported that he was on-station and steady with the target once more. Twelve minutes later, they executed a separation burn from the Agena for the final time. By this stage, the success had been so great that they could afford to be jocular; Conrad even asking Flight Director Glynn Lunney to send up a tanker to refuel them for more rendezvous.

The ambitious mission came to a close with a fully-automatic re-entry; unlike previous returns, in which the command pilots had flown their spacecraft down from 120 km using the Gemini's offset centre-of-gravity to generate lift for changes in direction, Conrad would not use his hand controller in conjunction with computer directions. Gemini XI would follow computer commands automatically, a technique derisively nicknamed 'chimp mode' by the astronauts. Seventy hours and 41 minutes after launch, at 8:24 am on 15 September, partway through their 44th orbit, the retrorockets fired. Conrad disengaged his hand controller and put the system onto autopilot. This performed admirably and Gemini XI splashed down safely, within 4.6 km of the helicopter carrier Guam, at 8:59 am.

Half an hour after hitting the Atlantic, Conrad and Gordon were aboard the Guam, almost exactly three days since leaving Cape Kennedy. Rendezvous, docking and their record-breaking altitude boost had been, all three, successfully achieved; yet EVA issues remained. Buzz Aldrin, in training to conduct his own excursion on Gemini XII in November, became one of the first astronauts to work underwater in 'neutral buoyancy' to prepare himself for working in space. His was supposed to be one of the most complex to date, completing what Gene Cernan had been unable to do in June: flying the Air Force's AMU backpack. Ironically, had the AMU flown on Gemini XII, Aldrin would not, and almost certainly he would not have gone on to become the second man on the Moon.

CONQUEST OF THE 'HIGH GROUND'

Bob Crippen, Dick Truly, Gordo Fullerton, Hank Hartsfield, Bob Overmyer, Karol 'Bo' Bobko and Don Peterson: to most space aficionados, their names will forever be connected to the early test flights of the Shuttle. However, prior to their NASA days – and more than a decade before each would make an orbital voyage – all seven almost found fame as the first men to inhabit a long-term space station. Yet they were not, insisted their sponsor, the Air Force, 'astronauts', but rather 'aerospace research pilots', with an agenda that remains largely classified to this day. "When the manned space programme started," remembered Peterson in a 2002 oral history for NASA, "the Cold War was in full swing. We were scared to death and there was this feeling that space was the high ground; that is, if you conquered space, you had command of Earth. The idea that the Russians might be ahead of us was pretty frightening, so there was strong public and government support for the manned space programme and an unlimited budget."

The most visible example of the seemingly bottomless moneypit available for space exploration during this period was, of course, Project Apollo, but in May 1966, more than $1.5 billion was pledged by President Lyndon Johnson for a space-based outpost known as the Manned Orbiting Laboratory (MOL). Had this actually reached fruition, it would have provided Peterson, his six colleagues and a handful of others with their first space missions and made them the first men ever to launch into the heavens from the United States' west coast.

Ideas for a military space station can be traced back to June 1959, when preliminary plans were laid for a two-man laboratory to support a range of biomedical, scientific and engineering tasks in the microgravity environment of low-Earth orbit. Within three years, these sketches had crystallised into a formal proposal for three separate cylindrical modules, launched separately atop Titan II boosters and joined together in space to form a triangular structure. Crews would then be ferried to and from MOL using an Air Force variant of NASA's Gemini spacecraft, launched from Vandenberg Air Force Base in California.

At one stage, it was envisaged that the laboratory may operate in tandem with Dyna-Soar – the reusable winged craft which, in a different life, Neil Armstrong might have flown – but by December 1963 it seemed inevitable that the latter would be cancelled in favour of MOL. Eitherway, the perceived 'militarisation' of space caused concern for many observers, including James Haggerty of the Army-Navy-Air Force Journal and Register, who described MOL as "an ominous harbinger of a reversal in trend, an indication that the services may play a more prominent role in future space exploration at NASA's expense. "Whether you label it a development platform, satellite or laboratory, it is clearly intended as a beginning for space station technology," continued Haggerty. "It is also clearly the intent of this [Johnson] administration that, at least in the initial stages, space station development shall be under military rather than civil cognisance."

Moreover, despite its official emphasis on biomedical research and evaluating humanity's effectiveness in space, President Johnson rather tellingly announced in August 1965 that the ultimate aim of MOL was to "relate that ability to the defence of America". The dedication of the outpost to exclusively military activities was further underlined by the head of the Air Force's aerospace medicine group, Stan White, at a meeting with NASA representatives in May 1966, when he called for greater exploitation by the military of the agency's biomedical data. This, White argued, would relieve MOL's pilots of having to conduct such experiments, which the Air Force regarded as a burden to their own research. One of the most important military investigations was reconnaissance and surveillance, employing large optics, powerful cameras and side-looking radar.

It had already been realised from early spy satellites that having trained military observers available in orbit with specialised equipment would permit the real-time selection of ground-based targets and the acquisition of images through gaps in cloud cover. Furthermore, the return and interpretation of Corona reconnaissance satellite images typically took weeks or months, a lapse that the Air Force could not afford. With this in mind, the central element of MOL's surveillance payload was a telescope dubbed 'Dorian', fitted with a 1.8 m-wide mirror and supposedly capable of resolving ground-based objects the size of a softball.

Other surveillance instruments included high-resolution optical and infrared cameras, the side-looking synthetic-aperture radar, built by the Navy, with a resolution of 7.5 m – later cancelled because it was too large and heavy to be easily placed into orbit – and an electronic intelligence antenna. Much of the Dorian hardware, analysts have speculated, was probably later employed on KH-9 Big Bird and KH-11 Kennan reconnaissance satellites and may offer hints that both the CIA and Air Force doubted that MOL would ever fly.

For the 'aerospace research pilots' selected to travel to the outpost, there were also doubts, but in Hank Hartsfield's mind they centred on budgetary matters, as more money was siphoned away from MOL to finance the escalating conflict in Vietnam. Every year, with depressing predictability, Hartsfield recalled, the project would have its funding cut and be forced to lay off contractors, placing MOL further and further behind schedule ... and doubling its cost. In fact, by the time Stanley White's team met with NASA at Brooks Air Force Base in Texas to discuss the sharing of biomedical data, MOL had already guzzled $2.2 billion of taxpayers' money. The station, in its final configuration, was quite different from the three-module triangular structure planned in 1962: it took the form of a 12 m-long cylinder, powered by solar arrays or fuel cells, and comprised a transfer tunnel from the Gemini spacecraft, a laboratory divided into 'working' and 'living' quarters and an equipment section filled with oxygen and other tanks. One early plan actually called for a tunnel in the base of the laboratory, extending to an aft docking collar, which would enable two MOLs to be linked together in orbit. Such plans were never realised and had vanished from the Air Force's radar long before the project was cancelled in June 1969.

An interesting aspect of MOL was that the Gemini would have been attached 'backwards' to the outpost: rather than docking 'nose-first', as Apollo did with Skylab, it was through a hatch in the base that crews would have passed to reach their 30-day home in orbit. Gemini was very cramped and enabling pilots to unstrap, turn around and get through a 60 cm-wide hatch behind them would have been tricky. As a result, the Air Force tilted the seats slightly apart, as well as completely redesigning the spacecraft's instrument displays. However, it was not the size of the Gemini that caused concern; rather, it was the hatch in its heat shield – the very component upon which the crew's lives would depend during their fiery re-entry through the atmosphere.

The spacecraft and its pilots would have ridden into orbit already attached to MOL, neatly sidestepping the complications of rendezvous or docking, but at the end of their two-to-four-week mission, after closing the hatch, they would have had to hope that the tiniest – yet potentially deadliest – gap around the edge of the heat shield would not admit hot gases and tear the Gemini apart. As a result, it was decided to attach a MOL mockup, built from the propellant tank of a Titan II rocket, to NASA's Gemini 2 spacecraft and launch them on an unmanned test of the new hatch. On 3 November 1966, the unusual combo lifted-off from Pad 40 at Cape Canaveral and instantly made history as Gemini 2 became the first 'used' spacecraft to be refurbished and reflown. Originally launched in January 1965, it had been modified by the Air Force with a MOL-specification hatch in its base. Following a 33-minute suborbital flight, it separated from the MOL mockup and began its fiery plunge to Earth. "It came through with flying colours," exulted Hartsfield. "There was no heating problem or any burn-through. It proved the concept."

However, one of the key objectives of MOL was that it would operate in polar orbit, which would have resulted in higher-energy re-entries; consequently, the heat shield's diameter was increased to stick out from the base of the Gemini. Other changes from the standard NASA version of the spacecraft were that its OAMS thrusters were removed and its orientation managed instead by several forward-mounted reaction-control thrusters. Unlike the 'civilian' Gemini, designed to operate for periods of up to a fortnight with two men aboard, the systems of the Air Force variant were intended for longer-term, untended 'storage'.

After reaching orbit, the men's first task would be to shutdown the spacecraft and begin their long mission aboard MOL. They would reawaken the slumbering Gemini's systems just before re-entry, undock from the outpost and commence their descent. To this end, the spacecraft had only a 14-hour 'loiter' capability in its thrusters and life-support systems after separating from MOL. Naturally, the two groups of astronauts – NASA and Air Force – prepared in similar ways to fly different versions of the same machine and Hartsfield later provided some insight into his first experience as a trainee spacefarer. "We got some of the routine survival training," he said. "We'd had water survival [and] we'd gone down to Panama for jungle survival. We were getting that kind of training to get ready to go, because we were going to fly out of Vandenberg into a high-inclination orbit, which meant you covered a pretty good piece of the world. If you had to abort, you could almost go anywhere – jungle or polar regions."

Even three decades later, both he and Don Peterson were reluctant to talk much about their specific tasks and restricted their descriptions to 'generic' issues. "Of course," recalled Peterson, "we were flying a capsule in those days, so we were going to land in the water. The Earth is two-thirds water and you might come down someplace that you hadn't planned, [so] you had to be able to stay afloat and alive maybe for several days until [the Air Force] could get to you. Finding people in those days wasn't nearly as good as it is now."

To this day, it is unclear which MOL missions Hartsfield or Peterson would have flown, although the first – targeted for February 1972 at the time of the project's cancellation – would have been conducted, appropriately, by two Air Force pilots: Jim Taylor and Al Crews. Further two-man teams would then have been despatched at nine-month intervals for roughly 30-day orbital stays until the fifth and final manned mission in February 1975. At least one MOL flight, it was expected, would carry two naval officers, probably Bob Crippen and Dick Truly. The pilots, like the project itself, were supposed to be highly classified. However, there was a problem. One of them, Bob Lawrence, would have become the first African-American spacefarer when MOL finally flew and, naturally, the media grew to recognise him. "The rest of us were unknowns," said Don Peterson, "and I could travel on false IDs and nobody had any idea who I was. But [the Air Force] worried because [the press] knew him on sight and it becomes much harder to run a secret programme when one of your guys is a high interest to the media."

Tragically, Lawrence died in an aircraft crash in 1967, two years before MOL was cancelled. Both Peterson and Hartsfield are convinced that, had he lived, he would have gone on to join NASA and probably would have flown the Shuttle ...

After unstrapping and curling and twisting themselves around and through the hatch in the base of their Gemini, the MOL crews would first have drifted along a tunnel surrounded by cryogenic storage tanks for helium, hydrogen and oxygen to supply the station's atmosphere and fuel cells. Indeed, following the Apollo 1 disaster in January 1967, it was intended that MOL would have an atmosphere of 31 per cent helium and 69 per cent oxygen, pressurised at 0.34 bars, to reduce the risk of fire or detrimental medical effects on the pilots. The only adverse effect, it seemed, would have been some unusual chatter from the crew as they climbed to orbit: after breathing pure oxygen into their space suits, helium would also have been steadily pumped into the Gemini to better acclimatise them to the MOL environment. One can imagine that there would have been a few light-hearted smirks and chuckles among flight controllers as they listened in to the squeaky voices of two high-on-helium space explorers ...

After entering the outpost, and floating between the cryogenic storage tanks, the crew would have found themselves in its pressurised section. This was organised into two 'stories', each furnished with eight bays nicknamed 'the birdcage'. Providing further hints as to the kind of work the men would have done, these bays would have contained biochemical test consoles, an experiment airlock, a glovebox for liquids handling, a motion chair on rails, a physiology console to monitor their health and an Earth-facing viewport.

By the end of the Sixties, however, with a launch seemingly getting closer, the reality was that MOL was drifting further into oblivion. Members of the workforce complained that, no matter how hard they worked, the project always appeared to be at least a year away from launch. "General Bleymaier was the commander," recalled Don Peterson, "and he finally went to [President Richard] Nixon and said

'Either fund this programme or kill it, because we're burning time and money and we're not making progress because we don't have enough funds'." After cuts to the number of technical personnel, followed by woefully inadequate budgetary allocations, Nixon finally cancelled MOL on 10 June 1969 to save an unspent $1.5 billion of its estimated total price tag. The Outer Space Treaty, signed two years earlier, had already imposed enough restrictions on the Air Force to effectively demilitarise many of their proposed activities in Earth orbit and the Vietnam conflict continued to soak up more funds.

By now, MOL had swallowed close to three billion dollars, without even a single, full-scale unmanned flight. For the pilots, its cancellation was devastating. "We all thought it was going to come to fruition," said Bo Bobko in a 2002 oral history. "It was a surprise that it was just cancelled one day. I can remember I had a classmate from the Air Force Academy [who] had come to the MOL programme and it was his first day and they called everybody down to the auditorium and [told us that it had] been cancelled."

Hank Hartsfield, who was travelling to Huntington Beach in California, was also astonished, particularly in light of the work accomplished thus far. "The crew quarters were built, the training building was built, the pad was 90 per cent complete," he remembered. "It broke our hearts when it got cancelled. I won't forget the day. I was on my way to a meeting, listening to the news and they announced the cancellation. When I got to [prime contractor] Douglas and walked in, it was like walking into a morgue. It caught them completely by surprise. They heard it on the radio, like I did, or they came to work and found they didn't have a job. It was massive layoffs. People were getting pink slips almost immediately on the contractor force ... a very unhappy day."

For the Air Force pilots, the next step seemed to be to volunteer for Vietnam, although they were excluded from flying in combat. "We had a two-year duty and travel restriction," said Hartsfield, "because of the classified things we'd been exposed to. We couldn't be put in an environment where we could get captured." Several of the MOL pilots did return to active duty, but seven – Hartsfield, Peterson, Bobko, Crippen, Truly, Overmyer and Fullerton – were hired in September 1969 as NASA's seventh group of astronauts. After a wait of more than a decade, each flew at least one Shuttle mission and, in Crippen's case, as many as four. In fact, the final irony of MOL seems to be that, had Challenger not exploded in January 1986, Crippen would have commanded Shuttle mission STS-62A – the first-ever manned space launch from the United States' west coast, just as he had been trained to do almost two decades earlier ...

DR. RENDEZVOUS

Not long after his return from Gemini IX-A, astronaut Gene Cernan was summoned to Deke Slayton's office and posed an unusual question.

"Geno, how soon can you be ready to fly again?"

"Just say the word, Deke. When?"

"Right now. Would you be willing to jump from backup to prime? Fly [Gemini] XII with Lovell?"

The year 1966 had certainly been a dramatic one for Cernan. When it began, he and Tom Stafford confidently looked forward to flying Gemini XII – the last manned mission in the series – themselves. Then, with awful suddenness, the deaths of Elliot See and Charlie Bassett in February pushed them from backup to prime crew on Gemini IX. Following his return from his first spaceflight, Cernan had been given a 'dead-end' slot, with Gordo Cooper, to back up Jim Lovell and Buzz Aldrin on Gemini XII. Now, with barely two-thirds of the year gone, Slayton was offering to break his own crew-rotation system, bumping Aldrin from the mission. Cernan's first question to Slayton was a simple one. Why?

The reason was the Astronaut Manoeuvring Unit (AMU) – the Air Force-built rocket armchair which Cernan was originally detailed to test in June 1966 – whose military sponsors were pushing strongly to fly again on Gemini's last mission. Without giving much away, Slayton told Cernan simply that he was the best man to fly the AMU, which was probably true, but a number of contributory factors centred on Buzz Aldrin himself: a man of mathematical and engineering genius, the first astronaut to possess a doctorate, an unquestioned expert in the field of space rendezvous ... and a constant worry to Slayton. Aldrin had already raised eyebrows during Gemini IX-A, specifically those of Bob Gilruth and Chris Kraft, when he advocated having Cernan cut the lanyards of The Blob. Although not an outrageous suggestion, Slayton acquiesced, Aldrin's advice was a little too adventurous in light of NASA's limited EVA expertise.

Gilruth, Kraft and Slayton were not the only ones with worries about Aldrin. In his autobiography, Cernan hinted strongly that Aldrin's intelligence was tempered by a seeming inability to stick to one topic: he had a tendency to fly off at tangents and drastically re-engineer everything, at a time when NASA had little time to do so. Astronauts and their wives would roll their eyes when Aldrin collared them, even over coffee, and engaged them in hours-long discussions of the intricacies of celestial navigation and mechanics. Coupled with reports of his performance in the Gemini simulators, it was Slayton's judgement that the AMU test flight should be entrusted to Cernan, rather than Aldrin.

In his own defence, Aldrin would blame the decision on problems experienced by both Cernan and Dick Gordon on their EVAs: exhaustion, fogged-up visors and a difficulty in performing even simple tasks. "An urgent meeting of senior officials concerned with the Gemini XII EVA," Aldrin wrote, "was held at the end of September and ... they decided arbitrarily that I stood a poor chance of putting the innovative AMU backpack to good use. They felt the risks outweighed the benefits."

Despite the risks to his colleague's career, Cernan accepted Slayton's invitation on the spot – "when Deke asked if you would take a mission, there was only one answer" – and would have flown Gemini XII had not the decision been made that an AMU test was too risky. Gemini XII's EVAs would focus instead on less dramatic evaluations of a spacewalker's performance outside the pressurised confines of his spacecraft. Edwin Eugene Aldrin Jr, nicknamed 'Dr Rendezvous' behind his back, retained his place on the mission. Born in Glen Ridge, New Jersey on 20 January 1930, the son of an Army Air Corps pilot and a mother whose maiden name, propitiously, happened to be 'Moon', Aldrin's development, even into adulthood, was very much guided by his father.

Naturally, in light of his father's career, the man who would someday fly Gemini XII and become the second person to walk on the Moon was brought up with aviation in his blood. He first flew aboard an aircraft with his father in 1932, when he was barely two years old. (As a child, he earned the nickname 'Buzz' from his young sister, who, unable to pronounce 'brother', called him her 'buzzer'.) Graduation was followed by enrolment in a military 'poop school' – aimed at preparing him for the Naval Academy at Annapolis – although Aldrin sought the Military Academy at West Point. Despite his father's outspoken preference for the Navy, which he considered "took care of its people better", his son persisted and eventually won his reluctant approval.

When Aldrin graduated third in his class from West Point in 1951, his father's immediate reaction was a question: who had finished first and second? He was not accepted for a coveted Rhodes postgraduate scholarship and instead entered the Air Force, earning his pilot's wings later that same year after initial training in Bryan, Texas. During the conflict in Korea, Aldrin was attached to the 51st Fighter Wing, flying F-86 Sabres, and by the time hostilities ended in the summer of 1953 he had no fewer than 66 combat missions in his military logbook. Just a month before the end of the war, one of Aldrin's gun-camera photographs – a Russian pilot ejecting from his stricken MiG – ended up in Life magazine.

In total, Aldrin returned from Korea having shot down three MiGs. Back in the United States, he became a gunnery instructor at Nellis Air Force Base in Nevada and in 1955 was accepted into Squadron Officer School in Montgomery, Alabama. At around the same time, he met and married Joan Archer and shortly thereafter became the father of a son, James. Professionally, his military career prospered: he was assigned as an aide to General Don Zimmerman, the dean of the new Air Force Academy, then moved to Germany in 1956, flying the F-100 Super Sabre as part of the 36th Fighter-Day Wing, stationed in Bitburg. During this time, he became a father twice more: to Janice and Andrew.

Before pursuing his next ambition of test pilot school, Aldrin, like his Bitburg flying comrade Ed White, sought to gain further education and was accepted into the Massachusetts Institute of Technology on military detachment for a doctorate of science degree in astronautics. His 259-page ScD thesis, completed in 1963, just months before his selection as an astronaut, was entitled 'Line of Sight Guidance Techniques for Manned Orbital Rendezvous'. He chose the topic, he later wrote, because he felt it would have practical applications for the Air Force and aeronautics, although it also drew the attention of NASA, which was by now looking at lunar-orbital rendezvous for its Apollo effort.

Aldrin dedicated his thesis to future efforts in human exploration, wistfully remarking "if only I could join them in their exciting endeavours". By this time, of course, his application for the 1963 astronaut class was already being processed; he had tried to gain admission the previous year, seeking a waiver for his lack of test-piloting experience, but this time achieved success. A concern about his liver function, thanks to a bout of infectious hepatitis, did not prevent Aldrin from becoming one of the 14 astronauts named to the world that October.

Assigned to work on mission planning, his early days saw him focusing his attention on rendezvous and re-entry techniques ... and, gradually, as each Gemini crew was named, he became increasingly frustrated that he was receiving no flight assignment. At one stage, he even approached Deke Slayton to stress his confidence in his own abilities – that his qualifications and understanding of orbital rendezvous far exceeded those of anyone else in the office – and was politely told that his comments would be noted.

Shortly thereafter, in early 1966, Aldrin and Jim Lovell were assigned as the backup team for Gemini X. His heart sank. Taking into account Slayton's three-flight rotation system for backup-to-prime crews, there would be no Gemini XIII to which Aldrin and Lovell could aspire. It was, in effect, a 'dead-end' assignment. "Apparently, petitioning Deke – an arrogant gesture by 'Dr Rendezvous' – had not been well-received by the stick-and-rudder guys in the Astronaut Office," Aldrin wrote. "By being direct and honest rather than political, I'd shafted myself." All that changed on the last day of February, when See and Bassett were killed and their Gemini IX backups were pushed into prime position. In mid-March, Lovell and Aldrin were named as the new Gemini IX backups, with a formally-unannounced (but anticipated) future assignment as the prime Gemini XII crew.

For Aldrin, whose Nassau Bay backyard bordered that of the Bassetts, it was a devastating way to receive his long-desired flight assignment. Three weeks after the accident, he and Joan visited Jeannie Bassett to tell her the news. "I felt terrible," he wrote, "as if I had somehow robbed Charlie Bassett of an honour he deserved." Jeannie responded with quiet dignity and characteristic grace: her husband, she explained, felt that Aldrin "should have been on that flight all along ... I know he'd be pleased".

THE END

As the Gemini XII flight hardware – Lovell and Aldrin's spacecraft home, their Titan booster, their Atlas-Agena – was readied for launch during 1966, it was accompanied by an impending deadline to terminate the project and press on with Apollo. Indeed, when the two men walked out to Pad 19 on launch morning, 11 November, they would wear placards reading 'The' and 'End' on their backs. By this point, the corroded dome of the second-stage fuel tank previously assigned to Gemini X had been repaired and was delivered from Martin's Denver to Baltimore plants in late January. Seven months later, on 12 August, the entire Titan for Gemini XII was approved by Martin and by mid-September both of its stages were at Cape Kennedy.

It was at around this time that the Air Force's AMU test was deleted from Gemini XII and Gene Cernan lost his final opportunity to test-fly it in orbit. Persistent problems with mastering EVA techniques on previous missions had, NASA management concluded, made it inadvisable to proceed with such an ambitious endeavour and Gemini XII would instead focus on perfecting the 'fundamentals' with just basic extravehicular tasks. During his time outside, Aldrin would remove, install and tighten bolts with the power tool whose evaluation had been denied both Dave Scott and Dick Gordon, as well as operating connectors and hooks, stripping patches of Velcro and cutting cables.

To physically condition themselves, Aldrin and Cernan spent a considerable amount of time underwater in the neutral buoyancy tank, just outside Baltimore. They wore carefully-ballasted suits, Aldrin said later, to completely neutralise their buoyancy and approximate microgravity conditions as closely as possible. "Eventually," he wrote, "I mastered the intricate ballet of weightlessness. Your body simply had to be anchored, because if it wasn't, flexing your pinkie would send you ass-over-teakettle. And you don't want to do that dangling at the end of an umbilical cord 160 miles above Earth."

Processing of the Atlas booster and Gemini's final Agena target ran in tandem with that of the Titan. Designated Agena '5001', it was actually the non-flying version of the target delivered to Cape Kennedy in the summer of 1965, which had since been upgraded and made space-capable. "Getting a docking target took a bit of juggling," wrote Deke Slayton. The Air Force formally accepted it for advanced processing early in September 1966 and by the end of October it had been mated atop its Atlas and installed on Pad 14. For Lovell and Aldrin, their scheduled launch just a few days after Halloween had spawned an interesting orange-and-black embroidered crew patch, together with a crescent Moon offering a nod to the impending Apollo project.

Plans to launch on 9 November were abandoned when a malfunctioning power supply in the Titan's secondary autopilot reared its head and Lovell and Aldrin were recycled to fly two days later. The morning of the 11th dawned fine and clear and the Agena set off promptly that afternoon at 2:08 pm. (During insertion into space, an anomaly was noted in the target's propulsion system and plans to boost Gemini XII into a higher orbit were abandoned.) Strapped inside their tiny cabin, both astronauts could clearly hear the Atlas' thunderous roar. Ninety-eight minutes later, at 3:46:33 pm, it was their turn.

"There was no noise at first," Aldrin wrote, "but then a growing rumble began as the spacecraft rolled through its pre-programmed manoeuvre, twisting to the proper south-east launch trajectory." Steadily, the Titan accelerated, "like a subway train", Aldrin recalled, and as they climbed ever higher the sky turned to dark blue and eventually to black. Inside their space suits, both men felt their limbs rise and their toes lift to touch the tops of their boots. It felt almost as if they were stretching their feet, but not quite. They were weightless.

Once established in their 160 × 270 km orbit, Lovell and Aldrin set to work ploughing through their checklists, preparing for rendezvous and docking with the Agena some three orbits – and a little over four hours – into the mission. At around 5:11 pm, they made their first attempt at radar contact with the target and were surprised when the computer responded with the desired digits. "Houston," radioed a jubilant Aldrin, "be advised we have a solid lock-on ... two hundred thirty-five point fifty nautical miles."

However, the astronauts' success proved short-lived. As they circularised their

Jim Lovell (left) and Buzz Aldrin at breakfast on launch morning.

orbit to align themselves 'behind' and 'below' the Agena, above North America, Gemini XII's radar began giving intermittent readings. It was at this stage that Aldrin's years of rendezvous work came to the fore: he broke out the intricate charts and reverted to what he called the 'Mark One Cranium Computer' – the human brain. In his autobiography 'Men from Earth', Aldrin vividly described the hours-long effort: as Lovell piloted Gemini XII, he laboured over the charts, barely able to see the closely-printed data, occasionally aware of the passage of orbital daytime into nighttime and vice-versa.

It paid off. A little over four hours into the mission, Lovell eased the spacecraft's nose into the Agena's docking collar and announced, somewhat nonchalantly, "Houston, we are docked". The response from the ground, delivered with similar excitement, was a simple "Roger". A potentially serious obstacle – the failure of a critical piece of equipment, the rendezvous radar – had been overcome by human brainpower and flying abilities. Should a similar contingency occur during a rendezvous situation in orbit around the Moon, Lovell and Aldrin's work had at least proved that workarounds could be achieved. They had also used barely 127 kg of their fuel supply in one of the project's most economical rendezvous.

For the fourth time in eight months, a Gemini was securely linked to an Agena and Lovell and Aldrin became the second crew to practice undocking and redocking exercises. One attempt by Lovell during orbital darkness caused the docking latches to 'hang up' – producing a rather disturbing grinding sound – but he was nevertheless able to rock Gemini XII free without damage. A few minutes later, they switched roles and Aldrin redocked them onto the target.

Original plans, laid out before launch, had called for a reboost to high altitude, but this had to be abandoned eight minutes after the Agena lifted-off when its engine suffered a momentary decay in thrust chamber pressures and a drop in turbine speed. Instead, the astronauts were directed to turn their attention to solar eclipse photography; this task had been a scheduled part of their mission had they launched on 9 November, but the two-day delay caused it to be dropped. Now that the Agena reboost had been cancelled, it was reinstated, thanks to the input of Gemini XII's experiments advisory officer James Bates.

The inclusion of Bates' recommendation marked a shift in operations, with the scientists' representative, for the first time, being allowed to participate as one of the flight control team in the main Mission Control room. Moreover, it was determined that the Agena's secondary propulsion system had enough power to orient the spacecraft for an eight-second photographic pass at the proper time. At 10:51 pm, a little over seven hours into the mission, Lovell duly fired the target's smaller engines to reduce the combination's speed by 13 m/sec. The adjustment was successful and, after their first sleep period, the astronauts were advised to perform a second firing. Sixteen hours after launch, they reported seeing the eclipse "right on the money", cutting a swath across South America from north of Lima down to the southernmost tip of Brazil.

At first, it had seemed to the disgruntled crew that the second Agena burn might throw out the remainder of their schedule and adversely affect the start of Aldrin's first EVA. It did not, and at 11:15 am on 12 November, some 20 minutes before orbital sunset, Aldrin cranked open his hatch and pushed his helmeted head outside. "The hatch rose easily," he wrote, "and I rose with it, floating above my seat, secured to the spacecraft by short oxygen inflow and outflow umbilical hoses." Years later, he would vividly describe the immensity of the Universe all around him, remember the absence of any sense of speed and the recall the distinct curvature of Earth.

Aldrin quickly set to work on his first task, dumping a small bag containing used food pouches, and watched it slowly tumble away like a top, straight 'above' him. Next he moved on to quickly attach cameras onto brackets to photograph star fields on ultraviolet film and retrieved a micrometeoroid package, which he passed inside to Lovell. Unlike Cernan and Gordon, he did not overheat, thanks partly to regularly-scheduled rest breaks of two minutes apiece, and he returned inside Gemini XII at 1:44 pm after two and a half hours.

His real work had yet to begin. The mission's second period of EVA, which got underway at 10:34 am the following morning, required Aldrin to move away from the spacecraft on a 9 m tether. He set up a movie camera to allow flight controllers to monitor his performance, then moved to Gemini XII's nose and affixed a waist restraint strap to the docking adaptor. Next, Aldrin removed a tether from the Agena's nose and snapped it onto the Gemini, connecting the two vehicles for a gravity gradient exercise scheduled for later in the mission. He then manoeuvred himself towards the rear of the spacecraft, using flatiron-shaped handholds fitted with Velcro patches, and slipped his boots into a pair of foot restraints nicknamed 'golden slippers'. These, coupled with two small waist tethers, kept him anchored securely and Aldrin was able to satisfactorily complete a number of tool-handling and dexterity tests.

"Back in the buoyancy pool in Maryland," he wrote later, "I had torqued bolts and cut metal dozens of times – what I used to call 'chimpanzee work' – and I had no problem with these chores in space. Someone even put a bright yellow paper Chiquita Banana sticker at my busy box." He was even able to wipe Lovell's window (who asked him to change the oil, too) before returning to the cabin after two hours and six minutes outside. Back on Earth, Aldrin would claim quite openly that he had personally solved many of the problems of EVA, arousing criticism among the other astronauts, including Gene Cernan, who felt that his tasks were nowhere near as difficult as theirs. "Quite frankly," Cernan wrote, "we said he was only working a monkey board. Draw your own conclusions."

Shortly after Aldrin's return inside Gemini XII, the two men completed their evaluation of the tether by undocking from the Agena. The tether tended to remain slack, although they believed that slow gravity gradient stabilisation was achieved. "Within minutes," wrote Aldrin, "the two vehicles had stabilised without the aid of thrusters." After two full orbits thus connected, they finally fired an explosive squib to jettison the tether at 7:37 pm on 13 November.

Aldrin's record-breaking five and a half hours of cumulative extravehicular experience concluded the following day, the 14th, when he ventured outside at 9:52 am for a second stand-up period, lasting 55 minutes. He dumped unneeded equipment overboard, together with a sack containing his umbilical tether and two rubbish bags – hurled in lazy arcs over his shoulder – and took one last lingering

Aldrin during one of his three periods of EVA.

look at Earth below him: the vast land mass of Indochina ... and thought of his friend, Sam Johnson, with whom he had undergone flight training, and who was at that very moment a prisoner of war somewhere in North Vietnam.

Lovell and Aldrin's four-day mission had brought Project Gemini to a spectacular conclusion and had satisfactorily demonstrated rendezvous, docking, gravity gradient tethered operations and the ability of skilled human pilots to calculate a rendezvous with sextants and charts and a slide rule and pencil. Such human skills, using, in Aldrin's own words, the Mark One Cranium Computer, had relaxed managers' concerns about the viability of astronauts being able to perform a manual rendezvous, if necessary, in orbit around the Moon.

Gemini XII's problems were comparatively minor. Four of its 16 thrusters failed during the course of the mission and two its six fuel cells went dead, obliging flight controllers to instruct Lovell and Aldrin to drink more than their planned rations of water. This would make room for the excess fuel-cell water, which otherwise threatened to flood the spacecraft's power system. Whenever they drank water or used it to prepare their food, the red warning light blinked off, and in this way they nursed the fuel cell through 80 hours of flight.

A re-entry controlled completely by the computers brought Gemini XII into the Atlantic, barely 4.8 km from its target impact point, at 2:21 pm on 15 November. Within half an hour of splashdown, Lovell and Aldrin were safely aboard the aircraft carrier Wasp. The only unexpected event during re-entry had come at the onset of peak G loads, when a pouch containing books, filters and equipment broke free from the sidewall and landed on Lovell's lap. By this time, both men had unstowed the D-rings for their ejection seats and Lovell fought the urge to catch the pouch, lest he accidentally grab and pull the ring. "I didn't want to see myself punching out right at this high heating area," he said later.

With the safe return of Lovell and Aldrin to Earth, many of the procedures needed to get to the Moon and back had been thoroughly tested. Extravehicular suits had been used for extended periods of time and five astronauts had completed useful tasks outside. Unlike Alexei Leonov's swim in the void 20 months earlier, they had actually begun to demonstrate an astronaut's ability to really work in space. It provided the closest analogue yet attained of what working on the lunar surface might be like. Rendezvous, despite its complexity, had been completed with seemingly effortless ease by six Gemini command pilots ... and Lovell and Aldrin's work had shown it could be done without the aid of radar.

The radiant Moon above Cape Kennedy in the early winter of 1966–67 seemed considerably brighter than normal, as an altogether different kind of space vehicle geared up for its first manned shakedown cruise. Sitting on Pad 34 was a far larger rocket – the Saturn 1B – topped with the Apollo 1 spacecraft. In February 1967, astronauts Gus Grissom, Ed White and Roger Chaffee would evaluate the machine that would carry Americans to the Moon. Forget 1970, said many within NASA; it was becoming increasingly likely that a lunar landing might be achieved two years ahead of schedule. For ten euphoric weeks from mid-November 1966, the Moon was within humanity's grasp. Then, on the fateful Friday evening of 27 January 1967, all such dreams dissolved.

6

Disaster, Recovery, Triumph

"FIRE"

In a barren, disused area of Cape Canaveral stands a gaunt, concrete-and-steel hulk which once formed the launch platform of Pad 34. Four decades ago, it served as the starting point for the first Apollo mission, an 11-day trek into low-Earth orbit to demonstrate the capabilities of the spacecraft that would one day deliver men to the Moon. Today, overgrown with bushes, weeds and a few wild pepper trees, it slowly decays in the salty air. A faded 'Abandon in Place' sign adorns one of its legs. Near its base are a pair of plaques, memorialising a far darker and more tragic event. The first reads simply: 'Launch Complex 34 Friday 27 January 1967 1831 Hours' and dedicates itself to the first three Apollo astronauts, Virgil 'Gus' Grissom, Ed White and Roger Chaffee. The second pays tribute to their 'ultimate sacrifice' that January evening, long ago. Nearby are three granite benches, one in honour of each man.

Every year, without fail, their families are invited by NASA to visit the spot and reflect upon the disaster which befell them that Friday. It is a time of year which has become synonymous with tragedy in America's space programme; the losses of Challenger and Columbia having occurred within the very same week in 1986 and 2003. Indeed, 27 January 1967 marked a line in the sand, as the seeming good fortune and achievement of Project Gemini gave way to the stark and brutal realisation that reaching the Moon before the end of the decade was by no means assured and, technically and literally, our closest celestial neighbour remained a long way off. It would be a bad year for both the United States and the Soviet Union and by its close no fewer than six spacefarers would be dead: three killed in Apollo 1, a cosmonaut during his return to Earth and two astronauts in aircraft and car accidents.

Gene Cernan, for whom 1966 ended brightly with assignment to the Apollo 2 backup crew, recounted that 1967 was such a rotten year that his wife, Barbara, did not even feel able to write her usual Christmas letters to friends. It should have been quite different. NASA's plans called for as many as three Apollo missions, the first (led by Grissom) employing a spacecraft design known as 'Block 1', capable only of reaching Earth orbit. The others, commanded, respectively, by Jim McDivitt and Frank Borman, would utilise a more advanced Block 2 type, which possessed the navigational, rendezvous and docking equipment needed for lunar expeditions. "There were hundreds of differences between the two," wrote Deke Slayton, "the major one being that Block 1 vehicles didn't have the docking tunnel that would allow you to dock with a lunar module."

Each mission, though, would find common ground in that it would evaluate what Grissom had already described as a machine infinitely more complex than Mercury or Gemini; a machine which came in two parts. The 'command module', firstly, was a conical structure, 3.2 m high and 3.9 m across its ablative base, and would provide its three-man crews with 5.95 m^3 of living and work space whilst aloft. Brimming with reaction controls, parachutes, propellant and water tanks, cabling, instrumentation and controls, it would truly live up to its name as the command centre for future voyages to the Moon. Both it and the second section, the 'service module', were built by North American Aviation of Downey, California – today part of Boeing – and the latter consisted of an unpressurised cylinder, 7.5 m long and 3.9 wide, housing propellant tanks, fuel cells, four 'quads' of manoeuvring thrusters, an S-band communications antenna, oxygen and water stores and the giant Service Propulsion System (SPS) engine. The latter, some 3.8 m long and fed by a propellant mixture of hydrazine and unsymmetrical dimethyl hydrazine with an oxidiser of nitrogen tetroxide, would provide the impulse for inserting Apollo into and removing it from lunar orbit. It was a critical component without a backup and this demanded that its design be as simple as possible: its propellants, pushed into the combustion chamber by helium, were 'hypergolic', meaning that they would burn on contact, with no need for fuel pumps or an ignition system. The propellants would flow so long as the valves were held open and the valves were designed to be extremely reliable. The command and service modules would remain connected throughout a mission, with the latter jettisoned just minutes before re-entry.

Evaluating this machine for the first time was Grissom's responsibility. He and his crew were tasked with an 'open-ended' mission, lasting anywhere from six orbits to 14 days. In his book on lost and forgotten Apollo missions, spaceflight historian Dave Shayler wrote that it was certainly the crew's desire to fly for as long as practical – to extract as much valuable engineering data from the spacecraft as they could – but admitted that the Block 1 vehicle was almost certainly incapable of supporting operations for longer than 14 days. It will never be known what duration Apollo 1 might have achieved, but it seems doubtful that it could have seriously threatened the 14-day record established by Frank Borman and Jim Lovell a year earlier. Indeed, in the weeks before 27 January, Gus Grissom joked darkly that as long as the crew returned from orbit alive, he would consider the mission a success. His fatalistic outlook and lack of confidence in Block 1 was shared by many of his fellows within the astronaut corps.

Their launch, according to Flight Director Chris Kraft at the Apollo News Media Symposium, held in mid-December 1966, was targeted for late February or early March of the following year. As the event drew nearer, this date was refined. By the end of January, it was set for 21 February. Internally, the mission was known as

The first Apollo crew: Gus Grissom, Ed White and Roger Chaffee.

'Apollo-Saturn 204' (AS-204) and Grissom, White and Chaffee would have become the first humans to ride Wernher von Braun's mighty Saturn 1B booster, a two-stage behemoth which, in just 11 minutes and 30 seconds, would have injected their 20,400 kg spacecraft into an orbit of 136 × 210 km. (The '204' indicated that the launch vehicle was the fourth production unit of the Saturn 1B type. Despite this nomenclature, the crew had successfully pushed to rename their flight 'Apollo 1'.)

Grissom and his backup, Wally Schirra, light-heartedly dubbed the Saturn 1B "a big maumoo". Its first stage, known as 'S-IB', was 25.5 m tall and 6.6 m wide and would have boosted Apollo 1 to an altitude of 68 km under the combined thrust of its eight H-1 engines. Next, the 'S-IVB' second stage, fitted with a single J-2 engine, would have completed the climb into orbit. The latter, built by the Douglas Aircraft Company, would also form the third stage of the Moon-bound Saturn V booster. Minutes after achieving Earth orbit, Ed White, the senior pilot of Apollo 1, would have unstrapped and headed into the command module's lower equipment bay to begin setting up cameras and scientific hardware. Two and a half hours into the flight, he and pilot Roger Chaffee would have returned to their seats as Grissom separated Apollo 1 from the S-IVB.

Under Grissom's control, the spacecraft would have maintained tight formation with the spent stage as White and Chaffee employed a battery of cameras to record the venting of its residual liquid oxygen and hydrogen propellants. Much of this work would be crucial for subsequent, Moon-bound missions, which would involve a third Apollo component, the lunar module, housed inside the upper part of the S-IVB and extracted by the command and service module shortly after launch. No rendezvous manoeuvre with the stage was planned on Apollo 1, however, but several tests of the SPS engine were timetabled. In total, it would have been fired on eight occasions: three times by Grissom and White and twice by Chaffee, with the astronaut responsible for 'flying' each burn stationed in the commander's seat. The other crew members would respectively fulfil 'navigation' and 'engineering' roles. The first pair of SPS firings would have occurred during Apollo 1's second day in orbit, with the remainder being executed at roughly 50-hour intervals throughout the rest of the mission.

Additionally, the crew was assigned one of the largest complements of scientific, photographic and medical experiments ever carried into orbit. During their long flight, they would have operated instruments to monitor aerosol concentrations in the command module's cabin, carried out synoptic terrain photography of diverse targets ranging from the coast of Africa to the Mississippi River Mouth and Oyster Bay in Jamaica to the South China Sea and observed a range of meteorological and marine phenomena, from cloud eddies to dust storms and smog-laden cities to ocean currents. Other photographic tasks and medical investigations – an in-flight exerciser, a photocardiogram to monitor heart function and an otolith 'helmet' to track the effects of weightlessness on the balance mechanism of the inner ear – would have filled much of their time. Moreover, they would have been required to evaluate every aspect of Apollo itself, from the performance of its rudimentary 'toilet' to its general habitability.

Had their mission gone ahead and run to its maximum length, the three men would have returned to Earth on 7 March, separating from their service module a few minutes before plunging, base-first, into the upper fringes of the atmosphere. The command module would then have borne the brunt of re-entry heating and finally, beneath a canopy of three red-and-white parachutes, would have splashed into the Pacific some 330 hours after launch. Assuming nothing untoward happened, the flight of Apollo 1 would clear the way for the first mission of the Block 2 spacecraft and, eventually, for a lunar landing, possibly as soon as late 1968.

Thanks to Wally Schirra, there would only be one manned Block 1 flight. Originally, NASA wanted to virtually duplicate Grissom's mission with Apollo 2, featuring Schirra and rookie astronauts Donn Eisele and Walt Cunningham. However, the man who had commanded the world's first space rendezvous in December 1965 wanted nothing to do with it. "I argued it made no sense to do a repeat performance," he wrote, "and I succeeded in getting the mission scrubbed." (In fact, Walt Cunningham, in a 1999 oral history for NASA, speculated that Deke Slayton himself hoped to command Apollo 2, but was overruled. Slayton made no reference to this in his autobiography.) Regardless of Schirra's involvement, the hands of fate had already turned on Apollo 2 by October 1966, when the propellant tanks inside one of the service modules exploded during a ground test. Although it had not been assigned to Apollo 2, it made sense to keep downstream flights on schedule, replace it with Schirra's service module and cancel the 'repeat performance'. Whether he liked it or not, Schirra was told on 15 November that Apollo 2 was no more.

In winning his battle to get rid of the second Block 1 mission, Schirra had shot himself in the foot because he, Eisele and Cunningham were reassigned in early December as Grissom's new backup team. Schirra was reluctant to accept the change – he wanted a mission of his own that would provide a challenge and certainly did not want to serve on another backup crew – but Slayton and Grissom eventually talked him around. (He would complain to Tom Stafford, however, that Slayton had "screwed him".) Meanwhile, astronauts Jim McDivitt, Dave Scott and Rusty Schweickart, who had served as Apollo 1's backups since March 1966, would now become the prime crew of a 'new' Apollo 2, which would test the Grumman-built lunar module in Earth orbit on a mission involving two Saturn 1Bs. McDivitt, to be fair, had been working on the lunar module for more than a year and Scott had rendezvous experience, so it made sense to assign them, rather than Schirra's team, to this mission. Next up would be Frank Borman, Mike Collins and Bill Anders on Apollo 3, tasked with the first manned flight of the Saturn V to a high Earth orbit, reaching an record-breaking apogee of 6,400 km. Had the 27 January tragedy not occurred, these missions might have followed Apollo 1 in the summer and autumn of 1967, setting the stage for a lunar landing the year after.

Schirra's wish to cancel the second Block 1 flight was not just based on a whimsical reluctance to fly what he perceived to be a pointless mission. It also had much to do with his lack of confidence in the Block 1 design, which many astronauts considered sloppy and unsafe. "The craft was like an old friend," Time magazine told its readers on 3 February 1967. It was nothing of the sort. Since his assignment

The Apollo 1 command and service module during testing, with the Service Propulsion System (SPS) engine and thruster quads clearly visible. Apollo 1 would have been the first and only flight of the 'Block 1' design – a design disliked and distrusted by both Gus Grissom and Wally Schirra.

to Apollo 1, Gus Grissom had spent months overseeing poor performance and low standards on the part of North American in their efforts to get the Block 1 hardware ready for space. Unlike Gemini, where they could approach James McDonnell himself if issues arose, the North American set-up was far larger, more impersonal and little heed was paid to the astronauts' concerns. It "was a slick, big-time bunch of Washington operators," wrote Tom Stafford, "compared to the mom-and-pop operation at McDonnell." Some astronauts felt the technicians were more worried about their free time than with building a Moonship. Even NASA's Apollo manager Joe Shea remarked that, after receiving the two-billion-dollar contract, North American had thrown a party and made hats with 'NASA' printed on them, albeit with the 'S' replaced by a dollar sign ...

In fact, as 1966 wore on, the number of problems with 'Spacecraft 012' – the command and service module assigned to Grissom's mission – was so high that technicians were having trouble tending to them all. Riley McCafferty, responsible for updating the simulators, noted that at one point more than a hundred modifications awaited implementation. Plans to ship Spacecraft 012 to Cape Kennedy in August 1966 were postponed by three weeks when problems arose with a water glycol pump in the command module's environmental control system. Upon arrival in Florida, more than half of the engineering work assigned to North American remained incomplete, together with other serious deficiencies: a leaking SPS engine, coolant problems, computer software that "never quite worked right" and faulty wiring.

It did not inspire confidence in Grissom's crew, who presented Shea with a half-joking photograph of themselves bowed in prayer over their spacecraft. "It's not that we don't trust you," they told him, "but this time we've decided to go over your head!" Repairs and rework eliminated any hope of launching Apollo 1 before the end of the year and a February 1967 target became unavoidable. On 22 January, just before leaving his Houston home for the last time, Grissom was so angry with the problems that he plucked a grapefruit-sized lemon from a tree in his backyard, flew it to the Cape in his baggage and hung it over the hatch of the Apollo simulator.

Things had seemed quite different the previous March, when Grissom, White and Chaffee, "the coolest heads in the business," according to Bob Gilruth, had been named as the first Apollo crew. Together with their backups, McDivitt, Scott and Schweickart, they had closely monitored the spacecraft's progress throughout 1966, spending virtually every waking hour in Downey, renting rooms in a nearby motel and even sleeping in bunks near the production line from time to time. "Gus and Jim were the first two guys available to move from Gemini to Apollo," wrote Deke Slayton, "which is why I assigned them. Gus would keep an eye on the command and service module, while Jim would start following development of the lunar module."

To be fair, North American had faced immense technical challenges of its own. One of these was NASA's mandate that the command module should operate a pure oxygen atmosphere – a dangerous fire hazard, admittedly, but infinitely less complex than trying to implement an oxygen-nitrogen mixture which, if misjudged, could suffocate the crew before they even knew about it. In space, the cabin's atmosphere would be kept at a pressure of about a fifth of an atmosphere, but for ground tests would be pressurised to slightly above one atmosphere. This would eliminate the risk of the spacecraft imploding, but at such high pressures there remained the danger that anything which caught fire would burn almost explosively. At an early stage, North American had objected to the use of pure oxygen aboard Apollo, but NASA, which had employed it without incident on Mercury and Gemini, overruled them.

The selection of pure oxygen was not made lightly. NASA engineers had long been aware that a two-gas system, providing an approximately Earth-like atmosphere of some 20 per cent oxygen and 80 per cent nitrogen, pressurised to one bar, would reduce the risk of fires. Further, having a gaseous mixture of this type would avoid many of the physiological effects of pure oxygen, such as eye irritation, hearing loss and a clogging of the chest. However, at the time, the complexities of building a system which could mix and monitor these gases would have added to the spacecraft's weight, making it intolerably heavy. Other complications included the astronauts' space suits, which were pressurised to 0.31 bars. "To walk on the Moon," wrote Deke Slayton, "you needed to get out of the spacecraft . . . and with a mixed-gas system you'd have to pre-breathe for hours, lowering the pressure and getting the nitrogen out of your system so you didn't get the bends. Of course, if there was a real emergency and you had to use the suit, you'd really have been in trouble."

Other worries surrounded the craft's hatch, a complex affair which actually came in two cumbersome segments: an inner piece which opened into the command module's cabin, overlaid by an outer piece. North American had proposed a single-piece hatch, fitted with explosive bolts, which could be swung open easily in an emergency; NASA, however, argued that this might increase the risk of it misfiring on the way to the Moon. By adopting an inward-opening hatch, cabin pressure would keep it tightly sealed in flight . . . but notoriously tough to open on the ground. These two factors – a pure oxygen atmosphere and an immovable hatch – coupled with a mysterious ignition source would spell death for the Apollo 1 crew.

That Grissom was beginning to feel a strong sense of foreboding about the mission is reflected in his propitious comments in the weeks preceding the 27 January test – most famously, his declaration that "if we die, we want people to accept it . . . this is a risky business". He had already told colleagues that, if an accident did occur in the programme, it would probably involve him. Also, for the first time, he began to take the frustrations of work home with him. "When he was home," recalled his wife, Betty, "he normally did not want to be with the space programme. He would rather be just messing around with the kids, but he was uptight about it." Others, including Walt Cunningham, added that Spacecraft 012 "just wasn't as good as it should have been for the job of flying the first manned Apollo mission".

It was with an air, perhaps, of scepticism and contempt that Grissom, White and Chaffee, clad in their white space suits, crossed the gantry at Pad 34 and eased themselves into their seats early on the fateful afternoon of 27 January. All three were described by NASA secretary Lola Morrow – who had herself nicknamed Apollo as 'Project Appalling' – as unusually subdued and in no mood for the so-called 'plugs-out' test. By this point, the backup crew had changed to Wally Schirra,

Donn Eisele and Walt Cunningham, who had themselves spent the previous night aboard the spacecraft for a 'plugs-in' test, with the vehicle still entirely dependent upon electrical power from the ground and the hatch open. Schirra liked Spacecraft 012 no more than did Grissom. After emerging from his own test, he took his friend to one side and told him to "get outta there" if he sensed even the slightest glitch. "I didn't like it."

Communications between the spacecraft and the nearby blockhouse, manned by rookie astronaut Stu Roosa, caused problems almost from the outset. Grissom was so frustrated that he had even asked Joe Shea, at breakfast that morning, to climb into the cabin with them and witness, first-hand, from a manager's perspective, how bad the problems were. Shea weighed up the pros and cons of rigging up a headset and somehow squeezing himself in shirtsleeves into Apollo 1's lower equipment bay, beneath the astronauts' footrests, but ultimately decided against it. Even Deke Slayton, who was based in the blockhouse that day, considered sitting in with them, but decided to remain where he was and help monitor the test.

Grissom was first aboard, taking the commander's seat on the left side of the cabin, and almost as soon as he had hooked his life-support umbilicals into the oxygen supply, he noticed a peculiar odour. It smelled, he said, like sour buttermilk, and technicians were promptly scrambled to the spacecraft to take air samples. Nothing untoward was found and Roger Chaffee entered next, taking his place with the communications controls on the right-hand side. Finally, Ed White slid into the centre couch. With all three men aboard, the command module's inner and outer hatches were closed and sealed and finally the boost cover to protect it from the exhaust of the Saturn 1B's escape tower was secured. Next, pure oxygen was steadily pumped into the cabin.

Throughout the afternoon, a multitude of niggling problems disrupted and delayed things. A high oxygen flow indicator periodically triggered the master alarm and spotty communications between Apollo 1 and Roosa grew so bad that Grissom barked in exasperation: "How are we gonna get to the Moon if we can't talk between two or three buildings?" One such problem, which arose at 4:25 pm, turned out to be caused by a live microphone that could not be turned off. The NASA test conductor, Clarence Chauvin, recalled that communications were so poor that his team could barely hear the astronauts' voices. Eventually, the simulated countdown was put on hold at 5:40 pm. Forty minutes later, after more communications headaches, controllers prepared to transfer the spacecraft to internal fuel cell power ... and the countdown was held again.

Suddenly, with no warning, controllers noticed the crew's biomedical readings jump, indicating increased oxygen flow in their space suits. At the same time, around 6:30:54 pm, other sensors recorded a brief power surge aboard Apollo 1. Ten seconds later came the first cry from the spacecraft. It was Roger Chaffee's voice: "Fire!"

Seated in the blockhouse, Deke Slayton glanced over at a monitor which showed Apollo 1's hatch window, normally a dark circle, but now lit up, almost white. In quick succession came more urgent calls from the spacecraft. "We've got a fire in the cockpit!" yelled Chaffee, adding "let's get out ... we're burning up" and finally uttering a brief scream. His words described the desperate bid by the astronauts to save themselves. Downstairs, on the first floor of Pad 34, technician Gary Propst, watching a monitor in the first seconds after the call, could clearly see Ed White, his arms raised above his head, fiddling with the hatch. Propst could not understand why the men did not simply blow the hatch, little realising that there was no way for them to do this.

Instead, White had to laboriously use a ratchet to release each of the six bolts spanning the circumference of the inner hatch. Years later, Dave Scott recalled that, during training, he and White had weightlifted the hatch over their heads whilst lying supine in their couches. Now, in the few seconds he had available before being overcome, White barely had chance to even begin loosening the first bolt. It would have made little difference. As fire gorged its way through Apollo 1, the build-up of hot gases sealed the hatch shut with enormous force. No man on Earth could have opened the hatch under such circumstances. In fact, even under the best conditions in the simulator, the inner hatch alone took 90 seconds to remove and none of the crew – even the super-fit White – had ever done it inside two minutes during training.

Subsequent investigations would determine that the fire had started somewhere beneath Grissom's seat, perhaps in the vicinity of some unprotected and chafed wires and, once sparked in the pure oxygen atmosphere, fed itself hungrily and soon exploded into a raging inferno. Other readily-combustible products – Velcro pads, nylon netting, polyurethane padding and paperwork – added fuel to the flames. The astronauts themselves had even taken a block of styrofoam into the spacecraft to relieve the pressure on their backs throughout the test; it had exploded like a bomb in the pure-oxygen environment. "At such pressure, and bathed by pure oxygen," wrote Grissom's biographer Ray Boomhower, "a cigarette could be reduced to ashes in seconds and even metal could burn."

At length, pressures inside the cabin exceeded Apollo 1's design limit and the spacecraft ruptured at 6:31:19 pm, filling Pad 34's white room with thick black smoke and flames. By now, the pure oxygen had been guzzled by the fire and poisonous fumes had long since asphyxiated the three men. Outside, just a couple of metres away, North American's pad leader Don Babbitt sprang from his desk and barked orders to lead technician Jim Greaves to get them out of the command module. It was hopeless. The waves of heat and pressure were too intense and repeatedly drove the men back. "The smoke was extremely heavy," Babbitt recalled. "It appeared to me to be a heavy thick grey smoke, very billowing, but very thick." In fact, none of the pad staff could see far beyond their noses and had to physically run their hands over the outside of the boost cover to find holes into which they could insert tools to open the hatch.

The effects of the smoke were so bad that no fewer than 27 technicians were treated that night by Cape Kennedy's dispensary for inhalation and Babbitt had to order Greaves out of the white room at one stage, lest he pass out. Eventually, with the assistance of more technicians and the arrival of firefighters, the hatch was opened and the would-be rescuers gazed at a hellish scene of devastation within. By the flickering glow of a flashlight, they could see nothing but burnt wiring, firefighter Jim Burch recounted, and it took a few seconds before they finally realised that the

The burnt-out remains of Apollo 1.

unreal calmness meant only one thing: Grissom, White and Chaffee were dead. It was 6:37 pm, five and a half minutes after the first report of fire. Choking over the phone to Slayton, Babbitt could not find words to describe what he saw.

Flight surgeon Fred Kelly, who arrived on the scene with Slayton shortly thereafter, was equally shocked. He observed that it would probably take hours to remove the three men from the spacecraft, because the intense heat had caused everything to melt and fuse together. Moreover, there remained the risk that the heat could accidentally set off the Saturn 1B's escape tower and the pad was cleared of all personnel. Not until the small hours of the following morning, 28 January, were the bodies removed. Grissom had detached his oxygen hose, probably in an effort to help White with the hatch, while Chaffee, in charge of communications, remained securely strapped into his seat. The autopsy team set to work immediately and found that none of the astronauts had suffered life-threatening burns; all had died from asphyxia when their oxygen hoses burned through and their space suits filled with poisonous carbon monoxide.

It was a devastating blow to the astronaut corps and to NASA as a whole. Deke Slayton described it as his "worst day" and Frank Borman, whom the agency appointed as its astronaut representative on the internal review board, admitted that he, Max Faget and Slayton "went out and got bombed" in the hours after the accident. "I'm not proud to say it," continued the normally-teetotal Borman, "but ... we ended up throwing glasses, like a scene out of an old World War One movie." The wives of the three dead men – Betty Grissom, Pat White and Martha Chaffee – were also angry and would sue North American for its shoddy spacecraft. Each was awarded hundreds of thousands of dollars in compensation in 1972.

Three days after the accident, in flag-draped coffins, the bodies of the astronauts were met at Andrews Air Force Base in Washington, DC, by an honour guard and by representatives of NASA and President Lyndon Johnson's administration. Grissom and Chaffee were laid to rest in Arlington National Cemetery, whilst White's family insisted on his interment at the Military Academy at West Point.

The one saving grace of the tragedy was that it happened when it did, that it stopped NASA's incessant 'go' fever in its tracks at a point from which there might be some recourse. If Grissom, White and Chaffee had been on their way to the Moon when such a fire erupted, it could conceivably have spelled the end of Apollo. Moreover, astronauts Tom Stafford, John Young and Gene Cernan, who were in Downey on the night of the disaster, preparing to back up Jim McDivitt's Apollo 2 mission, felt the deaths of three men on the ground had probably saved six or more lives later on. By extension, the fact that investigators had the burnt-out spacecraft to examine meant that the cause could be pinpointed and rectified. A fire on the way to the Moon would have eliminated any chance of finding out what had happened.

MOONSHIP

Project Apollo, which brought about the deaths of Grissom, White and Chaffee and which also enabled the steps of Neil Armstrong and Buzz Aldrin on the Moon, was born very soon after NASA's own creation late in 1958. At that time, it was expected that exploration of the Solar System would be one arena in which the abilities of men, rather than machines, would be required. A fundamental obstacle, however, was the distinct absence of large boosters capable of fulfilling such roles and in mid-December of that year, newly installed Administrator Keith Glennan listened as Wernher von Braun, Ernst Stuhliner and Heinz Koelle presented the capabilities of existing hardware and stressed the need for a new 'family' of rockets. Landing men on the Moon, for the first time, was explicitly discussed as a long-term objective and, indeed, Koelle suggested a preliminary timeframe for achieving this as early as 1967.

Von Braun's vision for the new family of rockets was that, first and foremost, their engines should be arranged in a 'cluster' formation, directly carrying an aviation concept into the field of spacegoing rocketry. The famed missile designer also discussed propellants and the idea of employing different combinations for different stages ... then broached the subject of precisely how such enormous boosters could deliver a manned payload to the lunar surface. Von Braun had five methods in mind: one involving a 'direct ascent' from Earth to the Moon, the other four involving some sort of rendezvous and docking of vehicles in space. In whatever form the mission took, the rocket would need to be enormous, comprising, he said, "a seven-stage vehicle" weighing "no less than 6.1 million kg". Alternatively, he suggested flying a number of smaller rockets to rendezvous in Earth orbit and assemble a 200,000 kg lunar vehicle, which could then depart for the Moon. Aside from the immense practical problems of building and executing such a plan were the very real unknowns, Stuhlinger added, of how men and machines could operate in a weightless environment, with concerns of temperature, radiation, micrometeorites and corrosion an ever-present hazard.

Glennan's focus at the time was, of course, Project Mercury, although in testimony before Congress early in 1959 he and his deputy, Hugh Dryden, admitted that there was "a good chance that within ten years" a circumnavigation of the Moon might be achieved, although not a landing, and that similar projects may be underway in connection with Venus or Mars. In support of NASA's long-term aims, Glennan requested funding to begin developing the cornerstone for such epic ventures – the booster itself – and presented President Dwight Eisenhower with a report on four optional 'national space vehicle programmes': Vega, Centaur, Saturn and Nova. Although the first and last of these scarcely left the drawing board, the others would receive developmental funding and von Braun's team, which had championed a rocket known as the Juno V, gained backing to develop it further under the new name 'Saturn'.

In April 1959, Harry Goett, later to become director of NASA's Goddard Space Flight Center in Greenbelt, Maryland, was called upon to lead a research steering committee for manned space exploration. The major conclusions of his panel were that, after Project Mercury had sent a man into orbit, the agency's goals should encompass manoeuvring in space, establishing a long-term manned laboratory, conducting a lunar reconnaissance and landing and eventually surveying Mars or Venus. "A primary reason," remarked Goett of the choice of the Moon as a major target, "was the fact that it represented a truly end objective which was self-justifying and did not have to be supported on the basis that it led to a subsequent more useful end."

Elsewhere, efforts to begin developing the Saturn were gathering pace. Its challenges, though, were both huge and staggering, with propellant weights alone for a direct-ascent rocket producing a vehicle of formidable scale; indeed, even the prospects for constructing a lunar spacecraft in Earth orbit would require more than a dozen 'smaller' launches and the added complexity of rendezvous, docking and assembly operations. At this early stage, the problems of being able to store cryogenics for long periods in space, to have a throttleable lunar-landing engine and takeoff engine with storable propellants and auxiliary power systems were first identified.

Unfortunately, midway through Dwight Eisenhower's second term in office, and with the emphasis of his administration on balancing the budget "come hell or high water", it proved impossible for Glennan to formally commit NASA to a long-term lunar effort. Instead, small groups at the agency's field centres began springing up, including one within the Space Task Group, which considered a second-generation manned vehicle capable of re-entering the atmosphere at speeds almost as great as those needed to escape Earth's gravitational pull. "The group was clearly planning a lunar spacecraft," wrote Courtney Brooks, James Grimwood and Loyd Swenson, and by the autumn of 1959 sketches of a lenticular re-entry vehicle had emerged and crystallised to such a point that its designers even applied for it to be patented.

Early January of the following year finally brought approval from Eisenhower for NASA to accelerate development of von Braun's Saturn and offered the first hint of political support for manned space efforts beyond Project Mercury. Within weeks, Glennan's request to Abe Silverstein, director of the Office of Space Flight Programs, to encourage advanced design teams at each NASA field centre and within the aerospace industry began to bear fruit: von Braun's team proposed a Saturn-based lunar exploration design and J.R. Clark of Vought Astronautics offered a brochure entitled 'A Manned Modular Multi-Purpose Space Vehicle'. At this point, of course, Project Mercury had yet to accomplish its first manned mission; however, regardless of their limited chances of receiving presidential or congressional approval, the proposals continued.

Other efforts focused on exactly how the spacecraft and other hardware could be delivered to the lunar surface in the most economical way. In May 1960, NASA's Langley Research Center sponsored a two-day conference on rendezvous, with several techniques discussed, although it was recognised that they would be unlikely to bear fruit until the agency secured funding for a flight test programme.

It was at around this time that the decision was made over naming the spacecraft which would bring about the most audacious engineering and scientific triumph in the history of mankind. The name 'Apollo', formally conferred upon the programme on 28 July 1960 by Hugh Dryden, would honour the Greek god of music, prophecy, medicine, light and – perhaps above all – progress. "I thought the image of the god Apollo riding his chariot across the Sun," wrote Abe Silverstein, who had consulted a book on mythology to come up with the name, "gave the best representation of the grand scale of the proposed programme."

The scope of Apollo, Bob Gilruth and others revealed to more than 1,300 governmental, scientific and industry attendees at a planning session in August, was for a series of Earth-orbital and circumlunar expeditions as a prelude for the first manned landing on the Moon. Guidelines for the design of the spacecraft would be fourfold: it would need to be compatible with the Saturn booster under development by von Braun's team, it had to be able to support a crew of three men for a period of up to a fortnight and it needed to encompass the lunar or Earth-orbital needs of the project, perhaps in conjunction with a long-term space station. By the end of October, three $250,000 contracts were awarded to teams led by Convair, General Electric and the Martin Company for initial studies.

In spite of this apparent brightening of the lunar project's chances, Glennan himself remained unconvinced that Apollo was ready to move beyond the feasibility stage and felt a final decision would have to await the arrival of the new president in January 1961. By this point, Glennan was estimating Apollo to cost around $15 billion and felt that the Kennedy administration needed to spell out, clearly, and with no ambiguity, its precise reasons for pursuing the lunar goal, be they for international prestige or scientific advancement. At around the same time, Hugh Dryden and Bob Seamans directed George Low to head a Manned Lunar Landing Task Group, detailed to draft plans for a Moon programme, utilising either direct-ascent or rendezvous, within cost and schedule guidelines, for use in budget presentations before Congress. When Low submitted his report in early February, he assured Seamans that no major technological barriers stood in the way and that, assuming continued funding of both the Saturn and Apollo, a manned lunar landing should be achievable between 1968–70.

Moreover, Low's committee was considerably more optimistic than Glennan in terms of cost estimates: they envisaged spending to peak around 1966 and total some seven billion dollars, reasoning that by that time the Saturn and larger Nova-type boosters would have been built and an Earth-circling space station would probably be in existence. It stressed, however, that manned landings would require a launch vehicle capable of lifting between 27,200 and 36,300 kg of payload; the existing conceptual design, dubbed the 'Saturn C-2', could boost no more than 8,000 kg towards the Moon. Low's group advised either that several C-2s needed to be refuelled in space or an entirely new and more powerful booster awaited creation. Both approaches seemed realistic, the committee concluded, with Earth-orbital rendezvous probably the quickest option, yet still requiring the technologies and techniques to refuel in space.

Of pivotal importance in the subsequent direction of Apollo was the new president, John Fitzgerald Kennedy, who had already appointed a group before his inauguration to assess the perceived American-Soviet 'missile gap' and investigate ways in which the United States could pull ahead technologically. The group was headed by Jerome Wiesner of the Massachusetts Institute of Technology – later to become Kennedy's science advisor – and it advocated, among other points, that NASA's goals needed to be both redefined and sharpened. Another key figure, and long-time ally of NASA, was Vice-President Lyndon Johnson, who pushed strongly to appoint James Webb, a man with immense experience in government, industry and public service, to lead NASA. On 30 January 1961, Webb's appointment as the agency's second administrator was authorised by Kennedy.

It was Webb who would guide NASA through the genesis of Apollo; indeed, his departure from the agency would come only days before the project's first manned launch in October 1968. His importance to America's space heritage and the respect in which he continues to be held will be recognised, just a few years from now, by the launch of the multi-billion-dollar James Webb Space Telescope (JWST), successor to Hubble. Yet Webb's background was hardly scientific or in any way related to space exploration: a lawyer by profession, he directed the Bureau of the Budget and served as Undersecretary of State for the Truman administration, but throughout the Sixties he would prove NASA's staunchest and most fierce champion.

Also championing the agency's corner was President Kennedy himself, who, only weeks before Yuri Gagarin's flight, raised its budget by $125 million above the $1.1 billion appropriations cap recommended by Eisenhower. Much of this increase was funnelled into the Saturn C-2 development effort and, specifically, its giant F-1 engine. Built by Rocketdyne, the F-1 – fed by a refined form of kerosene, known as 'Rocket Propellant-1' (RP-1), together with liquid oxygen – remains the most powerful single-nozzled liquid engine ever used in service. Although it experienced severe teething troubles during its development, particularly 'combustion instability', it would prove impeccably reliable and the cornerstone to a lunar landing capability.

By this time, Convair, General Electric and the Martin Company had submitted their initial responses to NASA, none of which overly impressed the agency's auditors; indeed, recounted Max Faget, all three had stuck rigidly with the same shape as the Mercury capsule. Some theoreticians had already predicted that a Mercury-type design would be unsuitable for Apollo's greater re-entry speeds and Space Task Group chief design assistant Caldwell Johnson had begun investigating the advantages of a conical, blunt-bodied command module.

Early in May 1961, after more adjustments and rework, the contractors offered their final proposals to NASA. Convair envisaged a three-component Apollo system, its command module nestled within a large 'mission module'. Notably, it would return to Earth by means of glidesail parachute and develop techniques of rendezvous, docking, artificial gravity, manoeuvrability and eventual lunar landings. General Electric offered a semi-ballistic blunt-bodied re-entry vehicle, with an innovative cocoon-like wrapping to provide secondary pressure protection in case of cabin leaks or micrometeoroid punctures. Martin, lastly, proposed the most ambitious design of all. Conical in shape, its Apollo was remarkably similar to the design ultimately adopted, although it featured a pressurised shell of semi-monocoque aluminium alloy coated with a composite heat shield of superalloy and a charring ablator. Its three-man crew would sit in an unusual arrangement, with two abreast and the third behind, in a set of couches which could rotate to better absorb the G loads of re-entry and enable better egress.

All three contractors spent significantly more than the $250,000 assigned by NASA, with Martin's study topping three million dollars, requiring the work of 300 engineers and specialists and taking six months to complete. In their seminal work on the development of Project Apollo, Brooks, Grimwood and Swenson pointed out that, had times been less fortunate, NASA may have been obliged to spend months evaluating the contractors' reports before making a decision. However, it was at this time that Yuri Gagarin rocketed into orbit and John Kennedy pressed Lyndon Johnson to find out how the United States could beat the Russians in space. On 25 May 1961, before a joint session of Congress, he made the lunar goal official ... and public.

In the wake of Kennedy's speech, one of the key areas into which the increased funding would be channelled was a new booster idea called 'Nova'; this was considered crucial to achieving a lunar landing by the direct-ascent method. At this stage, although NASA was "studying" orbital rendezvous as an alternative to direct-ascent, Hugh Dryden explained that "we do not believe ... that we could rely on [it]". More money and increased urgency for Apollo was not necessarily a good thing: both Webb and Dryden felt that decisions over direct-ascent or orbital rendezvous and liquid or solid propellants would have been better made two years further down the line.

Nonetheless, rendezvous as an option was steadily coming to the fore, with a realisation that it could provide a more attractive alternative to the need for enormous and unwieldy boosters, instead allowing NASA to use two or three advanced Saturns with engines that were already under development. Although Earth-orbital rendezvous was considered safer, a lunar-orbit option would require less propellant and could be done with just one of von Braun's uprated Saturn 'C-3' rockets.

The Apollo spacecraft which would fly missions to the Moon was also taking shape. Max Faget, the lead designer of the Space Task Group, set the diameter of its base at 4.3 m and rounded its edges to fit the Saturn for a series of test flights. These rounded edges also simplified the design of an ablative heat shield which would be wrapped around the entire command module. Encapsulating the spacecraft in this way provided additional protection against space radiation, although on the downside it entailed a weight penalty. Others, including George Low, saw merits in both blunt-bodied and lifting-body configurations and suggested that both should be developed in tandem. Most within the Space Task Group, however, felt that a blunt body was the best option.

Notwithstanding these issues, in August 1961 NASA awarded its first Apollo contract to the Massachusetts Institute of Technology, directing it to develop a guidance, navigation and control system for the lunar spacecraft. Two months later, five aerospace giants vied to be Apollo's prime contractor, with the Martin Company ranked highest in terms of technical approach and a very close second in technical qualification and business management. In second place was North American Aviation, whom the NASA selection board recommended as the most desirable alternative. On 29 November 1961, word quickly leaked out to Martin that its scores had won the contest to build Apollo, but proved premature; the following day, it was announced by Webb, Dryden and Seamans that North American would be the prime contractor, in light, it seemed, of their long-term association with NASA and NACA and their spaceflight experience. The choice of North American, whose fees were also 30 per cent lower than Martin, would in many minds return to haunt NASA in years to come.

Rumour quickly abounded that it was politics, and not technical competency, which had won North American the mammoth contract. Astronaut Wally Schirra would recount that he felt the decision was made because companies in California had yet to receive their fair share of the space business, while others pointed to the company's lobbyist Fred Black, who had developed a close relationship with Capitol Hill insider Bobby Gene Baker, a protégé of Vice-President Lyndon Johnson.

As North American and NASA hammered out their contractual details, the nature of Apollo's launch vehicle remained unclear, as, indeed, was its means of reaching the Moon. It was likely that the production of large boosters capable of accomplishing a direct-ascent mission would take far longer than the development of smaller vehicles. The attractions of rendezvous were also becoming clearer as a means of meeting Kennedy's end-of-the-decade deadline. At around this time, Bob Gilruth wrote that "rendezvous schemes may be used as a crutch to achieve early planned dates for launch vehicle availability and to avoid the difficulty of developing a reliable Nova-class launch vehicle".

As the debate over the launch vehicle continued, it was recognised that, in whatever form it took, it would be enormous and would demand a correspondingly enormous launch complex. Under consideration were Merritt Island, north of Cape Canaveral, together with Mayaguana in the Bahamas, Christmas Island, Hawaii, White Sands in New Mexico and others. Only White Sands and Merritt Island proved sufficiently economically competitive, flexible and safe to undergo further study. The final choice: a 323 km^2 area of land on Merritt Island for a site later to become known (after the assassination of President John Kennedy) as the Kennedy Space Center. One of the most iconic structures to be built here in the mid-Sixties, and associated forever with the lunar effort, was the gigantic Vehicle (originally 'Vertical') Assembly Building (VAB), used to erect and test the Saturn rockets. Standing 160 m tall, 218 m long and 158 m wide, it covered 32,400 m^2 and to this day remains the world's largest single-story building.

Elsewhere, a site near Michoud in Louisiana was picked for the Chrysler Corporation and Boeing to assemble the first stages of the Saturn C-1 and subsequent variants. In October 1961, NASA purchased 54 km^2 in south-west Mississippi and obtained easement rights over another 518 km^2 in Mississippi and Louisiana for a static test-firing site for the large booster, prompting around a hundred families, including the entire community of Gainsville, to sell up and relocate. It was around the same time that the decision to move the Space Task Group – now superseded by the Manned Spacecraft Center – from Virginia to Houston, Texas, was made.

On the morning of 27 October 1961, shortly after 10:06 am, the maiden mission in support of Apollo got underway with the test of the Saturn 1 (originally C-1) rocket from Pad 34 at Cape Canaveral. Although the vehicle was laden with dummy upper stages, filled with water, its performance was satisfactory, but its 590,000 kg of thrust was woefully insufficient to send men to the Moon and back. Still, it marked the first of ten Saturn 1s launched, which, by the time of its last flight in July 1965, had carried a 'boilerplate' command and service module into orbit. Most engineers envisaged the lunargoing Saturn would need at least four or even five F-1 engines in its first stage. This would permit an Earth-orbit or lunar-orbit rendezvous mode to deliver a payload to the Moon's surface. Despite continuing interest in a large, direct-ascent Nova, employing as many as eight F-1s, the decision was taken on 21 December to proceed with a rocket known as the Saturn C-5 (later the Saturn V), capable of supporting both Earth-orbital and lunar-orbital rendezvous missions.

However, direct ascent was still considered by many as the safest and most natural means of travelling to the Moon, sidestepping the dangers of finding and docking with other vehicles in space. Yet procedures for exactly how a lander might be brought onto the lunar surface remained sketchy, with some suggesting the bug-like spacecraft touching down vertically on deployable legs or horizontally on skids. An Air Force-funded study, begun in 1958 and called 'Lunex', had already addressed a direct-ascent method of reaching the Moon. However, Wernher von Braun doubted it was possible to build a rocket large enough to accomplish such a mission and favoured rendezvous with smaller vehicles. Before coming to NASA, von Braun's team had proposed a mission known as 'Project Horizon', which justified the need for a lunar base for military, political and, lastly, scientific purposes. He felt that only Saturn was powerful enough to complete such a mission and one of his conditions upon joining NASA was that its development should continue.

Against this backdrop came the appearance of the lunar-orbital rendezvous plan, whereby a craft would descend to the Moon's surface and, after completing its mission, return to rendezvous with a 'mother ship'. The landing crew would then transfer to the orbiting mother ship and return to Earth. Since 1959, in fact, this idea had been recognised as the best technique to reduce the total weight of the spacecraft. Many within NASA, however, were terrified by the prospect of attempting rendezvous so far from home. Proponents, on the other hand, considered it relatively simple, with no concerns about weather or air friction, lower fuel requirements and no need for a monster Nova rocket. It was NASA engineer John Houbolt who finally convinced Bob Seamans to place it on an equal footing with direct ascent and Earth-orbital rendezvous when a decision came to be made. By July 1962, the decision had been made: lunar-orbital rendezvous would be adopted, employing a separate lander in addition to the command ship.

At the same time, the first steps to actually design the lander got underway, with early plans ranging from short-stay missions involving one man for a few hours to seven-day expeditions with crews of two. One design took the form of an open, Buck Rogers-like 'scooter' with landing legs, which the fully-suited astronaut would manoeuvre onto the surface. As these plans crystallised, the paucity of knowledge of the lunar surface material, and the effect of exhaust gases on its rocks and dust, made it imperative that astronauts could 'hover', brake their spacecraft and select an appropriate landing spot.

North American, which had already been awarded the contract to build the command and service modules, strongly opposed the lunar-orbital rendezvous mode, partly because it wanted its spacecraft to perform the landing. (Indeed, in August 1962, cartoons adorned its factory walls, depicting a somewhat disgruntled Man in the Moon looking suspiciously at an orbiting command and service module and declaring "Don't bug me, man!") With this in mind, North American made a strong bid to build the lander, which NASA rejected on the basis that the company already had its hands full with the development of the main spacecraft. By September 1962, 11 companies had submitted proposals to build the lander and in November the Grumman Aircraft Engineering Corporation of Bethpage, New York, was chosen for the $388 million contract. Although each bidder was judged technically and managerially capable, Grumman had spacious design and manufacturing areas, together with clean-room facilities to assemble and test the lander.

The decision to proceed with lunar-orbital rendezvous eliminated the requirement for the Apollo command module to land on the Moon, but created a new problem: the need for a form of docking apparatus by which it could link up with Grumman's lander. The need was quickly identified for a series of Earth-orbital missions to demonstrate and qualify the command module's systems before committing them to lunar sorties; the result was the Block 1 and 2 variants, the second of which provided the docking hardware and means of getting to the Moon. By mid-1963, North American had begun work on an extendable probe atop the command module, which would fit into a dish-shaped drogue on the lunar lander.

As the design of the command module moved through Block 1 and 2 variants, so the lunar module itself was changing into its final form: a two-part, spider-like 'bug' which would deliver astronauts to the Moon's surface and back into orbit. Its four-legged descent stage would be equipped with the world's first-ever throttleable rocket engine, whilst the ascent stage, housing the pressurised cabin, would have a fixed-thrust engine to boost the crew back into lunar orbit. The organic appearance of the lunar module produced something which Brooks, Grimwood and Swenson described as "embodying no concessions to aesthetic appeal ... ungainly looking, if not downright ugly". Operating within Earth's atmosphere, obviously, would be unnecessary and aerodynamic streamlining was ignored by the Grumman designers. However, when the time came for the ascent stage to liftoff from the lunar surface, its exhaust in the confined space of the inter-stage structures – 'fire-in-the-hole' – could produce untoward effects, perhaps tipping the vehicle over. Clearly, many problems remained to be solved.

Shape-wise, the ascent stage was originally spherical, much like that of a helicopter, with four large windows for the crew to see forward and 'down'. This design was ultimately discarded when it became clear that the windows would need extremely thick panes and strengthening of the surrounding structure. Two smaller windows were chosen instead, but the need for visibility remained very real, eliminating the spherical cabin design in favour of a cylindrical one with a flat forward bulkhead cut away at various planar angles. The windows became small, flat, triangular panels, canted 'downwards' so that the crew would have the best possible view of the landing site.

Changing from a spherical to a cylindrical cabin, though, meant that Grumman's engineers could not easily weld the structure. By May 1964, they had decided to weld areas of critical structural loads, but rivets would be employed where this was impractical. The interior of the 4,930 kg ascent stage cabin, with a volume of 60 m^3, made it the largest American spacecraft yet built and NASA pressed Grumman to make its instruments as similar to those in the command module as possible. As it evolved, the astronauts became an integral part of it, with Pete Conrad working on the design perhaps more so than anyone else. He was instrumental in implementing electroluminescent lighting inside the lunar module, as well as the command module, reducing weight and power demands.

Another crucial change in the design of the lunar module was the removal of seats, which were seen as too heavy and restrictive in view of the fact that the astronauts would be clad in bulky space suits. Bar stools and metal cage-like structures were considered, but the brevity of the lunar module's flight and moderate G loads eventually rendered them totally unnecessary. Moreover, standing astronauts would have a better view through the windows and the eliminated worry about knee room meant that the cabin could be reduced in size. Instead of seats, restraints would be added to hold the astronauts in place and prevent them from being jostled around during landing.

The hatch, through which the astronauts would exit and re-enter from the lunar surface, was changed from circular to square to make it easier for their pressurised suits and backpacks to fit through. At the base of the 10,334 kg descent stage were five legs, later reduced to four as part of a weight-versus-strength trade-off, and 91 cm footpads with frangible probes to detect surface impact. Keeping the lander's weight down was of pivotal importance, to such an extent that NASA paid Grumman $20,000 for every kilogram they could shave off. Even the weight of the astronauts helped determine which of them would fly the lunar module and which would not.

Inside the third stage of the Saturn V launch vehicle, the lander's legs would be folded against the structure of the descent stage and extended in space. In addition to its ascent and descent engines, the lunar module possessed 16 small attitude-control thrusters, clustered in quads, pointing upwards, downwards and sideways around the ascent stage for increased manoeuvrability. The ascent engine, built by Bell, was a key component which simply had to work to get the astronauts away from the lunar surface; as a result, it was the least complicated device, with a pressure-fed fuel system employing hypergolic propellants. The descent engine was more challenging, since it had to be throttleable: Rocketdyne, its builder, used helium injection into the propellant flow to decrease thrust while maintaining the same flow rate.

As the command, service and lunar modules took shape, the launch vehicles for the Earth-orbital (Saturn 1B) and lunar (Saturn V) missions also approached completion. The two-stage Saturn 1B – Gus Grissom and Wally Schirra's "big maumoo" – underwent its first test on 26 February 1966 and also marked the first 'real' flight of a 'production' Apollo command and service module. The rocket's S-IB stage had arrived at Cape Kennedy in mid-August of the previous year, followed by the S-IVB a month later. By the end of October, the rocket's instrument unit and the command and service module for the mission, designated 'Apollo-Saturn 201' (AS-201), were in Florida. After numerous delays, including lower-than-allowable pressures in the S-IVB, the flight got underway at 11:12 am. The S-IB carried the Saturn to an altitude of 57 km, whereupon the S-IVB took over and boosted AS-201 to an altitude of 425 km.

After raising its own apogee to 488 km, the command and service module's SPS engine was ignited to accelerate its return to Earth. Splashdown came at 11:49 am, half an hour after launch, and the undamaged spacecraft was hauled aboard the recovery vessel Boxer. Despite problems, AS-201 proved that the Apollo spacecraft was structurally sound and that its heat shield could survive a high-speed re-entry. However, its SPS had not performed as well as expected; firing, but only operating correctly for about 80 seconds, after which its pressure fell by 30 per cent due to helium ingestion into its oxidiser chamber. Managers, obviously, did not want such an event to occur during a return from the Moon. The SPS problem had to be rectified. Further, the effects of microgravity on the propellants in the S-IVB, which would be needed to perform the translunar injection burn, needed to be better understood.

Consequently, a decision was taken to reverse the plan of unmanned Saturn 1B launches for the remainder of the year. The six-hour AS-203 mission, not planned to carry a command and service module, was shifted ahead of AS-202 and launched on 5 July. It satisfactorily demonstrated that the S-IVB's single J-2 engine could indeed restart in space and that the propellants behaved exactly as predited. Seven weeks later came AS-202, during which the SPS was fired four times without incident, demonstrating its quick-restart capabilities, and the heat shield was tested. Its 90-minute mission cleared the way for Apollo 1, still internally dubbed 'AS-204', at the end of the year.

WITCH HUNT

The fire in the AS-204 spacecraft on 27 January 1967 left plenty of blame to go around and both North American, whose workmanship was seen as shoddy, and NASA, who had overseen them and given their seal of approval, were savaged by the media, by the public and by lawmakers alike. The media, indeed, were making up their own stories. On 10 February, for example, Time magazine cited the New York Times as having quoted an unidentified official who claimed that Grissom, White and Chaffee had screamed repeatedly for help in those frantic seconds. Their bodies, the official added, had been incinerated ...

Fearing that Congress could pull the plug on Apollo with immediate effect, the agency set to work on the night of the disaster on its own internal review, with an eight-man panel headed by Langley Research Center director Floyd Thompson. Although Olin Teague, chair of the House Space Subcommittee, was keen for NASA to complete its work, others within the Senate were impatient and called for a hearing on 27 February. There, Administrator Jim Webb was verbally grilled, with representatives condemning "the level of incompetence and carelessness" as "just unimaginable". Recriminations took an uglier turn when Senator Walter Mondale probed Webb for details of something called 'The Phillips Report'.

Apollo's programme manager, a retired Air Force general named Sam Phillips, had strongly criticised North American's performance as prime contractor for over a year. He considered their relationship with NASA to be quarrelsome and disagreeable and had established a 'tiger team' to inspect the situation. This had

Seated before a Senate hearing, NASA's senior management were verbally grilled and NASA's "carelessness" and "incompetence" were particularly attacked. From left to right are Bob Seamans, Jim Webb, George Mueller and Sam Phillips.

left him with serious concerns, so much so that on 16 December 1965 he wrote a scathing memo to North American chairman Lee Atwood, placed the company on notice to improve and told George Mueller, NASA's associate administrator for the Office of Manned Space Flight, that he had "lost confidence" in the prime contractor. Now, in the spring of 1967, Jim Webb revealed that he had never been made privy to the contents of Phillips' report.

Others, including North American inspector Thomas Baron, had since 1965 condemned the level of poor workmanship they saw at Cape Kennedy, together with infractions of cleanliness and safety rules. Although Baron's judgements were refuted by North American in its congressional testimony, they cannot have helped to quieten those who were looking for blame. Some, including the writer Erik Bergaust in his 1968 book 'Murder on Pad 34', even implied that NASA had blood on its hands for racing recklessly with the decade and killing the three men in the process.

Against this backdrop of public and media fury, the Thompson board worked for ten weeks, assisted by 1,500 technicians, and traced all possible sources of fire in Apollo's 30 km of electrical wiring and even re-enacted the blaze in a command module mockup. Additionally, Spacecraft 014, the Block 1 vehicle originally assigned to Wally Schirra's Apollo 2 mission, was shipped from Downey to Cape Kennedy for systematic dismantling and inspection alongside the burnt-out Spacecraft 012. Cabin pressures, the investigators found, had soared from the normal test pressure of 1.15 bars, slightly above sea-level equivalent, to 2.0 bars, rupturing the spacecraft's hull, but it was Bureau of Mines expert Robert van Dolah who revealed the damning truth: an escape hatch, capable of being opened in a couple of seconds, might have saved the astronauts. Thompson's report was published on 9 April and ran to 3,300 pages. It found no definitive cause for the fire, but suspected an unexplained arc on wiring beneath Grissom's left footrest, which spurted to another object and ignited the 100 per cent oxygen atmosphere.

The report cited "deficiencies in command module design, workmanship and quality control", including uncertified and highly-flammable materials in the cabin, as having contributed to the tragedy. Additionally, it revealed that many safety checks simply were not done, nor was there enough fire-suppression equipment at Pad 34. "It was," wrote Deke Slayton, "about as scathing a document as you'd ever see from a government agency towards itself."

Days later, Thompson and others found themselves testifying before the House and quickly discovered that even pro-Apollo congressmen were fiercely unsympathetic. Some lawmakers even went so far as to suggest reviewing the business of selecting contractors for the lunar effort. At one stage, responding to a question from Congressman John Davis of Georgia, North American's John McCarthy raised the possibility that Grissom himself might have inadvertently started the fire by kicking a batch of loose wires. Although Slayton admitted that McCarthy's comment was only raised in response to a question, he wrote that "it really pissed me off ... because there were no grounds for the story – it was pure speculation, not to mention physically impossible".

The effect of the fire elsewhere in the space agency was equally dramatic. Bob Gilruth, who had become a virtual father figure to many of the astronauts, broke down in tears upon learning of the tragedy. In his autobiography, Wally Schirra recalled taking him out on for a spin on his Cal 25 sailboat a few months after the accident and, whilst manning the tiller, Gilruth fell asleep. "Maybe it was the first chance he'd had to relax, to realise he had to push ahead and forget the tragedy," Schirra wrote. "Gilruth was carrying a tremendous load." So too was Joe Shea, the man who might have been inside the command module, sitting in precisely the spot where the fire started that terrible evening. He took the fire very badly, shifting into overdrive in an impossible personal crusade to solve Apollo's problems ... and, in doing so, drove himself to the brink of a breakdown. Eventually, he was moved to NASA Headquarters, then left to work for Raytheon. Years later, Shea would wonder if he could have snuffed out the fire ... and convinced himself, with 70 per cent certainty, that he could have successfully smothered it.

It was the straight-talking Frank Borman who summed up what should happen in testimony on 17 April. "Let's stop the witch hunt," he told Congress, "and get on with it." Getting on, though, would involve more than a year and $75 million-worth of changes to turn Apollo into a very different machine to that in which Grissom, White and Chaffee had died. Its cabin would now be pressurised with a mixture of 60 per cent oxygen and 40 per cent nitrogen, then steadily replaced with pure oxygen at partial pressure after launch as the nitrogen leaked out. No major structural reworking of the command module would be necessary. All flammable materials were to be removed and, crucially, a new 32 kg single-piece hatch was implemented, which opened outwards and could be sprung in just five seconds. Its mechanism, assisted by a cylinder of compressed nitrogen gas, could be opened with a little finger.

Elsewhere, aluminium plumbing, which melted at 580°C, was replaced by stainless steel, and coolant pipelines which could release flammable glycol when ruptured were 'armour-plated' with high-strength epoxy. Wire bundles were encased in protective metal panels and nylon netting and plastic containers were replaced by fire-retardant materials such as Teflon. Intricate 'Velcro maps' were created to limit the presence of this useful, but highly flammable, material and identify exactly where every piece of it would be located in the command module's cabin. Paperwork was kept to a minimum, to such an extent that the crews were barred from taking reading materials with them. "No books or magazines," wrote Wally Schirra. "Nor could we take anything made of paper to play with, such as cards or puzzles. We would find boredom a serious problem as we progressed through ten days in orbit."

The space suits to be worn by the astronauts had their nylon outer coatings replaced by beta cloth – an advanced fibreglass material produced by Owens-Corning Fibreglass Corporation – and supported by 14 layers of fire-resistant material. "We're paying a price for safety," Apollo 7 flight director Glynn Lunney told Time magazine. "The suits are bulkier, the fibreglass itches like hell and the seat belts are difficult to cinch down because they are so stiff, but you are seeing a spacecraft several hundred per cent improved." Further, an emergency venting system capable of reducing the cabin pressure in seconds provided an extra safeguard to snuff out fires. Overall, the changes increased Apollo's weight by 1,750 kg and placed it just beneath the Saturn V's total lifting capacity for lunar missions. As a result, parachutes were enlarged to permit safer splashdowns at greater weights, some redundant systems were eliminated and lead ballast was removed.

By extension, of course, the disaster which had befallen the command module could also afflict Grumman's lunar module and increased fervour was placed on reviewing its materials, too. Nylon-based items were replaced by beta cloth and 'booties' were installed over circuit breakers to lessen the risk of electrical shorts. This work on the lunar module – the machine which would actually set men on the Moon – refocused attention on the key question: would John Kennedy's dream ever be realised, within the decade, or at all? At the beginning of 1967, NASA had spent $23 billion on Project Apollo and many now questioned the need for America to go there. The continuing threat of the Soviet Union provided one reason: Leonid Brezhnev's increasingly regressive and repressive regime had, only a year before, consigned writers Yuli Daniel and Andrei Sinyavsky to hard labour for penning satirical, anti-Soviet texts. In some minds, it harked back to far darker times under Stalin.

When physicist Edward Teller was asked by Congress what he expected men to find on the Moon, he replied: "The Russians!" Even now, the sense of fear was as strong as ever. For their part, the Russians, mysteriously, had been conspicuously absent from the manned spaceflight business for almost two years by the time of the Apollo 1 fire, but their ambitions in Earth orbit were ready for a new resurgence. The death of Sergei Korolev and the appearance of a successor, Vasili Mishin, had pushed the Soviet Union's new spacecraft – Soyuz ('Union') – further and further behind schedule. Now, three months after the deaths of Grissom, White and Chaffee, it was ready to go. Or was it?

OUTER SPACE TREATY

On the evening that the Apollo 1 crew lost their lives, the astronaut office in Houston was unusually quiet. At one point, only Al Bean was on duty and it was he who received the first word from Cape Kennedy of the fire. Several other astronauts were at Downey, California, running through simulations and practice for their missions … and a select delegation was at the White House in Washington, DC. There, veteran astronauts Scott Carpenter, Gordo Cooper, Jim Lovell, Neil Armstrong and Dick Gordon witnessed the signing by President Johnson of a document popularly called 'The Outer Space Treaty'. Four decades later, the document has around a hundred signatories and a further two dozen who are partway through their ratification of it.

Officially, it is known as 'The Treaty on Principles Governing the Activities of States in the Exploration and Use of Outer Space, Including The Moon and Other Celestial Bodies'. Essentially, the document forms the basis for the earliest international space law and on the very day that Grissom, White and Chaffee died, it was opened for signing by the United States, Great Britain and the Soviet Union. Its 17 articles decree that signatories will refrain from the placement of nuclear weapons or weapons of mass destruction into Earth orbit, onto the Moon or onto any other celestial body. The treaty explicitly states that the Moon and other celestial bodies are to be used for peaceful purposes and forbids weapons-testing and military exercises or implacements on them. Moreover, it denies signatories the right to 'claim' a celestial resource, such as the Moon, as its own and declares all to be "province of mankind". It also assures the safe and cordial return of any astronauts or cosmonauts who make an unexpected landing within the borders of another nation.

The astronauts liked to call it the "non-staking-a-claim treaty" and as the afternoon wore into evening, they mingled with guests at the event, including ambassadors from the Soviet Union (Anatoli Dobrynin), Great Britain (Patrick Dean) and Austria (Kurt Walheim, later Secretary-General of the United Nations). In his biography of Armstrong, James Hansen noted the astronaut's recollection that the event ended at 6:45 pm and that, with the exception of Carpenter, the NASA delegation returned to the Georgetown Inn on Wisconsin Avenue. When they got to their rooms, they were greeted by flashing red lights on their answer machines. Something terrible had happened in Florida. A difficult year lay ahead.

FALL FROM HEAVEN

On 27 April 1967, an unusual communiqué was issued by the Soviet news agency, Tass. Days earlier, Vladimir Komarov – veteran of Voskhod 1 and the first cosmonaut to make two spaceflights – had been launched into orbit aboard the new Soyuz spacecraft. Within hours, however, euphoria had vanished into tragedy. In a handful of sentences, carefully crafted by the secretary of the Central Committee of the Communist Party, Dmitri Ustinov, it was revealed that Komarov's ship had "descended with speed" from orbit, "the result of a shroud line twisting". The result: "the premature death of the outstanding cosmonaut". Little more would be known in the western world for nearly three decades and only recently would details begin to trickle out. They would uncover a harrowing tragedy still shrouded in myth, mystery and rumour.

Soyuz was the brainchild of Sergei Korolev, the famous 'Chief Designer' of early Soviet spacecraft and rockets, with the original intention that it would support a series of lunar missions to rival the United States' Apollo effort. When it became increasingly clear that neither the Soyuz, nor an enormous booster rocket needed to reach the Moon, called the 'N-1', would be able to beat the Americans, the Soviet paradigm shifted to near-Earth missions: in 1971, they would establish the world's first space station in orbit. Soyuz would provide a ferry for missions which, by the end of the Seventies, would be routinely spending many months aloft. Four decades later, its basic design remains operational and, heavily modified, continues to transport cosmonauts and astronauts from a variety of nations to and from the International Space Station.

In his 1988 book about the early Soviet space programme, Phillip Clark traced the history of its development back to a three-part 'Soyuz complex' – a manned craft, a dry rocket block and a propellant-carrying tanker – which Korolev envisaged in the

Yuri Gagarin, Yevgeni Khrunov, Vladimir Komarov, Alexei Yeliseyev and Valeri Bykovsky during training for the Soyuz 1/2 joint mission. Note the EVA suits worn by Khrunov and Yeliseyev, providing clear evidence that an extravehicular transfer between the two spacecraft was probably planned.

early Sixties could be assembled in orbit for circumlunar missions. The first part, known as 'Soyuz-A', was closest in appearance to the spacecraft which actually flew. Measuring 7.7 m long, it comprised three sections: a cylindrical orbital module, a bell-shaped descent module to house the crew positions and a cylindrical instrument module for manoeuvring equipment, propellant and electrical systems. According to Korolev's early blueprints, Soyuz-A weighed around 6,450 kg, but unlike the eventual version it was not fitted with solar panels.

Supporting Soyuz-A were the 'dry' Soyuz-B rocket block and the propellant-carrying Soyuz-V tanker. Clark has hinted that a typical flight profile would have begun with the launch of a Soyuz-B, followed, at 24-hour intervals, by up to four Soyuz-Vs, which would dock, deliver their propellant loads, then separate. When the Soyuz-B had been fully fuelled, a manned Soyuz-A would be launched to dock onto the rocket block. "Mastering rendezvous and docking operations in Earth orbit may have been one of the primary objectives of the Soyuz complex," wrote Asif Siddiqi, "but the incorporation of five consecutive dockings in Earth orbit to carry out a circumlunar mission was purely because of a lack of rocket-lifting power in the Soviet space programme." Nonetheless, the sheer 'complexity' of the Soyuz complex seems to have foreshadowed its restructuring sometime in 1964 and effected a postponement of its maiden voyage until at least 1966. It was as a result of this setback, Clark explained, that the stopgap Voskhod effort was ultimately born.

When Voskhod began with such apparent promise – the world's first three-man cosmonaut crew, then the first-ever spacewalk – it surprised many in the western world, among them NASA's astronauts, when nothing more was heard from the Soviets until April 1967. "They hadn't flown in over two years," wrote Deke Slayton, "which nobody could understand ... Some people were beginning to say there wasn't really a race to the Moon, and on the evidence you had to admit that possibility." It was Korolev's successor, Vasili Mishin, who spearheaded the abandonment of Voskhod, which many within the Soviet space programme felt was a diversion of resources from the more versatile Soyuz. "Given what we know about Voskhod," added Slayton, "it was the right decision."

By October 1969, seven manned Soyuz spacecraft would have rocketed into orbit. However, a key physical difference between these missions and the original Soyuz-A concept was that they employed a pair of rectangular solar panels, mounted on the instrument module, to generate electrical power. The total surface area of these wing-like appendages was 14 m^2, each measuring 3.6 m long and 1.9 m wide. The remainder of the craft's design was strikingly similar to Soyuz-A: a spheroid orbital module, 2.65 m long and 2.25 m wide, atop the beehive-shaped descent module, itself 2.2 m long and 2.3 m wide at its base. Beneath the descent module was the cylindrical instrument module, 2.3 m long and 2.3 m wide. In total, Soyuz was somewhat larger than Apollo's command module, yet smaller than the combined command and service module.

Its propulsion system, designated 'KTDU-35', consisted of a pair of engines operating from the same fuel and oxidiser supply. The primary engine had a specific impulse of some 2,750 m/sec, equivalent to around 280 seconds' burn time, and a thrust of 417 kg, with early reports speculating that the propulsion system was capable of lifting Soyuz to an altitude of 1,300 km. This led Clark to suggest that a propellant load of 755 kg would have been required. Propellants took the form of unsymmetrical dimethyl hydrazine and an oxidiser of nitric acid, loaded in tanks on the instrument module. Clark speculated that, for the first few Soyuz missions at least, a lower-than-full propellant supply of around 500 kg was probably carried.

Like Vostok and Voskhod before it, the spacecraft and its three-stage rocket – an uprated version of Korolev's Little Seven, including four tapering boosters strapped to its central core – were typically delivered to the launch pad horizontally aboard a railcar. The Soyuz' own propellants were fully loaded before attachment to the rocket's third stage, after which a payload shroud was installed and, following rollout, the entire combination was tilted into an upright position. Four cradling arms, nicknamed 'the tulip', supported the rocket at its base and a pair of towering gantries provided pre-launch servicing access. Cosmonauts entered the spacecraft through its orbital module and dropped down into their seats in the descent module.

Yet the development of this complex spacecraft had been mired in technical and managerial problems since the death of Sergei Korolev in January 1966. Indeed, only days before Soyuz 1 was launched, engineers are said to have reported no fewer than 200 design problems to party leaders, all of which were overruled by the political pressure of getting a cosmonaut back into space. Even Vladimir Komarov, the man who would fly Soyuz 1, is reputed to have said one night in March 1967 that he would not – could not – turn down the assignment, even though he knew the spacecraft was imperfect and his chances of returning alive were slim. His reason: Yuri Gagarin, the first man in space and the Soviet Union's most treasured hero, was Komarov's backup. When asked by Gagarin's KGB friend Venyamin Russayev why he could not simply resign from Soyuz 1, Komarov's response was simple. "If I don't make this flight, they'll send the backup pilot instead," he said slowly. "That's Yura, and he'll die instead of me. We've got to take care of him."

Russayev was so concerned by Komarov's admission that he spoke to one of his own superiors, Konstantin Makharov, whose department dealt with spaceflight matters relating to personnel. Makharov told him that he intended "to do something" and asked Russayev to pass on a letter to Ivan Fadyekin, the head of Department Three, who directed him instead to a close personal friend of Leonid Brezhnev himself, a KGB man named Georgi Tsinev. The letter consisted of a covering note from a team of the cosmonauts, led by Gagarin, together with a ten-page document detailing all 200 problems with Soyuz. "While reading the letter," Russayev was quoted by Jamie Doran and Piers Bizony as saying, "Tsinev looked at me, gauging my reactions to see if I'd read it or not." It seemed to Russayev that Tsinev knew of Soyuz' inadequacies, but was not interested in the details. "He was glaring at me very intently," Russayev continued, "watching me like a hawk, and suddenly he asked, 'How would you like a promotion up to my department?' He even offered me a better office.'" Russayev carefully declined the offer and Tsinev kept the document ... which was never seen again. Makharov was fired, without a pension; Fadyekin was demoted simply for reading the document; and the hapless Russayev was stripped of all space-related responsibilities. "I kept my head down like a hermit for the next ten years," he said later.

Against this backdrop, Soyuz' problems had become almost chronic, with difficulties involving its Igla docking system, its simulators, its space suits, its hatches, its parachutes and its environmental controls. At one stage, early in its development, over 2,000 defects awaited resolution. Further, a series of unmanned Soyuz test flights under the 'Cosmos' cover name suffered troubles of their own. Phillip Clark noted that, as the break in Soviet manned launches stretched through 1965 and 1966, it became "almost a sport" among analysts to find evidence that a future crewed spacecraft was undergoing trials. Certainly, the flight of Cosmos 133 on 28 November 1966 and that of Cosmos 140 in early February of the following year were strongly suggestive of bearing some link with Soyuz. The first suffered a malfunctioning attitude-control system, which caused rapid fuel consumption and unanticipated spinning. An inaccurate retrofire and the likelihood that it would land in China eventually forced flight controllers to issue a self-destruct command to Cosmos 133. It exploded early on 30 November.

Two months later, Cosmos 140 suffered similar attitude and fuel problems, but at least remained controllable ... for a while. Its control system malfunctioned during retrofire, producing a steeper-than-intended re-entry which burned a 300 mm hole into the heat shield. The only reason its parachutes successfully deployed was because of this burn-through; otherwise, they would have failed ... an ominous harbinger of what would befall Komarov in April. Clearly, a Cosmos 140-type event would have doomed a human occupant, but the descent module separated successfully, parachuted to Earth and crashed through the ice of the frozen Aral Sea. It was retrieved by divers in 10 m of water and, astonishingly, the results of its mission were deemed "good enough" for Komarov to take the helm of a future flight.

In his autobiography, Alexei Leonov remarked that the Cosmos 140 burn-through had been caused by a flawed design feature which was slightly different to that on a manned Soyuz and admitted that "there was no chance of the fault recurring". Still, today, it seems ludicrous to have even contemplated a manned mission with such unpromising test results and unforgiving hardware. Political pressure seems to have been the overriding impetus driving Soyuz' schedule. One Soviet heat shield engineer, Viktor Yevsikov, hinted in 1982 that "some launches were made almost exclusively for propaganda purposes ... the management knew that the vehicle had not been completely debugged: more time was needed to make it operational, but the Communist Party ordered the launch despite the fact that preliminary launches had revealed faults in the co-ordination, thermal control and parachute systems". The situation was so bad, admitted Yevsikov, that Vasili Mishin himself refused to sign the endorsement papers permitting Soyuz 1 to fly. He felt it was unready.

Mishin, despite being an excellent mathematician and fast-thinking engineer, was no Korolev. He had none of his predecessor's stature or clout and was not renowned for his diplomatic skills. "Lacking the political instincts of, say, a Wernher von Braun or a Sergei Korolev," wrote Asif Siddiqi, "he suffered dearly. Some would argue that so did the Soviet space programme in the coming years." Nonetheless, with little opposition, Mishin was named Chief Designer in May 1966 and, although he quickly asserted himself, his insistence on filling the cosmonaut corps with non-pilot engineers from the OKB-1 design bureau to fly the early Soyuz missions infuriated Nikolai Kamanin. In his diary, the latter fumed that Mishin placed no value in six years' worth of experience of his command's training of cosmonauts to fly space missions. Kamanin considered it absurd that Mishin wanted to prepare civilian engineers for Soyuz command positions, with no pilot training, no parachute experience, no medical screening and no centrifuge practice. Eventually, under pressure from Dmitri Ustinov, Mishin was forced in July 1966 to accept pilot-cosmonauts for Soyuz command positions, with OKB-1 engineers filling support roles. It was only the first of many stand-offs between he and Kamanin which would place their relationship at a very low ebb.

Mishin's desire to fly civilians into space had been shared by Sergei Korolev and, intermittently in the early Sixties, a few OKB-1 engineers had passed preliminary screening, but were never seriously considered by the Soviet Air Force. When eight military cosmonauts began training for the first Soyuz missions in September 1965, Korolev entrusted one of his engineers to explore the possibility of forming a parallel group of civilians. Eleven candidates passed initial tests at the Institute of Biomedical Problems and several months later, on 23 May 1966, Mishin signed an official order to establish the first non-military cosmonaut group. Candidates Sergei Anokhin, Vladimir Bugrov, Gennadi Dolgopolov, Georgi Grechko, Valeri Kubasov, Oleg Makarov, Vladislav Volkov and Alexei Yeliseyev seemed to have little hope of actually flying into space and the nomenclature used to describe them – 'cosmonaut-testers' – seemed to support the assumption that they would be of limited use.

Despite his doubts, Kamanin was finally appeased when Grechko, Kubasov and Volkov passed tests at the Air Force's Central Scientific-Research Aviation Hospital and arrived at the cosmonauts' training centre, Zvezdny Gorodok, on 5 September. Within two months, another pair, Yeliseyev and Makarov, had also arrived. All five, wrote Siddiqi, "were accomplished engineers", Grechko having worked on fuelling Korolev's R-7s and Makarov having been involved in Vostok, Voskhod and Soyuz development. Unfortunately, Anokhin, Bugrov and Dolgopolov did not pass the Air Force's screening and were never considered for positions on the early Soyuz missions.

For the others, however, a seat on a spaceflight seemed only months away. Military pilot Vladimir Komarov had long been pointed at Soyuz 1, owing to his expertise, but Mishin, naturally, wanted two civilian engineers on the three-man Soyuz 2 crew. Nikolai Kamanin opposed this move, feeling that the complexity of the early missions made it inadvisable. A compromise was reached, thanks to the chief of the Communist Party's Defence Industries Department, Ivan Serbin, who suggested flying an Air Force pilot (Yevgeni Khrunov) and an OKB-1 engineer (Alexei Yeliseyev) alongside Vostok 5 veteran Valeri Bykovsky on Soyuz 2. A few days later, on 21 November 1966, Komarov told a State Commission meeting at Tyuratam that he had been picked to fly Soyuz 1 and that Bykovsky, Khrunov and Yeliseyev would follow aboard Soyuz 2. It was a triumph for the civilians. Yet had Yeliseyev flown as planned on Soyuz 2, he would not only have become the first of Mishin's civilians to enter space, but would have also been the first of them to die during his descent to Earth ...

Over the years, western observers suspected that the Soyuz 1 mission had been pushed to fly prematurely and improperly as a political stunt in advance of the May Day celebrations, since 1967 coincided with the half-century anniversary of the Bolshevik Revolution. Additionally, Leonid Brezhnev was in Karlovy Vary in Czechoslovakia at the time, at a meeting of the Soviet bloc leadership; the propaganda value of a major space success, for him, would be incalculable. In a dispatch to the Washington Star newspaper, Moscow correspondent Edmund Stevens wrote that the space effort under Mishin was less able to resist political pressure than Korolev had been. (It was even suggested that Leonid Smirnov, chairman of the Military-Industrial Commission, had personally told Komarov, still sceptical about Soyuz' readiness, that the cosmonaut might as well remove all of his military decorations if he refused to fly the mission ...)

In the days preceding the manned shot, rumours hinted of a space spectacular to rival Gemini and Apollo: a joint mission involving not one Soyuz, but two, and perhaps featuring rendezvous, docking and even the spacewalking transfer of crew members from one vehicle to the other. Reuters, for example, revealed on 19 April 1967 that such stories were circulating with some excitement in Moscow. Three days later, western journalists in the Soviet capital were told that two spacecraft with five or six cosmonauts would be launched, beginning on 23 April. If all went well with the first mission, it seemed likely that Soyuz 2 would fly at 3:10 Moscow Time the next morning. Komarov would attempt a docking on Soyuz 2's first or second orbit and the two spacecraft would remain docked for perhaps three days. "There was speculation," Time magazine told its readers on 5 May, "that the second ship had a restartable engine that would push the joined ships as far out as 50,000 miles." This was obviously a false assumption, but it does highlight the uncertainty of exactly what the Soviets were up to.

Actually, the joint mission, and specifically the spacewalking transfer of cosmonauts between two spacecraft, had caused concern for months. The hatch in the Soyuz orbital module, for example, was barely 66 cm in diameter, scarcely wide enough for a fully-suited man to get outside and virtually impossible for him to get back inside. (The problems of space suits 'ballooning' had already been experienced by Alexei Leonov.) A redesign of the hatch, Mishin realised, would add months to the schedule and the decision was instead taken to modify the suits by moving their oxygen supplies from the cosmonaut's back to his waist. Enlarged hatches would then be implemented on later missions. Nikolai Kamanin was unimpressed. "I am personally not fully confident that the whole programme of flight will be completed successfully," he wrote, "although there are no sufficiently weighty grounds to object to the launch. In all previous flights we believed in success. Today, there is not such confidence in victory ... This can perhaps be explained by the fact that we are flying without Korolev's strength and assurances." It did not bode well for the four men assigned to fly the Soyuz 1/2 joint mission.

Photographs released over the years have shown Komarov training with Bykovsky, Khrunov and Yeliseyev, the latter pair clad in EVA-type suits, confirming that they would have attempted the risky Soyuz-to-Soyuz transfer. Others show Yuri Gagarin, Komarov's backup, assisting Khrunov with his helmet. In their biography of Gagarin, Jamie Doran and Piers Bizony pointed out that it was Korolev's death in January 1966 which refocused the First Cosmonaut on somehow getting himself back into space. His renewed self-discipline and vigour in completing an engineering diploma at the Zhukovsky Air Force Academy impressed Nikolai Kamanin sufficiently to assign Gagarin in October 1966 as Komarov's backup. However, despite his confidence, Kamanin noted in his diary that Gagarin's importance to the Soviet state made it unlikely he would ever fly again.

Years later, Soviet journalist Yaroslav Golovanov would recall Gagarin's behaviour in the hours before the Soyuz 1 launch as quite unusual. "He demanded to be put into the protective space suit," Golovanov was quoted by Doran and Bizony. "It was already clear that Komarov was perfectly fit to fly, and there were only three or four hours remaining until liftoff time, but he suddenly burst out and started demanding this and that. It was sudden caprice." Venyamin Russayev expressed his belief over the years that Gagarin was trying to elbow his way onto the mission to save Komarov from almost certain death in a botched spacecraft. Others have countered that, since Komarov was not meant to wear a space suit on Soyuz 1, Gagarin's antics were actually designed to encourage his comrade to take one as an additional safety margin. Alternatively, maybe Gagarin was simply trying to disrupt matters somehow. Whatever the reality, archived pre-launch footage of the cosmonauts from that fateful third week of April 1967 – an unhappy Komarov, a downcast Gagarin and a team of very dejected technicians – show that that the atmosphere at Tyuratam was one of tense pessimism.

Other official images of Komarov arriving at the launch site showed him quite differently: bedecked with flowers … as, indeed, were Bykovsky, Yeliseyev and Khrunov, also in attendance for their own mission a day later. Plans for the flights were still very much in flux. Disagreement flared over whether to dock automatically or manually, with Mishin favouring the former and Komarov expressing confidence that he could guide Soyuz 1 by hand to a linkup from a distance of 200 m. At length, the chair of the State Commission, Kerim Kerimov, supported an automatic approach to 50–70 m, followed by a manual docking, although his judgement was still hotly contested.

Nevertheless, at 3:35 am Moscow Time on 23 April, Soyuz 1 was launched and inserted into a satisfactory orbit of 201–224 km. Within moments of reaching space, the Soviets referred to his mission, by name, as 'Soyuz 1', clearly indicating that a 'Soyuz 2' would follow soon. Fellow cosmonaut Pavel Popovich told Komarov's wife, Valentina, that he was in orbit, to which she responded that "he never tells me when he goes on a business trip!" Four and a half hours into the mission, a bulletin announced that the flight was proceeding normally; as, indeed, did another report at 10:00 am. More than 12 hours then elapsed before any more news emerged from the Soviets, and when it did finally come, it was devastating. Not only had there been no Soyuz 2 launch, but, stunningly, Komarov had lost his life during re-entry.

Little information other than the basics were forthcoming in the terse final report. It alluded to Soyuz 1's "very difficult and responsible braking stage in the dense layers of the atmosphere" and concluded that the "tangling of the parachute's cords" had caused the spacecraft to fall "at a high velocity, this being the cause of the death of Colonel Vladimir Komarov". Twenty years later, Phillip Clark wrote of "persistent reports" that problems had been experienced during Soyuz 1's first few hours in orbit. Its left-hand solar array failed to deploy properly, depriving Komarov of more than half (some sources say as much as 75 per cent) of his electricity supply. Soyuz 1 would be forced to run on batteries for a shortened mission of around a day in orbit. The subsequent, unusual, lack of televised images from the cabin and no other reports of in-flight activities lent credence to notions that the flight was in deep trouble.

A backup telemetry antenna also failed, probably triggering intermittent reception, and problems with solar and ionic sensors prevented Komarov from achieving even basic control of his craft's orientation. (It later became clear that the Sun sensor had actually been contaminated by Soyuz' thruster exhausts.) Although the antenna failure was a minor annoyance, the solar sensor was more serious, because without it Soyuz 1 could not be properly oriented for rendezvous and docking. During his fifth orbit, the cosmonaut tried to use his periscope and Earth's horizon to reorient the craft, but found it virtually impossible to do so. The failure of the left-hand solar panel to open had also left Soyuz 1 in an asymmetric configuration, which made attitude control far more difficult. At one point, Komarov even knocked with his boots on the side of the spacecraft, to free a stubborn deployment mechanism for the panel, but without success. By this time, the Soyuz 2 launch – already hampered by heavy rain at Tyuratam, but now exacerbated by the ongoing problems in orbit – had been called off and the focus had shifted instead to ensuring Komarov's safe return to Earth.

Attempts to bring him home, Clark continued, were planned on the 16th, 17th and 18th orbits, with the first retrofire attempt called off, presumably because the spacecraft could not be properly stabilised. Indeed, Doran and Bizony have reported that, at one stage, Komarov complained with fury: "This devil ship! Nothing I lay my hands on works properly." Unlike the spherical Vostok, the underside of Soyuz' bell-shaped descent module was distinctly flattened and it had an offset centre of gravity to provide it with some aerodynamic 'lift' during re-entry. However, it also required far more precision as it began to enter the atmosphere and, with Soyuz 1's guidance system out of action, the cosmonaut could not keep it under control. When it began to spin, he attempted to fire his attitude-control thrusters to stabilise the situation, but their close proximity to the navigation sensors meant that he could not accurately align the spacecraft. In desperation, Komarov resorted to using the Moon to work out his alignment.

The first retrofire attempt apparently began at 2:56 am on 24 April, but the problems forced the automatic control system to inhibit it. A decision was made shortly thereafter not to make another attempt on the 17th circuit, but to use that pass over Russia to prepare him for re-entry on the next orbit. Sometime between 3:30–4:00 am, a Japanese station received signals from Soyuz 1 and Tass announced that a routine communications event was being held between mission controllers and Komarov. That 'event', according to some, was far from routine. In August 1972, a former National Security Agency analyst, under the pseudonym Winslow Peck (real name Perry Fellwock), reported being on duty at a monitoring station near Istanbul in Turkey on the morning of Komarov's death. According to Fellwock's report, the cosmonaut and ground controllers knew that the situation would produce fatal consequences and Komarov even spoke personally to his wife, Valentina, and to a tearful Soviet premier Alexei Kosygin. "He told [his wife] how to handle their affairs and what to do with the kids," wrote Fellwock. "It was pretty awful. Towards the last few minutes, he was falling apart . . . "

These and other harrowing, though unverified, reports imply that Komarov knew that the problems with Soyuz 1 were insurmountable. Unconfirmed stories over the years hinted that, when he finally began re-entry, he grumbled that "the parachute is wrong" and "heat is rising in the capsule". Evidently, the actual retrofire on his 18th orbit was far from perfect, in light of the asymmetrical shape of the spacecraft and the inability of the attitude-control thrusters to maintain proper orientation. Still, retrofire began at 5:59 am and ran for long enough to ensure entrance into the atmosphere. The Yevpatoriya control station in the Crimea picked up voice communications at 6:12 am, in which Komarov apparently advised them of the results of the retrofire and his loss of attitude, before entering a period of blackout as heated plasma surrounded the spacecraft.

During re-entry, the descent module should have separated from the remainder of the Soyuz – the orbital and instrument sections – about 12 minutes after retrofire. Parachute deployment should have begun 14 minutes later and touchdown some 39 minutes and 27 seconds after retrofire. Komarov's voice reappeared during re-entry, sometime between 6:18 and 6:20 am, and was described as calm and unhurried, in spite of the 8 G load imposed by what was effectively a steep, 'ballistic' descent. Notwithstanding these problems, Soyuz 1 might still have landed safely. Then its parachutes failed.

In his autobiography, fellow cosmonaut Alexei Leonov related being based in the control centre, participating in the recovery effort. He wrote that "the brake chute deployed as planned and so did the drag chute, but the latter failed to pull the main canopy out of its container. While the reserve chute was then triggered, it became entangled with the cords of the drag chute and also failed to open". Indeed, Soyuz 1's landing point – at 51.13 degrees North latitude and 57.24 degrees East longitude, some 65 km east of the industrial city of Orsk, in the southern Urals – was considerably farther west than normal and has been seen by many analysts as "consistent with a purely ballistic re-entry . . . and no parachute deployment". Locals in the Orsk area, who witnessed the final stages of the descent, confirmed that Soyuz 1's parachutes were simply turning, not filling properly with air . . .

Meanwhile, Soviet anti-aircraft radar installations detected the incoming descent module at 6:22 am and predicted its 'landing' two minutes later. Elsewhere, listening posts in Turkey are said to have intercepted Komarov's cries of rage and frustration as he plunged to his death, cursing the engineers and technicians who had launched him in a fault-ridden spacecraft. Whether this really happened will probably never be known with certainty. Travelling at more than 640 km/h, Soyuz 1 hit the ground like a meteorite, killing the cosmonaut instantly and completely flattening the descent module. Solid-fuelled rockets in its base – meant to cushion the touchdown – detonated on impact, causing the remains to burst into flames. The whole landing site was soon engulfed in smoke and the first helicopter pilot on the scene quickly judged that it was a fatal situation. "But he also knew he was on an open loop with Yevpatoriya and the Ministry of Defence satellite control centre in Moscow," wrote Deke Slayton. "All he said was 'the cosmonaut is going to need emergency medical treatment outside the spacecraft', at which point the lines were cut by somebody in the rescue units."

The misleading call for 'urgent medical attention' is an intriguing story in itself. Flight surgeons Oleg Bychkov and Viktor Artamoshin, members of the search and rescue group which found Soyuz 1, recounted later that their helicopter touched down 70–100 m from the point of impact. "Everybody rushed to the capsule," they wrote, "but only upon reaching it, realised that the pilot would no longer need help. Fire inside the spacecraft was spreading and its bottom completely burned through with streams of molten metal dripping down." The rescue team was equipped with coloured flares to signal the overflying aircraft about the situation on the ground. No code existed to denote the death of the cosmonaut, so they were forced to fire the flare which equated to Komarov needing medical aid. It was this misunderstood message which, tragically, kindled some hope that Vladimir Komarov had survived.

On the ground, the flames were so fierce that portable foam extinguishers proved insufficient and the would-be rescuers began shovelling heaps of dirt onto the capsule. The force of impact had already reduced it from its normal 2 m height to a tangled mess no more than 70 cm tall and it was during the frantic firefighting effort that Soyuz 1 literally collapsed, leaving a pile of charred wreckage and a couple of congealed pools of molten aluminium, topped by the circular entrance hatch. Nearby lay the three parachutes. Komarov's remains were "excavated" from what was left of his ship at 9:30 am and his death was pronounced as having been caused by multiple injuries to the skull, spinal cord and bones. Later eyewitness reports revealed that his 'body' took the form of a 'lump', 30 cm wide and 80 cm long, while Venyamin Russayev recounted that a heel bone was the only recognisable fragment left ...

By this time, Nikolai Kamanin himself was on the scene and it was he who telephoned Dmitri Ustinov, who in turn contacted Leonid Brezhnev. Five hours later, it was Ustinov who carefully edited Tass' communiqué on the subject of Komarov's death.

A government investigation, headed by V.V. Utkin of the Flight Research Institute of the Aviation Industry, revealed that Soyuz 1's parachute container had opened at an altitude of 11 km and had become 'deformed', squeezing the main canopy and preventing it from opening correctly. Although a small drogue had come out, the main parachute simply could not exit the container, and not just because of the deformation. The drogue was supposed to impart a force of 1,500 kg to pull out the main parachute, whereas it actually required upwards of 2,800 kg, perhaps a result of air pressure in the descent module pushing against the container. Such problems had never arisen in tests, Utkin's panel found, but attributed them to the abnormal and 'random' conditions surrounding the Soyuz 1 descent. Future missions, the panel decreed, would benefit from enlarged and strengthened parachute containers. The failure of the drogue to pull out the main parachute was compounded by its backup canopy. This quickly became entangled with the fluttering drogue, leaving nothing to arrest Komarov's meteoric fall to Earth.

Unofficially, gross negligence on the part of manufacturing technicians has also been blamed for Komarov's death. During pre-flight preparations, explained Asif Siddiqi, the Soyuz 1 and 2 spacecraft were coated with thermal protection materials and placed in a high-temperature test chamber. Both were evaluated with their parachute containers in place, but lacking covers. This resulted in the interiors of both containers becoming covered with a polymerised coating, which formed a very rough surface and directly prevented Soyuz 1's parachute from deploying. "Clearly," wrote Siddiqi, "the most chilling implication of this manufacturing oversight was that both Soyuz spacecraft were doomed to failure – that is, if Komarov had not faced any troubles in orbit and the Soyuz 2 launch had gone on as scheduled, all four cosmonauts would have died on return." None of this was mentioned in the official Soyuz 1 accident report.

As the Soviets, like the Americans, dug in for a lengthy period of self-criticism and introspection to make their craft spaceworthy, not another cosmonaut would venture aloft until October 1968. That cosmonaut, Georgi Beregovoi, would establish a new record as the oldest man yet to be launched into orbit, aged 47. He was also one of Yuri Gagarin's harshest critics – a senior Soviet Air Force officer, Second World War combat veteran and decorated test pilot, albeit unflown in space – who considered the First Cosmonaut to be "an upstart" and a bit-of-a-lad who was "too young to be a proper Hero of the Soviet Union". Their relationship in the months before Komarov's death grew so stormy that Gagarin even shouted that Beregovoi would never fly in space.

Seven months after Gagarin's untimely death in an aircraft crash, Beregovoi finally got his chance. It was he who would lay the ghost of Vladimir Komarov to rest and nurse Soyuz through its first successful manned mission.

SLOW RECOVERY

Within days of the publication of Floyd Thompson's damning report into the Apollo 1 fire, the first efforts were implemented to fulfil its recommendations. Of paramount importance was the redesign of the hatch, which would change from a complex two-piece device into a 'unified' single section. Although it was heavier than the hatch which had prevented Gus Grissom, Ed White and Roger Chaffee from escaping the inferno of Spacecraft 012, it could be opened in as little as five seconds and had a manual release for either internal or external operation. At the same time, fire and safety precautions were upgraded at Cape Kennedy and a slidewire was added to Pad 34's service structure to allow crews to rapidly descend to ground level.

By the beginning of May 1967, a sense pervaded NASA and North American that the first steps to recover from the fire were underway; so much so that George Mueller proposed an unmanned test flight of the gigantic Saturn V lunar rocket as soon as possible. A crewless demonstration of the improved Apollo system was definitely needed and, utilising a command and service module combo known as 'Spacecraft 017', was pencilled-in for the early autumn of that year. By that time, four manned missions had also been timetabled, one featuring the command and service module on its own, the other three inclusive of the lunar module, after which an attempt to actually touch down on the Moon might go ahead. Certainly, Time magazine told its readers on 19 May that unmanned Apollos were scheduled for September, October and December, followed by an inaugural manned mission in March 1968. NASA Headquarters were even more optimistic. Some managers suggested that a lunar landing could occur on the fourth manned Apollo flight, but their counterparts in Houston expressed more caution. Chris Kraft, for one, had warned George Low, who replaced Joe Shea to head the Apollo Spacecraft Program Office, that a lunar landing should not be attempted "on the first flight which leaves the Earth's gravitational field".

Others, including Mueller, wanted to skip the flight of a manned command and service module in Earth orbit entirely and press on with a complete 'all-up' test of the entire Apollo combination, including the lunar module. "Bob Gilruth got in the way of this one," wrote Deke Slayton. "For one thing, the Apollo CSM was a sufficiently complex piece of machinery that it needed a shakedown flight of its own. Why try to test two manned vehicles for the first time at the same time? We thought a CSM-only flight was the way to go before the fire and nothing we were going to learn was likely to change that." Moreover, the lunar module itself was running months behind schedule and a manned flight was not anticipated until at least the end of 1968. Mueller was finally persuaded to accept a command and service module flight in Earth orbit for the first manned Apollo mission.

Despite the increased optimism, concerns remained. The schedule for the first unmanned Apollo test atop the Saturn V – designated 'Apollo 4' or 'Apollo-Saturn 501' (AS-501) – was extremely tight. In particular, the Saturn's S-II second stage had undergone a difficult year of testing in 1966. Nonetheless, at the stroke of 7:00 am on 9 November 1967, the entire Cape Kennedy area received a jolt when the five F-1 engines of the Saturn V ignited with what Brooks, Grimwood and Swenson later described as "a man-made earthquake and shockwave ... the question was not whether the Saturn V had risen, but whether Florida had sunk!" Deke Slayton, who had come to the Cape to watch the behemoth fly, later recounted that he had "seen a lot of launches ... but nothing was ever as impressive as that first Saturn V. It just rose with naked power, lots of noise and light". Fellow astronaut Tom Stafford, also there, commented that Walter Cronkite's CBS News trailer almost shook itself to pieces. "Suddenly," added Mike Collins, "you realise the meaning of 7.5 million pounds of thrust – it can make the Cape Kennedy sand vibrate under your feet at a distance of four miles ... "

The merest mention of the name 'Saturn V' implies power. From a height, weight and payload-to-orbit standpoint, it remains the largest and most powerful rocket ever brought to operational status, although the Soviet Union's short-lived Energia had slightly more thrust at liftoff. It evolved from a series of rockets, originally dubbed the Saturn 'C-1' through 'C-5', of which NASA announced its intent to build the latter in January 1962. It would be, the agency revealed, a three-stage launcher with five F-1 engines on its first stage, five Rocketdyne-built J-2 engines on its second stage and a single J-2 on its third stage. These engines, when tested, had shattered the windows of nearby houses. It would be capable of delivering up to 118,000 kg into low-Earth orbit or up to 41,000 kg into lunar orbit. Early in 1963, the C-5 received a new name: Saturn V.

When a mockup of the rocket was rolled out to Pad 39A at Cape Kennedy on 25 May 1966, it amply demonstrated its colossal proportions. It stood 110.6 m tall and 10 m wide, only a few centimetres shorter than St Paul's Cathedral in London. It comprised an S-IC first stage, an S-II second stage and was topped by the S-IVB which would be restarted in space to boost the Apollo spacecraft towards the Moon on a so-called 'translunar injection' (TLI) burn. All three stages used liquid oxygen as an oxidiser. Fuel for the first stage was the RP-1 form of refined kerosene, while the S-II and S-IVB utilised liquid hydrogen. Eighty-nine truckloads of liquid oxygen and 28 of liquid hydrogen, together with 27 railcars filled with RP-1, were needed to fuel the Saturn V.

The S-IC first stage, built by Boeing, was 42 m tall and its five F-1 engines, arranged in a cross pattern, produced over 3.4 million kg of thrust to lift the Saturn to an altitude of 61 km. The four 'outboard' engines could be gimballed for steering during flight, whilst the centre one was fixed. The S-II, built by North American, was

Spectacular panoramic view of the Cape Kennedy landscape as 'Moon-fever' gripped NASA in mid-1966. Clearly visible are a Saturn V test vehicle, the gigantic Vehicle Assembly Building (VAB) and the Launch Control Center (LCC).

25 m tall and would make history as the largest cryogenic-fuelled rocket stage ever built. Finally, the Douglas Aircraft Company's 17.85 m-tall S-IVB would be used to place the Apollo spacecraft into Earth orbit, then restart a couple of hours later for a six-and-a-half-minute-long TLI burn. It also provided a 'garage' to house the lunar module.

The Apollo 4 spacecraft was an old Block 1 with many features of the upgraded Block 2 design, including an improved heat shield and the new unified hatch. The aim of its mission was to evaluate its structural integrity, its compatibility with the Saturn V and its ability to enter an elliptical orbit and re-enter the atmosphere to land in the Pacific. The mission ran perfectly: the Saturn V boosted the spacecraft into a 185 km parking orbit and, after two circuits of the globe, for the first time, its S-IVB third stage restarted to propel Apollo 4 to an apogee of more than 17,000 km. Next, the service module's SPS engine ignited, sending the spacecraft out to 18,000 km for a four-and-a-half-hour-long 'soak' in the little-known radiation and temperature environment of deep space. In doing so, Apollo 4 dipped its toe into the conditions that astronauts would one day experience as they traversed the 370,000 km translunar gulf.

Finally, with the command module's nose pointed Earthward, the SPS fired a second time to bring it home. The service module separated and the command module hit the upper atmosphere, just as it would on a lunar return, at 40,000 km/h. Nine hours after its launch, Apollo 4 hit the waves of the Pacific, near Hawaii, just 16 km from the primary recovery ship Bennington. As successful as the mission had been, a long road remained before an actual lunar landing could be accomplished. Certainly, an additional uncrewed flight was highly desirable to many within NASA, providing further confirmatory data that the enormous rocket was capable of delivering men safely to the Moon. One crucial vehicle which still needed an 'all-up' performance test was Grumman's lunar module, the first flight-ready version of which – designated 'LM-1' – was delivered to Cape Kennedy, three months late, at the end of June 1967.

By a strange twist, Apollo 5, which would consist solely of the lunar module, with no command and service module aboard, was assigned the Saturn 1B originally meant to carry Gus Grissom's crew into orbit. In the immediate aftermath of the fire, it had been destacked from Pad 34, checked for corrosion or damage and finally restacked on Pad 37 on 12 April 1967. With the lunar module installed in its nose, the 55 m rocket looked unusual, 'stubby' even, since it lacked the command and service modules and an escape tower. The LM-1, encased in the final stage of the Saturn, had an incomplete environmental control system and was not fitted with landing gear, since it was destined to burn up during re-entry into the atmosphere.

Loading propellants aboard the rocket proved troublesome, mainly due to procedural difficulties and minor irritations such as clogged filters and ground support equipment glitches, but a simulated launch demonstration ended success-fully on 19 January 1968. Three days later, at 5:48 pm, Apollo 5 set off and was inserted perfectly into orbit. Forty-five minutes into the flight, LM-1's attitude control thrusters pushed it away from the S-IVB and a lengthy checkout of its systems began. Two orbits later, its TRW-built descent engine – the world's first-ever

AS-501, the first flight of the Saturn V, topped by the Apollo 4 command and service module, is prepared in the VAB for its November 1967 launch.

The legless Apollo 5 lunar module is prepared for flight.

throttleable rocket, capable of slowing it down for landing on the Moon – was fired for 38 seconds, but was ended abruptly by the lunar module's guidance system when it sensed the vehicle had not accelerated fast enough. In response to the cutoff, flight controllers moved to an alternate plan: firing the descent engine on two further occasions, then igniting the ascent engine. With all primary tests done, LM-1 re-entered the atmosphere to destruction and its remains plunged into the Pacific, several hundred kilometres south-west of Guam, on 12 February. So successful, in fact, was Apollo 5 that a further unmanned test of the lunar module was considered unnecessary. Its next flight, atop the Saturn V, would be carried out with a crew aboard.

However, the lander still had many problems of its own. The instability of its Bell-built ascent engine, in particular, caused concern throughout 1967 and for much of 1968. Although both George Mueller and Sam Phillips felt that Bell had a good chance of solving the engine's fuel-injector problems, the agency nevertheless hired Rocketdyne to develop an alternate device. Despite difficulties in both cases,

Rocketdyne was ultimately chosen to outfit the lunar module's fuel injector. Other problems with the bug-like lander included windows blown out and fractured during high-temperature tests, broken wiring and stress corrosion cracks in its aluminium structural members; the latter led to the formation of a team to identify the cause and implement corrective actions. Grumman analysed more than 1,400 components and heavier alloys were employed for newer sections of the lunar module. Weight, too, posed an issue. In 1965, more than 1,100 kg had been shaved from the lunar module and NASA even offered incentives to Grumman to remove yet more unwanted bulk. The LM-1 flight had been good enough for NASA to cancel an unmanned LM-2 test, but LM-3 – the first mission to fly manned – would not be ready until at least the end of 1968.

Meanwhile, the performance of the Saturn V on the Apollo 4 mission fired up hopes that it could soon be entrusted with a human crew. Nonetheless, another test flight, that of Apollo 6, was still required ... and rightly so, for the rocket's second mission, AS-502, almost ended in a disaster. On 13 March 1967, the S-IC first stage arrived at Cape Kennedy and, inside the cavernous interior of the Vehicle Assembly Building, was mated to its S-II second stage in May. By February of the following year, topped by the S-IVB third stage and the Apollo 6 command and service module, it was rolled into wind-driven rain towards its destination: Pad 39A, today revered as one of the most famous and historic launch platforms in the world. Despite communications difficulties, which forced a two-hour halt, the stack arrived at the pad at 6:00 pm.

Aside from being a second unmanned test of the Saturn V, the Apollo 6 mission would put Spacecraft 020 through its paces on the final flight of the command and service modules before a human crew headed aloft on Apollo 7. Originally scheduled for launch in the first quarter of 1968, the flight was postponed several times. First, the tank 'skirt' on another service module split during structural tests, prompting an inspection and restrengthening of Apollo 6 to prevent a similar problem. Next, after rollout to the pad, water seepage was detected in the Saturn V's S-II second stage and some parts had to be replaced. Eventually, at 7:00 am on 4 April, the rocket thundered into the heavens, seemingly with perfection ... and then, things began to go wrong.

Throughout the first two minutes of its climb, the five F-1 engines burned steadily and normally, then experienced thrust fluctuations which caused the entire rocket to oscillate longitudinally like a pogo stick for around 30 seconds. Low-frequency modulations were recorded in the Apollo 6 command module, exceeding design criteria, but otherwise the first stage completed its work. However, the time soon came for the S-II second stage to exhibit problems: two of its five J-2 engines suddenly stopped, four minutes into a six-minute firing, requiring the others to burn for 59 seconds longer than planned to compensate for the abrupt power loss. The rocket did not tumble and explode, however, because the failed J-2s were adjacent to one another and the Saturn survived by gimballing its remaining 'good' engines. Still, the second stage did not achieve its desired velocity and ended up at a higher altitude than it should before its fuel was exhausted.

This meant that the S-IVB had to burn for correspondingly longer. It "was confusing to the computer guiding the S-IVB," wrote Deke Slayton, "which realised it was higher than it should be ... and slower. So while it added 29 seconds to the burn, it actually pointed itself down toward the centre of the Earth." At length, after a difficult ascent in which the S-IVB pitched itself back upwards and entered orbit firing backwards, Apollo 6 was inserted into a wild 178–367 km elliptical orbit, instead of a 160 km circular path. The Saturn's troubles, though, were still not over. An attempt to restart the S-IVB – just as it would be required to do in order to boost Apollo crews toward the Moon – failed when the third stage refused to ignite. "If this had been a manned flight," wrote Deke Slayton, "the escape tower on the Apollo would have been commanded to fire, pulling the spacecraft away from the Saturn for a parachute landing in the Atlantic."

An 'alternate' mission was now inevitable and the command and service module were duly separated from the S-IVB and the SPS engine burned for seven minutes, simulating a TLI manoeuvre and pushing the apogee of Apollo 6's looping elliptical orbit to 22,200 km. This gave it enough altitude to mimic a lunar-type return, but not enough velocity, and it splashed down in the Pacific, missing its impact point by 80 km. Ten hours after launch, the command module was hauled aboard the amphibious assault ship Okinawa. Despite a NASA press release which declared that preliminary data indicated the spacecraft had done its job well, many felt that, overall, the mission had not been a success. The Saturn V might need a third unmanned test before it could be flown with astronauts aboard.

In fact, pogo effects had been observed, to a lesser extent, during the Apollo 4 launch and its apparent cause was traced to a partial vacuum created in the fuel and oxidiser suction lines by the rocket engines. The condition, wrote Brooks, Grimwood and Swenson, produced a hydraulic resonance; in effect, the engine 'skipped' when bubbles caused by the partial vacuum reached the firing chamber. Engineers later determined that two of the Saturn V's engines had been inadvertently tuned to the same frequency, which probably made the problem worse. In future, all clustered engines were tuned to different frequencies to prevent any two or more of them from pulling the rocket off-balance and changing its trajectory.

As part of efforts to rectify the issue, Rocketdyne began retesting the F-1 engine in late May, injecting helium into the liquid oxygen feed lines to interrupt the resonating frequencies which had caused the unacceptable vibration levels. In four of the six tests, the 'cure' proved worse than the 'disease', by making the oscillations more pronounced. Attempts at NASA's Marshall Space Flight Center in Huntsville, Alabama, used the same technique, but produced quite different results; no oscillations were observed. Elsewhere, the cause of the J-2 failures proved more of a mystery. During tests, engineers discovered that frost forming on propellant lines when the engines fired at ground temperatures served as an extra protection against the fuel lines rupturing. However, frosting did not take place in the vacuum of space, pointing at a possible cause of the failure. The chances of American bootprints on the Moon before the end of 1969, it seemed, was still very much touch-and-go.

MY LAI

In the spring of 1968, as NASA wrung its hands over the Saturn V, the United States' strategy of attrition in Vietnam seemed to be failing as the ongoing conflict consumed ever more hundreds of lives and President Lyndon Johnson was being pressured by his generals to commit an additional 206,000 troops to the half-million-strong military force already in south-east Asia. At the end of January, a hammer blow struck the misguided sense of complacency that the Vietcong were little more than snipers and unable to mount major co-ordinated attacks. The so-called 'Tet Offensive', which ran in three devastating waves until September, was intended to strike military and civilian command centres throughout South Vietnam and spark uprisings among the population. Although it ultimately proved disastrous, militarily, for the Vietcong, the offensive was so vast (countrywide) and so well-organised (involving more than 80,000 troops) that it shocked both Johnson's failing administration and the American public. In March, citing conflict both abroad and at home, Johnson announced that he had no intention to "seek and ... will not accept the nomination of my party for another term as your president".

Against this backdrop of self-doubt and introspection came one event which has become infamous as perhaps the most notorious act of mass murder in American military history, involved the tiny South Vietnamese hamlet of My Lai. There, on 16 March, at least 300 – some reports say as many as 500 – unarmed civilians, including women and children, were raped, tortured, mutilated and massacred by American troops. Excuses have been banded around over the years, that such-and-such was reaching for a grenade, for instance, or even that the South Vietnamese peasantry were seen as 'inhuman', but the precise reasons for My Lai have never been divulged. When it reached the ears of the world a year later, the incident sparked outrage and condemnation and strengthened already simmering public discontent over an unpopular war.

The so-called 'Charlie Company' who gained notoriety for the massacre had arrived in South Vietnam three months earlier, just before the January outbreak of the Tet Offensive. My Lai and several neighbouring hamlets were suspected of harbouring Vietcong fighters and the wheels of a major American offensive were quickly set in motion. It would, the commanders urged, be an aggressive assault, involving the total destruction of the hamlets, the slaughter of livestock and even the pollution of wells. On the evening before the attack, Captain Ernest Medina of Charlie Company advised his men that nearly all civilians at My Lai would have left for market by early morning and only Vietcong sympathisers or fighters would remain. Differing opinions would materialise over the years over whether Medina specifically instructed his men to slaughter women and children ...

Certainly, upon reaching My Lai soon after dawn, no enemy fighters were found, but their presence was suspected and Lieutenant William Calley – the only military officer to be convicted of murder that day – began shooting at what he later described as a "suspected enemy position". Calley's actions lit the touchpaper for the murderous rampage that followed: the soldiers began attacking anything that moved, using rifle butts, bayonets and hand grenades to summarily execute young and old alike. At one point, it was said, Calley took a weapon from a soldier who refused to kill ... and used it to continue the massacre. When the bloodbath ended, My Lai was torched.

Warrant Officer Hugh Thompson, a helicopter pilot, witnessed much of this from the air and identified many of the soldiers committing the atrocities; his and others' testimony would prove crucial when the perpetrators were brought to trial. However, the real carnage might have gone unknown had Ron Ridenhour, a former member of Charlie Company, not sent a damning letter in March 1969 to newly-inaugurated President Richard Nixon, numerous congressmen, the Joint Chiefs of Staff, the Pentagon and the State Department, detailing the chain of events at My Lai. Eventually, in September 1969, William Calley was convicted of premeditated murder and 25 other officers were later charged with related crimes. At around the same time, Time, Life and Newsweek magazines broke the story ... and public support for the Vietnam War, already on shaky ground, vanished.

'THE WALLY, WALT AND DONN SHOW'

Despite the Saturn V's woes, the situation was somewhat brighter for Spacecraft 101, the command and service module assigned to Apollo 7, the first manned flight. Heavily refurbished and quite different from the machine which had claimed Grissom, White and Chaffee's lives, it finally arrived at Cape Kennedy on 30 May 1968. By this time, its crew had already been training for more than a year. Commander Wally Schirra, senior pilot Donn Eisele and pilot Walt Cunningham were personally announced before Congress by NASA Administrator Jim Webb on 9 May 1967, with Tom Stafford, John Young and Gene Cernan as their backups. Schirra wanted to call his spacecraft 'The Phoenix' – the mythical firebird of classical lore, said to end its 500-year lifespan on a pyre of flames, then return from the ashes – but NASA, fearing unpleasant reminders of Apollo 1, vetoed the idea. Schirra's crew was also granted a three-man 'support' team, drawn from a new pool of astronauts announced in 1966: Jack Swigert, Ron Evans and Bill Pogue. In effect, each Apollo crew would now comprise nine dedicated members, emphasising its complexity over previous Mercury and Gemini missions. In Schirra's words, the support crew's role was to "maintain a flight data file, develop emergency procedures in the simulators and prepare the cockpit for countdown tests". Interestingly, Schirra asked Swigert to devise techniques to handle a fuel tank explosion in space. Less than two years after Apollo 7, Swigert would find this task very helpful ...

From the time that their spacecraft arrived in Florida to the day of their launch on 11 October 1968, Schirra, Eisele and Cunningham would spend nearly 600 hours in the command module simulator, operating the 725 manual controls and responding to countless simulated emergencies and malfunctions. Moreover, they had occupied Spacecraft 101 during an altitude chamber test, had checked out the Chrysler-built slidewire to practice escaping from Pad 34 in the event of a pre-launch emergency, had crawled out of a mockup command module in the Gulf of Mexico and had pored over hundreds of pages of documentation and flight plans.

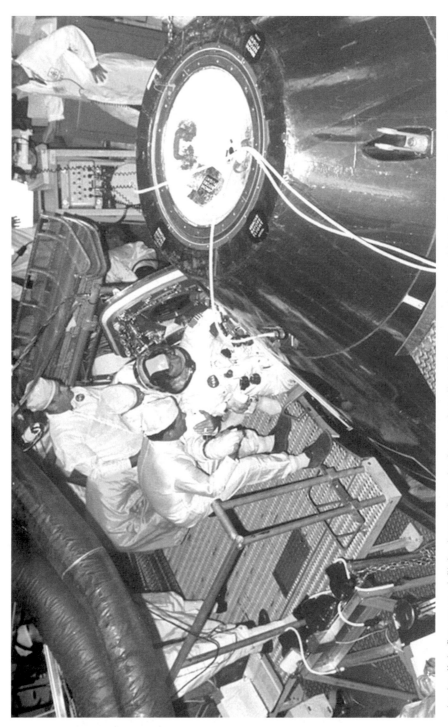

As Apollo's first manned flight draws closer, backup senior pilot John Young is inserted into the spacecraft prior to a test at North American's Downey plant.

North American's attitude towards the astronauts had changed dramatically, with the company now heartily opening its doors to their inspections, and Schirra, Eisele and Cunningham had not been shy in prowling the Downey plant to check on progress. This irritated Frank Borman, who had been attached to Downey as the astronaut office's representative. He considered the Apollo 7 crew to be obnoxious and felt that they were just causing disruptions and doing more harm than good. Borman even went so far as to complain to Deke Slayton, who spoke in turn to Schirra. However, it was understandable that the crew behaved as it did: they had seen poor workmanship from North American in the past and three of their comrades had died as a result of it. They wanted to make sure that they would be flying into space in the best possible ship. When Schirra saw an increasingly positive response and commitment from North American, from engineers to managers to the president, Lee Atwood, himself, he was appeased.

On one occasion, Schirra got into trouble himself ... albeit to his pleasure. One day in early 1968, he was visiting Downey and, dressed in the required clean-room garb, carefully tried to clamber aboard Spacecraft 101 without damaging any mechanical or electrical parts. "But it was a tight fit," he wrote, "and my knee landed on a bunch of wires. When it did, I felt a sharp slap on my face and I heard a woman's voice: 'Don't you dare touch those wires. Don't you know we lost three men?'" When she found out who Schirra was, the woman was embarrassed and apologetic, but for the commander of Apollo 7, she was just the person NASA needed on its contractor workforce. Schirra's new-found confidence was shared by John Young, who tracked Spacecraft 101's progress and concluded that it was "a pretty clean machine".

Elsewhere, fellow astronauts Jim Lovell, Stu Roosa and Charlie Duke simulated how quickly the command module could right itself if it flipped upside down – nose down – in the water (the so-called 'Stable 2' position) and for how long it might support them after splashdown. They experienced no difficulties in getting to the manual control switches which inflated three airbags and turned the ship into its upright 'Stable 1' orientation. Occasionally, water would splash into the cabin through a post-landing air vent, but the urine collection device proved more than adequate to vacuum this away and dump it overboard. Their consensus: not only was the command module seaworthy, but it could support the astronauts for a prolonged wait of up to two days until recovery. Other astronauts, including Joe Kerwin, Vance Brand and Joe Engle, spent weeks working in test vehicles and concluded that the ship's systems were virtually trouble-free.

In late July 1968, Schirra, Eisele and Cunningham spent nine hours inside Spacecraft 101 itself under conditions which simulated an altitude of 68.9 km. This gave them the opportunity to perform some of their actual mission tasks and evaluate their ability to work inside pressurised space suits. Technicians purged the cabin, using a mixture of 65 per cent oxygen and 35 per cent nitrogen, then 'dumped' the atmosphere so the men were obliged to rely upon their suits as the pressure dropped to almost zero. After an hour in a near-vacuum, the cabin was repressurised with pure oxygen, the normal atmosphere to be used in orbit. A few days later, the backup crew of Stafford, Young and Cernan repeated the exercise with similar success.

For Schirra, it was the altitude chamber testing which 'sold' Spacecraft 101 to him. "I said many times that we would not accept [it] until it had completed its run in the altitude chamber," he wrote, "similar to the launch pad test where fire killed Grissom, White and Chaffee." The success of these tests prompted him to chortle that Apollo 7 was now on a high-speed track and "the train is moving out".

Spacecraft 101 had begun its manufacturing cycle at North American in early 1966 and, by July of that year, had been assembled, wired, fitted with subsystems and was ready for testing. In the wake of the Apollo 1 fire, it went through a recertification and modification process, during which time its wiring was upgraded and its two-part hatch replaced with the newer unified version. In December 1967, it was finally ready for testing to resume and it passed its three-part acceptance review at Downey in May of the following year. No items were found which might prove "constraints to launch" and on the penultimate day of that month, North American shipped the spacecraft to Cape Kennedy. A flight readiness review in September confirmed that Spacecraft 101 was "a very good spacecraft". By that time, wrote Schirra, the crew was ready to go.

For Schirra, Eisele and Cunningham, the events surrounding the Apollo 1 fire and its aftermath were laced with irony. Initially assigned as the prime crew for the original Apollo 2, it had been Schirra's reluctance to duplicate Gus Grissom's flight that led him to request its cancellation. In doing so, he had been assigned Apollo 1 backup chores, with no 'prime' mission to aspire to. The events of 27 January 1967, in a roundabout way, landed Schirra with the kind of flight he really wanted: something new and challenging. During Apollo 7, he and his crew would spend 11 days in space – the second-longest manned mission to date – and were tasked with the comprehensive and systematic evaluation of the command and service module in Earth orbit.

In theory, Schirra's mission objectives could be 'achieved' in as little as three days, but according to Sam Phillips in a letter to Jim Webb it would be open-ended to 11 days in order "to acquire additional data and evaluate the aspects of long-duration space flight". The countdown, punctuated by three built-in holds to correct any last-minute problems, began on the evening of 6 October and proceeded without incident until ten minutes before launch on the 11th. At that point, thrust-chamber jacket chilldown was initiated for the Saturn's S-IVB second stage, but took longer than anticipated, forcing a hold of two minutes and 45 seconds. (After launch, analysis confirmed that the chilldown would have occurred without the hold, but waiting, in real time, was prudent in order to meet revised temperature requirements.) The countdown resumed at 10:56 am and the Saturn 1B lifted-off at 11:02:45 am, watched by more than 600 accredited journalists.

If the early stages of the ascent seemed laborious, Time magazine told its readers, they should not have been surprised: the booster weighed 590,000 kg, only slightly less than the 725,750 kg thrust of its first stage. Acceleration from the astronauts' point of view, therefore, was much calmer and less oppressive than the G loads experienced by previous Mercury and Gemini crews.

Inside the command module, the crew experienced a clear sense of movement, but only Eisele had a good view of the commotion that was going on outside. "We had a

Wally Schirra becomes the first man to launch into space three times, and aboard three different rockets, as the Saturn 1B carries its first crew into Earth orbit.

boost protective cover over the command module," Cunningham recalled later. "There's an escape rocket that you can use anytime until you get rid of it, and that's a little after a minute into the flight. Because that rocket puts out a plume, you had to have a cover over the command module so that you wouldn't coat the windows and you wouldn't be able to see anything out of the windows in the event you were coming down on a parachute during an abort. So, the only place you can see out is over Donn's head in the centre seat. There's a little round window, about six inches across, and he was the only one that could see out. We had no windows until the boost protective cover [was jettisoned]."

Two and a half minutes into the thunderous ascent, the eight H-1 engines of the Chrysler-built S-IB first stage burned out and it was released, allowing the S-IVB and its single J-2 engine to pick up the thrust and deliver Apollo 7 into orbit. A little under six minutes after launch, as he, Eisele and Cunningham became the first men ever to fly atop a load of liquid hydrogen rocket fuel, Schirra reported that the Saturn was "riding like a dream". On the ground, the situation was not quite so dreamy: for a minute or so, the Manned Spacecraft Center had suffered a power failure which temporarily knocked out its lights, control consoles, screens and instruments. Fortunately, generator power took over and no telemetred data was lost.

Despite the successful launch, Schirra would later admit to some anger. Months earlier, during a meeting in Downey, he had learned that Apollo 7 would fly with old Block 1-style couches, rather than the improved Block 2 type. The latter, it was realised, would offer better protection for the crew if they happened to inadvertently touch down on land. Schirra felt that if Block 1 couches had to be used on his flight, the mission rules dealing with wind speeds at launch needed to be revised, since an abort over Florida could push their command module back over land. He agreed to accept Block 1 couches, on condition that Apollo 7 would not launch if wind conditions were unfavourable. On 11 October 1968, winds at the Cape were around 40 km/h, considerably higher than the maximum-allowable 32 km/h needed to avoid a touchdown on land. It was felt that the Saturn 1B's record of reliability made it unlikely that an abort would occur in the early stage of ascent, but to Schirra, commanding the crew that would be flying the thing, it was black and white. "A mission rule had been broken," he wrote. "Needless to say, I was not the happiest guy in town."

Ten and a half minutes after launch, following little bumpiness and loads which never exceeded 1 G, Apollo 7 was inserted perfectly into an orbit of 227–285 km and the S-IVB duly shut down. Both stages of the Saturn performed to near-perfection. Two hours and 55 minutes into the flight, the spacecraft undocked from the S-IVB and pulsed its reaction controls twice to turn back in a simulated rendezvous approach which Moon-bound crews would use to pick up their lunar module. Although there was no lunar module housed inside the stage, it provided useful practice and Schirra brought his ship within 1.2 m of the spent S-IVB. Unfortunately, said Cunningham, one of the four adaptor panels had not fully deployed, due to a stuck retention cable, although they would be jettisoned explosively on subsequent flights to ensure lunar module extraction. It "had sort of

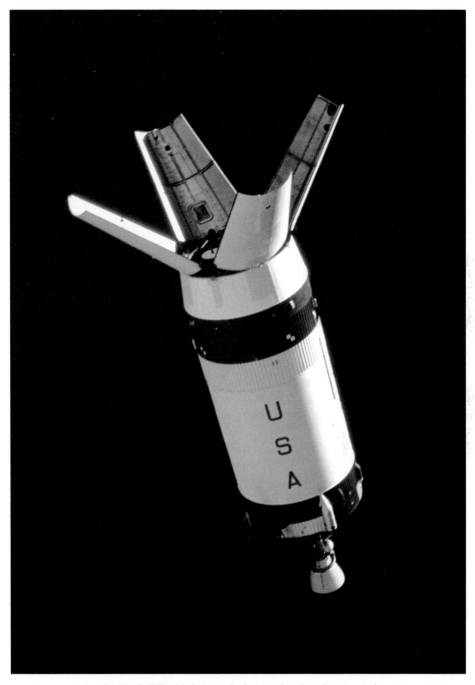

Apollo 7's S-IVB third stage during station-keeping operations.

bounced back," wrote Deke Slayton. "It posed no danger to the crew, but had this flight carried a lunar module, it might have been tough to get it out of there."

Elsewhere, the performance of the big SPS engine was highly successful. It was fortunate, indeed, that this was the case, for this was a component which simply 'had to work' or lunar crews would not be able to return home. During their mission, Schirra, Eisele and Cunningham oversaw no fewer than eight SPS firings, the first of which posed something of a surprise. In contrast to the exceptionally smooth Saturn 1B liftoff, the SPS noticeably jolted the astronauts, prompting Schirra to whoop "Yabadabadoo!" in imitation of Fred Flintstone. Eisele, too, said that the entire crew "got more than we expected" and that the additional boost literally plastered them back into their seats. Later SPS burns lasted anywhere from half a second to more than a minute in duration and simulated virtually everything from a return from the Moon to a rendezvous with a phantom lunar module.

Other systems aboard Apollo 7 performed equally well. Occasionally, one of the three electricity generating fuel cells would develop unwanted high temperatures, but load-sharing hookups prevented any power shortages. The astronauts complained about noisy fans in the environmental circuits and turned one of them off, but, when this did not help, switched off the other. Visibility through the windows was mixed, with sooty deposits noted shortly after the jettisoning of the Saturn 1B's escape tower and spots of water condensation seen at other times. Two days into the flight, however, Cunningham reported that most of the windows were in fairly good condition, although moisture was gathering between the inner panes in one case. A similar situation was seen by Schirra a few days later. Nonetheless, the windows proved adequate, particularly during the rendezvous and station-keeping with the S-IVB, when they were almost clear. Navigational sightings with a telescope and sextant on any of 37 pre-selected stars proved difficult if done too soon after a waste-water dump and, indeed, the astronauts typically had to wait several minutes for the frozen droplets to disperse. Eisele reported that unless he could see at least 40 or 50 stars at a time, it was tough to decide which part of the sky he was looking at.

On a more mundane level, the 'waste-management system' – a somewhat euphemistic term for Apollo's rudimentary toilet – proved adequate, if annoying. Its defecation bags, which contained a blue germicidal tablet to prevent bacterial and gas formation, could be sealed easily and stored in empty food containers in the command module's lower equipment bay. However, they were far from ideal, still produced unpleasant odours and took each astronaut 45–60 minutes to complete. Bill Anders, who flew Apollo 8, would later tell Andrew Chaikin that, since nothing in microgravity 'falls', it was necessary to "flypaper this thing to your rear end and then reach in there with your finger [the bags had 'fingers' for this] – and suddenly you were wishing you'd never left home!" To add insult to injury, the germicidal tablets then had to be kneaded into the contents of the defecation bag to ensure that they were fully mixed. Not surprisingly, many astronauts found themselves postponing their 'need to go' for as long as possible and particularly to wait until there was no work to do.

The urine-collection device took the form of a hose with a condom-like fitting on the end, which led, by way of a valve, to a vent on the outside of the command module, out of which would periodically spill a cloud of frozen droplets. (One Apollo astronaut, when asked what he thought was the most beautiful sight in space, responded "Urine dump at sunset!") During one toilet session, as Cunningham fitted the urine hose to his suit, he instinctively turned his back to the window for privacy. "Walt," Schirra asked with some humour in his voice, "who is out there?"

As for moving around in the new spacecraft, the astronauts turned, in Schirra's words, into spacegoing gymnasts. "You can move any place you want to fairly freely," said Cunningham, "and you certainly don't need strong handholds to take care of it." Exercise was important, though. At first, when they slept in their couches, their bodies curled up into foetal positions, giving them lower back and abdominal pains; these were relieved by working on a stretching device called an Exer-Genie, which relaxed their cramped and aching muscles.

Sleep brought mixed blessings, with Schirra complaining about the around-the-clock operations which disrupted their normal routine. Sometimes they might go to bed as early as 4:00 pm or as late as 4:00 am, he said, and a consensus was finally reached whereby Eisele kept watch on Apollo 7's systems whilst Schirra and Cunningham slept and vice versa. Two sleeping bags were mounted beneath the couches and the astronauts typically zipped themselves inside, although the incorrect positioning of restraint straps made them less than ideal. Cunningham, certainly, preferred to sleep in his couch with a shoulder harness and lap belt to keep still. However, if two crew members did this, they invariably disturbed their colleague who was awake. By the third day of the flight, thankfully, they had worked out a routine to get enough sleep.

All three men expressed unhappiness over their food, which tended to crumble and whose particles floated around the cabin. Following his Gemini VI-A mission three years earlier, Schirra requested taking coffee with him aboard Apollo, which he did. Not so fulfilling were the head colds which, first Schirra, and then Eisele and Cunningham, developed during the mission. To be fair, this caused severe discomfort, because it proved extremely difficult to clear the ears, nose and sinuses in microgravity. Mucus rapidly accumulated, filling their nasal passages and stubbornly refusing to drain from their heads; indeed, their only relief was to blow their noses hard, which proved painful on their ear drums. A little under a day into the mission, an irritable Schirra, already annoyed that Mission Control had added two thruster firings and a urine dump to their workload, cancelled the first planned television transmission from Apollo 7, "without further discussion". It was the first of many conflicts with ground staff.

Indeed, according to Schirra in his autobiography, Donn Eisele, watching over the spacecraft systems at the time, began the dispute with Mission Control over rescheduling the first television transmission early on 12 October. "When I awoke," Schirra wrote, "I could hear Eisele in an argument. I put on a headset and heard a ground controller say, rather insistently, that our first television transmission was on the agenda for that day." Schirra butted in and backed Eisele that they had enough to do on their second day in orbit, with engineering objectives, rendezvous practice and SPS preparations, without having to worry about the transmission. However, there was more to it than that. "We were scheduled to test the TV circuit later that day," explained Schirra, "and we'd test it before using it. It was an electrical circuit and I had not forgotten than an electrical short had resulted in the loss of the Apollo 1 crew."

In fact, Schirra had complained about the scheduling of the transmissions on the ground, before launch, "but hadn't been able to win the battle," wrote Deke Slayton. "He probably figured there wasn't much we could do to him while he was in orbit, and he was right, but it made my life kind of difficult." The commander's antics also upset the flight directors, including Chris Kraft and Glynn Lunney, when he began sarcastically criticising "the genius" who designed a particularly balky piece of equipment. "He might have been right," continued Slayton, "but it sure didn't endear him to the guys on the ground to have the astronaut implying they were idiots over the open line for everyone to hear." Eisele and Cunningham, despite their admonitions to the contrary, followed Schirra's lead. After one test which he perceived as pointless, Eisele, clearly annoyed, quipped that he "wanted to talk to the man, or whoever it was, that thought up that little gem". 'The man' turned out to be Flight Director Lunney himself. Going over the mission tapes and transcripts after the flight, Cunningham would conclude that he "never had any problem with the ground", although Deke Slayton felt that all three men "were pretty testy". One flight controller even muttered, only half-jokingly, about letting the Apollo 7 crew land in the middle of a typhoon . . .

Still, on the third day of the mission, the first of seven eagerly awaited transmissions began, marking the first live televised event from an American manned spacecraft. Each one took place as Apollo 7 passed between Corpus Christi in Texas and Cape Kennedy, the only two ground stations equipped to receive the transmissions. The crew opened the first telecast with a sign which read 'From the lovely Apollo room, high atop everything', then aimed their camera through the window as they passed above New Orleans and over Florida. Later transmissions included tours of the command module, demonstrations of the Exer-Genie and explanations of how food was prepared in space and how dried fruit juice was reconstituted with water. All in all, the telecasts were well-received and the astronauts enjoyed them, displaying cards with 'Keep Those Cards and Letters Coming In, Folks' . . . and offering Schirra a chance to gain revenge on Deke Slayton by asking him, live, if he was a turtle. Sitting in Mission Control, the gruff Slayton acknowledged that he had recorded his answer, switching off his microphone to utter the necessary answer: "You bet your sweet ass I am!" After the flight, these 'Wally, Walt and Donn Shows' proved so popular that the astronauts even won a special Emmy award.

By 12 October, the day that Schirra cancelled the first transmission, Apollo 7 had drifted about 130 km from the S-IVB. The crew's task was to re-rendezvous with it. This was not as straightforward as it had been on Gemini, since the command module lacked a rendezvous radar and the astronauts were unable to read their range and closing velocity to the target. However, wrote Schirra, "we made it through the rendezvous, with each of us ageing about a year", and Apollo 7 edged to within 20 m of the discarded stage. The manoeuvre proved quite traumatic, with no clear awareness of their closing motions, and the S-IVB itself was spinning throughout like an angry whale. Twenty metres was close enough, before moving away. The stage eventually re-entered the atmosphere on 18 October, its debris splashing into the Pacific Ocean.

Experimental work included synoptic terrain photography, employing a hand-held 70 mm Hasselblad 500C camera to monitor the Carolina bays in the United States and examine wind erosion in desert regions, tropical morphology and the origin of the African rift valley. Imagery was also acquired of Baja California, parts of Mexico and the Middle East in support of geological inquiries and geographical urban studies were aided by photographs of New Orleans and Houston. Islands in the Pacific, extensive coverage of northern Chile and Australia and other areas added to Apollo 7's photographic haul. Overall, that haul amounted to some 500 images, of which more than a third proved usable, although the need to change film magazines, filters, settings and keep cameras steady accounted for the improper exposure of many frames. Weather photography was also important, with numerous images of various cloud and meteorological structures, including the best-ever photographs of a tropical storm, in this case Hurricane Gladys and Typhoon Gloria. Pre-flight and post-flight X-rays of bones also contributed towards a demineralisation study, whilst sampling before launch and after splashdown helped to determine cellular changes in the astronauts' blood.

The final days of what both Wernher von Braun and Sam Phillips were lauding as "the perfect mission" were marred by the worsening head colds. Schirra's, indeed, had materialised barely 15 hours after launch, forcing him to admit that he had "gone through eight or nine Kleenexes", so he had to endure it for most of the 11-day mission. (He also took Actifed, to which he became so attracted that he helped sell it on television commercials after leaving the astronaut corps.) Years later, Walt Cunningham would blame Schirra's cold on a dove-hunting trip that the Apollo 7 crew took in a rainy Florida shortly before launch. "Wally was kind of a General Bull Moose complex," Cunningham said. "What's good enough for Bull Moose is good enough for the world. So, when Wally had a cold, everybody had to be miserable."

A head cold anywhere is miserable, but in the pressurised confines of a spacecraft, it proved much more so, and Eisele and Cunningham quickly succumbed. Physician Chuck Berry advised them to take aspirin and decongestant tablets and, as re-entry neared, they began to worry that the build-up of pressure whilst wearing their helmets might burst their eardrums. It was not an idle fear. During his days at test pilot school, Schirra had made a short flight in a propellor aircraft, with a head cold, and "almost busted an ear drum". The choice he now faced on Apollo 7 – not wearing a helmet or running the risk of lifelong hearing loss – was an easy one to make.

Deke Slayton explicitly ordered the crew to wear their helmets, but Schirra refused, agreeing only to keep them stowed in case of emergencies. There were, admittedly, contingency options in place for returning home suitless, perhaps in the event of contamination, but after almost 11 days it seemed unlikely that cabin pressure would fail during re-entry. Each man took a decongestant pill an hour before hitting the atmosphere and endured no major problems. As the command module's pressure was raised to conditions approximating normal sea-level, Schirra,

Eisele and Cunningham performed the Valsalva manoeuvre – holding their noses, closing their mouths and vigorously exhaling through their nostrils – to keep their middle ears equal to the increasing cabin pressure. In doing so, they avoided ruptured ear drums … but aroused the wrath of flight controllers. All three men would be "tarred and feathered" for their insubordination during the mission. Schirra had already announced his retirement from NASA and probably could not have cared less, but Eisele and Cunningham, who had followed their commander's lead, would never fly again.

WALT AND 'WHATSHISNAME'

In truth, Air Force Major Donn Fulton Eisele's NASA career was waning by the time Apollo 7 splashed down. Born in Columbus, Ohio, on 23 June 1930, Eisele had followed the classic path to become an astronaut: a bachelor's degree from the Naval Academy in 1952, a master's credential in astronautics from the Air Force Institute of Technology and graduation from the Aerospace Research Pilots' School at Edwards Air Force Base in California. Prior to his selection as an astronaut, along with Cunningham, in October 1963, Eisele served as a project engineer and test pilot at the Air Force's Special Weapons Center at Kirtland Air Force Base in New Mexico.

The easygoing Eisele's performance as an astronaut is hinted at by Deke Slayton in his autobiography, when he notes that his original intention was to "try out some of the guys who, frankly, I thought were weaker" on the Apollo 1 mission. "My original rotation had Donn Eisele and Roger Chaffee as the senior pilot and pilot, working for Gus," he continued. Had it not been for the fact that Eisele damaged his shoulder during a zero-G training flight aboard a KC-135 aircraft just before Christmas 1965, he might have been in the senior pilot's seat aboard Apollo 1, instead of Ed White. Instead, Slayton considered it easier to swap Eisele for White, the latter of whom was previously attached to Wally Schirra's original Apollo 2 crew.

Eisele quickly assumed the moniker 'Whatshisname', bestowed upon him by Schirra and Cunningham, when nobody seemed to be able to pronounce his surname. Phonetically, it ran EYE-SEL-EE, but when NASA Administrator Jim Webb tried to introduce the crew to President Lyndon Johnson, he mistakenly called him Donn 'Isell'. "From then on," Schirra wrote, "Donn was 'Whatshisname'".

Eisele's career, in addition to Apollo 7, was harmed by a particularly ugly divorce from his wife Harriet, the result of an affair which caused his work in the astronaut office to suffer. Indeed, the pressures of the job had led many astronauts to look elsewhere, outside the marital home, and after Eisele it would be John Young who would next go through a divorce. Unlike Eisele, however, Young did not allow his personal life to disrupt his work and remained devoted to the space programme. Stories would abound over the years that the funeral of one astronaut killed in the early Sixties – his name was never divulged – was attended not only by his wife and family … but also by his long-term mistress, discreetly escorted to the ceremony by a close and trusted friend.

In spite of the criticisms levelled at them in the wake of Apollo 7, both Eisele and Cunningham were at least considered for backup roles on future missions. The former had already been assigned to serve as the backup command module pilot on Apollo 10, the dress-rehearsal for the first lunar landing. For Tom Stafford, the commander of that mission's prime crew, however, Eisele's assignment was little more than "a temporary step into oblivion". Cunningham, on the other hand, would work for several years on the United States' space station project, Skylab, and even trained as backup commander for its first mission. He "wanted to fly again," wrote Deke Slayton. "In spite of the flight operations opinion that he shouldn't, I wasn't going to rule him out. But it was a numbers game." Cunningham, like Eisele, never flew again.

Ronnie Walter Cunningham was born on 16 March 1932 in Creston, Iowa, and came to be seen as one of 'the scientists' among the astronaut corps, owing to his credentials as a civilian physicist. He received bachelor's and master's degrees from the University of California at Los Angeles in 1960 and 1961, respectively, then began doctoral research, which he completed, save for his final thesis. However, his military experience certainly paralleled his scientific knowledge: he joined the Navy in 1951, began flight training and served on active duty, then as a reservist, with the Marine Corps. "In the Navy, in those days, you ran the risk of being assigned to torpedo bombers or transport pilots," Cunningham recalled, "and the Marine Corps guaranteed you that your first tour ... would be flying single-engine fighter planes." He remained a reservist throughout his astronaut career. Prior to selection as one of 'The Fourteen' in October 1963, Cunningham worked for the Rand Corporation, performing research in support of classified projects and problems relating to the magnetosphere.

"I was working on defence against submarine-launched ballistic missiles, trying to write in ... the crudest fashion the equations that would intercept a missile on the rise," Cunningham explained. "At the same time, I was doing my doctoral work on the Earth's magnetosphere. It was a tri-axial search coil magnetometer and we were trying to measure fluctuations in the Earth's magnetic field. It was during this period that I applied and got accepted at NASA. I never did finish the thesis."

As a non-test pilot, possessing an air of academia and a self-confessed irreverence to authority, Cunningham stood out among the Fourteen. He "seemed determined to be different from the rest of us," wrote his 1963 classmate Gene Cernan, "whether reading The Wall Street Journal while we busted our asses during a classroom lecture or driving a Porsche instead of a Corvette." He would also lose support through his criticisms, notably over the performance of Neil Armstrong and Dave Scott during the Gemini VIII emergency. When Cunningham claimed years later that he, Schirra and Eisele had been tarred and feathered for their antics on Apollo 7, Cernan would retort that it was "probably with good reason".

"101 PER CENT SUCCESSFUL"

It was not just the head colds that overshadowed Apollo 7. Schirra's morale had been decidedly more sombre in the weeks before launch and many of his colleagues wondered what had happened to the normally good-humoured, 'chummy' astronaut. Schirra, in truth, was simply burned-out from nine years in the astronaut business. He had lost his best friend and neighbour, Gus Grissom, and was tired by the constant grind and long hours demanded of him. "I had changed over the span of time that encompassed my three flights," Schirra wrote. "As the space programme had matured, so had I. I was no longer the boy in scarf and goggles, the jolly Wally of space age lore." He steadfastly refused to allow Apollo 7 to be jeopardised by what he perceived to be the influence of 'special interests' – scientific or political – and declared that he "would not be an affable fellow when it came to decisions that affected the safety of myself and my two mates".

Despite his intense focus on Apollo 7, the monotony of the last few days – SPS burns, navigational sightings, water dumps, photography, experiments – proved somewhat less than fun or challenging. The astronauts became less enamoured with seeing 16 sunrises and sunsets each day. Years later, despite Sam Phillips' assertion that the flight had been "101 per cent successful", Cunningham would state that both he and Eisele felt Apollo 7 could have achieved more. "The initial plan was no more than 60–75 per cent of what we should've had on it, because so many things got thrown off the flight, principally through Wally's efforts. We felt like we could've accomplished a whole lot more. It turns out that the last several days were fairly boring."

Yet there were spots of fun and games in the cabin, with Cunningham making a ring with his thumb and forefinger and Schirra shooting a weightless pen through it. They learned to catch cinnamon cubes in their mouths, even blasting them off-course with an air hose for added fun. However, in spite of those 'ho-hum' closing days, the mission opened up exciting possibilities for the next flight, which had, since August, been under consideration to go to the Moon and back. By the end of the first week of Apollo 7, NASA confidently predicted that Apollo 8, with its crew of Frank Borman, Jim Lovell and Bill Anders, could liftoff atop the giant Saturn V rocket as early as 5 December. That date, ultimately, was untenable, but indicated the growing confidence in Apollo and the success that Schirra's mission brought.

That success also established Apollo 7 as the second-longest manned mission to date. Its conclusion was brought about by the eighth and final SPS burn, lasting almost 12 seconds, executed at 6:41 am on 22 October as the spacecraft hurtled around the globe for the 163rd time. Four minutes later, the service module was jettisoned and at 6:55 am the command module hit the upper fringes of Earth's atmosphere to begin its re-entry. Shortly thereafter, descending beneath three beautiful red-and-white parachutes, it dropped into the Atlantic, just south-east of Bermuda, after a mission of ten days, 20 hours, nine minutes and three seconds. Apollo 7 splashed down at 7:11:48 am, just 3.5 km from its target point and barely 13 km from the recovery vessel, the aircraft carrier Essex, which incidentally had been involved in the quarantine of Cuban waters prior to the Bay of Pigs fiasco seven years before.

It was fortunate that Jim Lovell, Stu Roosa and Charlie Duke had practiced 'Stable 2' splashdown positions, because gusty winds and choppy waves caused Apollo 7 to quickly assume an upside-down orientation, but its flotation bags righted it within minutes. An hour after splashdown, following a tense loss of communications, Schirra and his team were aboard the Essex. Shortly thereafter, the command module – part of the vehicle which Cunningham called a "magnificent flying machine" – was also safely on the carrier's deck; it was too heavy to be hauled to safety with the men aboard. The astronauts, heavily bearded, weary and unsteady on their feet, had all lost weight, but their humour returned quickly. Deke Slayton, aboard the Essex, admitted to having "a few words in private" with Schirra, not so much about his own behaviour, but about his effect on Eisele and Cunningham. Others were less complimentary. Although he would later deny it to Cunningham's face, Chris Kraft is said to have announced that nobody from Apollo 7 would ever fly into space again. "I made the selections," admitted Slayton, "but I wasn't going to put anybody on a crew that Kraft's people wouldn't work with. Not when I had other guys."

In a year which had seen the steady rise of the hippie movement in America and protests ranging from civil rights to the war in Vietnam and outrage over the murders of Martin Luther King and Bobby Kennedy, NASA's public affairs officer Paul Haney knew precisely why the bearded Schirra and his crew had been so irritable. "Something happens to a man when he grows a beard," Haney joked. "Right away, he wants to protest!"

SOYUZ RESURGENT

For Yuri Gagarin, the first man in space, still in his early thirties, yet seemingly thwarted from ever flying again, the death of Vladimir Komarov cast a long shadow over his career. Shortly after the disaster, Nikolai Kamanin gathered the cosmonauts together and told him in no uncertain terms that his chances of another mission were virtually nil and that the next manned Soyuz would be flown instead by Georgi Timofeyevich Beregovoi, the oldest active pilot in the corps and a harsh critic of Gagarin, a man he considered to be an upstart. Meanwhile, retaining their slots on the 'passive' Soyuz, which would involve the ship-to-ship EVA, were Valeri Bykovsky, Yevgeni Khrunov and Alexei Yeliseyev. All four men began intensive training in November 1967.

Correcting Soyuz' chronic problems was entrusted to engineers at TsKBEM, the Scientific-Research Institute for Automated Devices, and the Gromov Flight-Research Institute. The TsKBEM was the 'new name' for the OKB-1. Their work led to a number of improvements, including changes to the operating schedule of the reserve parachute, and by September 1967 the Utkin commission declared it was satisfied that Soyuz could commence automated missions. In mid-October, Vasili Mishin announced that test flights would be launched with an 'active' vehicle sent aloft for three days, followed, if its health proved acceptable, by a 'passive' craft. The two would then automatically rendezvous, using their Igla radars, with docking not mentioned, but considered an option. The active craft, under the cover name of Cosmos 186, was successfully launched from Tyuratam at 12:30 pm Moscow Time on 27 October, entering an orbit of 209 × 235 km. Unlike Komarov's Soyuz, its solar panels deployed successfully and its Igla worked perfectly, but a malfunction in its solar-stellar attitude-control sensor prevented it from adjusting its orbit. Nevertheless, the second launch was given the go-ahead.

Enthused by Cosmos 186's success, Mishin was keen to attempt not only a rendezvous, but also a docking, and the vehicle named Cosmos 188 was duly launched to conduct this new mission at 12:12 pm on 30 October. Its launch vehicle's trajectory was precise: inserting it into a 200 × 276 km orbit and within 24 km of Cosmos 186. The latter then fired its engine 28 times under automatic command from its Igla and at 1:14 pm, barely an hour after the passive vehicle's launch, the pair had docked. Clear images appeared on Soviet television that evening, giving the outside world its first brief glance at the configuration of the Soyuz spacecraft.

Despite a small, 8.5 cm gap between the two craft, the Cosmos vehicles undocked after three and a half hours to commence their respective re-entries. Here the problems arose. Cosmos 186 suffered a failure of its solar-stellar sensor, which altered its descent trajectory into a purely ballistic fall from orbit; still, it was recovered safely. On the following day, 1 November, Cosmos 188 proved unable to perform a guided re-entry because of an incorrect attitude. It re-entered at an excessively steep angle, to such an extent that its self-contained package of explosives remotely destroyed the descent module, lest it land on foreign soil. (Had the explosives not fired, it was later concluded, Cosmos 188 would have landed about 400 km east of Ulan-Ude, just to the north of the Mongolian border, but still in the Soviet Union ...)

Early the following year, Yuri Gagarin and a handful of other cosmonauts defended their 'candidate of technical sciences' theses at the Zhukovsky Air Force Academy and the prospects of another flight into space seemed to brighten a little. On 27 March, he and test pilot Vladimir Seregin took off from the Chkalovskaya airfield, near Moscow, in an antiquated MiG-15UTI trainer. Shortly afterwards, Gagarin requested permission to alter his course ... and then, at 10:31 am, communications were lost.

"The weather was very bad that day," remembered fellow cosmonaut Alexei Leonov, who was overseeing parachute jumps from a helicopter near Kirzach airfield. "The cloud cover was low and it was raining hard. My team had performed just one jump when the weather deteriorated even further. The rain turned to sleet and conditions were so bad that I cancelled the session and requested permission to return to base." As he waited to learn if his request had been granted, Leonov heard two loud bangs from the distance, one of them clearly an explosion, the other a sonic boom, with barely a second or so between them. During the return to base, he was puzzled when the control tower kept radioing Gagarin's callsign. Leonov wondered if they were mistakenly calling him instead, but upon landing he was told that contact had been lost with Gagarin and Seregin. When Leonov described the explosions he had heard, a helicopter was hastily despatched to the last known location of Gagarin.

At length, as late afternoon gave way to a wintry twilight, the helicopter commander reported finding the wreckage of the MiG some 64 km from the airfield. Debris, he said, was scattered in a wooded area and the aircraft's engine was buried several metres underground. Search and rescue forces, who arrived shortly thereafter, would determine that the MiG-15 had hit the ground at over 700 km/h.

An upper jaw, identified as that of Seregin, was found and Soviet Air Force officials informed Leonid Brezhnev and Alexei Kosygin of the accident. As yet, however, they had no confirmatory evidence that Gagarin had also died. Early the following morning, a piece of cloth hanging from a birch tree offered the first proof: it was from Gagarin's flight jacket. Clearly, neither he nor Seregin had ejected. The men's remains – which Doran and Bizony described as "fingers, toes, pieces of ribcage and skull" – were both interred in the Kremlin Wall. The cause of the accident was hard to find. Theories included a bird strike, a collision with a hot-air balloon (the remains of which, in fact, were found close to the crash site) and even more outlandish notions that Gagarin was drunk or Seregin was taking pot-shots at wild deer from the MiG. Still others postulated that after angrily throwing a cognac in Leonid Brezhnev's face in the wake of Komarov's death, Gagarin had been imprisoned or confined to a mental asylum . . .

In December 1968, the official accident report pointed towards pilot error, but when the classified files were reopened two decades later it became more likely that Gagarin and Seregin did not have accurate altitude data and had flown into an area where a supersonic Sukhoi SU-15 jet was operating. Bizony and Doran noted that Seregin was told the cloud base was 1,000 m, when in fact it was nearer to 450 m. Witnesses would later confirm seeing both Gagarin's aircraft and the Sukhoi. "According to the flight schedule of that day," wrote Leonov, "the Sukhoi was prohibited from flying lower than 10,000 m. I believe now, and believed at the time, that the accident happened when the jet pilot violated the rules and dipped below the cloud cover for orientation . . . that he passed within 10 or 20 m of Yuri and Seregin's plane while breaking the sound barrier. The air turbulence overturned their jet and sent it into a fatal flat spin." In such a situation, and thinking they were higher than they actually were, neither Seregin nor Gagarin would have had much time to respond or eject.

It was Leonov who finally identified Gagarin's physical remains . . . from fragments of flesh removed from the crash site and placed into a metallic bowl. "A few days before," he wrote in his autobiography, "I had accompanied Yuri to the barber to have his hair cut. I had stood behind Yuri talking while the barber worked. When he came to trim the hairs at the base of Yuri's neck, he noticed a large, dark brown mole." Leonov had joked that the barber should be careful not to cut the mole, little realising that it would prove pivotal shortly thereafter in identifying the last mortal remains of Yuri Alexeyevich Gagarin. "Looking down at the fragments of flesh lying in that metal bowl," Leonov wrote, "I saw that one bore the mole." The first man to conquer space was dead at the age of just 34.

The day before Gagarin and Seregin died, the Soyuz State Commission, headed by Kerim Kerimov, met to discuss future plans. The parachute design for the spacecraft had been extensively overhauled and cleared to fly. At 1:00 pm on 14 April, with several cosmonauts, including Georgi Beregovoi, in attendance, Cosmos 212 was successfully launched into a 210 × 239 km orbit. Next day, at 12:34 pm, it was followed by Cosmos 213, which entered orbit just four kilometres from its target. Within an hour, and with Cosmos 212 leading the rendezvous, the two craft docked automatically, separated later that evening and each completed five-day independent missions. Cosmos 212 performed the first-ever guided Soyuz re-entry on 19 April, touching down in high winds near Karaganda in Kazakhstan. Its twin, after performing automated tasks in radiation sensing, micrometeoroid detection and photometry, landed near Tselinograd on 20 April.

By this time, many were looking to Beregovoi to fly the next manned Soyuz mission, planned as a four-day flight, with Boris Volynov, Yevgeni Khrunov and Alexei Yeliseyev aboard the second, 'passive' mission. They would fulfil the denied missions of Soyuz 1 and 2. But not yet. Trials of the spacecraft's backup parachute were not considered good enough to assign a human pilot and it was deemed likely that, with a crew of three cosmonauts aboard, it might rip during deployment. Vasili Mishin and the parachute's designer proposed reducing the three-man crew to two. Further, it seemed prudent, to avoid unnecessary risk, to dock the two Soyuz vehicles, but not yet to attempt a risky EVA transfer.

Mstislav Keldysh, head of the Soviet Academy of Sciences, was even more cautious, refusing to endorse a manned flight until further automated tests had been conducted. On 29 May 1968, Mishin suggested a compromise: a docking of two Soyuz vehicles in orbit, one of them unmanned, the other carrying a single cosmonaut. After the success of that flight, the next crews would be committed to the EVA transfer, perhaps as early as September. Dmitri Ustinov stepped in the way of this plan, demanding an additional automated flight, which caused the intended August date for the first mission to slip until October. On 10 June, the Soyuz State Commission convened and Kerim Kerimov approved a plan to launch the automated mission in July, followed by the joint manned mission in September and a full-scale docking and EVA in November or December. Ustinov added to this the proviso that the EVA should transfer not one, but two, cosmonauts between ships. One Soyuz, obviously, would land with a crew of three, necessitating the repair of the backup parachute ... and quickly.

Cosmos 238 was duly launched on 28 August, a month late because of problems with its parachutes, and apparently conducted at least one major orbital manoeuvre, before touching down four days later. All hurdles appeared to have been cleared and the joint mission, with an unmanned Soyuz 2 and Beregovoi aboard Soyuz 3, was scheduled for mid-October. This was slightly postponed due to pre-flight malfunctions and problems during testing of the spacecraft, but the State Commission met on 23 October and confirmed Beregovoi as the Soyuz 3 pilot. The new backups would be Vladimir Shatalov and Boris Volynov. Born on 15 April 1921 in Fedorovka in the Ukraine, Beregovoi was a fully-fledged colonel in the Soviet Air Force and, thanks to his exploits as a squadron commander in the Second World War, had long since been decorated with the coveted Hero of the Soviet Union medal.

He had joined the Air Force in 1941 and was rapidly assigned to a ground-attack unit, flying the Ilyushin Il-2 and completing 185 combat missions against Nazi Germany. Following the end of hostilities, he became a test pilot and flew more than 60 different types of aircraft, becoming deputy chief of the Air Force's flight testing department. He was accepted for cosmonaut training in 1962 and, as a fellow war veteran, had been looked upon favourably by Nikolai Kamanin. In their biography of Gagarin, however, Doran and Bizony suggested that after serving in a backup capacity for an unflown Voskhod mission, Beregovoi locked horns with the First

Cosmonaut over flight assignments ... and personally insulted the younger man over his limited flying experience and qualifications. According to onlookers, Gagarin threatened to do everything in his power to keep Beregovoi from flying in space. Even in the last days before Soyuz 3, serious concerns were raised over 47-year-old Beregovoi's suitability to carry out the mission. He had failed his pre-launch examination, receiving a 'bad' mark rather than the expected 'excellent', but instead of substituting him for Shatalov, he was given a second chance, which he passed with a respectable 'good'. After the flight, when asked if his advanced age was a factor in the indecision over whether he should fly, Beregovoi responded that his height – 1.8 m – was actually a deciding factor ...

At noon on 25 October 1968, the unmanned Soyuz 2 (it only received this name after Beregovoi's launch aboard Soyuz 3) lifted-off from Tyuratam, entering a perfect orbit of 183 × 224 km. Although the systems aboard the spacecraft appeared to be functioning normally, conservatism and scepticism over the reliability of the Igla system prompted suggestions that the attempt to dock with Beregovoi be dropped in favour of a simplified, two-part rendezvous, firstly to a distance of 30 km and then to 100–200 m. Next morning, at 11:34, Soyuz 3 set off and within minutes Beregovoi was in an orbit of 205 × 225 km, ready to exorcise the ghost of Vladimir Komarov and put Soyuz through its paces.

During his first orbit, ground controllers activated the Igla system, which guided Soyuz 3 towards its passive target and brought it to within 200 m. At this point, as an external television camera relayed pictures to Earth, Beregovoi took manual control to attempt a docking. As he closed in to around 50 m, Soyuz 3 inexplicably banked 180 degrees from the target, despite the cosmonaut's best efforts to counter it. Suspicion that the Igla had contributed to this failure was denied by the system's designer, Armen Mnatsakanyan, who claimed that "the cosmonaut had been confused by the light beacons [on Soyuz 2] and thereby [had manoeuvred in such a way] that a certain angle had been formed between the antennas of the [two] craft". This, it was concluded, had caused Soyuz 3 to turn away to one side. Mnatsakanyan's judgement: pilot error.

Years later, Asif Siddiqi would write that, indeed, upon realising that the two spacecraft were imperfectly aligned, Beregovoi should have gingerly stabilised his craft along a direct axis to the target. However, he used a stronger thruster firing to place Soyuz 3 into a completely incorrect orientation relative to the target. Soyuz 2's radar sensed this improper deviation and automatically turned its nose away to prevent docking, but Beregovoi tried to complete a fly around and a second approach. When the same thing happened again, it became clear that Soyuz 3's propellant load was running low, leaving barely enough for re-entry. Further docking attempts were immediately abandoned and the two craft drifted apart.

In spite of these problems, during his four days in space, Beregovoi demonstrated the basic habitability of Soyuz, even giving his terrestrial audience a televised tour of the descent and orbital modules during no fewer than three transmissions. It did not have quite the same impact as the Wally, Walt and Donn Show, but proved a close Soviet second. Clad in a woollen training top and a white helmet with microphones, he spoke of the sheer 'comfort' of Soyuz and that, although he did not 'need' a space suit for protection, one was carried aboard Soyuz 3 regardless. Beregovoi also participated in Earth observation studies, noting forest fires and thunderstorms close to the equator and conducting astronomical and Earth-resources photography.

Soyuz 2, meanwhile, suffered a failure of its astro-orientation sensor, but successfully re-entered and landed in Kazakhstan at 10:56 am on 28 October. This was soon followed by Beregovoi's own return, which aroused much anxiety, since it was the first Soviet manned re-entry since Komarov's death. An initial abortive retrofire was followed by a successful 145-second burn over the Atlantic Ocean early on 30 October. The descent module, thankfully, perfectly executed a guided return, hurtling over Africa and the Caspian Sea and hitting the snow-covered steppe near Karaganda with a firm thud at 10:25 am. The ebullient Beregovoi's first contact with a living being came when he was met by a bewildered local boy on a donkey.

Notwithstanding the problems with the Soyuz 2 docking attempt, the mission had been an extraordinary success, with all of the spacecraft's systems – including its Igla rendezvous device – performing as advertised. So too, apparently, did Beregovoi himself: in an oblique jab at the insubordination of Wally Schirra's Apollo 7 crew, Tass reported that the Soyuz 3 pilot had followed his instructions correctly and without complaint. Still, in his post-flight report to the State Commission the following day, Beregovoi actually had a number of complaints: the jettisoning of the launch vehicle's payload fairing was "unpleasant", he told them, and one of his viewports was fogged up, whilst others had dust between the panes. Moreover, the manual control system was "too sensitive" during his approach towards Soyuz 2. The completion of the Soyuz shakedown cruise, though, set the stage for a far more ambitious docking and spacewalking extravaganza planned for the Soyuz 4 and 5 missions in January 1969. Before that, however, a lunar launch window was scheduled to open for the Soviets at the beginning of December. Had they given up on their lunar dream, the Americans wondered, or was another space spectacular on the cards?

"NOT A PRIORITY"

With predictable ambiguity, in the dying months of 1968, the Soviets gave little away to either confirm or deny that they had any interest in competing with the United States to reach the Moon. Fortunately for them, their head of state had not committed them to the same audacious goal as John Kennedy had done for America seven years before. However, it is well known that the Soviets had a vigorous manned lunar effort ... and made equally vigorous attempts to hide it, although the intricacies will be discussed further in the next volume. On 14 October, indeed, Academician Leonid Sedov told the 19th Congress of the International Astronautical Federation in New York that "the question of sending astronauts to the Moon at this time is not an item on our agenda ... it is not a priority". A few weeks later, discussing Beregovoi's mission with the press, Mstislav Keldysh was forced to admit that Soyuz was not designed for a circumlunar flight and that journeys to the Moon were not being studied.

An N-1 rocket on the launch pad at Tyuratam in 1972.

For months, though, speculation had grown over the development of a large lunar rocket, known as the 'N-1', or, had it successfully flown, as the 'Raskat' ('Peal'). NASA Administrator Jim Webb, informed by CIA sources early in 1968 about the N-1, had proven so vocal about its existence that the booster was also colloquially known as 'Webb's Giant'. Its development had begun under Sergei Korolev in 1959 and it was originally intended to lift space stations into orbit or even deliver human crews to Mars with a nuclear-powered upper stage. Two years later, it received a small amount of funding and a government report established a schedule for its maiden launch sometime in 1965. The chance of the N-1 moving from paper to production took a dramatic turn with Kennedy's lunar speech in May 1961 and Korolev began laying plans for a Soviet mission to the Moon, featuring a new spacecraft called 'Soyuz'.

Several launches, it was realised, would be needed to loft the crew-carrying Soyuz, the lunar lander and additional engines and fuel. Powering the giant rocket would be a new engine known as the RD-270, built by Valentin Glushko, employing unsymmetrical dimethyl hydrazine and an oxidiser of nitrogen tetroxide; this hypergolic mixture reduced the need for a complex combustion system, but also yielded a reduced thrust when compared to a combination of, for example, kerosene and liquid oxygen. Korolev felt that a high-performance design demanded high-performance fuels and moved instead to work with Nikolai Kuznetsov's OKB-276 bureau, which suggested a 'ring' of NK-15 engines on the rocket's first stage. The interior of the ring would be open, with air piped into the hole through inlets near the top of the stage; this would then be mixed with the exhaust to augment the N-1's total thrust.

Eventually, in recognition of the difficulties of orbital rendezvous, Korolev opted for a direct-ascent means of reaching the Moon. The result was that the N-1 increased dramatically in size and was combined with a new lunar package called the 'L-3', which contained additional engines, an adapted Soyuz spacecraft (known as the 'Lunniy Orbitny Korabl' or 'LOK') and a lunar lander. Had the N-1 actually flown successfully, it would have been one of the largest launch vehicles ever built: at 105 m tall, it was slightly shorter than the Saturn V, yet produced a million kilograms more thrust from its first stage engines. It had five stages in total, three to boost it into orbit and two to support lunar activities. The lower three stages formed a truncated cone, some 10 m wide at the base, arranged to best accommodate the kerosene and liquid oxygen tanks.

Astonishingly, the first stage – known as 'Block A' – was powered by no fewer than 30 NK-15 engines, arranged in two rings, with one ring handling pitch and yaw and the others on gimballing mounts for roll manoeuvres. With an estimated thrust of 4.5 million kg, Block A would have easily out-thrust the Saturn V's first stage, although its use of kerosene offered poorer performance: it could place 86,000 kg into low-Earth orbit, compared to the Saturn V's 118,000 kg. The N-1's second stage would have comprised eight uprated NK-15V engines in a single ring and four smaller NK-21 engines in a square arrangement on its third stage.

The complexities of building plumbing to feed fuel and oxidiser into the clustered rocket engines proved an incredibly delicate and fragile task and contributed in part to the four catastrophic N-1 launch failures between 1969–72. The plumbing difficulties experienced during the construction of the Saturn V's first stage, which comprised just five engines, must have been exacerbated tenfold for the N-1's 30 engines. During the week that Apollo 8 circled the Moon, the first N-1 was being readied at Tyuratam for its first unmanned launch, scheduled for early in 1969.

As the N-1 proceeded through its torturous and ultimately unsuccessful development process, another rocket – the Proton – was meeting with greater success. It had been designed to loft a cosmonaut-carrying version of the Soyuz spacecraft, albeit without an orbital module, on a circumlunar trajectory. In readiness for this so-called 'Soyuz 7K-L1' variant, a number of unmanned missions were launched towards the Moon under the cover name of 'Zond'. Despite several dismal Proton failures, Zond 5 was launched on 14 September 1968, passing within 1,950 km of the lunar surface and splashing down in the Indian Ocean seven days later. In addition to demonstrating the spacecraft which might someday support two cosmonauts, Zond 5 took the first living creatures – a pair of steppe tortoises, wine flies, mealworms, plants, seeds and bacteria – to the vicinity of the Moon.

Eight weeks later, on 10 November, Zond 6 provided perhaps the biggest impetus to get Apollo 8 into lunar orbit before the end of the year, when it too carried a biological payload and numerous instruments to the Moon. It hurtled just 2,420 km past the lunar surface, acquiring high-resolution photographs, but its descent module depressurised during re-entry into Earth's atmosphere. This killed all of the biological specimens and would probably have resulted in the deaths of cosmonauts, too, had they been aboard. On the other hand, failure occurred because the spacecraft was held in an unusual attitude for a protracted time, owing to a problem that a pilot could have readily been able to overcome. Further, the seal suffered in the heat of the Sun and would not have been overly stressed if a crew had been aboard. To add insult to what had been a less-than-successful mission, only one negative was recovered from its camera container . . . and its parachutes deployed too early, causing it to crash land. Despite the depressurisation and parachute faults, either of which would have doomed a human crew, Tass' announcement a few days later explicitly stated that Zond 6's mission had been "to perfect the flight and construction of an automated variation of the manned spacecraft for flying to the Moon". It marked a stark contrast to Sedov's claim only weeks earlier that the Soviets had no interest in manned lunar missions . . .

In perhaps one of the bitterest ironies, Frank Borman, who would command Apollo 8, the first manned circumlunar expedition, became the first American astronaut to be invited to the Soviet Union afterwards. When he arrived, he was welcomed with a huge party, held in his honour at Moscow's Metropole Restaurant. Hundreds of guests flocked to shake his hand and have their photograph taken with him. One of them was Alexei Leonov. Borman congratulated Leonov on his Voskhod 2 spacewalk and the two chatted about the circumlunar mission and possible landing sites on the Moon. Most likely, Borman considered Leonov to be showing the interest typical of a space explorer. Little did he realise that Leonov had been training to complete the very same goal for which Borman was now being applauded.

Initially, in the months after Sergei Korolev's death, Leonov and a civilian cosmonaut, possibly Oleg Makarov, had been tipped to fly the first circumlunar Soyuz mission sometime in the summer of 1967. "We then expected to be able to accomplish the first Moon landing," wrote Leonov, "ahead of the Americans in September 1968." The Soviet plan was for one cosmonaut to descend to the surface in the LK ('Lunniy Korabl') landing craft, whose spidery appearance was uncannily similar to Grumman's lunar module, leaving his colleague in orbit aboard the main L-3 craft.

"To train for the extreme difficulties of a lunar landing," Leonov wrote, "we undertook exhaustive practice in modified Mi-4 helicopters. The flight plan of a lunar landing mission called for the landing module to separate from the main spacecraft at a very precise point in lunar orbit and then descend towards the surface of the Moon until it reached a height of 110 m from the surface, where it would hover until a safe landing area could be identified. The cosmonaut would then assume manual control of its descent . . . " In addition to flying the circumlunar mission, it is likely that Leonov would have hoped for such a landing mission of his own. For that reason, he qualified as a helicopter pilot and spent much of 1966 and 1967 using it to prepare himself as much as possible for a Moon landing.

Not only would the journey to the lunar surface be fraught with risk, but so too would the return to Earth. In fact, the Soviet plan to slow down from transearth velocities would be to enter the atmosphere, 'bounce' off and re-enter a second time,

The Zond 5 descent module after splashdown in the Indian Ocean.

as would Apollo. "The key to this difficult manoeuvre," continued Leonov, "was the angle of re-entry. If it was wrong, the spacecraft would be severely deformed, perhaps even destroyed ... Several times during training sessions, I sustained pressures of 14 G, the maximum to which a human being can be subjected on Earth. This put tremendous strains on every cell in my body and caused several haemorrhages at those points where my body was most severely compressed."

With the completion of Zond 6, western analysts knew that the next favourable 'window' for a Soviet circumlunar shot opened on 6 December 1968, just two weeks before the window for Apollo 8. It never happened. The parachute failure and the depressurisation issue were sufficiently serious to have killed a cosmonaut crew and Vasili Mishin did not want to risk one until Zond was proven by at least two further test flights. Of course, the Americans knew nothing of this and missing the December window would pose yet more questions as to what the Soviets were up to. Years later, Alexei Leonov would tell Tom Stafford that he and Oleg Makarov were prepared to take the risk and ride a Zond for seven days to the Moon and back. "We could have beaten the Apollo 8 crew," Leonov said sadly, "but Mishin was a blockhead."

MOONWARD BOUND

Since July 1969, Mike Collins has achieved fame as 'the other one' on the first lunar landing crew. A year before making that momentous flight, he might have been aboard Frank Borman's Apollo 9 mission, destined to perform a high-Earth-orbit test of the combined command and service module, complete with the lunar module. That mission changed significantly by the time it finally flew, renamed 'Apollo 8', in December 1968. For Collins, though, the most significant change was that he had gone from being the mission's senior pilot ... to sitting on the sidelines in Mission Control.

The original line-up for the Apollo lunar effort envisaged a seven-step process, labelled 'A' through to 'G'. First would come unmanned test flights ('A') of the command and service module, already achieved by Apollo 4 in November 1967 and Apollo 6 in April 1968. Next, the 'B' mission, completed by Apollo 5, would conduct an unmanned test of the lunar module. A manned 'C' flight, involving the command and service modules in Earth orbit, was originally assigned to Gus Grissom, but completed by Wally Schirra's Apollo 7 crew. Final strides towards the Moon focused on four increasingly more complex missions: 'D' (a manned demonstration of the entire Apollo system in Earth orbit), 'E' (repeating D, albeit in a high elliptical orbit with an apogee of 6,400 km), 'F' (a full dress-rehearsal in orbit around the Moon) and 'G' (the landing itself).

During the course of 1966, crews were assembled to support the first of these flights. When Wally Schirra removed himself from the original Apollo 2 crew and scuppered the 'duplicate' mission, a 'new' Apollo 2, destined to complete the D mission, was given to Jim McDivitt, Dave Scott and Rusty Schweickart for the manned lunar module flight in Earth orbit. Years later, Deke Slayton would note that McDivitt had been intimately involved with the lunar module's development for some considerable time and the rendezvous commitment of the mission necessitated a veteran command module pilot, Dave Scott. The command and service module and lunar module were to be launched on a pair of Saturn 1Bs. The E crew, targeted to fly the Saturn V for the first time on Apollo 3, was named as Frank Borman, Mike Collins and Bill Anders.

Numbering changed in the wake of the Apollo 1 fire, of course, but the crews remained more or less intact. In the months preceding Apollo 7, the McDivitt and Borman crews seemed on track to fly Apollos 8 and 9, which would respectively conduct the lunar module test flight and the high-orbit mission in late 1968 or early 1969. Since Borman's crew would be the third flight of the Saturn V, they were known internally as 'Apollo-Saturn 503' (AS-503). Not only would it be the first manned launch of the behemoth rocket, but also, wrote Collins, the S-IVB "would be reignited, just as if it were a lunar mission ... However, ours would be shut down early, causing us to stay in Earth orbit" with a 6,400 km apogee. "This little detail created all sorts of planning problems," Collins continued, "because one could only escape from this lopsided orbit at certain prescribed intervals and if one had troubles and was forced to return to Earth prematurely, it was entirely possible to end up landing in Red China."

As 1967 ended with the triumphant first flight of the Saturn V, an increased wave of optimism spread through NASA that the lunar landing could be accomplished, to such an extent that by August of the following year – before Apollo 7 had even flown – some managers were talking of expediting Borman's mission from high Earth orbit to a lunar distance. An already record-breaking apogee of six and a half thousand kilometres would be multiplied to almost three hundred and seventy thousand. Then, abruptly, in July 1968, Mike Collins was removed from the mission. One day, during a game of handball, he became aware that his legs did not seem to be functioning as they should, a phenomenon which progressively worsened: as he walked down stairs, his knees would buckle and he would feel peculiar tingling and numbness.

Eventually, and with a typical pilot's reluctance, Collins sought the flight surgeon's advice and was referred to a Houston neurologist. The diagnosis was that a bony growth between his fifth and sixth cervical vertibrae was pushing against his spinal column and relief of the pressure demanded surgery. A few days later, at the Air Force's Wilford Hall Hospital in San Antonio, Texas, Collins underwent an 'anterior cervical fusion' procedure, whereby the offending spur and some adjoining bone was removed and the two vertibrae fused together with a small dowel of bone from the astronaut's hip. Several months of convalescence followed, during which time Collins' backup, Jim Lovell, was assigned to his seat on AS-503 ... and something else happened. 'Apollo 9' would not be known as Apollo 9 anymore, but as 'Apollo 8' and, further, its destination had indeed changed: it would not just fly a basic circumlunar jaunt, but would actually go into orbit around the Moon.

A plan had been under consideration by George Low and Chris Kraft since April 1968 as a means of cutting out one of the seven steps and achieving a landing much sooner. They wanted to change the E mission into something called 'E-prime', moving from high Earth orbit to the vicinity of the Moon, but this quickly became untenable when it materialised that the lunar module would not be ready in time.

When it became evident that the first lunar module would not be available until early 1969, George Low came up with a radical idea: in place of the E mission would be a flight known as 'C-prime', which aimed to send a command and service module, without the lunar module, around the Moon in December 1968. Low knew from Rocco Petrone, director of launch operations at Cape Kennedy, that the lunar module would not be ready for December, but at a meeting in Bob Gilruth's office on 9 August it was felt that the navigation and trajectory teams, together with the astronauts and their training staff, could be ready for a Moon shot. The additional risk of actually entering lunar orbit would be beneficial in that it would provide empirical data on orbital mechanics and the formulation of better gravitational models.

Since Jim Webb and George Mueller were at a conference in Vienna at the time, it was left to Deputy Administrator Tom Paine, still sceptical after the Apollo 6 pogo problems and uncertain as to the reliability of the SPS engine, to conditionally approve it. Mueller, with some reluctance, also agreed, but Webb vehemently opposed the idea. He had been particularly lambasted after the Apollo 1 fire and did not want to be hauled over the coals if the Moon shot failed, particularly as the spacecraft had not even been tested with a human crew. On the other hand, of course, he intended to resign from NASA and was finally won over, with reservations, on 16 August. A few weeks later, Webb visited Lyndon Johnson in Washington to announce his resignation. With Webb gone, the prospects for C-prime brightened.

The plan was officially set in motion by NASA on 19 August by Sam Phillips, although some managers remained nervous about making such a bold move before Wally Schirra's shakedown flight of the command and service module. Officially, until that mission had flown successfully, the 'new' Apollo 8 would represent "an expansion of Apollo 7", but that "the exact content . . . had not been decided". The content of the mission may not have been decided, but the crew certainly had been. On 10 August, Deke Slayton told Jim McDivitt that the flight order was being switched: that his D mission with the lunar module would now become Apollo 9, preceded by Borman's C-prime expedition around the Moon. "Over the years," McDivitt recounted in Slayton's autobiography, "this story has grown to the point where people think I was offered the flight around the Moon, but turned it down. Not quite. I believe that if I'd thrown myself on the floor and begged to fly the C-prime mission, Deke would have let us have it. But it was never really offered." Offered or not, McDivitt acquiesced, he, Scott and Schweickart had been training for so long on the lunar module that they were the best-prepared and wanted to fly its maiden mission.

Privately, Frank Borman was pleased with his lot when he received command of C-prime. McDivitt's D mission "was a test-piloting bonanza," wrote Andrew Chaikin, "and Borman would have gladly traded places." Borman was at North American's Downey plant, working on tests of Spacecraft 104 – the command module for the E mission – when he was summoned to take a call from Deke Slayton. Shortly afterwards, he was back in Houston, in Slayton's office, hearing about the C-prime plan, together with disturbing CIA reports that the Soviets might be only weeks away from staging their own manned circumlunar flight. When Slayton asked Borman if he would command Apollo 8 to the Moon, it was essentially a question with only one answer.

Also pleased with the decision was Jim Lovell, Borman's senior pilot, who had been drafted in only weeks earlier to replace Mike Collins. The pair had, of course, already flown together on Gemini VII and were a good match. Lovell had been planning to take his family – his wife Marilyn and their three children – to Acapulco for Christmas 1968, but was now forced to tell her instead that his yuletide destination had a somewhat different, more exotic and far more extraterrestrial flavour. One evening, flying cross-country with Borman in a T-38 jet, he had sketched a design for Apollo 8's crew patch onto his kneeboard: a figure-eight emblem, with Earth in one circle and the Moon in the other. With Borman in command, it would be Lovell's job as senior pilot to oversee Apollo 8's navigation system, using the command module's sextant to make star sightings, verify their trajectory and track lunar landmarks. Lovell would be Apollo's final senior pilot. The next mission, Apollo 9, would feature a lunar module in addition to its command and service modules and the moniker of 'senior pilot' would be effectively superseded by that of 'command module pilot' (CMP). During subsequent Moon landing missions, the CMP would fly solo in lunar orbit, requiring them to have previous spaceflight experience. The 'pilot' position, in turn, would be replaced by that of 'lunar module pilot' (LMP), the man who would accompany the commander down to the Moon's surface.

One astronaut who was unhappy about the Apollo 9/8 change, though, was Dave Scott. As command module pilot of the D mission, he had nursed Spacecraft 103 – 'his' original ship – through testing and preparation at North American and was now being swapped for another vehicle, Spacecraft 104. On the opposite side of the coin, Bill Anders, on Borman's crew, was also unhappy. Since the beginning of 1968, he had been immersed in lunar module training and would tell Andrew Chaikin years later that he felt he had an 80 per cent chance of a seat on a landing crew. Now, with the change to their flight, the elimination of the E mission and the creation of C-prime, without a lunar module, Anders knew that he had gained the first circumlunar mission, at the expense of probably losing the chance to someday walk on the Moon.

OUTSIDER

When Bill Anders joined NASA in October 1963, he stood out: among the Fourteen, six astronaut candidates lacked test-piloting credentials, and he was one of these apparent 'outsiders'. As a child, he had never had aspirations of becoming a test pilot, but rather wanted to follow in the footsteps of his father, Arthur Anders, and become a career naval officer. William Alison Anders was born in Hong Kong on 17 October 1933 and, in his youth, was an active Boy Scout, receiving its second-highest rank. He earned a bachelor's degree from the Naval Academy in 1955, but opted for a commission in the Air Force, serving as a fighter pilot in an all-weather interceptor squadron in Iceland.

Anders' decision to move away from naval service was made in part by the extraordinary number of fatal accidents he saw as a midshipman: one aircraft, for example, landed on the carrier, missed the arresting net, hit a line of parked jets, careered off the deck and plunged into the sea. Anders accepted the risks of his new-found love of aviation, but preferred to face such risks in air-to-air combat, rather than whilst attempting to land. After Iceland, Anders' next step, with 1,500 hours in his flight logbook, was to apply for test pilot school. He was rejected, on the basis that the school was "pushing academics" and desired him to earn an advanced degree.

With test pilot school still at the back of his mind, Anders enrolled in the Air Force Institute of Technology at Wright-Patterson Air Force Base in Dayton, Ohio, graduating with a master's degree in nuclear engineering in 1962. He had wanted to study astronautical engineering, but places were full, although Anders simultaneously took night school classes in aeronautics at Ohio State University. Upon receipt of his master's qualification, Anders again tried for test pilot school, only to learn that it was not recruiting students, and he moved instead to become a T-33 instructor pilot at Kirtland Air Force Base in Albuquerque, New Mexico.

The following year, 1963, proved life-changing for Anders. He again applied for test pilot school, only to learn that now, under the new commandant, Chuck Yeager, it was looking for candidates with more flying experience. Unperturbed, he submitted his application and in June, whilst waiting to hear of the outcome, he learned that NASA were recruiting for its new class of astronauts. Test piloting qualifications, it turned out, were no longer mandatory and Anders had everything that the space agency needed: he was younger than 35, had an advanced degree and his logbook had now expanded to more than the 2,000 required hours.

When he was called to interview, Anders stressed his nuclear engineering work, aware that flights to the Moon would surely require an understanding of radiation hazards in cislunar space. That October, he was picked as one of the Fourteen ... and, ironically, received a letter rejecting him from test pilot school! In his early years as an astronaut, Anders supervised Apollo's environmental controls and his performance as the capcom during the Gemini VIII emergency in March 1966 quite possibly contributed to his assignment as Neil Armstrong's pilot on the Gemini XI backup crew. By the end of that year, Anders had drawn his first actual flight assignment: to Frank Borman's E mission, during which he hoped to get the opportunity to test-fly the lunar module in high Earth orbit. All that changed in August 1968 when Anders became a part of history: one of the first human explorers to visit the Moon.

C-PRIME

That August, around the time that Anders' E mission was beginning its metamorphosis into C-prime, Apollo command module 103 arrived at Cape Kennedy for testing. Its mission, unofficially, ranged from circumlunar to fully orbital, with around ten circuits of the Moon planned. During the translunar coast, to qualify the 'make-or-break' SPS engine, it would be test-fired for a few seconds. If the engine refused to work, the astronauts could still be brought home safely, thanks to a safety feature built into Apollo 8's trajectory design. Known as the 'free return', it would allow the crew to essentially loop around the Moon and use its gravitational influence to 'slingshot' them back to Earth without using the SPS. In fact, if Borman, Lovell and Anders did find themselves with a useless engine, they would only need to perform a couple of mid-course correction burns, using the service module's thruster quads, to keep them on track for home.

Aside from the chance of an SPS failure, a host of other concerns worried Borman. One of them surrounded Apollo 8's splashdown in the Pacific at the end of the six-day mission. To achieve a splashdown in daylight hours would require a trajectory design which included at least 12 lunar orbits. Borman, though, could not care less whether he landed in daylight or darkness. "Frank didn't want to spend any more time in lunar orbit than was absolutely necessary," wrote Deke Slayton, "and pushed for – and got – approval of a splashdown in the early morning, before dawn." Apollo 8 would stick at ten orbits. To understand Borman's reluctance to do more than was necessary is to understand part of his character and military bearing: he was wholly committed to The Mission, whatever it might be. On Gemini VII, he strenuously rejected any addition to the flight plan which might complicate his and Lovell's chances of fulfilling their primary objective: to spend 14 days in space. Now, on Apollo 8, Borman's mission was to reach the Moon and bring his crew home safely. Nothing else mattered.

All non-essential, 'irrelevant' requests irritated him. "Some idiot had the idea that on the way to the Moon, we'd do an EVA," he recounted years later in a NASA oral history. "What do you want to do? What's the main objective? The main objective was to go to the Moon, do enough orbits so that they could do the tracking, be the pathfinders for Apollo 11 and get your ass home. Why complicate it?"

The four months leading up to the mission were conducted at a break-neck pace. The lunar launch window opened on 21 December, at which time Mare Tranquilitatis (the Sea of Tranquility, a low, relatively flat plain tipped as a possible first landing site) would be experiencing lunar sunrise and its landscape would be thrown into stark relief, allowing Borman, Lovell and Anders to photograph and analyse it. In the final six weeks before launch, the Apollo 8 crew regularly put in ten-hour workdays, with weekends existing only to wade through piles of mail. At the end of November, outgoing President Lyndon Johnson threw them a 'bon voyage' party in Washington. Then, on the evening of 20 December, the legendary Charles Lindbergh, first to fly solo across the Atlantic, visited their quarters at Cape Kennedy. During their meal, the topic of conversation turned to the Saturn V rocket, which would burn over 18,000 kg of fuel in its first second of firing. Lindbergh was astounded. "In the first second of your flight tomorrow," he told them, "you'll burn ten times more fuel than I did all the way to Paris!"

Shortly after 2:30 am on launch morning, Deke Slayton woke them in Cape Kennedy's crew quarters and joined them for the ritual breakfast of steak and eggs. Also in attendance were chief astronaut Al Shepard and Apollo 8 backups Neil Armstrong and Buzz Aldrin. (The third backup crew member, Fred Haise, was busily setting switch positions inside the command module at Pad 39A.) Shortly thereafter, clad in their snow-white space suits and bubble helmets, they arrived at the brilliantly-floodlit pad, where their Saturn V awaited. First Borman, then Anders and finally Lovell took their seats in the command module, joining Haise, who had by now finished his job of checking switches. After offering them his hand in solidarity and farewell, Haise crawled out of the cabin and the heavy unified hatch slammed shut at 5:34 am. Years later, Bill Anders would tell Andrew Chaikin that, at one point, he glanced over at a window in the boost-protective cover and saw a hornet fluttering around outside. "She's building a nest," he thought, "and did she pick the wrong place to build it!"

As their 7:51 am launch time drew closer, a sense of unreal calm pervaded Apollo 8's cabin. With five minutes to go, the white room and its access arm rotated away from the spacecraft and, shortly thereafter, the launch pad's automatic sequencer took charge of the countdown, monitoring the final topping-off of propellants needed by the Saturn to reach space. Sixty seconds before launch, the giant rocket was declared fully pressurised and it transferred its systems to internal control. As the countdown ticked into the final dozen seconds, Borman, Lovell and Anders, despite being cocooned inside their space suits, could faintly hear the sound of fuel pouring into the combustion chambers of the five F-1 engines, a hundred and ten metres below. As the clock inside the command module read 'T-3 seconds', that faint sound was replaced by a distant, thunder-like rumbling and, at some point in the calamitous commotion that followed, the first Saturn V ever to be trusted with human passengers took flight.

"Liftoff," radioed Borman, gazing at the clock on his instrument panel. "The clock is running." After the mission, all three men would have their own recollections of what it was like to launch atop the biggest and most powerful rocket ever built, but Chaikin summed it up best when he quoted Bill Anders: they felt as if they were little more than helpless prey in the mouth of a giant, angry dog.

Forty seconds into the climb, the rocket broke through the sound barrier and G loads on the three astronauts climbed steadily – three, now four, and still climbing – but when they hit 4.5 the uncomfortable feeling of intense acceleration ended as the Saturn's S-IC first stage burned out and separated. "The staging," Borman recounted, "from the first to the second stage, as we went from S-IC cutoff to S-II ignition, was a violent manoeuvre: we were thrown forward against our straps and smashed back into the seat." So violent, in fact, was the motion that Anders felt he was being hurled headlong into the instrument panel. Seconds later, the now-unneeded escape tower and the command module's boost-protective cover were jettisoned, flooding the cabin with daylight as windows were uncovered. For Anders, his first glimpse of Earth from space – mesmerising clouds, vivid blue ocean and a steadily darkening sky – were electrifying.

A little under nine minutes after launch, the S-II finally expired and the S-IVB picked up the remainder of the thrust needed to achieve orbit. "The smoothest ride in the world" was how Borman would later describe riding the Saturn's restartable third stage, before it, too, shut down, at 8:02 am. Barely 11 minutes had passed since leaving Cape Kennedy and the Apollo 8 astronauts were in orbit. In less than three hours' time, assuming that their spacecraft checked out satisfactorily, they would relight the S-IVB for six minutes to begin the translunar injection, or TLI, burn and set themselves on course for the Moon. However, if Apollo 8 did not pass its tests with flying colours and the lunar shot was called off, they would be consigned to what had been uninspiringly termed 'the alternate mission': an Apollo 7-type jaunt for ten long days in Earth orbit, with little to do. Borman could think of nothing worse.

Indeed, at one stage, Lovell, working under one of the couches to adjust a valve, accidentally inflated his space suit's life vest and his commander gave him a dirty look. In true Frank Borman fashion, nothing would be permitted to interfere with or distract their attention from The Mission. At length, it was Capcom Mike Collins, who had been recovered from his neck surgery since early November and had even fruitlessly approached Deke Slayton with a view to staying on the crew, who gave them the news they so badly needed to hear: "Apollo 8, you are Go for TLI!"

Drifting high above the Pacific Ocean at the time, the astronauts knew that the burn would be entirely controlled by the computers and, with ten seconds to go, a flashing number '99' appeared on the command module's display panel. In essence, it asked them to confirm that they wanted to go ahead with the specific manoeuvre. Lovell punched the 'Proceed' button and at 10:38 am, some two hours and 47

Had the Saturn V risen or had Florida sunk? On 9 November 1967, the maiden flight of the mighty Saturn V got underway with "naked power, lots of noise and light". A little more than a year later, Apollo 8 would carry its first human crew.

minutes into the mission, the third stage ignited with a long, slow push. Although Borman kept a keen eye on his instruments in the event that he had to assume manual control, Collins relayed updates from the trajectory specialists that Apollo 8 was in perfect shape. It did not feel that way to Borman, who was convinced from the intense shaking and rattling that he might be forced to abort the burn. Steadily, as Anders watched the third stage's propellant temperatures and pressures, they turned from 'Earth-orbiting' astronauts to 'Moon-bound' adventurers. By the time the S-IVB finally shut down after five minutes and 18 seconds, their velocity had increased from 28,100 km/h to 37,300 km/h – the 'escape velocity' needed to reach the Moon. Frank Borman, Jim Lovell and Bill Anders were travelling faster than any human beings had ever flown before.

Surprisingly, though, with no outside point of reference, there was not the slightest sense of the tremendous speed at which Apollo 8 was moving. Then, when Borman separated the command and service module from the now-spent S-IVB and manoeuvred around to face the third stage, they saw for the first time the effect of TLI: their home world, Earth, was no longer a seemingly-flat expanse of land and sea and cloud 'below' them, but a planet, spherical, its curvature obvious in the black void. They could actually see it receding from them as they continued travelling outwards. At length, as their altitude increased, Earth grew so small that it seemed to fit neatly inside the frame of one of the command module's windows, then could be easily hidden behind a thumb. "Tell Conrad he lost his record," Borman radioed Collins. Jim Lovell promptly launched into a geography lesson and even asked Collins to warn the people of Tierra del Fuego to put on their raincoats, as a storm seemed to be approaching.

Manoeuvring Apollo 8 with its nose pointed toward Earth and the S-IVB had not been done simply for sightseeing: Borman's next task was to rendezvous with it, just as future crews would need to do in order to extract their lunar modules from the enormous 'garage' atop the S-IVB. After completing this demonstration, he pulled away for the final time and Apollo 8 set sail for the Moon. Five hours into the flight, after finally removing his space suit, Lovell set to work taking star sightings with the 28-power sextant and navigation telescopes. If they lost contact with Earth, he might have to measure the angles between target stars and the home planet and punch the data into the computer to figure out their position. He would do the same in lunar orbit, measuring craters and landmarks to help refine Apollo 8's flight path. Not for nothing was 'navigational expert' one of the senior pilot's main responsibilities.

Shortly after 6:00 pm, the first test firing of the SPS engine was performed, lasting just two seconds, which satisfied the astronauts and ground controllers that it could operate as advertised. As the first workday of Apollo 8 drew to a close, Lovell and Anders watched the instruments whilst Borman, unsuccessfully, tried to sleep.

Heading across the vast cislunar gulf, more than 370,000 km wide, the astronauts awakened the first sensations of space sickness. Borman, it seemed, suffered the most. A number of cases of gastroenteritis had plagued Cape Kennedy in the days before launch and it was suggested that this '24-hour intestinal flu' could have triggered the malady; alternatively, Borman had taken a Seconal tablet to help him sleep and blamed the medication for his discomfort. Upon awakening to begin his

The S-IVB recedes into the blackness of cislunar space as Apollo 8 heads for the Moon.

second day aloft, he suffered both vomiting and diarrhoea, but recovered sufficiently by the third day to tell Mission Control that "nobody is sick". Unknown to Borman, his "case of the 24-hour flu" had caused much consternation amongst the flight surgeons on the ground and even led to suggestions that the mission might have to be terminated. Fortunately, all three men were indeed fine and, even if they were ill, the SPS could not be fired to about-face them back to Earth. They were heading for the Moon, whether they liked it or not.

Strangely, since the Moon was barely a crescent to them at the time, none of the crew really saw it until shortly before their arrival. "I saw it several times in the optics as I was doing some sightings," admitted Lovell, but "by and large, the body that we were rendezvousing with – that was coming from one direction as we were going to another – we never saw ... and we took it on faith that the Moon would be there, which says quite a bit for ground control." As they headed towards their target, Apollo 8 slowly rotated on its axis in a so-called 'barbecue roll', to even out thermal extremes of blistering heat and frigid cold across its metallic surfaces.

Two hundred and twenty-three thousand kilometres from Earth, approximately two-thirds of the way to the Moon and 31 hours since launch, they began their first live telecast from Apollo 8. Borman had tried to have the camera removed from the mission, but had been overruled, and now found himself using it to film Jim Lovell in the command module's lower equipment bay, readying a dessert of chocolate pudding. Next there was a shot of Bill Anders, twirling his weightless toothbrush. "This transmission," Borman commenced for his terrestrial audience, "is coming to you approximately halfway between the Moon and the Earth. We have about less than 40 hours to go to the Moon ... I certainly wish we could show you the Earth. Very, very beautiful."

Unfortunately, a telephoto lens fitted to the camera by Anders did not work and when they switched back to the interior lens it resolved the home planet as little more than a white blob, giving away little of its splendour. Borman was disappointed that he had been unable to show viewers the "beautiful, beautiful view, with blue background and just huge covers of white clouds". Lovell closed out the transmission by wishing his mother a happy birthday, after which Borman placed Apollo 8 back into its barbecue roll, which took the high-gain antenna off Earth. A day later, their second telecast was somewhat better, allowing Lovell to describe for his spellbound audience the appearance of the western hemisphere: the royal blues of the deep ocean trenches, the varying browns of the landmasses, the bright whites of the cloud structures.

Lovell was an explorer at heart. His excitement in wanting to fly Apollo 8 was motivated equally as much, if not more so, by the simple urge to explore and see new sights and places than by a desire to carry out scientific investigations. The science was important, but Lovell's sentiment could perhaps be best tied to a statement made three years later by Apollo 15 commander Dave Scott: that going to the Moon was "exploration at its greatest". At one stage in the flight, Lovell turned to Borman and wondered aloud what alien travellers might think as they approached Earth. Would they believe it to be inhabited or not? Would they decide to land on the blue or the brown part of its surface?

"You better hope that we land on the blue part," deadpanned Anders.

By the afternoon of 23 December, almost 60 hours since their Saturn V left Earth, the gravitational influence of their home planet was finally overcome by that of the Moon. At this point, Apollo 8 was more than 300,000 km from Earth and just 62,600 km from its target and the spacecraft's velocity had slowed to 4,320 km/h as it moved farther into its gravitational 'well'. As they sailed towards Lunar Orbit Insertion (LOI), their trajectory was near-perfect: only two of four planned mid-course correction burns had been needed to keep Apollo 8 locked into its free return trajectory. At 3:55 am on Christmas Eve, Capcom Gerry Carr, a member of the Apollo 8 support crew, radioed Borman with the news that they were "go for LOI".

The three astronauts had still not seen the Moon, despite their close proximity to it, since their angle of approach caused it to be lost in the Sun's glare. At length, Carr asked them what they could see. "Nothing," replied Anders gloomily, adding "it's like being on the inside of a submarine". Less than an hour later, at 4:49 am, Apollo 8 passed behind the Moon, with Lovell telling Carr that "we'll see you on the other side". Eleven minutes later, moving at 9,300 km/h and 'backwards', they fired the SPS engine for four minutes to reduce their speed by 3,200 km/h and brake themselves into a 111 × 312 km orbit. The burn was flawless, although Lovell admitted that it was "the longest four minutes I ever spent". Had the engine burned too long or too short, they could have ended up either crashing into the Moon or vanishing into some errant orbit. Just to be sure, Borman hit the shutdown button as soon as the clock touched zero.

Back on Earth, a tense world – nearly a billion people were listening in, NASA estimated, scattered across 64 different countries – waited for word of their insertion into lunar orbit. If Apollo 8 had not fired the SPS, then Borman, Lovell and Anders would come back into communications range ten minutes sooner than planned. At length, right on time, following a 45-minute blackout, public affairs officer Paul Haney announced with joy: "We got it! We got it!" Fifteen minutes later, the astronauts' first close-range descriptions of the Moon came across more than three hundred thousand kilometres of emptiness: "The Moon," Lovell began, "is essentially grey; no colour; looks like plaster of Paris or sort of a greyish deep sand. We can see quite a bit of detail. The Sea of Fertility doesn't stand out as well here as it does back on Earth. There's not as much contrast between that and the surrounding craters. The craters are all rounded off. There's quite a few of them; some of them are newer. Many of them – especially the round ones – look like hits by meteorites or projectiles of some sort ... "

The lack of even the slightest vestiges of an atmosphere lent a weird clarity to what was, in effect, a scene of the utmost desolation, silence and stillness; the Moon was literally a world frozen in time. Only weeks earlier, the film of Arthur C. Clarke's '2001: A Space Odyssey' had premiered and even the astronauts imagined the lunar terrain to be composed of dramatic, sharp-edged mountains and jagged cliffs. Instead they were presented with an essentially dead place, seemingly ubiquitous in its dullness and blandness. Anders, tasked with the bulk of the lunar photography, had spent hours before launch with the only geologist-astronaut, Jack Schmitt, discussing the features of the surface, and had his own flight plan to plough

The lunar farside, never before seen directly by human eyes.

through, but found it hard because of dirty windows. In fact, only the command module's two small rendezvous windows remained reasonably clear.

For Anders, the far side of the Moon, never seen from Earth or ever by human eyes, resembled "a sand pile my kids have been playing in for a long time ... it's all beat up, no definition, just a lot of bumps and holes". He considered the lunar surface to be an unappealing place, albeit with "a kind of stark beauty" of its own, and all three men found pleasure in giving temporary names to some of the craters to honour their colleagues and managers: Low, Gilruth, Shea, Grissom, White, Webb, Chaffee, Kraft, See, Bassett and others. "These," said Borman, "were all the giants who made it work." At one stage during the excitement, when flight controller John Aaron noticed that the command module's environmental control system needed adjustment, they responded by naming a crater for him, too. (Before the flight, Lovell had even given his wife Marilyn a photograph of a mountain, near the eastern edge of the Sea of Tranquility, which he had unofficially named for her: Mount Marilyn.)

Four hours after entering orbit, another SPS burn, this time thankfully shorter at just 11 seconds, adjusted Apollo 8's path around the Moon into a near-perfect 111 km circle. Then, at 10:37 am on 24 December, their first glimpse of colour entered a Universe of endless blackness and greyness: the three astronauts became the first humans to witness 'Earthrise' from behind the lunar limb. Borman was in the process of turning the spacecraft to permit Lovell to take some sextant readings, when all at once Anders yelled: "Oh my God! Look at that picture over there." It would become a running, though light-hearted competition among the crew over who took the 'Earthrise Picture' which has since become world-famous: a shot of our home planet, a pretty blue-and-white marble, rising in the void above the Moon's grey-brown surface. With Lovell in attendance, it was Anders who, after fitting the colour magazine and aiming the telephoto lens, snapped one of the most iconic images of the Space Age. In perhaps no other image has the beauty, fragility and loneliness of Earth been captured with more meaning. Years later, Anders would win praise from environmentalists for his assertion that Apollo 8's goal was to explore the Moon ... and what it really did was discover the Earth!

The astronauts' intense workload during their 20 hours in orbit was getting the better of them, with tiredness causing them to make mistakes. On occasion, Lovell had punched the wrong code into the command module's computer, triggering warning alarms, and Anders was overcome with his own schedule: stereo imagery, dim-light photography and filter work. At length, clearly irritated that the timeline was too full, Borman snapped at Capcom Mike Collins that he was taking an executive decision for his two crewmates to get some rest. "I'll stay up and keep the spacecraft vertical," he told Collins, "and take some automatic pictures." With some difficulty, he had to force Lovell and Anders to pry their eyes away from the windows and get some sleep.

It seemed inevitable, after thousands of years of watching and wondering about the Moon, that humanity's first visit would be commemorated in a religious, spiritual or symbolic way. Before the launch, Borman, Lovell and Anders had discussed this issue at length with friends and concluded that they would read the story of Creation from the first ten verses of Genesis. During their ninth orbit, on their second live telecast from the Moon, they read it to a spellbound world, first Anders taking a part, then Lovell and finally Borman closing with "Good night, good luck, a Merry Christmas and God bless all of you ... all of you on the good Earth".

Eight minutes into Christmas morning, three days and 17 hours after launch, the return home got underway when the SPS engine was ignited to increase their speed by 3,800 km/h. As they rounded the Moon for the last time, Lovell told Capcom Ken Mattingly, who was just coming on duty in Houston, "Please be informed there is a Santa Claus". Mattingly replied that they were the best ones to know.

The return journey proved uneventful, with fogged windows, puddling water and clattering cabin fans creating mere annoyances. A final televised tour of Apollo 8 showed Anders preparing a freeze-dried meal ... and, when the camera stopped rolling, they found a real treat in their food locker: real turkey and real cranberry sauce, wrapped in foil with red and green ribbons. It was a far cry from the

One of the 20th century's most iconic images: Earthrise from Apollo 8.

Apollo 8's fiery return to Earth on 27 December 1968.

toothpaste tubes of Project Mercury and even better, perhaps, than Gus Grissom's corned beef sandwich. The turkey and cranberry sauce turned out to be their best meal of the entire flight, although Borman was annoyed that Deke Slayton had slipped three small bottles of brandy aboard as well. Why, if anything went wrong on the flight, the overzealous Borman fumed, the press and public would blame it on the 'drunk' astronauts. Lovell and Anders, who have admitted that they had no intention of touching the brandy, felt that Borman had gone a little too far. Christmas spirit returned, however, with festive presents: pairs of cufflinks and a man-in-the-Moon tie pin from Susan Borman and Marilyn Lovell and a gold 'figure 8' tie pin from Valerie Anders.

Only one minor trajectory correction burn was needed and early on 27 December, the astronauts fired pyrotechnics to jettison the service module and plunged into Earth's atmosphere at 34,900 km/h. During re-entry, which carried them over north-eastern China, then brought the command module in a long slanting path towards the south-east, Borman, Lovell and Anders were subjected to deceleration forces as high as 7 G. Splashdown came as Cape Kennedy clocks read 10:51 am, but still in pre-dawn darkness over the western Pacific, completing a mission of just over six days. At Mission Control in Houston, sheer pandemonium broke out, in the traditional American back-slapping way, and the smell of celebratory cigars scented the air for hours.

Among the cheering NASA throng was an overjoyed, though dejected Mike Collins. "For me personally, the moment was a conglomeration of emotions and memories," he wrote. "I was a basket case, emotionally wrung out. I had seen this flight evolve in the white room at Downey, in the interminable series of meetings at Houston ... into an epic voyage. I had helped it grow. I had two years invested in it – it was my flight. Yet it was not my flight; I was but one of a hundred packed into a noisy room."

A quarter of a world away, in the Pacific Ocean, some 1,600 km south-south-west of Hawaii, water came flooding through an open vent in the command module, drenching Borman and giving Anders the mistaken impression that the hull had cracked on impact. The ship overturned onto its nose, but quickly righted itself when Borman inflated the three airbags. It did not stop him from being sick. This time, Lovell and Anders, both of whom had served in the Navy, showed no mercy on their Air Force commander: "What do you expect from a West Point ground-pounder?"

Amidst the radio chatter from a rescue helicopter despatched by the aircraft carrier Yorktown came an age-old question which the whole world now wanted answered. "Apollo 8, is the Moon made from Limburger cheese?"

"Nope," replied Bill Anders. "It's made from American cheese!"

SAVED

A few weeks after Apollo 8 splashed down, amongst his mountains of fan mail, Frank Borman came across a telegram from a stranger which summed up the entire mission and the effect of the mission in three words. It read simply: "You saved 1968".

The Apollo 8 crew arrive aboard the recovery ship Yorktown.

In spite of Borman, Lovell and Anders' achievement, the telegram's sender was right in that the year had been a bad one, both in America and elsewhere. Israel and Palestine clashed in border disputes, three decades of 'Troubles' began in Northern Ireland, Soviet tanks rolled in Czechoslovakia to stifle the Prague Spring reforms, the Vietnam War seemed unwinnable and so unpopular that President Lyndon Johnson had glumly announced in March that he had no intention of running for re-election and the United States was left reeling by the murders of Martin Luther King and Senator Bobby Kennedy. Apollo 8, humanity's first journey to the Moon, had cast one of few rays of light over a desperately unhappy and violent year.

On 29 March, King had visited Memphis, Tennessee, in support of black sanitary works employees, who were striking for higher wages and better treatment. A few days later, he delivered his famous 'I've Been To The Mountaintop' speech, then checked into his room at the Lorraine Motel. At precisely 6:01 pm on 4 April, as he stood at his balcony, he was shot; the bullet passing through his right cheek, smashing his jaw, travelling down his spine and finally lodging somewhere in his shoulder. In spite of emergency surgery, the man who had fought tirelessly for civil rights in America was dead ... and his murder instantly sparked fury in as many as a hundred cities across the nation. Two months later, on 5 June, Senator Kennedy – brother of the murdered president, one-time attorney-general and having himself just won the California primary as part of his own presidential candidacy bid – was shot in the crowded kitchen passageway of the Ambassador Hotel in Los Angeles. Both killings provoked desperate outrage as two men who supported two of the strongest issues of the day – civil rights and ending the Vietnam War – were prematurely cut down.

As the last few days of blood-stained 1968 faded into history, however, the long road to the Moon had been won. The enormous technological challenges needed to navigate men and machines across a gulf of more than three hundred thousand kilometres of uncharted emptiness had been met. Nor was Apollo 8 simply a lucky shot: Frank Borman, Jim Lovell and Bill Anders had been guided into a precise orbit around our closest celestial neighbour and had taken a truly giant leap towards the small step which, seven months hence, would change humanity's view of itself forever. Completion of the first circumlunar mission was just the first part of John Kennedy's promise. Now, in the final few months of the decade, would come the most audacious task of all: achieving all that Apollo 8 had achieved and more, guiding the spidery, as-yet-untested lunar module down those last 111 km from orbit and planting American bootprints onto the Moon's dusty surface. The pieces were set. The machines, equipment and rockets were ready. So were the men. On 6 January 1969, Deke Slayton summoned Neil Armstrong, Mike Collins and Buzz Aldrin into his office. In just two words, he told them the news every astronaut had trained for years to hear: "You're it!"

'It', of course, meant that they were being tapped for the coveted lunar landing, tentatively pencilled-in for Apollo 11 in July. Armstrong and Aldrin had just come off their Apollo 8 backup duties and Slayton, perhaps, felt a pang of conscience for Collins, who had fought his way to full health and back onto flight status after his neck surgery. There would be a slight shift of roles, though. Following Jim Lovell's departure for the Borman crew, Aldrin had been promoted to senior pilot of the Apollo 8 backup team. It was a role later to become synonymous with the command module pilot, essentially a mission's second-in-command, but both Armstrong and Slayton had more confidence in Collins to fill this role. "I had a little difficulty putting Aldrin above Collins," Armstrong told James Hansen. "In talking with Deke, we decided, because the CMP had such significant responsibilities for flying the command module solo and being able to do rendezvous by himself and so forth, that Mike was best to be in that position." Thus, Aldrin missed out on being Apollo 11's second-in-command, but the alternative was far sweeter: if the schedule ran as planned, he would be the second man on the Moon.

Of course, we know today that Kennedy's goal was indeed met. Yet a lunar landing on Apollo 11 was by no means set in stone as 1969 dawned. (Mike Collins would later estimate the chance of success, in his mind at least, as no more than 50–50.) Still to be proven was Grumman's spidery lunar module, which Apollo 9 astronauts Jim McDivitt and Rusty Schweickart intended to test in Earth orbit late in February, as their colleague Dave Scott practiced rendezvous and docking in the command module. The space suit which astronauts would one day use to walk on the Moon would be put through its paces by Schweickart during a dramatic EVA, in which he would climb out of the lunar module's hatch and onto its porch. A couple of months later, Apollo 10 would do a full dress-rehearsal of the Moon landing ... 370,000 km away, in lunar orbit. Tom Stafford and Gene Cernan, old buddies from Gemini IX-A, would guide the lunar module to just 15 km above the surface, before firing their ascent engine to boost themselves back up to rendezvous with crewmate John Young. Only if both of these highly-complex missions, the details of which remained to be hammered-out, succeeded could Apollo 11 stand any chance of launching in mid-July.

The Soviets, too, were on the brink of re-entering the game with a vengeance. More than a year and a half after Vladimir Komarov's tragic death, the new Soyuz spacecraft was operating and, for them, 1969 would see no fewer than five manned missions: two in January which would feature their first spacewalk in almost four years, carried out, finally, by Yevgeni Khrunov and Alexei Yeliseyev, and a unique triple rendezvous involving seven cosmonauts in October, which the Soviets would laud as having laid the foundations of a long-term space station. However, the writing did seem to be on the wall as far as the their chances of getting a cosmonaut onto the lunar surface before the Americans was concerned; Nikolai Kamanin had long since written in his ubiquitous diary that he was convinced the United States would win the race. Yet the lure of the Moon and getting cosmonauts there would not fade from the Soviet psyche for some time and, indeed, 1969 would prove a make-or-break year for the enormous, temperamental N-1 booster. If that beast – even more powerful than the Saturn V – could somehow be tamed, made to work and entrusted with a human crew, a flag bearing the Hammer and Sickle might still end up sticking out of the lunar soil.

In addition to closing out the first decade of manned spaceflight, 1969 offered a starting point for the future: after the G mission, longer stays on the Moon were envisaged, running into the Seventies, with perhaps lunar bases and expeditions to

Mars thereafter. A revolutionary reusable spacecraft known as the Space Shuttle would begin its tumultuous development and it was hoped that, instead of simply visiting the heavens, men would actually come to live there. By the end of the Seventies, as the Shuttle prepared for its maiden launch and promised access to space that was cheaper than ever before, a total of six flags and hundreds of bootprints would dot half a dozen lunar landing sites and Soviet cosmonauts would routinely spend six months at a time in orbit, hosting guests from other nations in their orbiting stations.

Humanity had advanced enormously between the end of the Fifties and the close of the Sixties, in a thousand social, cultural, political and technological ways. It would have been impossible to imagine on the eve of Yuri Gagarin's orbital flight that within such a short span of time the techniques and tools of rendezvous, docking, spacewalking and reaching the Moon would have been tried, tested and mastered. Four per cent of the federal budget, in Apollo's case, had much to do with this speed and success, but it remains quite remarkable that the International Space Station has required two decades from conception to construction and at least 15 years will have passed by the time George W. Bush's vision of humans back on the Moon is realised sometime around 2020.

The Sixties were truly an inspirational, pivotal decade which shaped the future of space exploration. Lessons were learned which have much bearing on activities in orbit today and some continue to be relearned for the missions of the future. Yet they only represented the first few years of a human adventure which, to date, has spanned five decades. If the Sixties involved simply rising from Earth, as Socrates said, and reaching the top of the atmosphere to understand the world from which we came, then the Seventies would establish our first foothold in the heavens.

Bibliography

ARTICLES

'Man in Space.' Time, 5 September 1960.
'Freedom's Flight.' Time, 12 May 1961.
'Shepard and USA Feel A-OK.' Life, 12 May 1961.
'Astronaut's Story of the Thrust into Space.' Life, 19 May 1961.
'Saga of the Liberty Bell.' Time, 28 July 1961.
"I am Eagle." Time, 18 August 1961.
'Meditative Chimponaut.' Time, 8 December 1961.
'Nerveless?' Time, 23 February 1962.
'The Flight.' Time, 2 March 1962.
'Titov's Tour.' Time, 11 May 1962.
"Aurora 7. Do You Read Me?" Time, 1 June 1962.
'Suggestion to Astronauts: Look, Ma, No Hands.' Time, 8 June 1962.
'Duet in Space.' Time, 17 August 1962.
'The Heavenly Twins.' Time, 24 August 1962.
'Sweet Little Bird.' Time, 12 October 1962.
'Great Gordo.' Time, 24 May 1963.
'Under Whose Moon?' Time, 31 May 1963.
'Romanoff and Juliet.' Time, 21 June 1963.
'Women are different.' Time, 28 June 1963.
'Surprise With Troika.' Time, 23 October 1964.
'Here Comes Gemini.' Time, 19 February 1965.
'Adventure into Emptiness.' Time, 26 March 1965.
'The Flight of the Molly Brown.' Time, 2 April 1965.
'Molly's Laggard Lift.' Time, 30 April 1965.
'People' Section. Time, 14 May 1965.
'Closing the Gap.' Time, 11 June 1965.
'Towards the Moon.' Time, 18 June 1965.
'Tumult on Earth.' Time, 25 June 1965.

'The Fuel-Cell Flight.' Time, 27 August 1965.
'Flight To The Finish.' Time, 3 September 1965.
'Man is Moon-Rated.' Time, 10 September 1965.
'Journey to Inner Space.' Time, 17 September 1965.
'The Glitch and The Gemini.' Time, 5 November 1965.
'What Happened With Gemini 6.' Time, 19 November 1965.
'Inside While Outside.' Time, 26 November 1965.
'Far Out Date.' Time, 10 December 1965.
'Gemini's Week.' Time, 17 December 1965.
'The Moon in Their Grasp.' Time, 24 December 1965.
'Pictures of Success.' Time, 31 December 1965.
'Fountain of Youth.' Time, 7 January 1966.
'Rendezvous in St. Louis.' Time, 11 March 1966.
'Gemini's Wild Ride.' Time, 25 March 1966.
'The Lessons of Gemini VIII.' Time, 1 April 1966.
'Chasing an Angry Alligator.' Time, 10 June 1966.
'Down the Pickle Barrel.' Time, 17 June 1966.
'Fattening the Record Books.' Time, 29 July 1966.
'Of Glory and Cliches.' Time, 5 August 1966.
'Where's Ivan?' Time, 23 September 1966.
'The World is Round.' Time, 23 September 1966.
'How to Make Out with EVA.' Time, 30 September 1966.
'Two Steps Toward the Moon.' Time, 18 November 1966.
'And Now, Apollo.' Time, 25 November 1966.
'To Strive, To Seek, To Find, And Not To Yield.' Time, 3 February 1967.
'Inquest on Apollo.' Time, 10 February 1967.
'The Oxygen Question.' Time, 10 February 1967.
'Rescue Service for Astronauts.' Time, 10 March 1967.
'Death of a Cosmonaut.' Time, 5 May 1967.
'Premonition of Fire.' Time, 12 May 1967.
'Back To The Job.' Time, 19 May 1967.
'Fireproofing Apollo.' Time, 1 September 1967.
'A Chance To Be First.' Time, 11 October 1968.
'Testing Toward the Moon.' Time, 18 October 1968.
'Acrobats in Orbit.' Time, 25 October 1968.
'Perfection Plus 1%.' Time, 1 November 1968.
'Coloring the cosmos pink.' Time, 13 June 1983.
'My Father Nikita's Downfall.' Time, 14 November 1988.

BOOKS

Aldrin, Buzz and McConnell, Michael (1989) 'Men from Earth'. New York: Bantam.
Boomhower, Ray E. (2004) 'Gus Grissom: The Lost Astronaut'. Indianapolis: Indiana Historical Society Press.

Boynton, John H. (ed.) (1967) "Second United States Manned Three-Pass Orbital Mission (Mercury-Atlas 7, Spacecraft 18): Description and Performance Analysis'. Manned Spacecraft Centre, Houston, Texas.

Brooks, C., Grimwood, James M. and Swenson, Loyd S. (1979) 'Chariots of Apollo: A History of Manned Lunar Spacecraft'. NASA Headquarters, Washington, DC.

Brundin, Robert H. and Hohmann, B.A. (1961) 'Progress Report Atlas Booster Program for NASA Project Mercury'. Mercury Atlas Launch Vehicle Program Office.

Carpenter, Scott and Stoever, Kris (2002) 'For Spacious Skies'. New York: New American Library.

Cernan, Eugene and Davis, Don (1999) 'The Last Man on the Moon'. New York: St Martin's Press.

Chaikin, Andrew (1993) 'A Man on the Moon'. London: Penguin.

Chapman, S. (2007) 'From red star to rocket's glare: space travel, the early years'. Physics Education 42:4, pp. 335–344.

Clark, Phillip (1988) 'The Soviet Manned Space Programme'. London: Salamander.

Collins, Michael (1974) 'Carrying the Fire'. New York: Farrar, Straus and Giroux.

Conrad, Nancy and Klausner, Howard (2005) 'Rocketman'. New York: New American Library.

Doran, J. and Bizony, P. (1998) 'Starman: The Truth Behind the Myth of Yuri Gagarin'. London: Bloomsbury.

Erb, R. Bryan and Jacobs, Stephen. 'Entry Performance of the Mercury Spacecraft Heat Shield'. NASA Manned Spacecraft Center, October 1964.

Gemini 6 Press Kit. NASA Headquarters, Washington, DC, October 1965.

Gemini 7/6 Press Kit. NASA Headquarters, Washington, DC, November 1965.

Glenn, John and Taylor, Nick (1999) 'John Glenn: A Memoir'. New York: Bantam.

Gunston, W.T. 'Redstone: The Largest Missile in Service.' Flight, 23 May 1958.

Hacker, Barton and Grimwood, James (1977) 'On the Shoulders of Titans: A History of Project Gemini'. NASA Headquarters, Washington, DC.

Hammack, Jerome B., Smith, Norman F., Hodge, J.D., Slayton, Donald K., Augerson, William, and Donnelly, Paul C. Postlaunch Report for Mercury-Redstone No. 3 (MR-3). NASA Project Mercury Working Paper No. 192. Space Task Group, Langley Field, Virginia, 16 June 1961.

Hanson, James R. (2005) 'First Man: The Life of Neil Armstrong'. London: Simon & Schuster.

Hoffman, D.B. and Kontaratos, A.N. (1966) 'Astronaut and system performance during Gemini EVA'. Bellcom, Inc. paper.

Jones, Edward R. (1960) 'Predictions of man's vision in and from Mercury capsule'. Thirty-First Annual Meeting of the Aerospace Medical Association.

Laney, Charles C. Jr, (1964) 'Flashing-beacon experiment for Mercury-Atlas-9 (MA-9) mission'. NASA Langley Research Centre, Langley, Virginia.

Miller, F.E., Cassidy, J.L., Leveye, J.C. and Johnson, R.I. 'The Mercury-Redstone Project'. Saturn-Apollo Systems Office, George C. Marshall Space Flight Centre, Huntsville, Alabama, December 1964.

'NASA Outlines Astronaut Policy for New Space Flight Trainees.' NASA Headquarters Release, Washington, DC, 16 September 1962.

Pitts, David E. and Carter, Patricia C. (1966) High-altitude atmospheric measurements for the re-entries of Gemini VI and VII. NASA Technical Memorandum.

Portree, David S.F. and Treviño, Robert C. (1997) 'Walking To Olympus: An EVA Chronology'. History Office, NASA Headquarters, Washington, DC.

Post-Launch Report for Mercury-Atlas No. 1 (MA-1) (ed. Aleck C. Bond), Space Task Group, NASA, August 1960.

Results of the First United States Manned Orbital Space Flight. Manned Spacecraft Center, Houston, Texas, February 1962.

Results of the Third United States Manned Orbital Space Flight. Manned Spacecraft Center, Houston, Texas, October 1962.

Schaefer, Hermann J. and Sullivan, Jeremiah J. (1964) Measurements of the astronauts' radiation exposure with nuclear emulsion on Mercury missions MA-8 and MA-9. NASA-US Naval School of Aviation Medicine Joint Report.

Schroeder, Lyle C. and Russo, Francis P. (1968) Flight investigation and analysis of alleviation of communications blackout by water injection during Gemini 3 re-entry. Langley Research Centre, Hampton, Virginia.

Scott, David and Leonov, Alexei (2004) 'Two Sides of the Moon'. London: Simon & Schuster.

Seamans, Robert C., Jr. (2005) 'Project Apollo: The Tough Decisions'. Washington, DC: NASA History Division, Office of External Relations.

Shayler, David J. (2002) 'Apollo: The Lost and Forgotten Missions'. Chichester: Praxis.

Schirra, Wally with Billings, Richard (1988) 'Schirra's Space'. Annapolis, Maryland: Bluejacket.

Siddiqi, Asif A. (2000) 'Challenge to Apollo'. Washington, DC: NASA History Division, Office of Policy and Plans.

Slayton, Donald K. and Cassutt, Michael (1994) 'Deke'. New York: Forge.

Stafford, Thomas P. with Cassutt, Michael (2002) 'We Have Capture'. Washington, DC: Smithsonian Books.

Swenson, Loyd S., Jr., Grimwood, James M. and Alexander, Charles C. (1966) 'This New Ocean: A History of Project Mercury'. Office of Technology Utilisation, National Aeronautics and Space Administration, Washington, DC.

Technical Information Summary of Mercury-Atlas Mission No. 5 (Capsule No. 9), Space Task Group, Langley Air Force Base, Virginia, October 1961.

Thompson, Neal (2004) 'Light This Candle'. New York: Three Rivers Press.

'Training of Astronaut Candidates.' Manned Spacecraft Centre Release, Houston, Texas, 26 January 1963.

Voas, Robert B. (1960) A description of the astronaut's task in Project Mercury. Fourth Annual Meeting of the Human Factors Society, Boston, Massachusetts.

Warren, Carlos S. and Gill, William L. (1964) Radiation dosimetry aboard the spacecraft of the eighth Mercury-Atlas mission (MA-8). NASA Technical Note, Manned Spacecraft Centre, Houston, Texas.

Wilbur, Ted (1970) 'Once a Fighter Pilot.' Naval Aviator News.

Wolfe, Tom (1980) 'The Right Stuff'. London: Jonathan Cape.

Zygielbaum, J.L. (1961) 'The flight of Vostok 2.' Translation of Pravda Extra Edition, 7 August 1961.

Index

Printing: Mercedes-Druck, Berlin
Binding: Stein + Lehmann, Berlin